The Plastics Pipe Institute
Handbook of Polyethylene Pipe

The Plastics Pipe Institute
FIRST EDITION

ISBN # 0-9776131-0-0

Copyright © 2006
The Plastics Pipe Institute, Washington, DC

Contents

Foreword	**Handbook of Polyethylene Pipe**	
Chapter 1	**Introduction**	
	Features and Benefits of HDPE Pipe	6
	Ductility	10
	Visco-Elasticity	11
	Summary	12
	References	13
Chapter 2	**Inspections, Tests and Safety Considerations**	
	Scope	15
	Introduction	15
	Handling and Storage	16
	Receiving Inspection	16
	Product Packaging	16
	Checking the Order	18
	Load Inspection	18
	Receiving Report and Reporting Damage	19
	Unloading Instructions	19
	Unloading Site Requirements	19
	Handling Equipment	19
	Unloading Large Fabrications, Manholes and Tanks	20
	Pre-Installation Storage	21
	Exposure to UV and Weather	23
	Cold Weather Handling	23
	General Considerations During Installation	24
	Joining and Connections	24
	Field Joining	24
	Cleaning	24
	Field Fusion Joining	25
	Field Handling	25
	Inspection and Testing	26
	Pre-Construction	26
	During Construction	27
	Butt Fusion Joint Quality	27
	Soil Tests	28
	Pipe Surface Damage	28
	Deflection Tests	29
	Post Installation	30
	Leak Testing	30
	Pressure Testing Precautions	31
	References	32
	Test Pressure	32

	Test Duration	32
	Pre-Test Inspection	33
	Hydrostatic Testing	33
	Monitored Make-up Water Test	33
	Non-Monitored Make-Up Water Test	34
	Pneumatic Testing	35
	High Pressure Procedure	35
	Low Pressure Procedure	36
	Initial Service Testing	36
	Test Procedure	36
	Non-Testable Systems	36
	Considerations for Post Start-Up and Operation	36
	Disinfecting Water Mains	36
	Cleaning	37
	Squeeze-Off	37
	Tools	38
	Key Elements	39
	Procedure	40
	Routine or Emergency	42
	Repairs	42
	Damage Assessment	42
	Permanent Repairs	43
	Temporary Repair	44
	Conclusion	44
	References	44
Chapter 3	**Engineering Properties**	
	Scope	47
	Polyethylene Plastics	47
	History of Polyethylene	48
	Manufacture of Polyethylene	48
	Polymer Characteristics	49
	Density	50
	Crystallinity	51
	Molecular Weight	52
	Effect of Molecular Weight Distribution on Properties	55
	Mechanical Properties	57
	Establishing Long-Term Design Properties	57
	Tensile Creep Curves	59
	Tensile Creep or Apparent Modulus	60
	Stress Relaxation	62
	Simplified Representation of Creep and Stress-Relaxation Modulus	62
	Creep Recovery	62
	Creep Rupture	64
	Long-Term Hydrostatic Strength	65

Rate Process Method Validation	67
Fracture Mechanics	69
Cyclic Fatigue Endurance	70
Short-Term Mechanical Properties	71
Tensile Properties	71
Tensile Strength	73
Compressive Strength and Modulus	75
Flexural Strength and Modulus	76
Shear Properties	76
General Physical Properties	77
Impact Strength	77
Hardness	79
Abrasion Resistance	80
Permeability	81
Thermal Properties	81
Thermal Expansion and Contraction	81
Thermal Conductivity	82
Specific Heat	82
Glass Transition Temperature	83
Minimum/Maximum Service Temperatures	84
Deflection Temperature Under Load	84
Electrical Properties	85
Volume Resistivity	86
Surface Resistivity	86
Arc Resistance	86
Dielectric Strength	86
Dielectric Constant	87
Dissipation Factor	87
Static Charge	87
Flammability and Combustion Toxicity	88
Flammability	88
Combustion Toxicity	88
Chemical Resistance	88
Immersion Testing	89
Chemical Resistance Factors	96
Environmental Stress Crack Resistance (ESCR)	99
Compressed Ring ESCR Test	100
PENT Test	101
Aging	101
Weatherability	101
Stabilization	102
Toxicological Properties	103
Health Effects	103
Biological Resistance	104
Conclusion	104
References	105

Chapter 4	**Polyethylene Pipe and Fittings Manufacturing**	
	Introduction	107
	Pipe Extrusion	108
	Raw Materials Description	108
	Extrusion Line	109
	Raw Materials Handling	110
	Drying	111
	Extrusion Principles	111
	Extruders	113
	Breaker Plate/Screen Pack	113
	Die Design	113
	Pipe Sizing Operation	115
	Cooling	117
	Pullers	117
	Take-off Equipment	118
	Saw Equipment and Bundling	118
	Fittings Overview	118
	Injection Molded Fittings	118
	Fabricated Fittings	121
	Thermoformed Fittings	122
	Electrofusion Couplings	122
	Quality Control/Quality Assurance Testing	124
	Workmanship, Finish, and Appearance	124
	Dimensions	124
	Pressure Tests	125
	Physical Property Tests	125
	Quality Assurance Summary	126
	Summary	126
	References	126
Chapter 5	**Specifications, Test Methods and Codes for Polyethylene Piping Systems**	
	Introduction	129
	Properties and Classification of Polyethylene Materials	129
	Material Selection and Specification	131
	ASTM D-3350 "Standard Specification for Polyethylene Plastics Pipe and Fittings Materials"	131
	Thermal Stability	131
	Polyethylene Grade - D 3350	133
	PPI Designations	134
	Test Methods and Standards for Stress Rating, Dimensioning, Fittings and Joining of Polyethylene Pipe Systems	134
	Pressure Rating of Polyethylene Pipe	134
	Design Factors And Hydrostatic Design Stress	137
	Dimensioning Systems	138
	Diameters	138
	Standard Dimension Ratio	139

	Standard Specifications for Fittings and Joinings	141
	General	141
	Codes, Standards and Recommended Practices for Polyethylene Piping Systems	142
	Plastics Pipe Institute (PPI)	143
	ASTM	144
	ISO	145
	NSF International	145
	AWWA	146
	Plumbing Codes	147
	Other Codes and Standards	147
	Factory Mutual	148
	Conclusion	148
	References	149
	Appendix 1- Major Standards, Codes and Practices	150
	General	150
	Gas Pipe, Tubing and Fittings	152
	Water Pipe, Tubing and Fittings	153
	Installation	155
Chapter 6	**Design of Polyethylene Piping Systems**	
	Introduction	157
Section 1	**Design for Flow Capacity**	158
	Pipe ID for Flow Calculations	158
	Pipe Diameter for OD Controlled Pipe	158
	Pipe Diameter for ID Controlled Pipe	159
	Pressure Rating for Pressure Rated Pipes	159
	Fluid Flow in Polyethylene Piping	162
	Head Loss in Pipes	162
	Pipe Deflection Effects	165
	Head Loss in Fittings	166
	Head Loss Due to Elevation Change	167
	Pressure Flow of Water – Hazen-Williams	168
	Pipe Flow Design Example	169
	Surge Considerations	170
	Surge Pressure	170
	Pressure Class	172
	Working Pressure Rating	173
	Water Pressure Pipe Design Example	175
	Controlling Surge Effects	176
	Pressure Flow of Liquid Slurries	176
	Particle Size	177
	Solids Concentration and Specific Gravity	177
	Critical Velocity	178
	Compressible Gas Flow	182
	Empirical Equations for Low Pressure Gas Flow	183

	Gas Permeation	183
	Gravity Flow of Liquids	185
	Manning	185
	Comparative Flows for Slipliners	187
	Flow Velocity	188
	Pipe Surface Condition, Aging	189
Section 2	**Buried PE Pipe Design**	190
	Introduction	190
	Calculations	190
	Installation Categories	191
	Design Process	192
	Vertical Soil Pressure	193
	Earth Load	193
	Live Load	195
	Surcharge Load	202
	Installation Category #1: Standard Installation-Trench or Embankment	206
	AWWA Design Window	221
	Installation Category #2: Shallow Cover Vehicular Loading	222
	Installation Category #3: Deep Fill Installation	224
	Radial Earth Pressure Example	227
	Ring Deflection of Pipes Using Watkins-Gaube Graph	227
	Watkins – Gaube Calculation Technique	230
	Moore-Selig Equation for Constrained Buckling in Dry Ground	231
	Critical Buckling Example	232
	Installation Category #4: Shallow Cover Flotation Effects	232
	Design Considerations for Ground Water Flotation	232
	Unconstrained Pipe Wall Buckling (Hydrostatic Buckling)	236
	Ground Water Flotation Example	238
Section 3	**Thermal Design Considerations**	
	Introduction	240
	Unrestrained Thermal Effects	240
	End Restrained Thermal Effects	240
	Above Ground Piping Systems	243
	Buried Piping Systems	243
	Conclusion	244
	References for Section 1	245
	References for Section 2	245
	Appendix	247
Chapter 7	**Underground Installation of Polyethylene Piping**	
	Introduction	261
	Flexible Pipe Installation Theory	262
	Deflection Control	263

Acceptance Deflection	265
Pipe Design Considerations	265
Pipe Embedment Materials	265
Terminology of Pipe Embedment Materials	266
Classification and Supporting Strength of Pipe Embedment Materials	267
Strength of Embedment Soil	267
Embedment Classification Per ASTM D-2321	268
Use of Embedment Materials	270
Class I and Class II	270
Migration	271
Cement Stabilized Sand	272
Class III and Class IVA	272
Class IVB and Class V	273
Compaction of Embedment Materials	273
Density Requirements	273
Compaction Techniques	273
Compaction of Class I and II Materials	275
Compaction of Class III and IV Materials	276
Density Checks	276
Trench Construction	276
Trench Width	276
Trench Length	277
Stability of the Trench	278
Stability of Trench Floor	279
Stability of Trench Walls	280
Portable Trench Shield	281
Installation Procedure Guidelines	282
Trench Floor Preparation	282
Backfilling and Compaction	284
Backfill Placement	285
Proper Burial of HDPE Fabricated Fittings	286
Inspection	288
References	288
Appendix 1	288
Simplified Installation Guidelines for Pressure Pipe	288
Simplfied Step-by-Step Installation	289
Trenching	289
De-watering	289
Bedding	289
Pipe Embedment	290
Pressure Testing	290
Trench Backfill	290
Appendix 2	290
Guidelines for Preparing an Installation Specification General Requirements	290

	General Requirements	290
	Scope	291
	Quality Assurance	291
	Shop Drawings	292
	Storage	292
	Products	292
	General	292
	High-Density Polyethylene (HDPE) Pipe	292
	Execution	293
	Pipe Laying	293
	Connections	296
	Pipe Tunnels and Casing	297
	Inspection and Testing	298
	Cleanup	300
	Appendix 3	300
	Basic Soil Concepts For Flexible Pipe Installation	300
	Soil Classification	300
	Fine Grain Soil (Clay and Silt)	300
	Coarse Grain Soils	301
	Methods of Measuring Density	301
	Comparison of Installation of Rigid and Flexible Pipe	302
Chapter 8	**Above-Ground Applications for Polyethylene Pipe**	
	Introduction	305
	Design Criteria	306
	Temperature	306
	Pressure Capability	306
	Low Temperature Extremes	309
	Expansion and Contraction	309
	Chemical Resistance	310
	Ultraviolet Exposure	310
	Mechanical Impact or Loading	311
	Design Methodology	311
	Pressure Capability	311
	Expansion and Contraction	313
	Installation Characteristics	316
	On-Grade Installations	316
	Free Movement	316
	Restrained Pipelines	318
	Supported or Suspended Pipelines	321
	Support or Suspension Spacing	322
	Anchor and Support Design	325
	Pressure-Testing	325
	Conclusion	327
	References	327
	Equations	328

Chapter 9   **Polyethylene Joining Procedures**

Introduction	329
General Provisions	329
Thermal Heat Fusion Methods	329
Butt Fusion	330
Optional Bead Removal	331
Saddle/Conventional Fusion	331
Socket Fusion	333
Documenting Fusion	336
Heat Fusion Joining of Unlike Polyethylene Pipe and Fittings	336
Mechanical Connections	337
Mechanical Compression Couplings for Small Diameter Pipes	337
Stab Type Mechanical Fittings	338
Mechanical Bolt Type Couplings for Large Diameter Pipes	338
Flanged Connections	339
Polyethylene Flange Adapters and Stub Ends	340
Flange Gasket	341
Flange Bolting	341
Flange Assembly	341
Special Cases	342
Mechanical Flange Adapters	342
Mechanical Joint (MJ) Adapters	343
Transition Fittings	344
Mechanical Joint Saddle Fittings	344
Repair Clamps	345
Other Applications	346
Restraining Polyethylene Pipe	346
Squeeze-off	347
Summary	348
References	349

Chapter 10   **Marine Installations**

Introduction	351
The Float-and-Sink Method – Basic Design and Installation Steps	
Selection of an Appropriate Pipe Diameter	354
Determination of the Required SDR	354
Determination of the Required Weighting, and of the Design and the Spacing of Ballast Weights	355
Maximum Weighting that Allows Weighted Pipe To Be Floated into Place	355
Determining the Maximum Weighting That Still Allows PE Pipe To Float	357
Determining the Required Minimum Weighting for the Anchoring of a Submerged Pipe in its Intended Location	357

Ensuring that the Required Weighting Shall Not Be Compromised by Air Entrapment	359
Determining the Spacing and the Submerged Weight of the Ballasts To Be Attached to the Pipe	360
Design and Construction of Ballast Weights	361
Selection of an Appropriate Site for Staging, Joining and Launching the Pipe	363
Preparing the Land-to-Water Transition Zone and, When Required, the Underwater Bedding	364
Assembly of Individual Lengths of Pipe Into Long Continuous Lengths	364
Mounting the Ballasts on the Pipe	365
Launching the Pipeline into the Water	366
Submersion of the Pipeline Using the Float-and-Sink Method	369
Completing the Construction of the Land-to-Water Transition	372
Post-Installation Survey	373
Other Kinds of Marine Installations	373
Winter Installations	373
Installations in Marshy Soils	373
Water Aeration Systems	374
Dredging	374
Temporary Floating Lines	375
Conclusion	375
References	376
Appendix A-1	376
Derivation of the Equation for the Determining of the Buoyant Force Acting on a Submerged PE Pipe	376
Appendix A-2	378
Water Forces Acting on Submerged PE Piping	378
Appendix A-3	383
Some Designs of Concrete Ballasts	383

Chapter 11 Pipeline Rehabilitation by Sliplining with Polyethylene Pipe

Introduction	389
Design Considerations	390
Select a Pipe Liner Diameter	391
Determine a Liner Wall Thickness	391
Non-Pressure Pipe	391
Pressure Pipe	395
Other Loading Considerations	395
Determine the Flow Capacity	395
Design the Accesses	398
Develop the Contract Documents	401
The Sliplining Procedure	401
Other Rehabilitation Methods	410
Swagelining	410
Rolldown	411

	Titeliner	411
	Fold and Form	411
	Pipe Bursting	411
	Pipe Splitting	411
	Summary	411
	References	412
Chapter 12	**Horizontal Directional Drilling**	
	Introduction	413
	Background	413
	Polyethylene Pipe for Horizontal Directional Drilling	414
	Horizontal Directional Drilling Process	414
	Pilot Hole	414
	Pilot Hole Reaming	415
	Drilling Mud	415
	Pullback	416
	Mini-Horizontal Directional Drilling	416
	General Guidelines	416
	Safety	417
	Geotechnical Investigation	417
	Geotechnical Data for River Crossings	418
	Summary	418
	Product Design: DR Selection	419
	Design Considerations for Net External Loads	420
	Earth and Groundwater Pressure	422
	Stable Borehole - Groundwater Pressure Only	422
	Borehole Deforms/Collapse With Arching Mobilized	423
	Borehole Collapse with Prism Load	424
	Combination of Earth and Groundwater Pressure	425
	Live Loads	425
	Performance Limits	426
	Performance Limits of HDD Installed Pipe	426
	Live Loads	425
	Time-Dependent Behavior	427
	Ring Deflection (Ovalization)	427
	Ring Deflection Due to Earth Load	428
	Ring Deflection Limits (Ovality Limits)	429
	Unconstrained Buckling	430
	Wall Compressive Stress	432
	Installation Design Considerations	433
	Pullback Force	434
	Frictional Drag Resistance	434
	Capstan Force	435
	Hydrokinetic Force	436
	Tensile Stress During Pullback	437
	External Pressure During Installation	439

	Resistance to External Collapse Pressure During Pullback Installation	440
	Bending Stress	440
	Thermal Stresses and Strains	441
	Torsion Stress	441
	References	442
	Appendix A	442
	Design Calculation Example for Service Loads (Post-Installation)	442
	Example 1	442
	Example 2	444
	Example 3	445
	Solution	445
	Appendix B	447
	Design Calculations Example for Pullback Force	447
	Example 1	447

Chapter 13   HVAC Applications

Introduction	453
Ground Source Heat Pump Systems	453
Types of Ground Heat Exchangers	454
Pipe Specifications and Requirements	455
Pipe Joining Methods	458
Pipe Installation	458
Pressure Testing Ground Heat Exchanger	459
Solar Applications	460
Collector Technologies	461
Precautions	462
Installation	462
Vacuum Systems	463
Critical Buckling Under Vacuum	463
References	465

Chapter 14   Duct and Conduit

Introduction	467
Conduit Specifications	467
Applications	468
Advantages of PE Conduit	468
Installation	469
Features	469
Material Selection	470
Physical Properties	470
Cell Classification	470
Other Important Physical Properties	471
Stabilization	472
Colorants for Conduit	472

Design Considerations	473
Conduit vs. Pipe	473
Cable Dimension Considerations	473
Conduit Wall Determination	474
Installation Method vs. Short-Term and Long-Term Stress	475
Below Ground Installations	476
Open Trench/Continuous Trenching	476
Direct Plow	477
Conduit Network Pulling	478
Horizontal Directional Bore	479
Installation Methods	480
General Considerations	480
Mechanical Stress	480
Pulling Tension	480
Bending Radii	481
Underground Installation	481
Trenching Methods	481
Open Trench/Continuous Trench	481
Plowing	483
Plowing Variations	483
Directional Bores	484
Installation into Existing Conduit	484
Above Ground/Aerial	485
Installation	485
Joining Methods	486
Introduction	486
General Provisions	487
Mechanical Fittings	487
Barbed Mechanical Fittings	488
Threaded Mechanical Fittings	488
Compression Fittings	488
Expansion Joints	488
Heat Fusion	489
Butt Fusion Joining	489
Socket Fusion Joining	489
Electrofusion Joining	490
Repair Operations	490
Cable Installation	490
Pulling Cable into Conduit	490
Cable Blowing or Jetting	491
Cable Installed by the Conduit Manufacturer (Cable-in-Conduit)	492
Friction in Conduit Systems	492
Definitions	492
Friction Reduction	493
Field Effects of Friction	494

Placement Planning	495
Special Applications	496
Corrugated Duct	496
Bridge Structures	497
Underwater	497
Premise (Flame Retardant)	497
Electrical/Building Code (Conduit Entry Issues)	499
Armored (Rodent and Mechanical Protection)	499
Multi-Cell Conduit	499
Summary	499
References	500
Appendix A	500
Calculation of Frictional Forces	500
Calculations of Pulling Tensions	500

Glossary   509

    Organizations and Associations   526

Index   535

Handbook of Polyethylene Pipe

Foreword

Polyethylene piping is a comparatively new piping product. Alternate piping products such as concrete, steel, cast or ductile iron, and even PVC, have long been established within the engineering community. Polyethylene piping, on the other hand, has been utilized for a variety of piping applications for a mere 40-50 years. Despite this relatively short history, the engineering community has embraced the overall toughness and durability of HDPE pipe and the latitude afforded by the variety of installation methods that can be employed using polyethylene pipe to expand its use at a quickening rate.

Today, we see polyethylene piping systems operating in a broad array of installations; from pressure-rated potable water and natural gas lines to gravity sewers, from industrial and chemical handling to telecommunications and electrical ducting; from oil and gas production to marine installations and directional drilling.

This text has been developed to assist designers, installers and owners as they continue to look at the diversity of piping applications and help them recognize that polyethylene pipe is but one more tool available to them in the design and rehabilitation of our nation's pipeline infrastructure.

The material presented in this text has been written in a manner that is easily understood, with an emphasis on organization to provide the reader with ease of reference. It is only our efforts to be as comprehensive as possible with respect to the subject matter that have resulted in such an extensive publication.

The overall work consists of essentially three fairly discreet sections: Introductory Information, Design Considerations and Applications. Each chapter within these sections is written on somewhat of a stand-alone basis. That is, each chapter can be considered a thorough treatment of the subject material of note and the need to reference back and forth between chapters has been minimized to the extent possible. Each chapter is also annotated and industry recognized references, on which specific points are developed, are included at the end of each chapter.

The first section includes basic introductory information which reviews the origins and growth of the polyethylene pipe industry in North America, engineering properties of polyethylene, and safety considerations in the transport, handling and installation of polyethylene piping systems.

The second section, or design section, consists primarily of design considerations and includes chapters on pipe design, joining procedures, and basic information on buried and above-ground installations.

The final section of this text is comprised of a set of chapters that provide the reader with detailed information regarding design considerations, installation techniques and operation of polyethylene pipe in a variety of specific applications, such as directional drilling, marine, conduit, HVAC.

The overall work concludes with an extensive glossary and, of course, an index to provide ease of reference for specific topics of interest. The organization of the subject matter should allow the reader to quickly reference a specific area of interest or, moreover, for the college educator to utilize specific sections of the handbook within the context of a semester curriculum.

The Plastics Pipe Institute

This handbook has been developed as a result of a task group initiative within the Plastics Pipe Institute (PPI). Founded in 1950, the PPI is the major trade association representing all segments of the plastics piping industry. PPI is dedicated to the advancement of polyethylene pipe systems by:

- Contributing to the development of research, standards and design guides
- Educating designers, installers, users and government officials
- Collecting and publishing industry statistics
- Maintaining liaisons with industry, educational and government groups
- Providing a technical focus for the plastics piping industry
- Communicating up-to-date information through our website www.plasticpipe.org

The PPI develops and maintains a substantial body of engineering information and case studies related to the uses of polyethylene pipe in both potable water service, natural gas distribution and other applications.

PPI's engineering staff and members are experts in the design of polyethylene pipe systems and are available to answer your questions regarding HDPE pipe or any other PPI listed piping systems. Technical seminars are also conducted regularly for engineers and specifiers.

This handbook has been developed by the PPI as a service to the industry. The information in this handbook is offered in good faith and believed to be accurate at the time of its preparation, but is offered without any warranty, expressed or implied, including warranties of merchantability and fitness for a particular purpose. Additional information may be needed in some areas, especially with regard to unusual or special applications. In these situations, the reader is advised to consult the manufacturer or material supplier for more detailed information. A list of member companies is available from PPI.

PPI intends to revise this handbook from time to time, in response to comments and suggestions from member companies and from users of the handbook. To that end, please send suggestions for improvements to PPI. Information on other publications can be obtained by contacting PPI directly or by visiting the web site.

The Plastics Pipe Institute, Inc.
202-462-9607
www.plasticpipe.org

CHAPTER 1

Introduction

Since its discovery in 1933, polyethylene has grown to become one of the world's most widely used and recognized thermoplastic materials.[1] The versatility of this unique plastic material is demonstrated by the diversity of its use. The original application for polyethylene was as a substitute for rubber in electrical insulation during World War II. Polyethylene has since become one of the world's most widely utilized thermoplastics. Today's modern polyethylene resins are highly engineered for much more rigorous applications such as pressure-rated gas and water pipe, landfill membranes, automotive fuel tanks and other demanding applications.

Figure 1 Joining Large Diameter HDPE Pipe with Butt Fusion

Polyethylene's use as a piping material was first developed in the mid 1950's. In North America, its original use was in oil field production where a flexible, tough and lightweight piping product was needed to fulfill the needs of a rapidly developing oil and gas production industry. The success of polyethylene pipe in these installations quickly led to its use in natural gas distribution where a coilable, corrosion-free piping material could be fusion joined in the field to assure a "leak-free" method of transporting natural gas to homes and businesses. Polyethylene's success in this critical application has not gone without notice and today it is the material of choice for the natural gas distribution industry. Sources now estimate that nearly 95% of all new gas distribution pipe installations in North America that are 12" in diameter or smaller are polyethylene piping.[2]

The performance benefits of polyethylene pipe in these original oil and gas related applications have led to its use in equally demanding piping installations such as potable water distribution, industrial and mining pipe, force mains and other critical applications where a tough, ductile material is needed to assure long-term performance. It is these applications, representative of the expanding use of polyethylene pipe that are the principle subject of this handbook. In the chapters that follow, we shall examine all aspects of design and use of polyethylene pipe in a broad array of applications. From engineering properties and material science to fluid flow and burial design; from material handling and safety considerations to modern installation practices such as horizontal directional drilling and/or pipe bursting; from potable water lines to industrial slurries we will examine those qualities, properties and design considerations which have led to the growing use of polyethylene pipe in North America.

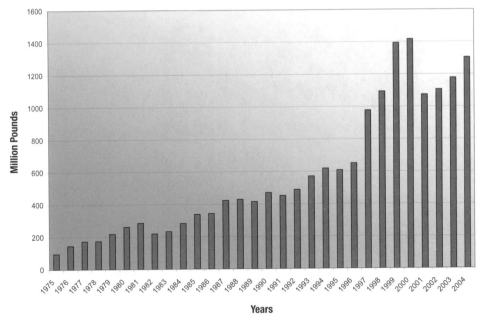

Figure 2 Historical Growth in North American HDPE Pipe Shipments[3]

Features and Benefits of HDPE Pipe

When selecting pipe materials, designers, owners and contractors specify materials that provide reliable, long-term service durability, and cost-effectiveness.

Solid wall polyethylene pipes provide a cost-effective solution for a wide range of piping applications including gas, municipal, industrial, marine, mining, landfill, and electrical and communications duct applications. Polyethylene pipe is also effective for above ground, buried, trenchless, floating and marine installations. According to David A. Willoughby, P.O.E., "…one major reason for the growth in the use of the plastic pipe is the cost savings in installation, labor and equipment as compared to traditional piping materials. Add to this the potential for lower maintenance costs and increased service life and plastic pipe is a very competitive product."[4]

Natural gas distribution was among the first applications for medium-density polyethylene (MDPE) pipe. In fact, many of the systems currently in use have been in continuous service since 1960 with great success. Today, polyethylene pipe represents over 95% of the pipe installed for natural gas distribution in diameters up to 12" in the U.S. and Canada. PE is the material of choice not only in North America, but also worldwide. PE pipe has been used in potable water applications for almost 50 years, and has been continuously gaining approval and growth in municipalities. PE pipe is specified and/or approved in accordance with AWWA, NSF, and ASTM standards.

Some of the specific benefits of HDPE pipe are discussed in the parargraphs which follow.

- **Life Cycle Cost Savings** For municipal applications, the life cycle cost of HDPE pipe can be significantly less than other pipe materials. The extremely smooth inside surface of HDPE pipe maintains its exceptional flow characteristics, and butt fusion joining eliminates leakage. This has proven to be a successful combination for reducing total system operating costs.

- **Leak Free, Fully Restrained Joints** HDPE heat fusion joining forms leak-free joints as strong as, or stronger than, the pipe itself. For municipal applications, fused joints eliminate the potential leak points that exist every 10 to 20 feet when using the bell and spigot type joints associated with other piping products such as PVC or ductile iron. As a result of this, the "allowable water leakage" for HDPE pipe is zero as compared to the water leakage rates of 10% or greater typically associate with these other piping products. HDPE pipe's fused joints are also self-restraining, eliminating the need for costly thrust restraints or thrust blocks while still insuring the integrity of the joint and the flow stream. Notwithstanding the advantages of the butt fusion method of joining, the engineer also has other available means for joining HDPE pipe and fittings such as electrofusion and mechanical fittings. Electrofusion fittings join the pipe and/or fittings together using embedded electric heating elements. In some situations, mechanical fittings may be required to facilitate joining to other piping products, valves or other system appurtenances. Specialized fittings for these purposes have been developed and are readily available to meet the needs of most demanding applications.

- **Corrosion & Chemical Resistance** HDPE pipe will not rust, rot, pit, corrode, tuberculate or support biological growth. It has superb chemical resistance and is the material of choice for many harsh chemical environments. Although unaffected by chemically aggressive native soil, installation of PE pipe (as with any piping material) through areas where soils are contaminated with organic solvents
(oil, gasoline) may require installation methods that protect the PE pipe against contact with organic solvents. Protective installation measures that assure the quality of the fluid being transported are typically required for all piping systems that are installed in contaminated soils.
- **Fatigue Resistance and Flexibility** HDPE pipe can be field bent to a radius of 30 times the nominal pipe diameter or less depending on wall thickness (12" HDPE pipe, for example, can be cold formed in the field to a 32-foot radius). This eliminates many of the fittings otherwise required for directional changes in piping systems. The long-term durability of HDPE pipe has been extremely well researched. HDPE has exceptional fatigue resistance and when, operating at maximum operating pressure, it can withstand multiple surge pressure events up to 100% above its maximum operating pressure without any negative effect to its long-term performance capability.
- **Seismic Resistance** The physical attributes that allow HDPE pressure pipe to safely accommodate repetitive pressure surges above the static pressure rating of the pipe, combined with HDPE's natural flexibility and fully restrained butt fusion joints, make it well suited for installation in dynamic soil environments and in areas prone to earthquakes or other seismic activity.

Figure 3 Butt Fused HDPE Pipe "Arced" for Insertion into Directional Drilling Installation

- **Construction Advantages** HDPE pipe's combination of light weight, flexibility and leak-free, fully restrained joints permits unique and cost-effective installation methods that are not practical with alternate materials. Installation methods such as horizontal directional drilling, pipe bursting, sliplining, plow and plant, and submerged or floating pipe, can save considerable time and money on many installations. At approximately one-eighth the weight of comparable steel pipe, and with integral and robust joining methods, installation is simpler, and it does not need heavy lifting equipment. Polyethylene pipe is produced in straight lengths up to 50 feet and coiled in diameters up through 6". Coiled lengths over 1000 feet are available in certain diameters. Polyethylene pipe can withstand impact better than PVC pipe, especially in cold weather installations where other pipes are more prone to cracks and breaks.

- **Durability** Polyethylene pipe installations are cost-effective and have long-term cost advantages due to the pipe's physical properties, leak-free joints and reduced maintenance costs. The polyethylene pipe industry estimates a service life for HDPE pipe to be, conservatively, 50-100 years provided that the system has been properly designed, installed and operated in accordance with industry established practice and the manufacturer's recommendations. This longevity confers savings in replacement costs for generations to come. Properly designed and installed PE piping systems require little on-going maintenance. PE pipe is resistant to most ordinary chemicals and is not susceptible to galvanic corrosion or electrolysis.

Figure 4 HDPE Pipe Weighted and Floated for Marine Installation

- **Hydraulically Efficient** For water applications, HDPE pipe's Hazen Williams C factor is 150 and does not change over time. The C factor for other typical pipe materials such as PVC or ductile iron systems declines dramatically over time due to corrosion and tuberculation or biological build-up. Without corrosion, tuberculation, or biological growth HDPE pipe maintains its smooth interior wall and its flow capabilities indefinitely to insure hydraulic efficiency over the intended design life.

- **Temperature Resistance** PE pipe's typical operating temperature range is from -40°F to 140°F for pressure service. Extensive testing at very low ambient temperatures indicates that these conditions do not have an adverse effect on pipe strength or performance characteristics. Many of the polyethylene resins used in HDPE pipe are stress rated not only at the standard temperature, 73° F, but also at an elevated temperature, such as 140°F. Typically, HDPE materials retain greater strength at elevated temperatures compared to other thermoplastic materials such as PVC. At 140° F, polyethylene materials retain about 50% of their 73°F strength, compared to PVC which loses nearly 80% of its 73° F strength when placed in service at 140°F.[5] As a result, HDPE pipe materials can be used for a variety of piping applications across a very broad temperature range.

The features and benefits of HDPE are quite extensive, and some of the more notable qualities have been delineated in the preceding paragraphs. For more specific information regarding these qualities and the research on which these performance attributes is based, the reader is encouraged to reference the information which is presented in the remaining chapters of this handbook.

Many of the performance properties of HDPE piping are the direct result of two important physical properties associated with HDPE pressure rated piping products. These are ductility and visco-elasticity. The reader is encouraged to keep these two properties in mind when reviewing the subsequent chapters of this handbook.

Ductility

Ductility is the ability of a material to deform in response to stress without fracture or, ultimately, failure. It is also sometimes referred to as strainabilty and it is an important performance feature of PE piping, both for above and below ground service. For example, in response to earth loading, the vertical diameter of buried PE pipe is slightly reduced. This reduction causes a slight increase in horizontal diameter, which activates lateral soil forces that tend to stabilize the pipe against further deformation. This yields a process that produces a soil-pipe structure that is capable of safely supporting vertical earth and other loads that can fracture pipes of greater strength but lower strain capacity.

With its unique molecular structure, HDPE pipe has a very high strain capacity thus assuring ductile performance over a very broad range of service conditions. Materials with high strain capacity typically shed or transfer localized stresses through deformation response to surrounding regions of the material that are subject to lesser degrees of stress. As a result of this transfer process, stress intensification is significantly reduced or does not occur, and the long-term performance of the material is sustained.

Materials with low ductility or strain capacity respond differently. Strain sensitive materials are designed on the basis of a complex analysis of stresses and the potential for stress intensification in certain regions within the material. When any of these stresses exceed the design limit of the material, crack development occurs which can lead to ultimate failure of the part or product. However, with materials like polyethylene pipe that operate in the ductile state, a larger localized deformation can take place without causing irreversible material damage such as the development of small cracks. Instead, the resultant localized deformation results in redistribution and a significant lessening of localized stresses, with no adverse effect on the piping material. As a result, the structural design with materials that perform in the ductile state can generally be based on average stresses, a fact that greatly simplifies design protocol.

To ensure the availability of sufficient ductility (strain capacity) special requirements are developed and included into specifications for structural materials intended to operate in the ductile state; for example, the requirements that have been established for "ductile iron" and mild steel pipes. Similar ductility requirements have also been established for PE piping materials. Validation requirements have been added to PE piping specifications that work to exclude from pressure piping any material that exhibits insufficient resistance to crack initiation and growth when subjected to loading that is sustained over very long periods of time, i.e. any material which does not demonstrate ductility or strainability. The PE piping material validation procedure is described in the chapter on Engineering Properties of Polyethylene.

Visco-Elasticity

Polyethylene pipe is a visco-elastic construction material.[6] Due to its molecular nature, polyethylene is a complex combination of elastic-like and fluid-like elements. As a result, this material displays properties that are intermediate to crystalline metals and very high viscosity fluids. This concept is discussed in more detail in the chapter on Engineering Properties within this handbook.

The visco-elastic nature of polyethylene results in two unique engineering characteristics that are employed in the design of HDPE water piping systems, creep and stress relaxation.

Creep is the time dependent viscous flow component of deformation. It refers to the response of polyethylene, over time, to a constant static load. When HDPE is subjected to a constant static load, it deforms immediately to a strain predicted by the stress-strain modulus determined from the tensile stress-strain curve. At high loads, the material continues to deform at an ever decreasing rate, and if the load is high enough, the material may finally yield or rupture. Polyethylene piping materials are designed in accordance with rigid industry standards to assure that, when used in accordance with industry recommended practice, the resultant deformation due to sustained loading, or creep, is too small to be of engineering concern.

Stress relaxation is another unique property arising from the visco-elastic nature of polyethylene. When subjected to a constant strain (deformation of a specific degree) that is maintained over time, the load or stress generated by the deformation slowly decreases over time. This stress relaxation response to loading is of considerable importance to the design of polyethylene piping systems.

As a visco-elastic material, the response of polyethylene piping systems to loading is time-dependent. The effective modulus of elasticity is significantly reduced by the duration of the loading because of the creep and stress relaxation characteristics of polyethylene. An instantaneous modulus for sudden events such as water hammer can be as high as 150,000 psi at 73°F. For slightly longer duration, but short-term events such as soil settlement and live loadings, the short-term modulus for polyethylene is roughly 110,000 to 120,000 psi at 73° F, and as a long-term property, the modulus is reduced to something on the order of 20,000-30,000 psi. As will be seen in the chapters that follow, this modulus is a key criterion for the long-term design of polyethylene piping systems.

This same time-dependent response to loading also gives polyethylene its unique resiliency and resistance to sudden, comparatively short-term loading phenomena. Such is the case with polyethylene's resistance to water hammer phenomenon which will be discussed in more detail in subsequent sections of this handbook.

Summary

As can been seen from our brief discussions here, polyethylene piping is a tough, durable piping material with unique performance properties that allow for its use in a broad range of applications utilizing a variety of different construction techniques based upon project needs. The chapters that follow offer detailed information regarding the engineering properties of polyethylene, guidance on design of polyethylene piping systems, installation techniques as well as background information on how polyethylene pipe and fittings are produced, and appropriate material handling guidelines. Information such as this is intended to provide the basis for sound design and the successful installation and operation of polyethylene piping systems. It is to this end, that members of the Plastics Pipe Institute have prepared the information in this handbook.

References:
1. *The History of Plastics.* (2005, May). www.americanplasticscouncil.org.
2. Mruk, S. (1985, November). Plastic Pipe in Gas Distribution Twenty-Five Years of Achievement, *Gas Industries.*
3. Shipments of Polyethylene Pipe, PPI Statistics Report. (2003). Plastics Pipe Institute, Washington DC.
4. Willoughby, D. A. (2002). *Plastic Piping Handbook*, McGraw-Hill Publications, New York.
5. PVC Pipe – Design and Installation. (2004). *AWWA Manual M-23*, American Water Works Association, Denver.
6. PE Pipe – Design and Installation. (2005). *AWWA Manual M-55*, American Water Works Association, Denver.

CHAPTER 2

Inspections, Tests and Safety Considerations

Scope

Once a polyethylene piping system has been selected and designed for an application, the design is implemented by securing the pipe, fittings and other necessary appurtenances, installing the system, and placing it in service. Piping installation involves setting various parts, people, and machines in motion to obtain, assemble, install, inspect and test the piping system. Whenever machinery, piping parts, and personnel are engaged in piping system construction, safety must be a primary consideration.

This chapter presents some of the inspections, tests and safety considerations related to installing polyethylene piping, placing an installed system into service, and operating a polyethylene piping system.

This chapter does not purport to address all of the product applications, inspections, tests, or construction practices that could be used, nor all of the safety practices necessary to protect persons and property. It is the responsibility of the users of this chapter, and the installers, inspectors and operators of piping systems to establish appropriate safety and health practices, and to determine the applicability of regulatory limitations before any use, installation, inspection, test or operation.

Introduction

Generally, piping system installation begins with obtaining the pipe, fittings, and other goods required for the system. Assembly and installation follow, then system testing and, finally, release for operation. Throughout the installation process, various inspections and tests are performed to ensure installed system quality, and that the system when completed is capable of functioning according to its design specifications. In the selection, design, and installation of polyethylene piping systems, professional engineering services and qualified installers should be used.

Handling and Storage

After the piping system has been designed and specified, the piping system components must be obtained. Typically, project management and purchasing personnel work closely together so that the necessary components are available when they are needed for the upcoming construction work.

Receiving Inspection

Few things are more frustrating and time consuming than not having what you need, when you need it. Before piping system installation begins, an important initial step is a receiving inspection of incoming products. Construction costs can be minimized, and schedules maintained, by checking incoming goods to be sure the parts received are the parts that were ordered, and that they arrived in good condition and are ready for installation.

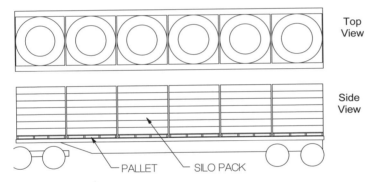

Figure 1 Typical Silo Bulk Pack Truckload

Polyethylene pipe, fittings, and fabrications are usually shipped by commercial carriers who are responsible for the products from the time they leave the manufacturing plant until they are accepted by the receiver. Pipe and fabricated fittings and structures are usually shipped on flatbed trailers. Smaller fittings may be shipped in enclosed vans, or on flatbed trailers depending upon size and packaging. Molded fittings are usually boxed, and shipped by commercial parcel services.

Product Packaging

Depending on size, polyethylene piping is produced in coils or in straight lengths. Coils are stacked together into silo packs. Straight lengths are bundled together in bulk packs or loaded on the trailer in strip loads. Standard straight lengths for conventionally extruded pipe are 40' long; however, lengths up to 60' long may be produced. Profile extruded pipes are typically produced in 20' lengths. State

Chapter 2
Inspections, Tests and Safety Considerations

transportation restrictions on length, height and width usually govern allowable load configurations. Higher freight costs will apply to loads that exceed length, height, or width restrictions. Although polyethylene pipe is lightweight, weight limitations may restrict load size for very heavy wall or longer length pipe.

Extruded profile pipe lengths are usually shipped on standard 40' flatbed trailers. Pipes are commonly packaged in strip loads. Pipes 96" ID (2438 mm ID) and 120" ID (3048 mm ID) will exceed 8' overall width, and are subject to wide load restrictions.

Figures 1 through 3 are general illustrations of truckload and packaging configurations. Actual truckloads and packaging may vary from the illustrations.

Small fittings are packaged in cartons which may be shipped individually by package carriers. Large orders may be palletized and shipped in enclosed vans. Large fittings and custom fabrications may be packed in large boxes on pallets, or secured to pallets.

Occasionally, when coiled pipe silos and boxed fittings are shipped together, fitting cartons are placed in the center of the silo packs. Tanks, manholes, and large fittings and custom fabrications are usually loaded directly onto flatbed trailers.

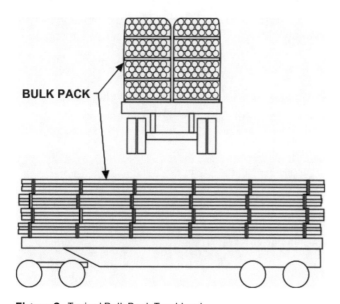

Figure 2 Typical Bulk Pack Truckload

Checking the Order

When a shipment is received, it should be checked to see that the correct products and quantities have been delivered. Several documents are used here. The **Purchase Order** or the **Order Acknowledgment** lists each item by its description, and the required quantity. The incoming load will be described in a **Packing List** which is attached to the load. The descriptions and quantities on the Packing List should match those on the Purchase Order or the Order Acknowledgment.

The carrier will present a Bill of Lading that generally describes the load as the number of packages the carrier received from the manufacturing plant. The Order Acknowledgment, Packing List, and Bill of Lading should all be in agreement. Any discrepancies must be reconciled between the shipper, the carrier, and the receiver. The receiver should have a procedure for reconciling any such discrepancies.

Load Inspection

There is no substitute for visually inspecting an incoming shipment to verify that the paperwork accurately describes the load. Products are usually identified by markings on each individual product. These markings should be checked against the Order Acknowledgment and the Packing List. The number of packages and their descriptions should be checked against the Bill of Lading.

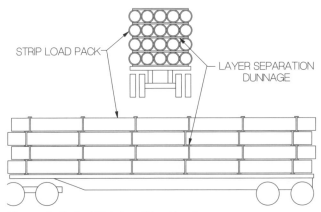

Figure 3 Typical Strip Load Truckload

This is the time to inspect for damage which may occur anytime products are handled. Obvious damage such as cuts, abrasions, scrapes, gouges, tears, and punctures should be carefully inspected.

Receiving Report and Reporting Damage

The delivering truck driver will ask the person receiving the shipment to sign the Bill of Lading, and acknowledge that the load was received in good condition. Any damage, missing packages, etc. should be noted on the bill of lading at that time.

Shipping problems such as damage, missing packages, document discrepancies, incorrect product, etc. should be reported to the product supplier immediately. Shipping claims must be filed as soon as possible as required by trade practice.

Unloading Instructions

Before unloading the shipment, there must be adequate, level space to unload the shipment. The truck should be on level ground with the parking brake set and the wheels chocked. Unloading equipment must be capable of safely lifting and moving pipe, fittings, fabrications or other components.

WARNING: Unloading and handling must be performed safely. Unsafe handling can result in damage to property or equipment, and be hazardous to persons in the area. Keep unnecessary persons away from the area during unloading.

WARNING: Only properly trained personnel should operate unloading equipment.

Unloading Site Requirements

The unloading site must be relatively flat and level. It must be large enough for the carrier's truck, the load handling equipment and its movement, and for temporary load storage. Silo packs and other palletized packages should be unloaded from the side with a forklift. Non-palletized pipe, fittings, fabrications, manholes, tanks, or other components should be unloaded from above with lifting equipment and wide web slings, or from the side with a forklift. The lifting slings/cables must be capable of safely carrying the weight of the product being unloaded. Damaged slings/cables must not be used.

Handling Equipment

Appropriate unloading and handling equipment of adequate capacity must be used to unload the truck. Safe handling and operating procedures must be observed.

Pipe must not be rolled or pushed off the truck. Pipe, fittings, fabrications, tanks, manholes, and other components must not be pushed or dumped off the truck, or dropped.

Although polyethylene piping components are lightweight compared to similar components made of metal, concrete, clay, or other materials, larger components can be heavy. Lifting and handling equipment including cables and slings must have adequate rated capacity to lift and move components from the truck to temporary storage. Equipment such as a forklift, a crane, a side boom tractor, or an extension boom crane may be used for unloading.

When using a forklift, or forklift attachments on equipment such as articulated loaders or bucket loaders, lifting capacity must be adequate at the load center on the forks. Forklift equipment is rated for a maximum lifting capacity at a distance from the back of the forks. (See Figure 4.) If the weight-center of the load is farther out on the forks, lifting capacity is reduced.

Before lifting or transporting the load, forks should be spread as wide apart as practical, forks should extend completely under the load, and the load should be as far back on the forks as possible.

WARNING: During transport, a load on forks that are too short or too close together, or a load too far out on the forks, may become unstable and pitch forward or to the side, and result in damage to the load or property, or hazards to persons.

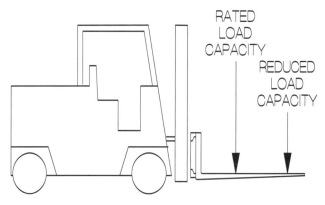

Figure 4 Forklift Load Capacity

Lifting equipment such as cranes, extension boom cranes, and side boom tractors, should be hooked to wide web choker slings that are secured around the load or to lifting lugs on the component. Only wide web slings should be used. Wire rope slings and chains can damage components, and should not be used. Spreader bars should be used when lifting pipe or components longer than 20'.

WARNING: Before use, inspect slings and lifting equipment. Equipment with wear or damage that impairs function or load capacity should not be used.

Unloading Large Fabrications, Manholes and Tanks

Large fabrications, manholes and tanks should be unloaded using a wide web choker sling and lifting equipment such as an extension boom crane, crane or lifting boom. The choker sling is fitted around the manhole riser or near the top of the tank. Do not use stub outs, outlets, or fittings as lifting points, and avoid placing slings where they will bear against outlets or fittings. Larger diameter manholes and tanks are typically fitted with lifting lugs.

WARNING: All lifting lugs must be used. The weight of the manhole or tank is properly supported only when all lugs are used for lifting. Do not lift tanks or manholes containing liquids.

Pre-Installation Storage

The size and complexity of the project and the components will determine pre-installation storage requirements. For some projects, several storage or staging sites along the right-of-way may be appropriate, while a single storage location may be suitable for another job.

The site and its layout should provide protection against physical damage to components. General requirements are for the area to be of sufficient size to accommodate piping components, to allow room for handling equipment to get around them, and to have a relatively smooth, level surface free of stones, debris, or other material that could damage pipe or components, or interfere with handling. Pipe may be placed on 4-inch wide wooden dunnage, evenly spaced at intervals of 4 feet or less.

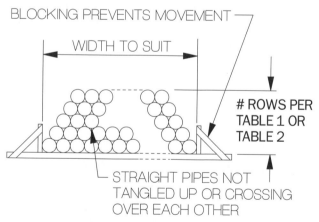

Figure 5 Loose Pipe Storage

TABLE 1
Suggested Jobsite Loose Storage Stacking Heights for Conventionally Extruded Pipe Lengths[1]

Pipe Size	Suggested Stacking Height* - Rows	
	DR Above 17	DR 17 & Below
4	15	12
5	12	10
6	10	8
8	8	6
10	6	5
12	5	4
14	5	4
16	4	3
18	4	3
20	3	3
22	3	2
24	3	2
26	3	2
28	2	2
30	2	2
32	2	2
36	2	1
42	1	1
48	1	1
54	1	1
63	1	1

* Stacking heights based on 6' for level terrain and 4' for less level terrain.

Pipe received in bulk packs or strip load packs should be stored in the same package. If the storage site is flat and level, bulk packs or strip load packs may be stacked evenly upon each other to an overall height of about 6'. For less flat or less level terrain, limit stacking height to about 4'.

Before removing individual pipe lengths from bulk packs or strip load packs, the pack must be removed from the storage stack, and placed on the ground.

Individual pipes may be stacked in rows. Pipes should be laid straight, not crossing over or entangled with each other. The base row must be blocked to prevent sideways movement or shifting. (See Figure 5, Table 1, and Table 2.) The interior of stored pipe should be kept free of debris and other foreign matter.

TABLE 2

Suggested Jobsite Loose Storage Stacking Heights for Extruded Profile Pipe[2]

Pipe Size	Suggested Stacking Height* Rows
18	4
21	3
27	2
30	2
33	2
36	1
42	1
48	1
54	1
60	1
66	1
72	1
84	1
96	1
120	1

*Suggested stacking heights based on 6' for level terrain and 4' for less level terrain.

Exposure to UV and Weather

Polyethylene pipe products are protected against deterioration from exposure to ultraviolet light and weathering effects. Color and black products are compounded with antioxidants, thermal stabilizers, and UV stabilizers. Color products use sacrificial UV stabilizers that absorb UV energy and are eventually depleted. In general, non-black products should not remain in unprotected outdoor storage for more than 2 years; however, some manufacturers may allow longer unprotected outside storage. Black products contain at least 2% carbon black to protect the material from UV deterioration. Black products with and without stripes are generally suitable for unlimited outdoor storage and for service on the surface or above grade.[11, 12, 13, 14]

Cold Weather Handling

Temperatures near or below freezing will affect polyethylene pipe by reducing flexibility and increasing vulnerability to impact damage. Care should be taken not to drop pipe, or fabricated structures, and to keep handling equipment and other things from hitting pipe. Ice, snow, and rain are not harmful to the material, but may make storage areas more troublesome for handling equipment and personnel. Unsure footing and traction require greater care and caution to prevent damage or injury.

Walking on pipe can be dangerous. Inclement weather can make pipe surfaces especially slippery.

WARNING: Keep safety first on the jobsite; do not walk on pipe.

General Considerations During Installation

Joining and Connections

For satisfactory material and product performance, system designs and installation methods rely on appropriate, properly made connections. An inadequate or improperly made field joint may cause installation delays, may disable or impair system operations, or may create hazardous conditions.

Polyethylene piping products are connected using heat fusion, electrofusion, thermal welding, and mechanical methods such as gasketed bell-and-spigot joints, flanges, and compression couplings. Joining and connection methods will vary depending upon requirements for internal or external pressure, leak tightness, restraint against longitudinal movement (thrust load capacity), gasketing requirements, construction and installation requirements, and the product.

Connection design limitations and manufacturer's joining procedures must be observed. Otherwise, the connection or products adjacent to the connection may leak or fail, which may result in property damage or hazards to persons.

The tools and components required to construct and install joints in accordance with manufacturer's recommendations should always be used. However, field connections are controlled by, and are the responsibility of, the field installer.

Field Joining

All field connection methods and procedures require that the component ends to be connected must be clean, dry, and free of detrimental surface defects before the connection is made. Contamination and unsuitable surface conditions usually produce an unsatisfactory connection. Gasketed joints may require appropriate lubrication.

Cleaning

Before joining, and before any special surface preparation, surfaces must be clean and dry. General dust and light soil may be removed by wiping the surfaces with clean, dry, lint-free cloths. Heavier soil may be washed or scrubbed off with soap and water solutions, followed by thorough rinsing with clear water, and drying with dry, clean, lint-free cloths.

WARNING: Before using chemical cleaning solvents, the potential risks and hazards to persons should be known by the user, and appropriate safety precautions must be taken. Chemical solvents may be hazardous substances that may require special handling and personal protective equipment.

The manufacturer's instructions for use, and the material safety data sheet (MSDS) for the chemical should be consulted for information on risks to persons and for safe handling and use procedures. Some solvents may leave a residue on the pipe, or may be incompatible with the material. See PPI Technical Report TR-19, Thermoplastics Piping for the Transport of Chemicals for additional information on chemical compatibility of polyethylene materials.[4]

Field Fusion Joining

Heat fusion joining requires specialized equipment for socket, saddle, or butt fusion, or electrofusion. Heat fusion joining may be performed in any season. During inclement weather, a temporary shelter should be set up over the joining operation to shield heat fusion operations from rain, frozen precipitation and cold winds. It is strongly recommended that installers consult with the manufacturer's recommended joining procedures before installation begins.

WARNING: Most heat fusion equipment is not explosion-proof. The fusion equipment manufacturer's safety instructions must be observed at all times and especially when heat fusion is to be performed in a potentially volatile atmosphere.

WARNING: When installing large diameter polyethylene pipe in a butt fusion machine, do not bend the pipe against an open fusion machine collet or clamp. The pipe may suddenly slip out of the open clamp, and cause injury or damage.

Field Handling

Polyethylene pipe is tough, lightweight, and flexible. Installation does not usually require high capacity lifting equipment. See "Handling and Storage" for information on handling and lifting equipment.

WARNING: To prevent injury to persons or property, safe handling and construction practices must be observed at all times. The installer must observe all applicable Local, State, and Federal Safety Codes, and any safety requirements specified by the owner or the project engineer.

Pipe up to about 8" (219 mm) diameter and weighing roughly 6 lbs per foot (20 kg per m) or less can usually be handled or placed in the trench manually. Heavier, larger diameter pipe will require appropriate handling equipment to lift, move and lower the pipe. Pipe must not be dumped, dropped, pushed, or rolled into a trench.

WARNING: Appropriate safety precautions must be observed whenever persons are in or near a trench.

Coiled lengths and long strings of heat-fused polyethylene pipe may be cold bent in the field. Field bending usually involves sweeping or pulling the pipe string into the desired bend radius, then installing permanent restraint such as embedment around a buried pipe, to maintain the bend.

WARNING: Considerable force may be required to field bend the pipe, and the pipe may spring back forcibly if holding devices slip or are inadvertently released while bending. Observe appropriate safety precautions during field bending.

These paragraphs have attempted to convey the primary safety and handling considerations associated with joining and connecting polyethylene pipe. For a more thorough discussion on the joining methods used with polyethylene pipe, the reader is referred to the joining procedures chapter in this Handbook, and PPI's TR-33 Generic Fusion Procedures.

Inspection and Testing

Pre-Construction

Inspections and tests begin before construction. Jobsite conditions dictate how piping may be installed and what equipment is appropriate for construction. Soil test borings and test excavations may be useful to determine soil bearing stress and whether or not native soils are suitable as backfill materials.

In slipline rehabilitation applications, the deteriorated pipeline should be inspected by remote TV camera to locate structurally deteriorated areas, obstructions, offset and separated joints, undocumented bends, and service connections. In some cases, a test pull, drawing a short section of slipliner through the line, may be conducted to ensure that the line is free of obstructions.

The installer should carefully review contract specifications and plans. It is important that the specifications and plans fit the job. Different piping materials require different construction practices and procedures. These differences should be accurately reflected in the contract documents. Good plans and specifications help protect all parties from unnecessary claims and liabilities. Good documents also set minimum installation quality requirements, and the testing and inspection requirements that apply during the job.

All incoming materials should be inspected to be sure that sufficient quantities of the correct products for the job are at hand, and that they arrived in good condition, ready for installation.

During Construction

Tests and inspections performed during construction include butt fusion joint quality tests, soil compaction and density tests, pipe deflection tests, pressure tests, and other relevant inspections. Fusion joint qualification and inspection guidelines for butt, socket and saddle fusions should be obtained from the pipe or fitting manufacturer.

Butt Fusion Joint Quality

Visual inspection is the most common joint evaluation method for all sizes of conventionally extruded polyethylene pipe. Visual inspection criteria for butt fusion joints should be obtained from the pipe manufacturer. Computer controlled ultrasonic inspection equipment is available for 12" IPS and smaller pipes with walls 1" or less in thickness. Equipment for larger diameters and wall thickness is being developed. X-ray inspection is generally unreliable because x-ray is a poor indicator of fusion quality.

When butt fusion is between pipe and molded fittings, the fitting-side bead may exhibit shape irregularities which are caused by the fitting manufacturing process. A slightly irregular fitting-side bead may not indicate an improper joint, provided that the pipe-side bead is properly shaped, and the v-groove between the beads is correct. Contact the pipe or fitting manufacturer if assistance is required.

Fusion joining may be destructively tested to confirm joint integrity, operator procedure, and fusion machine set-up. A field-performed destructive test is a bent strap test.

The bent strap test specimen is prepared by making a trial butt fusion, usually the first fusion of the day, and allowing it to cool to ambient temperature. A test strap that is at least 6" or 15 pipe wall thicknesses long on each side of the fusion, and about 1" or 1-½ wall thicknesses wide, is cut out of the trial fusion pipe. (See Figure 6.) The strap is then bent so that the ends of the strap touch. Any disbondment at the fusion is unacceptable, and indicates poor fusion quality. If failure occurs, fusion procedures and/or machine set-up should be changed, and a new trial fusion and bent strap test specimen should be prepared and tested. Field fusion should not proceed until a test joint has passed the bent strap test.

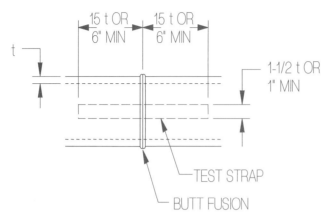

Figure 6 Bent Strap Test Specimen

WARNING: A test strap from thick wall pipe may require considerable effort to bend. Further, the test strap may spring back if the ends are inadvertently released while bending. Appropriate personnel safety precautions should be observed.

Soil Tests

During buried pipe installation, work should be checked throughout the construction period by an inspector who is thoroughly familiar with the jobsite, contract specifications, materials, and installation procedures. Inspections should reasonably ensure that significant factors such as trench depth, grade, pipe foundation (if required), quality and compaction of embedment backfill, and safety are in compliance with contract specifications and other requirements. To evaluate soil stability, density and compaction, appropriate ASTM tests may be required in the contract specifications.

Pipe Surface Damage

Surface damage may occur during construction handling and installation. Significant damage may impair the future performance of the pipeline. The following guidelines may be used to assess surface damage significance.

For polyethylene pressure pipelines, damage or butt fusion misalignment should not exceed 10% of the minimum wall thickness required for the pipeline's operating pressure.[5] Deep cuts, abrasions or grooves cannot be field repaired by hot gas or extrusion welding. Excessive damage may require removal and replacement of the damaged pipe section, or reinforcement with a full encirclement repair clamp. Severely misaligned butt fusions (>10% wall offset) should be cut out and redone.

If damage is not excessive, the shape of the damage may be a consideration. Sharp notches and cuts should be dressed smooth so the notch is blunted. Blunt scrapes or gouges should not require attention. Minor surface abrasion from sliding on the ground or insertion into a casing should not be of concern.

Deflection Tests

Buried flexible pipes rely on properly installed backfill to sustain earthloads and other loads. Proper installation requires using the backfill materials specified by the designer, and installing the pipe as specified by the designer.

Large diameter extruded profile pipes, and larger diameter, high DR conventionally extruded pipes are inherently flexible. Pipe deflection can be used to monitor the installation quality. Improperly embedded pipe can develop significant deflection in a short time, thus alerting the installer and the inspector to investigate the problem. Inspection should be performed as the job progresses, so errors in the installation procedure can be identified and corrected.

Figure 7 Determining Initial Deflection

Initial deflection checks of extruded profile pipe may be performed after embedment materials have been placed and compacted. The inside diameter of the pipe is measured after backfill materials have been placed to the pipe crown, and compacted. This is D1. Then final backfill materials are placed and compacted, and the pipe inside diameter is measured again at the exact location where the prior measurement was taken. This is D2. (See Figure 7.)

Percent initial deflection is calculated using the following:

$$\%Deflection = \left(\frac{D1 - D2}{D1}\right)100$$

Where D1 and D2 are as defined above and depicted in Figure 7.

Another method to measure deflection is to pull a pre-sized mandrel through the pipe. The mandrel should be sized so that if the pipe exceeds allowable deflection, the mandrel is blocked. (Calculations for mandrel sizing must include some allowance for fusion joint bead size and its reduction of the internal clearance.)

To properly size the mandrel, the allowable vertical diameter of the pipe must be established. It is necessary to account for pipe ID manufacturing tolerances and any ovality that may occur during shipping. Pipe base ID dimensions and tolerances should be obtained from the manufacturer. The maximum mandrel diameter is calculated as follows:

$$D_M = D - \left(\frac{Dy}{100}\right)$$

WHERE
DM = maximum mandrel diameter, in
D = base pipe ID, in

$$D = D_i - \sqrt{A^2 + B^2}$$

Di = nominal pipe ID, in
A = ID manufacturing tolerance, in
B = shipping ovality, in

$$B = 0.03 \, D_i$$

y = allowable deflection, percent

Deflection tests of conventionally extruded pipe may be performed in the same manner. However, conventionally extruded pipe is manufactured to a controlled outside diameter, so the inside diameter is subject to the combined tolerances of the outside diameter and the wall thickness.

Post Installation

Leak Testing

The intent of leak testing is to find unacceptable faults in a piping system. If such faults exist, they may manifest themselves by leakage or rupture.

Leakage tests may be performed if required in the Contract Specifications. Testing may be conducted in various ways. Internal pressure testing involves filling the test section with a nonflammable liquid or gas, then pressurizing the medium. Hydrostatic pressure testing with water is the preferred and recommended method. Other test procedures may involve paired internal or end plugs to pressure test individual joints or sections, or an initial service test. Joints may be exposed to allow inspection for leakage.

Liquids such as water are preferred as the test medium because less energy is released if the test section fails. During a pressure test, energy (internal pressure) is applied to stress the test section. If the test medium is a compressible gas, then the gas is compressed and absorbs energy while applying stress to the pipeline. If a failure occurs, both the pipeline stress energy and the gas compression energy are suddenly released. However, with an incompressible liquid such as water as the test medium, the energy release is only the energy required to stress the pipeline.

This testing methodology may not apply to gas distribution polyethylene piping systems. The municipal gas utility should be contacted to obtain their protocol for pressure testing of gas distribution pipelines. Additional guidance for testing gas pipelines may also be found in ANSI/GPTC Z380, "Guide for Gas Transmission and Distribution Piping Systems.

WARNING: Pipe system pressure testing is performed to discover unacceptable faults in a piping system. Pressure testing may cause such faults to fail by leaking or rupturing. This may result in failure. Piping system rupture may result in sudden, forcible, uncontrolled movement of system piping or components, or parts of components.

WARNING: Pipe Restraint - The pipe system under test and any closures in the test section should be restrained against sudden uncontrolled movement from failure. Test equipment should be examined before pressure is applied to insure that it is tightly connected. All low pressure filling lines and other items not subject to the test pressure should be disconnected or isolated.

WARNING: Personal Protection - Take suitable precautions to eliminate hazards to personnel near lines being tested. Keep personnel a safe distance away from the test section during testing.

Pressure Testing Precautions

The piping section under test and any closures in the test section should be restrained or otherwise restricted against sudden uncontrolled movement in the event of rupture. Expansion joints and expansion compensators should be temporarily restrained, isolated or removed during the pressure test.

Testing may be conducted on the system, or in sections. The limiting test section size is determined by test equipment capability. If the pressurizing equipment is too small, it may not be possible to complete the test within allowable testing time limits. If so, higher capacity test equipment, or a smaller test section may be necessary.

If possible, test medium and test section temperatures should be less than 100°F (38°C). At temperatures above 100°F (38°C), reduced test pressure is required. Before applying test pressure, time may be required for the test medium and the test section to temperature equalize. Contact the pipe manufacturer for technical assistance with elevated temperature pressure testing.

References
The following reference publications provide pressure testing information:

ASME B31.1 Power Piping, Section 137, "Pressure Tests."[6]

PPI TR-31 Underground Installation of Polyolefin Piping, Section 7, "System Testing."[1]

ASTM F 1417, Standard Test Method for Installation Acceptance of Plastic Gravity Sewer Lines Using Low-Pressure Air.[7]

Uni-Bell PVC Pipe Association Standard, Uni-b-6-90 Recommended Practice for Low-Pressure Air Testing of Installed Sewer Pipe.

The piping manufacturer should be consulted before using pressure testing procedures other than those presented here. Other pressure testing procedures may or may not be applicable depending upon piping products and/or piping applications.

Test Pressure
Test pressure may be limited by valves, or other devices, or lower pressure rated components. Such components may not be able to withstand the required test pressure, and should be either removed from, or isolated from, the section being tested to avoid possible damage to, or failure of, these devices. Isolated equipment should be vented.

- For continuous pressure systems where test pressure limiting components or devices have been isolated, or removed, or are not present in the test section, the maximum allowable test pressure is 1.5 times the system design pressure at the lowest elevation in the section under test.
- If the test pressure limiting device or component cannot be removed or isolated, then the limiting section or system test pressure is the maximum allowable test pressure for that device or component.
- For non-pressure, low pressure or gravity flow systems, consult the piping manufacturer for the maximum allowable test pressure.

Test Duration
For any test pressure from 1.0 to 1.5 times the system design pressure, the total test time including initial pressurization, initial expansion, and time at test pressure, must not exceed eight (8) hours. If the pressure test is not completed due to leakage,

equipment failure, etc., the test section should be de-pressurized, and allowed to "relax" for at least eight (8) hours before bringing the test section up to test pressure again.

Pre-Test Inspection

Test equipment and the pipeline should be examined before pressure is applied to ensure that connections are tight, necessary restraints are in-place and secure, and components that should be isolated or disconnected are isolated or disconnected. All low pressure filling lines and other items not subject to the test pressure should be disconnected or isolated.

Hydrostatic Testing

Hydrostatic pressure testing is preferred and is strongly recommended. The preferred testing medium is clean water. The test section should be completely filled with the test medium, taking care to bleed off any trapped air. Venting at high points may be required to purge air pockets while the test section is filling. Venting may be provided by loosening flanges, or by using equipment vents. Re-tighten any loosened flanges before applying test pressure.

Monitored Make-up Water Test

The test procedure consists of initial expansion, and test phases. During the initial expansion phase, the test section is pressurized to the test pressure, and sufficient make-up water is added each hour for three (3) hours to return to test pressure.[1]

TABLE 3
Test Phase Make-up Amount

Nominal Pipe Size, in	Make-Up Water Allowance (U.S. Gallons per 100 ft of Pipe)		
	1 Hour Test	2 Hour Test	3 Hour Test
1-14	0.06	0.10	0.16
1-1/2	0.07	0.10	0.17
2	0.07	0.11	0.19
3	0.10	0.15	0.25
4	0.13	0.25	0.40
5	0.19	0.38	0.58
5-3/8	0.21	0.41	0.62
6	0.3	0.6	0.9
7-1/8	0.4	0.7	1.0
8	0.5	1.0	1.5
10	0.8	1.3	2.1
12	1.1	2.3	3.4
13-3/8	1.2	2.5	3.7
14	1.4	2.8	4.2
16	1.7	3.3	5.0
18	2.0	4.3	6.5
20	2.8	5.5	8.0
22	3.5	7.0	10.5
24	4.5	8.9	13.3
26	5.0	10.0	15.0
28	5.5	11.1	16.8
30	6.3	12.7	19.2
32	7.0	14.3	21.5
34	8.0	16.2	24.3
36	9.0	18.0	27.0
42	12.0	23.1	35.3
48	15.0	27.0	43.0
54	18.5	31.4	51.7
63	–	–	–

After the initial expansion phase, about four (4) hours after pressurization, the test phase begins. The test phase may be one (1), two (2), or three (3) hours, after which a measured amount of make-up water is added to return to test pressure. If the amount of make-up water added does not exceed Table 3 values, leakage is not indicated.

Non-Monitored Make-Up Water Test

The test procedure consists of initial expansion, and test phases. For the initial expansion phase, make-up water is added as required to maintain the test pressure for four (4) hours. For the test phase, the test pressure is reduced by 10 psi. If the pressure remains steady (within 5% of the target value) for an hour, no leakage is indicated.

Pneumatic Testing

WARNING: Compressed air or any pressurized gas used as a test medium may present severe hazards to personnel in the vicinity of lines being tested. Extra personnel protection precautions should be observed when a gas under pressure is used as the test medium.

WARNING: Explosive Failure - Piping system rupture during pneumatic pressure testing may result in the explosive, uncontrolled movement of system piping, or components, or parts of components. Keep personnel a safe distance away from the test section during testing.

Pneumatic testing should not be used unless the Owner and the responsible Project Engineer specify pneumatic testing or approve its use as an alternative to hydrostatic testing.

Pneumatic testing (testing with a gas under pressure) should not be considered unless one of the following conditions exists:

- when the piping system is so designed that it cannot be filled with a liquid; or
- where the piping system service cannot tolerate traces of liquid testing medium.

The testing medium should be non-flammable and non-toxic. The test pressure should not exceed the maximum allowable test pressure for any non-isolated component in the test section.

Leaks may be detected using mild soap solutions (strong detergent solutions should be avoided), or other non-deleterious leak detecting fluids applied to the joint. Bubbles indicate leakage. After leak testing, all soap solutions or leak detecting fluids should be rinsed off the system with clean water.

High Pressure Procedure

For continuous pressure rated pipe systems, the pressure in the test section should be gradually increased to not more than one-half of the test pressure, then increased in small increments until the required test pressure is reached. Test pressure should be maintained for ten (10) to sixty (60) minutes, then reduced to the design pressure rating, and maintained for such time as required to examine the system for leaks.

Low Pressure Procedure

For components rated for low pressure service, the specified rated test pressure should be maintained for ten (10) minutes to one (1) hour, but not more than one (1) hour. **Test pressure ratings must not be exceeded.**

Leakage inspections may be performed during this time. If the test pressure remains steady (within 5% of the target value) for the one (1) hour test time, no leakage is indicated.

Pressure testing of gravity-flow sewer lines should be conducted in accordance with ASTM F 1417, Standard Test Method for Installation Acceptance of Plastic Gravity Sewer Lines Using Low-Pressure Air.[7]

Initial Service Testing

An initial service test may be acceptable when other types of tests are not practical, or where leak tightness can be demonstrated by normal service, or when initial service tests of other equipment are performed. An initial service test may apply to systems where isolation or temporary closures are impractical, or where checking out pumps and other equipment affords the opportunity to examine the system for leakage prior to full-scale operations.

Test Procedure

The piping system should be gradually brought up to normal operating pressure, and held at operating pressure for at least ten (10) minutes. During this time, joints and connections should be examined for visual evidence of leakage.

Non-Testable Systems

Some systems may not be suitable for pressure testing. These systems may contain non-isolatable components, or temporary closures may not be practical. Such systems should be carefully inspected during and after installation. Inspections such as visual examination of joint appearance, mechanical checks of bolt or joint tightness, and other relevant examinations should be performed.

Considerations for Post Start-Up and Operation

Disinfecting Water Mains

Applicable procedures for disinfecting new and repaired potable water mains are presented in standards such as ANSI/AWWA C651, *Disinfecting Water Mains*.[8] ANSI/AWWA C651 uses liquid chlorine, sodium hypochlorite, or calcium hypochlorite to chemically disinfect the main. Disinfecting solutions containing chlorine should not exceed 12% active chlorine, because greater concentration can chemically attack and degrade polyethylene.

Warning/Caution: After disinfection, all strong concentrations of disinfection solution must be purged from the main pipeline, lateral pipelines, and even service lines. All strong solutions must be flushed with purified potable water prior to isolation or commissioning. Leave no disinfection solution in stagnant or deadend pipe runs.

Cleaning

Pipelines operating at low flow rates (around 2 ft/sec or less) may allow solids to settle in the pipe invert. Polyethylene has a smooth, non-wetting surface that resists the adherence of sedimentation deposits. If the pipeline is occasionally subject to higher flow rates, much of the sedimentation will be flushed from the system during these peak flows. If cleaning is required, sedimentation deposits can usually be flushed from the system with high pressure water.

Water-jet cleaning is available from commercial services. It usually employs high-pressure water sprays from a nozzle that is drawn through the pipe system with a cable.

Pressure piping systems may be cleaned with the water-jet process, or may be pigged. Pigging involves forcing a resilient plastic plug (soft pig) through the pipeline. Usually, hydrostatic or pneumatic pressure is applied behind the pig to move it down the pipeline. Pigging should employ a pig launcher and a pig catcher.

A pig launcher is a wye or a removable spool. In the wye, the pig is fitted into the branch, then the branch behind the pig is pressurized to move the pig into the pipeline and downstream. In the removable pipe spool, the pig is loaded into the spool, the spool is installed into the pipeline, and then the pig is forced downstream.

WARNING: A pig may discharge from the pipeline with considerable velocity and force. The pig catcher is a basket or other device at the end of the line designed to receive the pig when it discharges from the pipeline. The pig catcher provides a means of safe pig discharge from the pipeline.

WARNING: Soft pigs must be used with polyethylene pipe. Scraping finger type or bucket type pigs may severely damage a polyethylene pipe and must not be used.

Commercial pigging services are available if line pigging is required.

Squeeze-Off

Squeeze-off (or pinch-off) is a means of controlling flow in smaller diameter polyethylene pipe and tubing by flattening the pipe between parallel bars. Flow control does not imply complete flow stoppage in all cases. For larger pipes, particularly at higher pressures, some seepage is likely. If the situation will not allow seepage, then it may be necessary to vent the pipe between two squeeze-offs.

Polyethylene gas pipe manufactured to ASTM D 2513 is suitable for squeeze-off; however, squeeze-off practices are not limited to gas applications. Squeeze-off is applicable to polyethylene PE3408 and PE2406 pressure pipe up to 24" IPS, and up to 100 psi internal pressure. Larger sizes and higher pressures may be possible, but suitable commercial equipment is not widely available, so there is limited experience with larger sizes or higher pressures.

WARNING: Squeeze-off is applicable ONLY to PE2406 and PE3408 polyethylene pipe and tubing. The pipe or tubing manufacturer should be consulted to determine if squeeze-off is applicable to his product, and for specific squeeze-off procedures.

Tools[9]

Squeeze-off tools should have:

- parallel bars that are shaped to avoid pipe damage,
- mechanical stops to prevent over-squeeze pipe damage,
- safety mechanisms to prevent accidental release, and
- a mechanism that controls the rate of closure, and the rate of release.

Typical bar shapes are single round bars, twin round bars, or flat bars with rounded edges. Other bar shapes may also be suitable as long as edge radius requirements are met. See Table 4.

Positive mechanical stops between the bars are essential to prevent over-squeeze and pipe damage. The stops limit bar closure to 70% of twice the maximum wall thickness.[9] For DR or SDR sized polyethylene pipe ONLY, (not tubing or schedule sized pipe), stop distance may be determined by:

$$g = 1.568 \left(\frac{D}{DR} \right)$$

WHERE
g = stop gap, in
D = pipe outside diameter, in
DR = pipe dimension ratio

TABLE 4
Squeeze Tool Bar Radius

Pipe Diameter	Minimum Bar Radius
≤ 0.750	0.50
> 0.750 ≤ 2.375	0.63
> 2.375 ≤ 4.500	0.75
> 4.500 ≤ 8.625	1.00
> 8.625 ≤ 16.000	1.50

Consult the pipe manufacturer for stopgap dimensions for tubing sizes.

Typical squeeze-off tools use either a manual mechanical screw or hydraulic cylinders. In either case, a mechanism to prevent accidental bar separation is an essential safety feature of the tool.

Key Elements

Closing and opening rates are key elements to squeezing-off without damaging the pipe. It is necessary to close slowly and release slowly, with slow release being more important. The release rate for squeeze-off should be 0.50 inches/minute or less as specified in ASTM F1041[10] or ASTM F1563.[9] The pipe must be allowed sufficient time to adjust to the high compressive and tensile stresses applied to the pipe's inside wall during squeeze-off.

Research work performed under contract to the Gas Research Institute indicates that the greatest damage potential is during release, especially with heavier wall pipes. Flattening places high compressive stress on the inside wall at the outer edges of the squeeze. Then releasing and opening applies high tensile stress to the same area. The material must be given ample time to accommodate these stresses. Opening too fast may cause excessive strain, and may damage the inside wall.

Lower temperatures will reduce material flexibility and ductility, so in colder weather, closure and opening time must be slowed further.

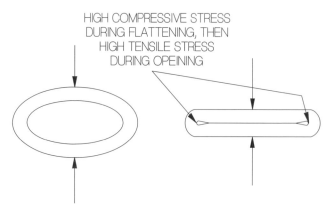

Figure 8 Squeeze-off Stresses

Testing of PE2406 and PE3408 polyethylene piping has shown that when proper procedures and tools are used, squeeze-off can be performed without compromising the expected service life of the system. However, pipe can be damaged during squeeze-off:

- if the manufacturer's recommended procedures are not followed, or
- if the squeeze is held closed too long, or
- from static electric discharge, or
- by altering or circumventing the closure stops, or
- by squeezing-off more than once in the same location.

WARNING: Pipe damaged during squeeze-off could leak or fail at the squeeze-off point. Pipe known or suspected to have been damaged during squeeze-off should be removed from the system, or should be reinforced at the squeeze-off point using a full encirclement clamp.

Procedure

From the installation of the squeeze-off tool to its removal, the total time should not exceed 8 hours. Excessive time may damage the pipe.

1. Select the correct size squeeze tool for the pipe being squeezed. Squeeze bar closure stops must be in place, and must be the correct size for the pipe's diameter and wall thickness or DR.

2. Fit the tool on the pipe so the pipe is centered in the tool, and the tool is square to the pipe. The squeeze-off tool must be at least 3 pipe diameters, or 12 inches, whichever is greater, away from any butt fusion, or any socket, saddle, or mechanical fitting.

WARNING: Static electricity control - When pipe conveying a compressed gas is being flattened, the gas flow velocity through the flattened area increases. High velocity, dry gas, especially with particles present in the flow, can generate a static electric charge on pipe surfaces which can discharge to ground. Before flattening the pipe, the tool should be grounded and procedures to control static charge build-up on pipe surfaces should be employed. Grounding and static control procedures should remain in place for the entire procedure.

3. Operate the bar closing mechanism and, at a controlled rate, flatten the pipe between the bars.

3.1 For 3" IPS and larger pipe, pause at least 1 minute when the pipe is flattened halfway and another minute when 3/4 closed. For all pipe sizes, pause 1 minute when the pipe inside walls make contact.

3.2 After pausing 1 minute when the pipe inside walls make contact, continue closing at about half the prior closing rate until the tool bars contact the closure stops.

3.3 If temperatures are near freezing or lower, closure rates should be halved and pauses should be doubled.

4. If necessary, engage the accidental release prevention mechanism.

5. Perform the necessary work downstream of the squeeze-off.

WARNING: Venting may be required for 100% shut-off. Squeeze-off may not stop all flow. If 100% shut-off is required, it may be necessary to install two squeeze-off tools at two points along the line, and vent between them. Any work performed must be downstream of the second squeeze-off. Do not remove or alter the closure stops, or place anything (rags, sticks, etc.) between the bars and the pipe.

6. When work is complete, disengage the accidental release mechanism (if required), and open the squeeze-off tool bars at a controlled rate no faster than the Step 3 closure rate. Opening must include a 1 minute pause at the wall contact point, and 1 minute pauses at 1/4 open (3/4 closed) and 1/2 open points for 3" IPS and larger pipes.

7. Open the bars, and remove the squeeze-off tool.

8. Identify the squeezed-off area by wrapping tape around the pipe, or installing a full encirclement clamp over the area.

WARNING: Do not squeeze off more than once in the same place. Doing so may damage the pipe.

Additional information on squeeze-off may be found in ASTM F 1041, *Standard Guide for Squeeze-off of Polyolefin Gas Pressure Pipe and Tubing*.[10]

Routine or Emergency?

Squeeze-off procedures may be used for routine, scheduled changes to piping systems, or as an emergency procedure to control gasses or liquids escaping from a damaged pipe. For scheduled piping changes, the above procedure should be followed, and if followed, the pipe's service life is not expected to be compromised.

However, an emergency situation may require quickly flattening the pipe and controlling flow because the escaping fluid may be an immediate hazard of greater concern than damaging the pipe.

WARNING: If an emergency situation requires rapid flattening, then the pipe or tubing will probably be damaged.

Repairs

Repair situations may arise if a polyethylene pipe has been damaged. Damage may occur during shipping and handling, during installation, or after installation. Damage may include scrapes or abrasions, breaks, punctures, kinks or emergency squeeze-off. Permanent repair usually involves removing and replacing the damaged pipe or fitting. In some cases, temporary repairs may restore sufficient serviceability and allow time to schedule permanent repairs in the near future.

Refer to TN-35, Repair of HDPE Pipelines, at www.plasticpipe.org for more information.

Damage Assessment

Damaged pipe or fittings should be inspected and evaluated. Pipe, fittings, fabrications or structures with excessive damage should not be installed. Damage that occurs after installation may require that the damaged pipe or component be removed and replaced.

WARNING: Scrapes or gouges in pressure pipe cannot be repaired by filling-in with extrusion or hot air welding. The damaged section should be removed and replaced.

WARNING: Improperly made fusion joints cannot be repaired.

Improper butt fusions must be cut out and re-done from the beginning. Poorly joined socket or electrofusion fittings must be removed and replaced. Poorly joined saddle fittings must be removed by cutting out the main pipe section, or, if the main is undamaged, made unusable by cutting the branch outlet or chimney off the saddle fitting, and installing a new saddle fitting on a new section of main.

WARNING: Broken or damaged fittings cannot be repaired and, as such, should be removed and replaced.

WARNING: Kinked pipe must not be installed and cannot be repaired.

Kinked pipe must be removed and replaced.

WARNING: Pipe damaged during an emergency squeeze-off cannot be repaired.

Squeeze-off damaged pipe must be removed and replaced.

Permanent Repairs

For buried large diameter polyethylene pipe that has been poorly backfilled, excessive deflection may be correctable using point excavation to remove backfill, then reinstalling embedment materials in accordance with recommended procedures.

Where replacement is required, any joining method appropriate to the product and service requirements is generally acceptable. Butt and socket fusion joining procedures require that one of the components move longitudinally. However, constrained installations, such as buried pipes, may not allow such movement.

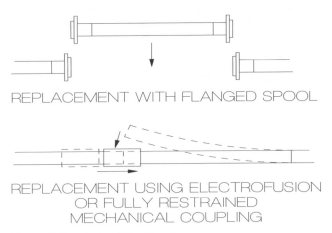

REPLACEMENT WITH FLANGED SPOOL

REPLACEMENT USING ELECTROFUSION OR FULLY RESTRAINED MECHANICAL COUPLING

Figure 9 Constrained Pipe Repair

Permanent repair of constrained pipe typically employs techniques that do not require longitudinal movement of one or both pipe ends. Techniques include deflecting one pipe end to the side, using a mechanical or electrofusion coupling, or installing a flanged spool. See Figure 9. Typical methods for joining repair pipe sections include flanges, electrofusion couplings, and fully restrained mechanical couplings.

To repair using a flanged spool, cut out, remove and discard the damaged pipe section. Install flanges on the two pipe ends. Measure the distance between the flange sealing surfaces, and prepare a flanged pipe spool of the same length. Install the flanged spool.

Repair using an electrofusion coupling or a fully restrained mechanical coupling is limited to pipe sizes for which such couplings are available. Mechanical or electrofusion coupling repairs are made by deflecting one pipe end to the side for the coupling body to be slipped on. The pipe ends are then realigned, and the coupling joint fitted up. To allow lateral deflection, a length of about 10 times the pipe outside diameter is needed.

Temporary Repair

Until permanent repairs can be effected, temporary repairs may be needed to seal leaks or punctures, to restore pressure capacity, or to reinforce damaged areas. Temporary repair methods include, but are not limited to, mechanical repair couplings and welded patches.

Conclusion

A successful piping system installation is dependent on a number of factors. Obviously, a sound design and the specification and selection of the appropriate quality materials are paramount to the long-term performance of any engineered installation. The handling, inspection, testing, and safety considerations that surround the placement and use of these engineered products is of equal importance.

In this chapter, we have attempted to provide fundamental guidelines regarding the receipt, inspection, handling, storage, testing and repair of polyethylene piping products. While this chapter cannot address all of the product applications, test and inspection procedures, or construction practices, it does point out the need to exercise responsible care in planning out these aspects of any job site. It is the responsibility of the contractor, installer, site engineer or other users of these materials to establish appropriate safety and health practices specific to the job site and in accordance with the local prevailing codes that will result in a safe and effective installation.

References

1. Plastics Pipe Institute, *Handbook of Polyethylene Pipe*, chapter on Underground Installation of Polyethylene Pipe, Washington, DC.
2. Gilroy, H.M. (1985). Polyolefin Longevity for Telephone Service, Antec Proceedings.
3. Plastics Pipe Institute, *Handbook of Polyethylene Pipe*, chapter on Polyethylene Joining Procedures, Washington, DC.
4. Plastics Pipe Institute, TR-19, Thermoplastics Piping for the Transport of Chemicals, Washington, DC.
5. American Gas Association. (1994). *AGA Plastic Pipe Manual for Gas Service*, AGA.
6. American Society of Mechanical Engineers, B31.1, Section 137.
7. American Society for Testing and Materials. (1998). ASTM F1417, Standard Test Method for Installation Acceptance of Plastic Gravity Sewer Lines Using Low Pressure Air, West Conshohocken, PA.
8. American Water Works Association. (1992). AWWA Standard for Disinfecting Water Mains, Denver, CO.
9. American Society for Testing and Materials. (1998). ASTM F1563, Standard Specification for Tools to Squeeze-off Polyethylene (PE) Gas Pipe or Tubing, West Conshohocken, PA.

10. American Society for Testing and Materials. (1998). ASTM F1041, Standard Guide for Squeeze-off of Polyethylene Gas Pressure Pipe or Tubing, West Conshohocken, PA.
11. American Society for Testing and Materials.(1998). ASTM D2104, Standard Specification for Polyethylene (PE) Plastic Pipe, Schedule 40, West Conshohocken, PA.
12. American Society for Testing and Materials. (1998). ASTM D2239, Standard Specification for Polyethylene (PE) Plastic Pipe (SIDR-PR) Based on Controlled Inside Diameter, West Conshohocken, PA.
13. American Society for Testing and Materials. (1998). ASTM D2447, Standard Specification for Polyethylene (PE) Plastic Pipe, Schedules 40 and 80, Based on Outside Diameter, West Conshohocken, PA.
14. American Society for Testing and Materials. (1998). ASTM D3035, Standard Specification for Polyethylene (PE) Plastic Pipe (DR-PR) Based on Controlled Outside Diameter, West Conshohocken, PA.

Chapter 3

Engineering Properties

Scope
This chapter reviews the history, structure and the fundamental properties of polyethylene (PE) pipe. A basic understanding of the physical and chemical nature of polyethylene and of its engineering behavior is very important for the proper design and installation of this material.

Polyethylene Plastics

Plastics are solid materials that contain one or more polymeric substances which can be shaped by flow. Polymers, the basic ingredient of plastics, compose a broad class of materials that include natural and synthetic polymers. Nearly all plastics are made from the latter. In commercial practice, polymers are frequently designated as resins. For example, a polyethylene pipe compound consists of polyethylene resin combined with colorants, stabilizers, anti-oxidants or other ingredients required to protect and enhance properties during fabrication and service.

Plastics are divided into two basic groups, thermoplastics and thermosets, both of which are used to produce plastic pipe.

Thermoplastics include compositions of polyethylene, polypropylene, polybutylene and polyvinyl chloride (PVC). These can be re-melted upon the application of heat. The solid state of thermoplastics is the result of physical forces that immobilize polymer chains and prevent them from slipping past each other. When heat is applied, these forces weaken and allow the material to soften or melt. Upon cooling, the molecular chains stop slipping and are held firmly against each other in the solid state. Thermoplastics can be shaped during the molten phase of the resin and therefore can be extruded or molded into a variety of shapes, such as pipe flanges or valves.

Thermoset plastics are similar to thermoplastics prior to "curing," a chemical reaction by which polymer chains are chemically bonded to each other by new cross-links. The curing is usually done during or right after the shaping of the final product. Cross-linking is the random bonding of molecules to each other to form a giant three-dimensional network. Thermoset resins form a permanent insoluble and infusible shape after the application of heat or a curing agent. They cannot be re-melted after they have been shaped and cured. This is the main difference between thermosets and thermoplastics. As heat is applied to a thermoset part, degradation occurs at a temperature lower than the melting point. The properties of thermosetting resins make it possible to combine these materials with reinforcements to form strong composites. Fiberglass is the most popular reinforcement, and fiberglass-reinforced pipe (FRP) is the most common form of thermoset-type pipe.

History of Polyethylene

The Imperial Chemical Company (ICI) in England first invented polyethylene in 1933.[24] ICI did not commercialize the production of polyethylene until 1939 when the product was used to insulate telephone cables and coaxial cables, the latter being a very important element in the development of radar during World War II. The early polymerization processes used high-pressure (14,000 to 44,000 psi) autoclave reactors and temperatures of 200° to 600° F (93° to 316° C). The polyethylene that came from these reactors was called "high pressure polyethylene." It was produced in a free radical chain reaction by combining ethylene gas under high pressure with peroxide or a trace amount of oxygen.

The original process was dangerous and expensive, so other safer and less expensive processes were developed. Polyethylene produced at low pressure was introduced in the 1950's. These methods also afforded greater versatility in tailoring molecular structures through variations in catalysts, temperatures, and pressures.

Manufacture of Polyethylene

Polymers are large molecules formed by the polymerization (i.e. the chemical linking) of repeating small molecular units. To produce polyethylene, the starting unit is ethylene, a colorless gas composed of two double-bonded carbon atoms and four hydrogen atoms (see Figure 1).

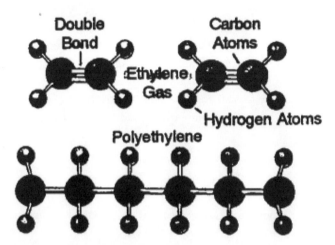

Figure 1 Manufacture of Polyethylene

There are currently three primary low-pressure methods for producing polyethylene: gas-phase, solution and slurry (liquid phase). The polymerization of ethylene may take place with various types of catalysts, under varying conditions of pressure and temperature and in reactor systems of radically different design. Ethylene can also be copolymerized with small amounts of other monomers such as butene, propylene, hexene, and octene. This type of copolymerization results in small modifications in chemical structure, which are reflected in certain differences in properties, such as density, ductility, hardness, etc. Resins that are produced without comonomer are called homopolymers.

Regardless of process type, the chemical process is the same. Under reaction conditions, the double bond between the carbon atoms is broken, allowing a bond to form with another carbon atom as shown in Figure 1. Thus, a single chain of polyethylene is formed. This process is repeated until the reaction is terminated and the chain length is fixed. Polyethylene is made by the linking of thousands of monomeric units of ethylene.

Polymer Characteristics

Polyethylene resins can be described by three basic characteristics that greatly influence the processing and end-use properties: density, molecular weight and molecular weight distribution. The physical properties and processing characteristics of any polyethylene resin require an understanding of the roles played by these three major parameters.

Density

The earliest production of polyethylene was done using the high-pressure process which resulted in a product that contained considerable "side branching." Side branching is the random bonding of short polymer chains to the main polymer chain. Since branched chains are unable to pack together very tightly, the resulting material had a relatively low density, which led to it being named low-density polyethylene (LDPE).

As time passed and polyethylenes of different degrees of branching were produced, there was a need for an industry standard that would classify the resin according to density. The American Society for Testing of Materials (ASTM) established the following classification system, still in use today. It is a part of ASTM D1248, *Standard Specification for Polyethylene Plastics Molding and Extrusion Materials*[2,5].

Type	Density
I	0.910 - 0.925 (low)
II	0.926 - 0.940 (medium)
III	0.941 - 0.959 (high)
IV	0.960 and above (high, homopolymer)

Type I is a low-density resin produced mainly in high-pressure processes. Also contained within this range are the linear-low-density polyethylenes (LLDPE), which represent a recent development in the polyethylene area using low-pressure processes.

Type II is a medium density resin produced either by low- or high-pressure processes.

Types III and IV are high-density polyethylenes. Type III materials are usually produced with a small amount of a comonomer (typically butene or hexene) that is used to control chain branching. Controlled branching results in improved performance in applications where certain types of stresses are involved. Type IV resins are referred to as homopolymers since only ethylene is used in the polymerization process, which results in least-branched and highest-possible-density material. Figure 2 depicts the various molecular structures associated with each type of polyethylene.

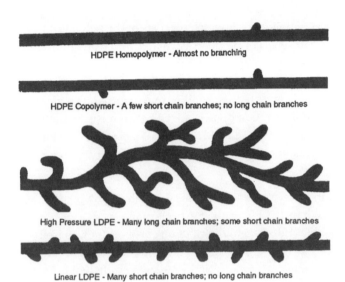

Figure 2 Chain Structure of Polyethylene

Crystallinity

The amount of side branching determines the density of the polyethylene molecule. The more side branches, the lower the density. The packing phenomenon that occurs in polyethylene can also be explained in terms of crystalline versus non-crystalline or amorphous regions as illustrated in Figure 3. When molecules pack together in tight formation, the intermolecular spacing is reduced.

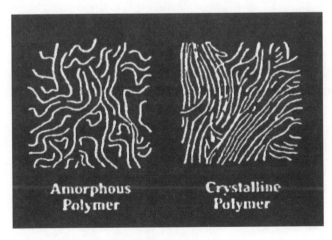

Figure 3 Crystallinity in Polyethylene

Polyethylene is one of a number of polymers in which portions of the polymer chain in certain regions align themselves in closely packed and very well ordered arrangements of polyhedral-shaped, microscopic crystals called spherulites. Other portions of the polymer chain lie in amorphous regions having no definite molecular arrangement. Since polyethylene contains both crystalline and amorphous regions, it is called a semicrystalline material. Certain grades of HDPE can consist of up to 90% crystalline regions compared to 40% for LDPE. Because of their closer packing, crystalline regions are denser than amorphous regions. Polymer density, therefore, reflects the degree of crystallinity.

As chain branches are added to a polyethylene backbone through co-polymerization, the site and frequency of chain branches affect other aspects of the crystalline/amorphous network. This includes the site and distribution of spherulites, as well as the nature of the intermediate network of molecules that are between spherulites. For example, using butene as co-monomer results in the following "ethyl" side chain structure[8]:

$$(-CH_2-CH_2-CH_2-CH_2-CH_2-)n$$
$$\quad\quad\quad\quad | $$
$$\quad\quad\quad CH_2$$
$$\quad\quad\quad\quad | $$
$$\quad\quad\quad CH_3$$

or using hexene results in this "butyl" side chain:

$$(-CH_2-CH_2-CH_2-CH_2-CH_2-)n$$
$$\quad\quad\quad CH_2$$
$$\quad\quad\quad CH_2$$
$$\quad\quad\quad CH_2$$
$$\quad\quad\quad CH_3$$

If two polymers were produced, one using ethyl and the other butyl, the polymer that contained the butyl branches would have a lower density. Longer side branching reduces crystallinity and therefore lowers density. For high-density polyethylene, the number of short chain branches is on the order of 3 to 4 side chains per 1,000 carbon atoms. It only takes a small amount of branching to affect the density.

Resin density influences a number of physical properties. Characteristics such as tensile yield strength and stiffness (flexural or tensile modulus) are increased as density is increased.

Molecular Weight

The size of a polymer molecule is represented by its molecular weight, which is the total of the atomic weights of all the atoms that make up the molecule. Molecular weight exerts a great influence on the processability and the final physical and mechanical properties of the polymer. Thermoplastics for piping systems are of high molecular weight (over 100,000) but not so high as to hamper shaping during manufacture or subsequent operations such as heat fusion.

Molecular weight is controlled during the manufacturing process. The amount of length variation is usually determined by catalyst, conditions of polymerization, and type of process used. During the production of polyethylene, not all molecules grow to the same length. Since the polymer contains molecules of different lengths, the molecular weight is usually expressed as an average value.

There are various ways to express average molecular weight, but the most common is the number average (Mn) and weight average (Mw). The definitions of these terms are as follows:

M_n = Total weight of all molecules ÷ Total number of molecules
M_w = (Total weight of each size) (respective weights) ÷ Total weight of all molecules

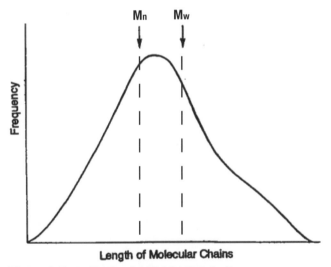

Figure 4 Typical Molecular Weight Distribution

Figure 4 illustrates the significance of these terms and includes other less frequently used terms for describing molecular weight.

Molecular weight is the main factor that determines the durability-related properties of a polymer. Long-term strength, toughness, ductility, and fatigue-endurance improve as the molecular weight increases. The current grades of highly durable materials result from the high molecular weight of the polymer.

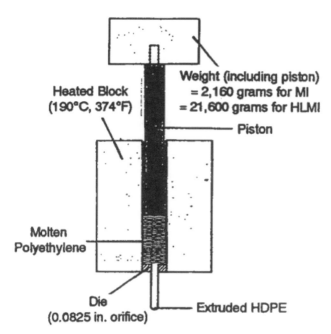

Figure 5 The Melt Index Test (per ASTM D1238)

Molecular weight affects a polymer's melt viscosity or its ability to flow in the molten state. The standard method used to determine this "flowability" is the melt flow rate apparatus, which is shown in Figure 5. ASTM D1238, *Standard Test Method for Flow Rates of Thermoplastics by Extrusion Plastometer*[2], is the industry standard for measuring the melt flow rate. The test apparatus measures the amount of material that passes through a certain size orifice in a given period of time when extruded at a predetermined temperature and under a specified weight The melt flow rate is the calculated amount of material that passes through the orifice in ten minutes.
The standard nomenclature for melt flow rate, as described in ASTM D1238, lists the test temperature and weight used. A typical designation is condition 190/2.16 that indicates the test was conducted at a temperature of 190°C while using a 2.16-kg weight on top of the piston. Other common weights include: 5 kg, 10 kg, 15 kg and 21.6 kg.

The term "melt index"(MI) is the melt flow rate when measured under a particular set of standard conditions – 190°C/2.16 kg. This term is commonly used throughout the polyethylene industry.

Melt flow rate is a rough guide to the molecular weight and processability of the polymer. This number is inversely related to molecular weight. Resins that have a low molecular weight flow through the orifice easily and are said to have a high melt flow rate. Longer chain length resins resist flow and have a low melt flow rate. The melt flow rates of these very viscous (stiff) resins are very difficult to measure under the common conditions specified by this test. Therefore, another procedure is used where the weight is increased to 21.6 kg from the 2.16 kg weight used in the normal test procedure. This measurement is commonly referred to as the High Load Melt Index (HLMI) or 10X scale. There are other melt flow rate scales that use 5 kg, 10 kg or 15 kg weights.

There are various elaborate analytical techniques for determining molecular weight of a polymer. The melt flow rate gives a very quick, simple indication of the molecular weight. The more sophisticated methods include Gel Permeation Chromatography (GPC). The essence of GPC is to dissolve the polymer in a solvent and then inject the solution into a column (tubing). The column contains a porous packing material that retards the movements of the various polymer chains as they flow through the column under pressure. The time for the polymer to pass through the column depends upon the length of the particular polymer chain. Shorter chains take the longest time due to a greater number of possible pathways. Longer chain molecules will pass more quickly since they are retained in fewer pores. This method measures the distribution of the lengths of polymer chains along with the average molecular weight.

Effect of Molecular Weight Distribution on Properties

The distribution of different sized molecules in a polyethylene polymer typically follows the bell shaped normal distribution curve described by Gaussian probability theory. As with other populations, the bell shaped curve can reflect distributions ranging from narrow to broad. A polymer with a narrow molecular weight distribution (MWD) contains molecules that are nearly the same in molecular weight. It will crystallize at a faster, more uniform rate. This results in a part that will have less warpage.

A polymer that contains a broader range of chain lengths, from *short* to *long* is said to have a broad MWD. Resins with this type of distribution have good Environmental Stress Crack Resistance (ESCR), good impact resistance and good processability.

Polymers can also have a bimodal shaped distribution curve which, as the name suggests, seem to depict a blend of two different polymer populations, each with its particular average and distribution. Resins having a bimodal MWD contain both very short and very long polyethylene molecules, giving the resin excellent physical properties while maintaining good processability. Figure 6 shows the difference in these various distributions.

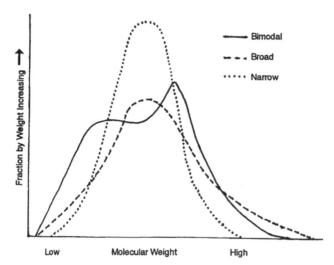

Figure 6 Molecular Weight Distribution

MWD is very dependent upon the type of process used to manufacture the particular polyethylene resin. For polymers of the same density and average molecular weight, their melt flow rates are relatively independent of MWD. Therefore, resins that have the same density and MI can have very different molecular weight distributions. The effects of density, molecular weight, and molecular weight distribution on physical properties are summarized in Table 1.

TABLE 1
Effects of Changes in Density, Melt Index, and Molecular Weight Distribution

Property	As Density Increases, Property	As Melt Index Increases, Property	As Molecular Wt. Distribution Broadens, Property
Tensile Strength (@ Yield)	Increases	Decreases	—
Stiffness	Increases	Decreases Slightly	Decreases Slightly
Impact Strength	Decreases	Decreases	Decreases
Low Temperature Brittleness	Increases	Increases	Decreases
Abrasion Resistance	Increases	Decreases	—
Hardness	Increases	Decreases Slightly	—
Softening Point	Increases	—	Increases
Stress Crack Resistance	Decreases	Decreases	Increases
Permeability	Decreases	Increases Slightly	—
Chemical Resistance	Increases	Decreases	—
Melt Strength	—	Decreases	Increases
Gloss	Increases	Increases	Decreases
Haze	Decreases	Decreases	—
Shrinkage	Increases	Decreases	Increases

Mechanical Properties

Establishing Long-Term Design Properties

In the case of metal piping, the conventional tensile test is relied upon to define basic mechanical properties such as elastic strength, proportional limit and yield strength. These are important for defining and specifying the pipe material. They are also basic constants for use in the many design equations that have been developed based upon elastic theory, where strain is always assumed to be proportional to stress. With plastics there is no such proportionality. The relationship between stress and strain is greatly influenced by duration of loading (e.g., rate of straining in a tensile test), temperature and environment. As depicted in Figure 3.7, the stress/strain response for polyethylene is profoundly dependent on the tensile test conditions. In addition, the stress-strain response is curvilinear. Even though near the origin there might appear to be an essentially linear response, in reality there is never a zone of true proportionality between stress and strain.

Accordingly, plastics have no true elastic constants, such as elastic modulus or proportional limit, nor do they have sharply defined yield points.

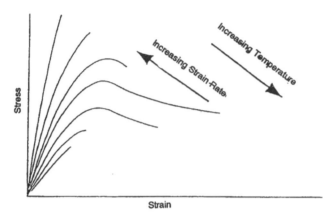

Figure 7 Schematic of Tensile Stress - Strain Response of a Thermoplastic Exhibiting Ductility at Intermediate Strain Rates

The values of moduli derived from tensile tests only represent the initial portion, either the secant or the tangent, of the stress-strain curve and only for the particular conditions of the test (see *Tensile Properties*). The primary value of the modulus, yield strength and other short-term properties of plastics is for defining and classifying materials. Strength and stiffness values that have been determined by means of short-term tests are not suitable constants for use in the large body of equations that have been derived on the assumption of elastic behavior. However, most of these equations can be, and are, used with plastics provided their strength and rigidity are defined by property values that give consideration to their non-elastic behavior. Test methods and systems for developing and applying such information are described in ensuing paragraphs.

Since polyethylene is composed of both crystalline and amorphous areas, its mechanical behavior is complex. The crystalline regions primarily account for the elastic response to forces, whereas the amorphous regions account for the viscous fluid-like response. The overall mechanical response to applied forces is called "viscoelastic" since it lies between these two types of behavior.

Models have been developed to understand, quantify, and characterize viscoelastic behavior[19,21]. The spring is used to demonstrate ideal elastic behavior. The deformation of the spring is directly proportional to the force needed to pull the spring. The relationship between force and deformation is $F = Kx$, where x represents the distance pulled, commonly called strain, and K is the spring stiffness. This relationship is known as Hooke's Law. An elastic material returns to its original length when the load (force) is removed. This is true because the spring has stored the applied force (energy) and has returned practically all of it back to the material.

A dashpot can be used to represent an ideal viscous material. When a force is applied to pull the dashpot, the amount of deformation (strain) is independent of the force but proportional to the velocity (v) at which the force is applied. This is shown mathematically as $F \cong v$. The dashpot will not return to its original position once the force is released. This is true because the energy is not stored in the dashpot but is fully spent in deforming the purely viscous material.

Since polyethylene behaves as both an elastic and viscous material, its behavior can be modeled by combining springs and dashpots together into a very simple configuration known as the Maxwell Model[40] as shown in Figure 8. The springs represent the elastic behavior, each one having a different spring constant.
The dashpots, each containing a different viscosity fluid, represent the viscous behavior. By combining various numbers of springs and dashpots, the stress-strain relationships for different plastics can be approximated.

These models may be used to characterize the stress-strain relationship of plastics as a function of duration of loading, temperature, and environment. However, there are other methods that express the stress-strain and fracture strength which are more commonly used for engineering design. These are based on tensile creep, stress-relaxation, and stress-rupture data that have been obtained on the subject material.

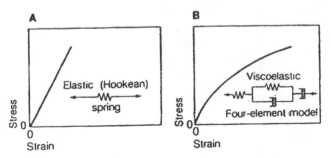

Figure 8 The Maxwell Model

Tensile Creep Curves
When a constant load is applied to a plastic part, it deforms quickly to an initial strain (deformation). It then continues to deform at a slower rate for an indefinite amount of time or until rupture occurs. This secondary deformation is termed creep. In ductile plastics, rupture is usually preceded by a stage of accelerated creep or yielding. In nonductile plastics, rupture occurs during creep. These typical responses are illustrated in Figure 9, which has been drawn on cartesian coordinate. As the stress level increases, so does the strain; however, it is not a linear relationship. A doubling of stress will not double the strain except at small strains or short times.

Temperature will also affect the strain. For a given stress level and time, a higher temperature will increase the strain.

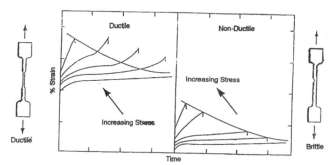

Figure 9 Schematic of Creep Rupture of Thermoplastics in Tension

A more practical way of representing tensile creep and creep-rupture information over the longer times of engineering interest is to plot the data on log-log coordinates. This method, which also facilitates graphical interpolation of data, is illustrated by Figure 10, which was developed for a high density polyethylene resin.

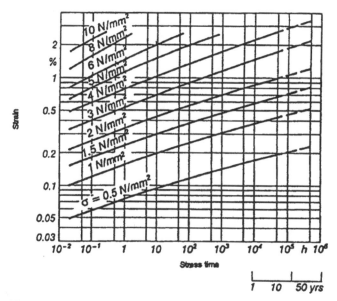

Figure 10 Strain-Time Curves at Constant Stress
Note: 1N/mm2 = 145 psi, Temperature = 73°4F (23°C)

Tensile Creep or Apparent Modulus

Any point on any creep curve gives a stress/strain ratio. The value of this ratio, E_c, is termed the creep modulus or apparent modulus. It is specific for the conditions where the stress is prescribed and the strain is free to vary. It is used in design equations in place of the tensile modulus.

The creep or apparent modulus is defined as the initial applied stress divided by the creep-strain at a given time and temperature, in units of lbs/in^2 (N/mm^2). The modulus decreases as the duration of loading increases.

When designing a pipeline for a 50-year life, the long-term tensile creep modulus of polyethylene should be used. This value will range from 20,000 psi to 30,000 psi depending on the type of polyethylene pipe material. As a comparison, the short-term modulus, derived from short-term tensile tests, is between 100,000 to 130,000 psi. Figure 11 represents typical tensile creep moduli data for polyethylene.

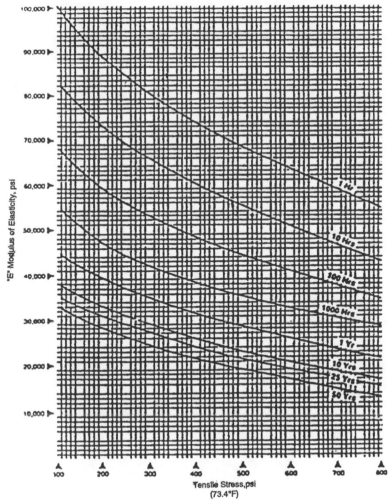

Figure 11 Tensile Creep Modulus versus Stress Intensity for a High-Density Polyethylene for Uniaxial Stress Conditions

Stress Relaxation

When a plastic part is deformed and maintained in that condition, the stress developed in the material decays gradually with time. The decrease in stress that occurs under constant strain is called stress relaxation. Initially, stress relaxation occurs at the fastest rate and then steadily decreases. Given enough time, the stress level approaches an equilibrium value.

An increase in temperature will decrease the time required for a given amount of stress relaxation to occur.

An example of stress relaxation is the reduction in stress that occurs when a polyethylene pipe bends around a curve or conforms to a contour.

Stress-relaxation apparent moduli can also be derived from stress-relaxation data. The stress-relaxation modulus is required for design calculations in which the strain is prescribed and the stress is free to vary. However, for the purpose of engineering design, the numerical difference between the relaxation and the creep modulus is often small when the strain and elapsed times are matched. The two can therefore be used interchangeably for most engineering design.

Simplified Representation of Creep & Stress-Relaxation Modulus

To simplify calculations, the creep and stress relaxation modulus for a certain time duration of loading is often represented as a fraction. The fraction is the long-term creep modulus divided by the short-term modulus (obtained by the tensile test). For such a simple representation, which is shown in Figure 12, it is assumed that the modulus is independent of stress intensity over the range of engineering stress for which this approximation applies. The consequence of this simplification is usually small and acceptable for most design.

Creep Recovery

Once the stress is removed at the end of a deformation test, the plastic will gradually return to its original dimension, but sometimes not completely. Incomplete recovery can occur even if the applied stress was below the yield point, established from the short-term tensile tests. The extent of recovery will depend upon the magnitude of the applied stress, the length of time over which the initial stress was applied and the properties of the material. At short-term creep or low-stress conditions, the recovery period can be rapid, but at long-term creep or high-stress conditions, the recovery can be quite slow. Figure 13 shows this phenomenon.

Duration of Uninterrupted Loading (Hours)	Approximate Ratio of Creep to Short-Term Modulus
1	0.80
100	0.52
10,000	0.28
438,000 (50 years)	0.22

Approximate Ratio of Creep Modulus to Short-Term Modulus as a Function of Loading Duration, for 73.4°F (23°C)

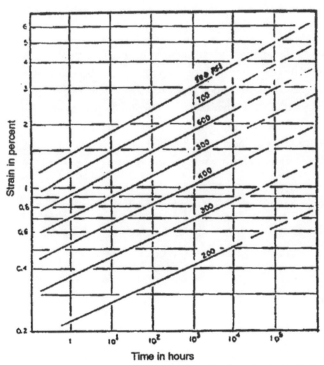

Figure 12 Tensile Creep Response for High-Density Polyethylene Pipe Material

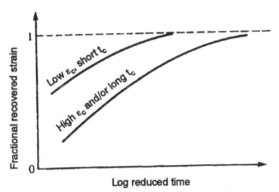

Figure 13 Creep Recovery as a Function of Time

Creep Rupture

Creep Rupture Characteristics

The relationship between tensile load and lifetime is described by the creep-rupture envelope of tensile creep curves. Each composition has a characteristic envelope for a given set of conditions of temperature and environment.

The stress versus lifetime characteristics of thermoplastic materials intended for pressure piping are determined by means of long-term pressure tests conducted on pipe specimens. Such characterization is generally referred to as stress-rupture testing.

The pipe samples are pressurized, usually with water, and immersed in a water bath at a certain temperature. The time required for each pipe to fail is recorded. By testing pipe at various temperatures and hoop stresses, creep rupture curves are generated, as shown in Figure 14. Note that the time for pipe failure to occur increases as applied stress decreases. The applied stresses are below the yield stress of polyethylene (measured by the short-term tensile test). This is typical of all materials that exhibit creep behavior, including metals and ceramics at very high temperatures[15,22].

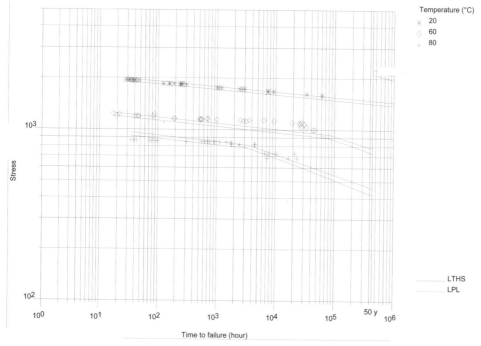

Figure 14 Typical Creep Rupture Curve for HDPE Pipe

When a pipe is pressurized, principal stresses are developed in both the axial and hoop (circumferential) directions. For a thick wall pipe, the fiber hoop stress is a function of the distance of the fiber from the inside pipe surface. The fiber at the inner surface is subject to a higher hoop stress than the outer wall. The axial stress is one-half the level of the average fiber hoop stress.

Long-Term Hydrostatic Strength

To correlate hoop stress with time-to-failure, the general practice is to calculate hoop stress by means of the relationship known as the ISO equation (denoting its adoption by the International Standards Organization):

$$S = p\,(OD-t)/2t$$

WHERE
S = hoop stress, psi
p = internal pressure, psi
t = minimum pipe wall thickness, inches
OD = outside pipe diameter, inches

This ISO equation is a form of a thin-wall vessel equation, which assumes that the fiber stress is constant between the inner and outer diameters. This assumption has been found to be a satisfactory representation for all plastic pipes. Stress-rupture testing is performed in accordance with the ASTM D1598[2] specification, *Time to Failure of Plastic Pipe Under Constant Internal Pressure*. Data obtained by D1598 testing is plotted on a log-log plot of stress versus time-to-failure. If the data falls along a straight line, then the *best least-squares straight line* is determined mathematically and extrapolated to the 100,000 hour intercept to forecast the long-term hydrostatic strength (LTHS). The extrapolation procedure used is that of ASTM D2837[3], *Obtaining Hydrostatic Design Basis for Thermoplastic Pipe Materials*. Each hydrostatic design basis (HDB) includes a range of the material's LTHS in a preferred stress category. These stress categories and the ranges of calculated LTHS are listed in Table 2.

The Hydrostatic Stress Board of the Plastics Pipe Institute issues recommendations of HDB for thermoplastic piping materials based upon ASTM D2837 and the additional requirements given in PPI Technical Report, TR-3[35], *Policies and Procedures for Developing Recommended Hydrostatic Design Stresses for Thermoplastic Pipe Materials*. Pipe compounds that are awarded an HDB are listed in the periodically update PPI Technical Report, TR-4[36], *Recommended Hydrostatic Strengths and Design Stresses for Thermoplastic Pipe and Fittings Compounds*.

TABLE 2
Hydrostatic Design Basis Categories

Range of Calculated LTHS Values		Hydrostatic Design Basis	
psi	Mpa	psi	MPa
190 to < 240	1.31 to < 1.65	200	1.36
240 to < 300	1.65 to < 2.07	250	1.72
300 to < 380	2.07 to < 2.62	315	2.17
380 to < 480	2.62 to < 3.31	400	2.76
480 to < 600	3.31 to < 4.14	500	3.45
600 to < 760	4.14 to < 5.24	630	4.34
760 to < 960	5.24 to < 6.62	800	5.52
960 to < 1200	6.62 to < 8.27	1000	6.89
1200 to < 1530	8.27 to < 10.55	1250	8.62
1530 to < 1920	10.55 to < 13.24	1600	11.03
1920 to < 2400	13.24 to < 16.55	2000	13.79
2400 to < 3020	16.55 to < 20.82	2500	17.24
3020 to < 3830	20.82 to < 26.41	3150	21.72
3830 to < 4800	26.41 to < 33.09	4000	27.58
4800 to < 6040	33.09 to < 41.62	5000	34.47

The ASTM D2837 extrapolation method has been used since the 1960's as the primary requirement for qualifying thermoplastic pressure piping materials. The excellent field performance achieved by adopting this test has proven its value for pipe design. Because a few exceptions to this performance have been noted with some polyethylene materials, the industry has investigated improved methods for predicting the long-term behavior of polyethylene pipe.

The stress-rupture line for polyethylene can have a downturn or "knee" where the failure mode changes from ductile to brittle. Figure 14 shows these "knees." Ductile failure mode is characterized by areas in the material that have undergone substantial cold-drawing with a significant elongation in the immediate area of the rupture. The failure looks like a "parrot's beak." Brittle failures are characterized by little or no deformation in the rupture. They are typically referred to as brittle or slit failures due to the formation of cracks or small pin holes within the pipe wall. These types of failures are the result of the manifestation of a fracture mechanics mechanism, which involves crack formation, propagation and ultimate failure. This is the type of failure generally seen in the field.

The assumption of the ASTM D2837 test method is that a straight line described by at least 10,000-hour test data will continue as a straight line through at least 100,000 hours. However, it is now known that certain polyethylene compounds that have met this requirement still can show a downturn prior to 100,000 hours. Therefore, another test method was needed that would confirm or "validate" that the 73°F (23°C) extrapolated line is straight at least through 100,000 hours.

Rate Process Method Validation

Many naturally occurring rate processes follow the law discovered by Arrhenius and represented by[28]:

$$k = k_o e^{-ER/T}$$

This expression states that the reaction rate is a function of the absolute temperature. This law applies to experimental data from many types of processes, including estimating the slit mode failure time of plastic pipe as a function of temperature.

The new test method selected is based upon an activated rate process theory for the rupture of materials and is commonly referred to as the Rate Process Method (RPM) Validation[1,29,32,33]. A three-coefficient mathematical formula is utilized to fit brittle failure data obtained at different elevated temperatures. The equation used is:

$$\log t = A + B/T + C/T \log S$$

WHERE
t = time to failure, hours
T = absolute temperature, °K
S = hoop stress, psi
A, B, C = coefficients

This equation is then used to predict the onset of the 73°F (23°C) brittle-type failures from brittle-type data obtained at elevated temperatures. It has been applied to six different polyethylenes of known long-term field performance and has accurately differentiated between good- and poor-performance pipes. It is mandatory that all pressure pipe compounds have to pass the validation test before a PPI listing is granted.

An example of the RPM validation procedure is explained in the following steps with the help of Figure 15.

1. Plot the log-stress versus log time for the ductile failures at 73°F (23°C) according to ASTM D2837, using data up to 10,000 hours. Extrapolate the data out to 100,000 hours (line aa') and obtain the LTHS intercept (Point 1).

At least six pieces of pipe are tested at each of the following conditions.

2. Select an elevated temperature (90°C or lower) and hoop stress where brittle failures will occur in 100 to 500 hours as shown by Point II. This is known as Condition 1.

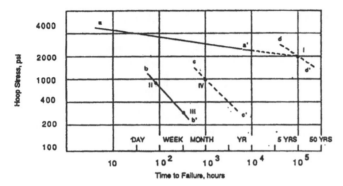

Figure 15 Hoop Stress vs Time to Failure

3. Using the same temperature, select a stress at least 75 psi lower than in Condition I. Failure times should range from 1,000 to 2,000 hours as shown by Point III. This is Condition II. The line (bb′) determined by points II & III will be used to determine the minimum failure time of the next test condition.

4. The underlying theory in ASTM D2837 assumes that the downturn or 'knee' will occur after 100,000 hours. Therefore, the worst case assumes that the 73°F knee will occur at 100,000 hours, which is indicated by line dd′. To confirm that the 73°F knee is at or beyond this worst case situation, select a temperature at least 15°C lower than Condition I but use the same stress as Condition I. This is known as Condition III and is indicated as Point IV and line cc′. The experimentally determined average log-failure time at Condition III is then compared to the time predicted by the RPM equation log t = A + B/T + (C/T)log S, where the coefficients A, B, and C are calculated using the Points I, II, and III.

5. If the experiment results in a test failure time that meets or exceeds this predicted RPM failure time for point IV, the hypothesis that the knee occurs at or beyond 100,000 hours has been confirmed independently and the ASTM D2837 procedure has been validated. If the actual failure time is less than the RPM predicted time, the pipe is disqualified and cannot be considered adequate for pressure pipe.

The RPM process can be used to study the effects of resin formulation or pipe processing changes in a shorter time period than having to retest under ASTM D2837. Another advantage to this method is that the long-term strength forecast is based on brittle-like failures that simulate long-term field failures. It can also be used as a Quality Control test for subsequent monitoring of pipe after the initial RPM rating is established.

Fracture Mechanics

The fundamental premise of fracture mechanics is that an object fractures under an applied stress due to the growth of cracks from flaws inherent in the object. These flaws may be material defects (contamination or an undesired multi-phase structure), manufacturing defects (voids or surface embrittlement), or post production damage (scratches, gouges, improper joining). They may be microscopic or macroscopic in size. Whatever the source or size, such flaws serve to intensify the nominal applied stress within their vicinity. At some point this intensified stress at the flaw will exceed the strength of the material and a crack will begin to grow. Ultimately, such growing cracks will lead to failure of the entire object.

Thus, the fracture resistance of a given structure or material will depend upon the level of stress applied to it, the presence and size of flaws, and the resistance of the material to crack initiation and growth. The objective of fracture mechanics is to provide quantitative relationships between some of these factors. There are two basic approaches to this problem, one method, originally proposed by Griffith[18], is concerned with the balance between strain energy stored within a stressed body and that released when a crack within that body extends by some amount. The second method, which will be presented here, deals with the mechanical environment near the tip of a flaw which will initiate a growing crack.

This latter method states that the amount of stress concentration at the tip of a crack (flaw) contained within a stress body can be characterized by a Stress Intensity Factor, K. For crack growth in the plane ahead of the flaw caused by some applied tensile stress, the value is denoted as K1 and is given by the equation:

$$K1 = (Y)(s)(\pi a)^{1/2}$$

WHERE
s = nominal applied stress, psi^p
π = length of the crack, inch
Y = a factor that accounts for the geometry of the specimen

Values of Y have been tabulated for a wide range of geometries[43,44]. Crack growth is assumed to occur when the value of $K1_i$, that exists in the stressed specimen exceeds some critical value, $K1_c$, that is characteristic of the material from which the specimen is made.

In its simplest form, this treatment assumes that the material behaves as a linear elastic solid up to the point where fracture occurs. However, it soon becomes apparent that, except for cases where fracture occurs at very low stress levels in comparison to the material yield stress, the fracture process does result in some plastic deformation. Due to the stress concentration effect, a volume of material near to the crack tip and at free surfaces of a specimen (if the crack extends to such surfaces) will yield, even if the bulk of the specimen is still at a stress below the yield stress. However, as long as this plastic deformation is constrained to a relatively small percentage of the total area ahead of the growing crack, linear elastic fracture mechanics can still be utilized.

In application, a material is evaluated by suitable laboratory testing to determine the values of K1 for which crack growth will occur. In brittle polymeric materials like polystyrene, crack growth (when it occurs) will be rapid and the specimen will fail immediately. In more ductile materials, like the pipe-grade polyethylenes, crack growth may proceed very slowly; e.g. at rates of $10^{-5} - 10^{-6}$ inches/hour. It is possible to observe both types of crack growth in the same material: i.e. under sufficient energy, a crack will grow slowly until its length is such that the value of K1 for the now longer crack exceeds the critical value $K1_c$ for the rapid crack growth in the material.

The phenomenon of rapid crack failure of polyethylene pipes has been studied extensively. Such failures are extremely rare, and methodologies exist to evaluate the potential conditions under which polyethylene could fail by such a mechanism[27].

In studies of slow crack growth behavior on resins that have been used for the production of polyethylene pipe, two types of behavior have been observed[31]. Some older polyethylene materials have exhibited slow crack growth (SCG) field failures. For these materials, it has been possible to determine crack incubation times (time for slow crack growth to commence at a given K1) and crack growth rates as a function of K1 and relate the pipe failure time to in-service stresses[12,45]. Newer polyethylene pipe resins are proving to be so extremely resistant to slow crack growth that more complex methods of non-linear analysis may be necessary[31].

Cyclic Fatigue Endurance

Fracture mechanics has also been utilized in evaluating fatigue fracture of polyethylene pipe[9]. For fatigue loading, the rate of crack growth is expressed as:

$$da/dn = D(\Delta K_i)^d$$

WHERE
da/dn = crack growth per fatigue cycle
D, d = material constants
ΔK_i = difference in stress intensity at the crack tip between the highest and lowest stresses imposed during each fatigue cycle

Testing of polyethylene pipes via fatigue loading appears to also be a reasonable method of assessing the relative resistance of the newest resins to slow crack growth[11]. Experimental evaluation of D and d in the above equation for a particular polyethylene pipe resin could permit prediction of field performance for pipe made from that resin if some estimate of applied stress and flaw size within the pipe can be made.

Each time a polyethylene pipe is pressurized, its circumference and length expand. For applications where the pressure is constant and below the pipe pressure rating, this small amount of expansion (strain) is not important. However, strain does become important when the pipe undergoes higher cyclic pressurization.

There is a maximum critical strain limit which, once exceeded, permanently changes the characteristics of the pipe. At high strain levels, microcracks can occur. Repeated straining that approaches the strain limit can cause the growth of flaws or microcracks that will propagate through the pipe wall, resulting in pipe failure. The effects of high strain levels are cumulative due to this non-reversible microcrack formation. As the critical strain limit is approached, there is a greater possibility that the pipe performance will be affected.

The typical pressure pipe in the field undergoes a strain of ½% to 1%. This is well below the critical strain limit of 6% to 7% for most polyethylenes. However, the actual value is a function of the material. Most field failures occur in the brittle mode, which indicate that crack propagation was the cause of the failure. Strain is not a factor in these cases. However, as the pressure increases so does the strain. Failures at high pressures (well above the pressure rating of the pipe) occur in the ductile mode, which is indicative of high strain conditions.

Predicting the service life under cyclic pressure conditions is not straightforward. It is a function of two parameters; Stress Level (Amplitude), and Frequency. By minimizing each of those items, the possibilities of failure by fatigue are reduced. See Figure 16 for a typical S-N curve. For more specific details on this subject, please see references[9,10,11].

Short-Term Mechanical Properties

Tensile Properties

Most short-term data is derived from a constant-speed tensile apparatus using test coupons as shown in Figure 17. The results from these tests are in the form of force and deformation data, which can then be transformed into stress-strain or elongation curves. The tests are usually conducted at a certain temperature, which is typically 73°F (23°C).

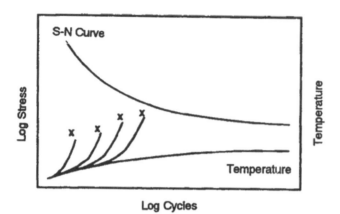

Figure 16 Typical S-N Curve, with Thermal Effects Which Sometimes Occur When Fatigue-Testing Plastics

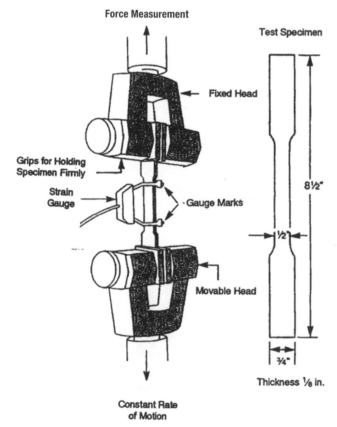

Figure 17 Typical Test Setup and Specimen

A material will deform whenever a force is applied[46]. The amount of deformation per unit length is called strain, and the force per cross-sectional unit area is called stress. At very low stress levels, strain is nearly proportional to stress and is reversible. Once the stress is removed, the material returns to its original dimension. The Modulus of Elasticity (Young's Modulus) is the ratio between stress and strain in this reversible region. This strain is also referred to as elastic strain since it is reversible. However, in practice, there is not a region of a true, reversible strain for plastics.

At higher stress levels, strain is no longer directly proportional to stress and it is not reversible when the stress is removed. The term used to describe strain in this region is called plastic strain. (This term is used for all materials, not just for polymeric materials.) Figure 18 illustrates stress-strain curves for polyethylene.

When testing plastics by using a typical constant crosshead rate tensile testing machine, two points should be noted that differ from results obtained from materials such as steel:

1. The stress-strain curve is usually not a straight line in the elastic region, as shown in Figure 18a. It is common to construct a secant line to a defined strain, usually 2% for polyethylene, and then read the stress level at that point[40].

2. The speed of the test will affect the elastic modulus. At slow speeds, the molecules have time to disentangle, which will lower the stress needed to deform the material and will lower the modulus. Conversely at higher crosshead speeds, the molecular entanglement requires a higher stress (force) for deformation and hence a higher modulus value as shown in Figure 18b. This is the reason the testing speed is specified in all test procedures.

It is necessary to know the exact conditions by which test data are obtained. Slight changes in conditions can drastically alter the test values.

Tensile Strength

The point at which a stress causes a material to deform beyond its elastic region is called the tensile strength at yield. The force required to break the test sample is called the ultimate strength or the tensile strength at break. The strength is calculated by dividing the force (at yield or break) by the original cross-sectional area. ASTM D638[2], *Standard Test Method for Tensile Properties of Plastics,* is used to determine the tensile properties of polyethylene pipe resins. Figure 19 shows the various inflection points for a typical stress-strain curve for polyethylene.

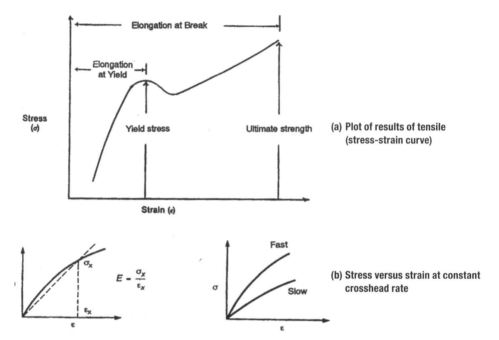

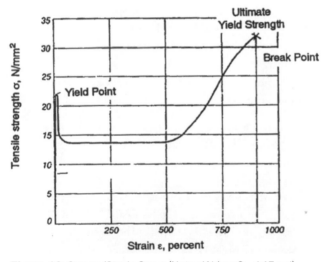

Figure 18 Stress vs Strain Curves Under Specified Conditions

Figure 19 Stress/Strain Curve (Note: 1N/mm2 = 145 psi)

Test specimens are usually shaped as a flat "dog-bone," but specimens can also be rod-shaped or tubular per ASTM D638. During the tensile test, polyethylene, which is a ductile material, exhibits a cold drawing phenomenon once the yield strength is exceeded. The test sample develops a "neck down" region where the molecules begin to align themselves in the direction of the applied load. This strain-induced orientation causes the material to become stiffer in the axial direction while the transverse direction (90° to the axial direction) strength is lower. The stretching or elongation for materials such as polyethylene can be ten times the original gauge length of the sample (1000% elongation). Failure occurs when the molecules reach their breaking strain or test sample defects, such as edge nicks, begin to grow and cause premature failure. Fibrillation, which is the stretching and tearing of the polymer structure, usually occurs just prior to rupture of a well-drawn sample.

Any stretching or compressing of a test specimen in one direction, due to uniaxial force (below the yield point), produces an adjustment in the dimensions at right angles to the force. A tensile force causes a small contraction, called lateral strain (Δd/d), to occur at right angles to the force, as shown in Figure 20. The ratio of lateral strain and tensile (longitudinal) strain (ΔL/L) is called Poisson's ratio (v).

$$v = \frac{\text{lateral strain}}{\text{longitudinal strain}} = -\frac{\Delta d/d}{\Delta L/L}$$

ASTM E132[6], *Standard Test Method for Poisson's Ratio at Room Temperature*, is used to determine this value. Poisson's ratio for polyethylene is between 0.40 and 0.45[20].

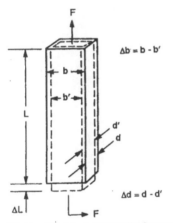

Figure 20 Loaded Tensile Strength Bar Showing Dimensional Change in Length and Width

Compressive Strength and Modulus

Compressive forces act in the opposite direction to tensile forces and can be measured on the same tensile testing machine. The sample is usually a solid rod rather than the tensile dog-bone specimen. The crosshead, instead of moving away from the test sample, moves into and compresses the sample. The deformation or strain is measured the same way as in a tensile test. Again, there is a region of the stress-strain curve in which the stress is proportional to the strain.

At small strains (up to 1–2%) the compressive modulus is about equal to the elastic modulus. However, at higher stress levels, the compression strain is lower than the tensile strain. Unlike the tensile loading, which results in a failure, stressing in compression produces a slow and infinite yielding which seldom leads to a failure. For this reason, it is customary to report compression strength as the stress required to deform the test sample to a certain strain. But even this is difficult to achieve. ASTM D695[2], *Standard Test Method for Compressive Properties of Rigid Plastics*, is used to determine this property.

Flexural Strength and Modulus

Flexural strength is the maximum stress in the outer fiber of a test specimen at rupture. The test is conducted using a specimen that is supported at each end with a load applied at the center. The distortion is measured as the load is increased. If the plastic does not break, as in the case of polyethylene, then the amount of stress is reported at a specified level of strain (usually at 2% or 5%).

Flexural strength is related to the density and, to a lesser extent, molecular weight. As the density increases, the polymer becomes stiffer since the molecules do not have as much space to move around one another. Also as the molecular weight increases, the entanglement of the molecules resists movement and therefore increases stiffness.

Using a tensile testing machine, a sample is bent while being held in a three- or four-point-contact holder. The amount of stress needed to deflect the outer surface of the sample a certain vertical distance (strain) is determined. Since most thermoplastics do not break in this test, the true flexural strength cannot be determined. Typically, the stress at 2% strain is used to calculate the flexural modulus. ASTM D790[2], *Standard Test Methods for Flexural Properties of Unreinforced and Reinforced Plastics and Electrical Insulating Materials* describes this test method.

Shear Properties

The best way to imagine shear stress is to slice a block of material into infinitesimally thin layers, as shown in Figure 21. If the block is subjected to a set of equal and opposite forces, Q, there is a tendency for one layer to slide past another one to produce a shear form of deformation or failure when the force is high enough. The displacement of one plane of molecules relative to another produces shear stresses.

The shear stress, γ, is defined as

γ = shear load / area resisting shear

   = Q/A

The shearing strain, e, is the angle of deformation which is measured in radians.

The shear modulus, G, is defined as the shear stress divided by the shear strain for stresses below the yield strength.

G = shear stress/shear strain

$= \gamma/\varepsilon$

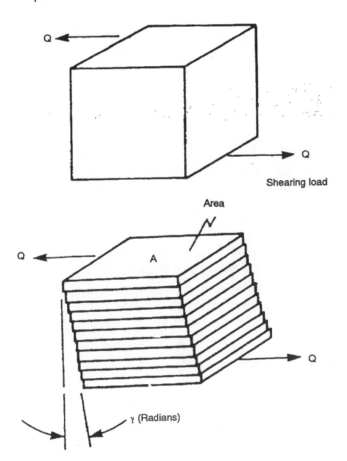

Figure 21 Shear Strain

General Physical Properties

Impact Strength

Impact strength measures the amount of energy that a material can absorb without breaking. The ability of a plastic part to absorb energy is a function of its shape and thickness and the molecular character of the resin. High-molecular-weight resins are very tough since they absorb more energy than lower molecular-weight resins.

The test results from these impact tests can only be used to rank the toughness or the notch sensitivity of similar materials. There are many factors that influence the results such as temperature, specimen orientation (compression or injection molded specimens give different results), and the shape and radius of the notch. The test results provide data for comparisons between the same types of resins (i.e., only polyethylenes to polyethylenes or acetals to acetals, etc.) under the same conditions of the test. Some very tough materials are notch-sensitive, such as nylons and acetals, and register very low notched impact values from this test.

There are several types of impact tests that are used today. The most common one in the United States is the notched Izod test. Izod specimens are tested as cantilever beams. The pendulum arm strikes the specimen and continues to travel in the same direction, but with less energy due to impact with the specimen. This loss of energy is called the Izod impact strength, measured in foot-pounds per inch of notch of beam thickness (ft-lb/in). The specimens can be unnotched or with the notch reversed with the results reported as unnotched or reverse notch Izod impact strength. (The impact test specimen is usually notched in order to have a controlled failure point.)

The Charpy impact test is widely used in Europe and is less common in the United States. The specimen is a supported beam, which is then struck with a pendulum. The loss of energy is measured in the same units as in Izod impact test. The specimens can be either notched or unnotched. Figure 22a illustrates the Izod and Charpy impact tests. ASTM D256[2], *Standard Test Methods for Impact Resistance of Plastics and Electrical Insulating Materials,* describes both test methods.

There is another type of test called the tensile impact strength test, which uses a swinging pendulum as shown in Figure 22b. It measures the amount of energy needed to break the specimen due to tensile impact loading.

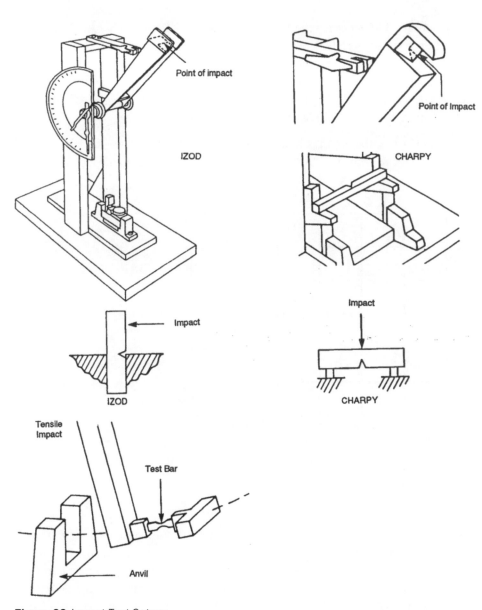

Figure 22 Impact Test Setups

Hardness

Hardness is the resistance of a material to penetration of its surface. It is related to the crystallinity and hence the density of the material. Figure 23 shows the relationship of density versus hardness. The typical hardness tests are either the Shore or the Rockwell. The hardness value depends on the shape, size, and time of the indenter used to penetrate the specimen. Depending upon the hardness of the material to be tested, each hardness test has several scales to cover the entire range of hardness. For polyethylene, the Shore D scale or Rockwell L scale is used. ASTM D785[2], *Standard Test Method for Rockwell Hardness of Plastics and Electrical Insulating Materials,* describes the Rockwell test, while ASTM D224[3], *Standard Test Method for Rubber Property-Durometer Hardness,* describes the Shore test.

For thermoplastics, the depth of indentation by the ball will be very dependent upon the amount of time that the specimen is under stress (due to the viscoelastic characteristics).

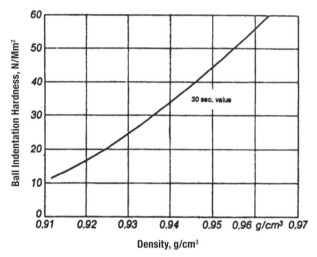

Figure 23 Ball Indentation of Hardness of Polyethylene as a Function of Density

Abrasion Resistance

The growth in the transportation of solids through hydraulic systems is increasing mainly due to the economic advantages of operating this type of system. More and more polyethylene pipe is being used to transport granular or slurry solutions, such as sand, fly ash and coal. The advantage of polyethylene in these applications is its wear resistance, which has been shown in laboratory tests to be three to five times longer than normal or fine-grained steel pipe at a typical velocity of under 15 ft/sec.

There are several factors that affect the wear resistance of a pipeline. The concentration, size and shape of the solid materials, along with the velocity, are the major parameters that will affect the wear resistance and thus affect the life of the pipeline. Some other factors include the angle of impingement and the type of flow characteristics (single- or two-phase flow)[41].

Permeability

The property of permeability refers to the passage of either gaseous or liquid materials through the plastic. Polyethylene has a low permeability to water vapor but it does exhibit some amount of permeability to certain gases and vapors. As a general rule, the larger the vapor molecule or the more dissimilar in chemical nature to polyethylene, the lower is the permeability.

The following gases are listed in order of decreasing permeability: sulfur dioxide, hydrogen, carbon dioxide, ethylene, oxygen, natural gas, methane, air and nitrogen.

Most of the permeability is through the amorphous regions of the polymer. It is related to density and, to a lesser extent, to molecular weight. An increase in density will result in a lower permeability. An increase in molecular weight will also slightly reduce the permeability. Table 3.3 shows permeation rate of methane and hydrogen through both MDPE and HDPE[1].

TABLE 3
Gas Permeation Rate Through Polyethylene

	Permeation Rate	
Resin	Methane	Hydrogen
Medium Density PE	4.2×10^{-3}	21×10^{-3}
High Density PE	2.4×10^{-3}	16×10^{-3}

Permeation from external reagents into the pipe can occur and should be properly addressed. Any pipe, as well as an elastomeric gasketed pipe joint, can be subjected to external permeation when the pipeline passes through contaminated soils. Special care should be taken when installing potable water lines through these soils regardless of the type of pipe material (concrete, clay, plastic, etc). The Plastics Pipe Institute has issued *Statement N - Pipe Permeation*[39] that should be studied for further details.

Thermal Properties

Thermal Expansion and Contraction

The coefficient of linear expansion for polyethylene is about 10 to 12×10^{-5} in./in./°F compared to steel at about 1×10^{-5} [42]. The typical method to determine this value is described in ASTM D696[2], *Standard Test Method for Coefficient of Linear Expansion of Plastics*. This means that an unconstrained polyethylene pipe will expand or contract at least ten times the distance of a steel pipe of the same length. The main concern to the piping engineer is the amount of internal stress generated during expansion and contraction movements. For constrained polyethylene pipe, the stresses developed due to this movement are substantially lower than that of a steel line. This is due to the lower modulus of elasticity of polyethylene as compared to steel. Polyethylene pipe that is properly anchored should not be adversely affected by normal expansion or contraction. The chaper on above ground applications of polyethylene pipe in this Handbook provides further information about pipe constraining techniques.

The equation to calculate expansion or contraction is

$$\Delta L = L\alpha(\Delta T)$$

WHERE
ΔL = Change in length
L = Original length
ΔT = Change in temperature
α = Coefficient of linear expansion

A 10°F rise or fall in temperature will cause a 100-foot length of an unconstrained polyethylene pipe to move 1.0 to 1.2 inches. The same length of steel pipe will move about 0.10 inch but will generate larger internal stresses than the polyethylene pipe.

Thermal Conductivity

The amount of heat that a polyethylene pipe can convey through its wall is a function of thermal conductivity. Polyethylene has a thermal conductivity of 2.4 Btu/in./ft²/hr/°F (0.43 W/m/°K) [42]. The amount of heat transmitted through a polyethylene wall is calculated by the following equation:

$$q = (k/x)(T_1-T_2)$$

WHERE
q = heat loss, BTU/hr/ft of length
k = thermal conductivity, BTU/in./ft²/hr/°F
x = wall thickness, inches
T_1 = outside temperature, °F
T_2 = inside pipe temperature, °F

This equation can be used to estimate the heat loss (or gain) from a polyethylene pipe.

The ASTM method commonly used to determine this value is ASTM C177[2], *Steady-State Heat Flux Measurements and Thermal Transmission Properties by Means of the Guarded-Hot-Plate Apparatus.*

Specific Heat

The specific heat is defined as the ratio of the heat capacity of the material to that of water (the specific heat of water is 1 cal/g/°C). The specific heat of polyethylene is a function of temperature that is shown in Figure 3.24[23].

For partially crystalline thermoplastics, a sharp maximum in the heat is observed in the melting point region. As the density increases, the specific heat maximum is higher and sharper. Once the polyethylene is in the molten state, the specific heat is independent of temperature and is the same value for all polyethylene resins.

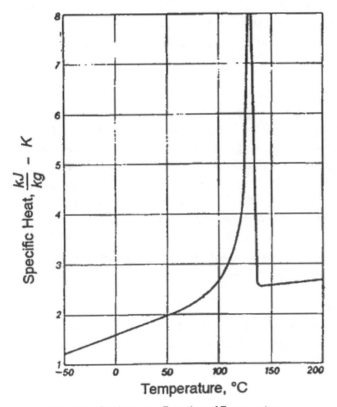

Figure 24 Specific Heat as a Function of Temperature

Glass Transition Temperature

The molecules that make up polyethylene are in constant motion. Even though the molecules are entangled, there is enough free space within the polymer to allow molecular movement. As the temperature falls below melting point, Tm, which is also called the crystallization temperature, there is a decrease in molecular movement and volume due to crystallization of part of the polymer. The structure now consists of crystalline regions separated by amorphous material. Within these latter regions, molecular motion still occurs. As temperature is decreased further, molecular movement and volume reduction in the non-crystalline areas continues. When the molecules cannot pack together any closer, any further decrease in temperature only allows molecular vibrations to occur. The temperature at which this occurs is called the glass transition temperature, Tg. Above the Tg temperature, polyethylene is flexible and ductile, and at temperatures below Tg it exhibits less ductility.

The following example illustrates the difference in polymer characteristics of polymers with different glass transition temperatures. Polyethylene has a Tg of -166°F (-110°C) and a Tm of +275°F (+135°C)[40]. As a comparison, polystyrene has a Tg of +212°F (+100°C) and a Tm of +230°F (+110°C), which accounts for its more brittle-like behavior at room temperature.

Minimum/Maximum Service Temperatures

Polyethylene has very good characteristics, such as impact strength, at low temperatures. It is the preferred material for operating temperatures below 0°F (-18°C). The highest permissible service temperature for a polyethylene pipe depends upon the duration and magnitude of stresses upon the pipe. Generally, +140°F (+60°C) is the typical maximum service temperature. However, some non-pressure applications can be used up to +176°F (+80°C). The use of polyethylene at high temperature necessitates reducing the working pressure in order to obtain the same service life as that of a lower temperature application. Refer to PPI Technical Note TN-11[34], *Suggested Temperature Limits For Thermoplastic Pipe Installation And For Non-Pressure Pipe Operation*, for further details.

Deflection Temperature Under Load

This test gives the temperature at which a plastic will deflect under a certain load. It is not intended to be a guide to high temperature service limits but to serve as a comparison of high temperature behavior of various materials.

Figure 25 shows a schematic of the typical apparatus used in this test. ASTM D648[2], *Standard Test Method for Deflection Temperature of Plastics Under Flexural Load*, is the standard method used to determine this value.

A 0.5 in. by 0.5 in. by 5.0 in. beam is immersed in a heat transfer liquid and heated until the beam deflects 0.01 in. with a maximum flexural stress load of 264 psi or 66 psi.

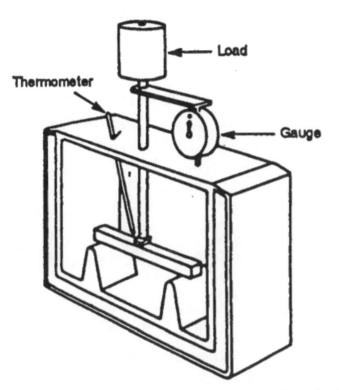

Figure 25 Test Apparatus for Deflection Temperature Under Load

Electrical Properties

Metals are very good electrical conductors due to their metallic crystal structure[13]. The outermost electrons are loosely bound to the atoms and can be easily broken free to move among the crystal lattice. This flow of free electrons accounts for the ability of metals to conduct large amounts of electrical current. In plastics and glass, the outer electrons are tightly bound to the atoms and are not available to move among the lattice. This accounts for the inability of plastics to conduct electricity; therefore they are called insulators. Table 4 lists the typical electrical properties of polyethylene.

TABLE 4
Selected Electrical Property Ranges for MDPE and HDPE

Property	Unit	Test Method	Value
Volume Resistivity	Ohms-cm	–	>10^{16}
Surface Resistivity	Ohms	–	>10^{13}
Arc Resistance	Seconds	ASTM D495	200 to 250
Dielectric Strength	Volts/mil	ASTM D 149, 1/8 in. Thick	450 to 1000
Dielectric Constant	–	D150 – 60 Hz	2.25 to 2.35
Dissipation Factor	–	D150 – 60 Hz	>0.0005

Volume Resistivity

Volume resistivity is the resistance to current leakage through an insulator. It is related to temperature, moisture in the insulator, and the type of the insulator. The units of measurement are ohm-centimeters (ohm-cm). Higher values indicate a better resistance to breakdown or leakage.

Also, a volume resistivity of 10^{10} ohm-cm normally forms the dividing line between conducting and non-conducting materials. Since polyethylene has a volume resistivity of 10^{16} ohm-cm, it will accumulate a static charge.

Surface Resistivity

Surface resistivity is the resistance to current leakage along the surface of the insulator, measured in ohms. It actually measures the ability of the current to flow over the surface of the material. This value is a function of surface conditions and is not a true property of the material. The higher the value, the better the resistance to leakage.

Arc Resistance

Arc resistance is the measurement of the breakdown of the surface of an insulator caused by an arc. If an electrical arc is imposed upon the surface, the current will flow along the path of least resistance. The test measures the time it takes for the breakdown to occur along the surface. Higher values indicate better resistance to breakdown.

Dielectric Strength

Dielectric strength is the voltage that an insulator can withstand before breaking down and allowing current to pass. The voltage just prior to breakdown divided by the sample thickness defines this property. It is expressed in terms of a voltage gradient, volts/mil. The higher the value, the better the insulator. The thinner the insulation thickness, the greater the dielectric strength.

Dielectric Constant

Molecules will become polarized when an electrical field is applied across an insulator. If the voltage potential is reversed, the polarization of insulator molecules will also become reversed. The ease with which polarization takes place is measured by a material constant called permittivity. The ratio of permittivity to the negligible permittivity in a vacuum is called the relative dielectric constant. The value changes with frequency, temperature, moisture level, and part thickness. This value is important when plastics are used in high frequency applications.

Dissipation Factor

If polarization of the sample occurs at a very high rate, a certain amount of energy will be dissipated in the form of heat. The dissipation factor is the ratio of energy dissipated at a certain frequency (usually 1 MHz) to that transmitted. A low value is important when plastics are used as insulators in high frequency applications such as microwave or radar equipment.

Static Charge

Since plastics are good insulators, they also tend to accumulate a static charge. A static charge is a result of either an excess or deficiency of electrons in the molecular structure of the polymer. Depending on the type of polymer, the static charge can be either positive or negative electrically.

Polyethylene pipe can acquire a static charge through friction. Sources of friction can be simply the handling of the pipe in storage, shipping, or installation. Friction can also be caused by the flow of gas containing dust or scale or by the flow of dry material through the pipe. These charges can be a safety hazard if there is leaking gas or an explosive atmosphere and should be dealt with prior to working on the pipeline.

Since polyethylene is electrically non-conductive, the static charge will remain in place until a grounding device discharges it. A ground wire will only discharge the static charge from its point of contact. The most effective method to minimize the hazard of static electricity discharge is to apply a film of water to the work area prior to handling. Please refer to the pipe manufacturer for further details.

There are special grades of electrically conductive polyethylenes that are used to prevent the build-up of static charges in explosion-proof areas. These resins usually contain 7% to 9% carbon black, which prevents an accumulation of a static charge by decreasing surface resistivity. For further information, please contact a polyethylene resin producer.

Flammability and Combustion Toxicity

Flammability

Polyethylene ignites on contact with a flame unless it contains a flame retardant stabilizer. Burning drips will continue to burn after the ignition source is removed. The flash ignition and self ignition temperatures of polyethylene are 645°F (341°C) and 660°F (349°C) respectively as determined by using ASTM D1929[3], *Standard Test Method for Ignition Properties of Plastics*. The flash point using the Cleveland Open Cup Method, described in ASTM D92[7], *Standard Test Method for Flash and Fire Points by Cleveland Open Cup*, is 430°F (221°C)[16].

During polyethylene pipe production, a certain amount of smoke may be generated. If smoke is present, it can be an irritant and as such should be avoided. Specific information and Material Safety Data Sheets (MSDS) are available from the polyethylene resin manufacturer.

Combustion Toxicity

The combustion of organic materials, such as wood, rubber, and plastics, can release toxic gases. The nature and amount of these gases depends upon the conditions of combustion. For further information on combustion gases, refer to *Combustion Gases of Various Building Materials* and *Combustion Toxicity Testing* from The Vinyl Institute[46,48].

The combustion products of polyethylene differ greatly from those of (poly)vinyl chloride (PVC). Polyethylene does not give off any corrosive gases such as hydrochloric acid, since it does not contain any chlorine in its polymer structure.

Chemical Resistance

Plastics are not subject to galvanic corrosion, as are metals, since they are nonconductors. However, plastics can be affected through direct chemical attack, strain corrosion or solvation. The extent of the resistance is a function of many items, including time, temperature and stress of contact.

Polyethylene is a non-polar high-molecular-weight paraffin hydrocarbon. It is very resistant to chemicals and other media such as salts, acids and alkalis. However, oils, fats and waxes will cause some slight swelling. Strong oxidizing agents tend to attack the molecules directly and lead to gradual deterioration of properties. Organic chemicals tend to be absorbed by the plastic through a process called solvation. The effects of solvation, which are very time-dependent, include swelling and softening of the polymer. Strain corrosion takes place under combined action of strain and a chemical environment. Another name for this phenomenon is environmental stress cracking (ESC).

Stress crack resistance increases as density decreases and also as the melt index (higher molecular weight) decreases. Copolymer type and placement on the polymer chain also greatly affect the ESC characteristics of a polymer. The following section describes various types of chemical resistance testing.

Immersion Testing

The chemical resistance information published by many polyethylene resin and pipe manufacturers was determined from immersion tests. These simple immersion chemical resistance tests use molded specimens that are immersed in the chemical media (under no stress) at two or three different temperatures. After a certain amount of time, the samples are removed, weighed, inspected and tested. The resistance is then determined by the amount of (or lack of) specimen swelling, weight loss and change in strength properties. The ratings are listed "generally resistant," "limited resistance," and "not resistant." Some test results are shown in Table 5. When using this type of data, special consideration must be made if the material is to be exposed to chemical, mechanical or thermal stresses. Consult PPI Technical Report TR19[38], *Thermoplastic Piping for the Transport of Chemicals*, for more complete information.

TABLE 5
Chemical Resistance of HDPE Pipe Resins
The following abbreviations are used in Table 5:

R Generally Resistant. (Swelling was less than 3% or observed weight loss less than 0.5%, and elongation at break was not changed significantly.)

C Has limited resistance only and may be suitable for some conditions. (Swelling between 3% and 8%, and/or observed weight loss between 0.5% and 5%, and/or elongation at break decreased less than 50%.)

N Is not resistant. (Swelling greater than 8% or observed weight loss greater than 5% and/or elongation at break decreased to greater than 50%.)

D Discoloration

The terms R to C, C to N, and R to N are used when disagreement exists in the literature.

Where no concentrations are given, the relatively pure material is indicated except in the case of solids, where saturated aqueous solutions are indicated.

* Indicates that this chemical resistance does not apply for weld joints.

(The complete Table 5 is included on the following pages.)

TABLE 5
Chemical Resistance of HDPE Pipe Resins

Chemical Name	Concentration	73°F	120°F	140°F
Acetaldehyde & Acetic Acid		R	—	—
Acetaldehyde (aqueous)	all	R to C	C	C to N
Acetamide		R	—	R
Acetic acid	100% (glacial)	R	C to N	C to N
Acetic acid	50%, 60%, 70%, 80%	R	C	R to C
Acetic acid	10%, 20%	R	—	R
Acetic acid vapor		R	R	—
Acetic anhydride	100%	R	—	C
Acetoacetic acid		R	—	—
Acetone, 100%		R to C	C	C
Acetophenone		R	—	—
Acetylene		R	—	—
Acids aromatic		R	—	R
Acronal® dispersons - usual commercial		R	—	C
Acrylic acid emulsions		R	—	R
Acrylonitrile	technically pure	R	—	R
Adiptic acid, saturated sol.		R	—	R
Adiptic acid ester		R	—	C
Aktivin® (chloramine)(aqueous)	1%	R	—	R
Alcohol, allyl		R	C	R
Alcohol, amyl	technically pure	R	C	R to C
Alcohol, benzyl		R to C	C	R
Alcohol, (n-butanol)		R	R	R
Alcohol (2-butanol)		R	R	—
Alcohol, ethyl		R	C	R to C
Alcohol, hexyl		R	R	R
Alcohol, isopropyl (2-propanol)		R	R to C	R to C
Alcohol, methyl		R	R	R to C
Alcohol, propyl (1-propanol)		R	R	R
Allyl acetone		R	—	R to C
Alums (aqueous)	all	R	R	R
Aluminum salts (chloride, flouride hydroxide, metaphosphate, sulphate)		R	R	R
Amino acids		R	—	R
Ammonia, gas		R	R	R
Ammonia, liquid		R	—	R
Ammonia (aqueous)		R	R	R
Ammonium salts (acetate, carbonate chloride, fluoride 10-25%, hydrosulphide, hydroxide, metaphosphate nitrate, phosphate, sulphate, sulphide, thiocyanate)		R	R	R
Amyl acetate	technically pure	R to C	C to N	C
Amyl chloride	100%	C	—	N
Amyl phthalate		R	—	C
Analine (aqueous)	all	R	C to N	C to N
Analine chlorhydrate		C	—	—
Analine hydrochloride (aqueous)	all	R to C	—	R
Analine dyes		C	—	—
Animal oils		R	—	R to C
Aniseed oil		C	—	N
Anisole		C	—	C to N
Antifreeze		R	—	R
Anthraquinone		C	—	—
Anthraquinone sulfonic acid		R	—	R
Antimony chloride, pentachloride		R	—	R
Antimony trichloride		R	R	R
Aqua Regia		N	N	N
Arsenic acid (anhydride)		R	—	R
Arsenic acid (aqueous)		R	R	R

Chemical Name	Concentration	73°F	120°F	140°F
Aryl sulfonic acid		R	—	—
Ascorbic acid		R	—	R
Asphalt		R	—	C
Aspirin		R	—	R
Barium salts	all	R	R	R
Barium hydroxide (aqueous)	all	R	R	R
Battery acid		R	—	R
Beater glue		R	—	R
Beer		R	R	R
Beet sugar liquor		R	—	—
Beeswax		R	C	C to N
Benzaldehyde (aqueous)	10%	R	—	C
Benzaldehyde in isopropyl alcohol	1%	R	—	R
Benzene	pure	C	C to N	C to N
Benzene Sulfonic acid	all	T ro C	—	R to C
Benzoic acid (aqueous)	all	R	R	R
Benzoyl chloride		C	—	N
Bichromate-sulphuric acid	concentrated	R	—	N
Bismuth salts		R	—	R
Bisulphite soution		R	—	R
Bitumen		R	—	C
Black liquor-paper		R	R	—
Bleach liquor (12.5% active chlorine)		R	N	N
Bleach liquor (5.5% active chlorine)		R	R	R
Bone oil		R	—	R
Borax		R	R	R
Boric acid (aqueous)	all	R	R	R
Boric acid methylester		R	—	C to N
Boron trifluoride		R	C	C
Brake fluid		R	—	R
Brandy - wine		R	—	—
Brine	saturated	R	R	R
Bromic acid		R	N	—
Bromine (gas)		N	N	N
Bromine (aqueous)		C	N	N
Bromine (liquid)		N	N	N
Bromochloromethane		N	—	—
Butanediol (aqueous)	all	R	R	R
Butadione		R	C	—
Butane tetrol (crythritol)		N	N	—
Butane, gas		R	R	R
Butanetriol (aqueous)	all	R	—	R
Butaxyl® (methoxylbutyl acetate)		R	—	C
Butter		R	—	R
Butyl acetate		R to C	C	C to N
n-Butyl acetate		R	—	C
Butyl acrylate		R	—	C
Butyl benzyl phthalate		R	—	R
Butylene glycol		R	—	R
Butylene		R	R	—
Butyl phenol		R to C	C	R
Butyric acid (aqueous)	all	R	C	C
Calcium chloride		R	R	—
Calcium salts (aqueous)		R	R	R to C
Camphor oil		N	—	N
Camphor (crystals)		R	C	C
Calcium hydroxide		R	R	R
Cane sugar liquors		R	R	R
Carbazole		R	—	R

TABLE 5
Chemical Resistance of HDPE Pipe Resins, continued

Chemical Name	Concentration	73°F	120°F	140°F
Carbolic acid		R	—	R
Carbolineum for fruit trees (aqueous)		R	—	C
Carbon bisulfide		C to N	N	N
Carbon dioxide (wet or dry)		R	R	R
Carbonic acid (aqueous)	all	R	R	R
Carbon monoxide		R	R	R
Carbon tetrachloride		C to N	N	N
Carnauba wax		R	—	R
Casein		R	C	—
Castor oil		R	C	R to C
Caustic potash (dry & solution)		R	R	R
Caustic soda (dry & solution)		R	R	R
Cellosolve		C	C	—
Cetyl alcohol (hexadecanol)		R	—	R
Cellosolve acetate		C	C	—
Chloral hydrate (aqueous)	all	R	R	RD
Chloramine		R	—	—
Chloroacetic acid		R to C	R to N	R to N
Chloric acid	20%	R	N	—
Chlorine, gaseous, dry		C to N	N	N
Chlorine, gaseous, moist		C to N	N	N
Chlorine, liquid		N	N	N
Chlorine, water		R to C	C	C to N
Chlorobenzene		C to N	C to N	N
Chlorocarbonic acid		R	—	C
Chlorobenzyl chloride		C	C	C
Chloroethanol	pure	R	—	RD
Chloroform	pure	C to N	C to N	N
Chloromethane	100%	C	—	N
Chloropicrin		R to C	—	N
Chlorosulphonic acid	100%	C to N	C to N	N
Chrome alum		R	R	R
Chrome anode mud		R	—	R
Chrome salts (aqueous)	all	R	—	R
Chromic acid	10%,30%,40%,50%	R	R to C	C to N
Chromic acid	80%	R	R to C	N
Chromium trioxide (aqueous*) up to 50%		R	—	—
Chromosulphuric acid		R	C	N
Cider		R	—	R
Citric acid (aqueous)	saturated	R	R	R
Clophen® A50 & A60		R	—	C to N
Coal-tar oil		RD	—	CD
Coconut oil		R	R to C	R to C
Cod liver oil		R	—	R to C
Coke oven gas (benzene free)		R	R	R
Coffee extract		R	—	R
Cognac		R	—	—
Cola concentrate		R	—	R
Copper salts (aqueous)		R	R	R
Copper chloride (aqueous)		R	R	R
Copper cyanide		R	—	R
Copper flouride (aqueous)		R	R	R
Copper nitrate (aqueous)		R	R	R
Copper sulphate (aqueous)	all	R	—	R
Corn oil		R	R to C	C
Corn syrup		R	R	R
Cranberry sauce		R	—	R
Coumarone resins		R	—	R
Creosote		R	R	RD
Cottonseed oil		R	C	R
Cresol	100%	R to C	R to N	CD
Cresol (aqueous)	diluted	R	—	RD

Chemical Name	Concentration	73°F	120°F	140°F
Crotonaldehyde	pure	R to N	N	C
Cresylic acid	50%	C	—	—
Cyclanone	usual commercial concentration	R	—	R
Crude oil		C	C	—
Cyclohexane		R	C	R to N
Cyclohexanol		R to C	C to N	N
Cyclohexanone		R to C	R to C	C to N
Decalin	pure	R	—	R
Detergents		R	R	R
Developer solutions (photographic)		RD	R	RD
Dextrin		R	R	R
Dextrose		R	R	R
Diazo salts		R	C	R
1,2-dibromoethane		C	—	N
Dibutyl ether		R to C	C	N
Dibutyl phthalate	pure	R	C	C
Dibutyl acetate		R	C	C
Dichloroacetic acid	pure	R	R	CD
Dichloroacetic acid	50%	R	—	CD
Dichloroacetic acid methyl ester		R	—	R
Dichlorobenzene		C	C to N	N
Dichloromethane		C	—	C
DDT (powder)		R	—	R
Dichloroethylene		C to N	C to N	N
Dichloropropane		C	—	N
Dichloropropene		C	—	N
Diesel fuel		R	C	C
Diethyl amine		C	C	—
Diethylene glycol		R	R	R
Diethyl ether		R to C	C	C to N
Di (2-ethylexyl) phthalate (DOP)		R	—	C
Diethyl ketone		R	.	C
Diglycolic acid (aqueous)	30%	R	R	R
Diisobutyl ketone	pure	R	R	C to N
Diisopropyl ether		R to C	—	N
Dimethylamine		R to C	C	C
Dimethyl formamide	pure	R	R	R to C
Dimethyl sulphoxide		R	R	R
Dioctyl phthalate		R to C	C	C to N
Dioxane 1,4		R	R	R
Diphenylamine		R	—	C
Diphenyl oxide		R	—	C
Disodium phosphate		R	R	R
Disodium sulphate		R	—	R
Dodecylbenzenesulphonic acid		R	—	C
Drinking water		R	—	R
Dyes		RD	—	RD
Electrolyte baths		R to C	—	C
Emulsifiers		R	—	R
Emulsions (photographic)		R	—	R
Emulsions (acrylic)		R	—	R
Ephetin (aqueous)	10%	R	—	R
Epichlorohydrin		R	—	R
Epsom salts	all	R	—	R
Esters, alphatic	pure	R	R	R to C
Ethane		R	—	R
Ether		R to C	C	C
Ethyl acetate	pure	R	C	C to R
Ethylbenzene	pure	C	—	—

TABLE 5
Chemical Resistance of HDPE Pipe Resins, continued

Chemical Name	Concentration	73°F	120°F	140°F
Ethyl chloride	pure	C	—	N
Ethyl ether		R to C	—	C to N
Ethylene		R	—	C
Ethyl esters		R	C	—
Ethyl halides		R	C	—
Ethylene diamine	pure	R	—	R
Ethylene diamine-tetraacetic acid		R	—	R
Ethylene dichloride		C to N	N	N
Ethylene chloride		C	—	C
Ethyl dibromide		C	—	N
Ethylene glycol		R	R	R
Ethylene oxide (gas)		R to C	C	R
2-Ethylhexanol		R	—	C
Euron® B		C	—	C
Euron® C		R	—	R
Fatty acids amides		R	—	C
Fatty acids		R	R	R to C
Fatty alcohols		R	—	C
Ferric chloride (aqueous)	all	R	R	R
Ferric and ferrous salts (aqueous)		R	R	R
Fertilizer salts (aqueous)	all	R	—	R
Film solutions		R	R	R
Fir wood oil		R	—	C
Fish solubles		R	—	R
Fluoboric acid		R	R	R to C
Fluorine, dry gas		C to N	N	N
Fluorine, wet gas		N	N	N
Fluorosilic acid	30%-40%	R	R	R
Formaldehyde	to 40%	R	R	R
Formamide		R	—	R
Formic acid (aqueous)	10%-50%	R	R to N	R
Formic acid (aqueous)	85%-100%	R	—	C
Freon - F11, 12, 113, 114		R to C	C	C to N
Freon - 21, F22		C	C	—
Fruit juices & pulp & fractose	all	R	R	R
Fuel oil		R	C	C
Furfural		C	C to N	C to N
Furfuryl alcohol		R	—	R to C
Gallic acid		R	—	—
Gas, coal, manufactured		R	R	R
Gas, natural, methane		R	R	—
Gasoline		R to C	C to N	C to N
Gelatin		R	R	R
Genantin®		R	—	R
Glucose		R	R	R
Glue		R	R	R
Glycerine (glycerol)(aqueous) to 100%		R	R	R
Gylcerol chlorohydrin		R	—	R
Glycine		R	—	R
Glycol		R	R	R
Glycolic acid (aqueous)	up to 70%	R	R	R
Glycolic acid butyl ester		R	—	R
Glysantin		R	—	R
Grisiron 8302		C	—	C
Grisiron 8702		R	—	R
Halothane		C	—	C to N
Heptane		R	C	C to N
Heating oil		C	C	—
Hexane		R	C	C

Chemical Name	Concentration	73°F	120°F	140°F
Hexanetriol		R	—	R
Hexanol		R	R	R
Honey		R	—	R
Hydraulic fluid		R	—	C
Hydrazine hydrate		R	—	R
Hydrobromic acid (aqueous)	up to 50%	R	R	R
Hydrobromic acid (aqueous)	100%	R	—	R
Hydrochloric acid	up to 100%	R	R	R
Hydrogen chloride gas wet & dry		R	—	R
Hydrocyanic acid	10%	R	R	R
Hydrfluoric acid	40%	R	R	C
Hydrofluosilicic acid (aqueous)	all	R	—	R
Hydrogen	100%	R	R	R
Hydrogen peroxide (aqueous)	10%	R	—	R
Hydrogen peroxide (aqueous)	30%	R	R	R
Hydrogen peroxide (aqueous)	50%	R	—	R
Hydrogen peroxide (aqueous)	90%	R	N	N
Hydrogen phosphide		R	—	R
Hydrogen sulphide	dry	R	R	R
Hydroquinone		RD	D	RD
Hydrosulphite	up to 10%	R	—	R
Hydroxylamine sulphate (aqueous)	12%	R	R	R
Hypochlorus acid		R	R	R to C
Ink		R	—	R
Iodine - in KI	3% (aqueous)	R	R	R
Iodine alcohol solution		C	C to N	N
Iodine (aqueous)	10%	C	C	—
Iron III chloride (aqueous)	all	R	—	R
Isobutyl alcohol		R	—	R
Isooctane		R to C	C	C
Isopropanol	pure	R	C	R to N
Isopropyl acetate	100%	R	—	C
Isopropyl ether	pure	R to C	C	N
Jam, jellies		R	—	R
Jet fuels, JP-4 & JP-5		R	C	—
Kerosene		C	C to N	C to N
Ketones		R to C	C	C to N
Kraft paper liquor		R	R	—
Labarraque's solution		R to C	—	
Lactic acid	10%-96%	R	R	R
Lactose		R	—	R
Lacquer thinners		C	C	—
Lanolin (wool fat)		R	—	R
Lard oil		R	R	—
Latex		R	—	R
Lauric acid		R	R	—
Lauryl chloride		R	R	—
Lauryl sulphate		R	R	—
Lead acetate (aqueous)	all	R	R	R
Lead salts		R	R	—
Lead tetraethyl		R	R	—
Lime		R	—	R
Lime sulphur		R	R	—
Lime water		R	—	R
Linseed oil		R to C	C	R to N
Liquor		R	R	R
Liqueur		R to C	R	N

TABLE 5
Chemical Resistance of HDPE Pipe Resins, continued

Chemical Name	Concentration	73°F	120°F	140°F
Liquid manure		R	—	R
Liquid paraffin		R	—	R
Liquid soaps		R	—	R
Lithium bromide		R	—	R
Lubricating oils		R to C	C	C
Lithium salts		R	R	—
Linoleic acid		R	R	—
Lysol		R	—	C
Machine oil		R	R	C
Magnesium salts		R	R	R
Magnesium carbonate		R	R	R
Magnesium chloride		R	R	R
Magnesium fluosilicate		R	—	R
Magnesium hydroxide		R	—	R
Magnesium iodide		R	—	R
Magnesium sulphate		R	R	R
Magnesium hydroxide		R	R	R
Magnesium nitrate		R	R	R
Maleic acid	50%-100%	R	R	R
Malic acid	50%	R	R	R
Manganese sulphate		R to C	R to C	R
Margarine		R	—	R
Mash		R	—	R
Mayonnaise		R	—	—
Menthol		R	R	C
Mercuric chloride		R	R	R
Mercuric cyanide		R	R	R
Mercurous nitrate		R	R	R
Mercuric salts		R	R	R
Mercury		R	R	R
Metallic soaps		R	R	R
Metallic mordants		R	—	—
Methacrylate		R	—	R
Methacrylic acid		R	—	R
Methane		R	R	—
Methanol	pure	R	R	R
Methyl acetate		C	C	—
Methyl bromide		C	C to N	N
Methyl cellosolve		C	C	—
Methyl chloride		C	C to N	N
Methyl chloroform		C	C	—
Methyl benzene		R	—	N
Methoxy butanol		R	—	C
Methoxybutyl acetate (Butozyl)		R	—	C
Methyl cyclohexane		C	C	N
Methyl cyclohexanone		R	C	—
Methyl methacrylate		R	C	R
Methyl salicylate		R	—	C
Methyl sulfate	50%	R	—	R
Methyl sulfuric acid		R	—	R
Methyl ethyl ketone		R to N	R	R
Methyl glycol		R	—	R
Methyl isobutyl ketone		R	—	C to N
4-Methyl2-pentanone		R	—	R to CD
Methyl propyl ketone		R	—	C
n-Methyl pyrrolidone		R	—	R
Methylene bromide		C	C	—
Methylene chloride*		C	C	C to N
Methylene iodide		C	C	—
Milk		R	R	R
Mineral oil		R to C	C	C to N

Chemical Name	Concentration	73°F	120°F	140°F
Molasses		R	R	R
Mixed acids (sulfuric & nitric)		N	N	—
Mixed acids (sulfuric & phosphoric)		R	C	—
Monochloroacetic acid		R	—	R
Monochloroacetic acid ethyl ester		R	—	R
Monochloroacetic acid methyl ester		R	—	R
Monochlorobenzene		C to N	C to N	N
Monoehtanolamine		—	—	—
Morpholine		R	—	R
Motor oil		R	R	R to C
Mowilith® polymer emulsions		R	—	R
Mustard		R	—	R
Nail varnish remover		R	—	C
Naphtha		R to C	C to N	C to N
Naphthalene		R	C	C
Nickel chloride		R	R	R
Nickel nitrate		R	R	R
Nickel salts		R	R	R
Nickel sulphate (aqueous)	all	R	—	R
Nicotine		R	—	R
Nicotine acid	diluted solution	R	—	R
Nitric acid	0-30%	R	R to C	R
Nitric acid	30-50%	R to C	C	N
Nitric acid	60%	C	N	N
Nitric acid	70%	C to N	N	N
Nitric acid	80%	N	N	N
Nitric acid	90%	N	N	N
Nitric acid	100%	N	N	N
Nitric acid fuming		N	N	N
Nitrobenzene		R to C	C	N
Nitrocellulose		R	—	—
Nitrotoluene		R	C	N
Nitrous acid		R	N	—
Nitrous oxide, gas		R	N	—
Nitroglycerine		R	C	—
Nitroglycol		—	—	—
Nitropropane		—	—	—
Nonyl alcohol		R	—	R
Octyl cresol		C	N	—
Oils and fats		R	R to C	C to N
Oils, vegetable		R to C	C	C
Oleic acid		R to C	C	C
Oleum		N	N	N
Olive oil		R	R	R
Optical brightners		R	—	R
Orange juice		R	—	R
Orthophosporic acid	50%	R	R	R
Orthophosporic acid	85%	R	R	C
Oxalic acid		R	R	R
Oxygen, gas		R	R	R
Ozone, gas		C	C	N
Palmitic acid	10%	R	R	R to C
Palmitic acid	70%	R	R	—
Palmityl alcohol		R	—	R
Paraffin		R to C	C	C
Palm kernel oil		R	—	R
Paraformaldehyde		R	—	R
Pentane		C	C	—
Pentanol		R	—	R

TABLE 5
Chemical Resistance of HDPE Pipe Resins, continued

Chemical Name	Concentration	73°F	120°F	140°F	Chemical Name	Concentration	73°F	120°F	140°F
Peppermint oil		R	—	—	Potassium nitrate		R	R	R
Peracetic acid		R	—	—	Potassium orthophosphate	saturated	R	—	R
Perchloric acid (aqueous)	up to 20%	R	R	R	Potassium perchlorate		R	R	R
Perchloric acid (aqueous)	20% to 50%	R	R	C	Potassium perborate		R	R	R
Perchloric acid (aqueous)	70%	R	R to C	N	Potassium permanganate	up to 25%	R	R	R
Perchloroethylene		C	C	N	Potassium persulphate (aqueous)	all	R	R	R
Perfume oils		C	—	C to N	Potassium salts		R	R	—
Petroleum (sour, refined)		R	C	C	Potassium sulphate		R	R	R
Petroleum ether		R	—	C	Potassium sulfide		R	R	R
Phenol		R	C	RD	Potassium sulfite		R	—	R
Phenolic resin molding materials		R	—	R	Potassium tetracyanocuprate		R	—	R
Phenylcarbinol		—	—	—	Potassium thiosulphate		R	—	R
Phenyethylalcohol		R	—	R	Propane, gas		R	R	R
Phenylhydrazine		C	C	C to N	Propargyl alcohol (aqueous)	7%	R	—	R
Phenylhydrazine hydrochloride		R to C	C	R	Propionic acid (aqueous)	all	R	—	R to C
Phenylsulphonate		R	—	R	Propylene dichloride	100%	C to N	—	N
Phenylsulphonate		R	—	R	Propylene glycol		R	R	R
Phosgene gas		C to N	C	—	Propylene oxide		R	—	R
Phosgene liquid		N	N	—	Prussic acid		R	R	R
Phosphorus oxycloride		R	R	C	Pseudocumene		C	—	C
Phosphorus pentoxide		R	R	R	Pyridine		R	C	C
Phosphorus trichloride		R	R	C	Pyrogallic acid		—	—	—
Phosphoric acid	50%	R	R	R	Pulp mill water (red & black liquor)		R	R	—
Phosphoric acid	80%-100%	R	—	CD					
Phosphorus, yrellow		—	—	—	Quinine		R	—	R
Phosphorus, red		—	—	—	Quinol (hydroquinone)		R	—	R
Phosphates (aqueous)	all	R	R	R					
Photographic developers		RD	—	RD	Rayon coagulating bath		R	—	R
Phthalic acid (aqueous)	50%	R	R	R	Rubber dispersions (latexes)		R	—	R
Phthalic acid ester		R	—	R to C					
Picric acid (aqueous)		R	R to C	C	Sargrotan®		R	—	C
Pineapple juice		R	—	R	Salinec acid (aqueous)		R	—	C
Pine-needle oil		R	—	C	Salicylic acid		R	R	R
Plating solution, metals (many types)		R	C	R	Saturated steam concentrate		R	—	R
Plasticizers		R	—	C	Sauerkraut		R	—	R
Polyester plasticizers		R	—	R to C	Salicylaldehyde		R	R	—
Polyester resins		C	—	R	Sea water		R	R	R
Polyglycols		R	—	R	Selenic acid		R	R	R
Potash		R	R	R	Sewage, residential		R	R	—
Potash aluminum (aqueous)		R	—	R	Silicic acid (aqueous)	all	R	R	R
Potassium alkyl xanthates		—	—	—	Silicone, oil		R	R to C	R to C
Potassium bicarbonate (aqueous)	all	R	—	R	Silver, acetate		R	R	R
Potassium bichromate	40%	R	—	R	Silver, cyanide		R	R	R
Potassium bisulphate (aqueous)	all	R	—	R	Silver, nitrate		R	R	R
Potassium borate (aqueous)	1%	R	R	R	Silver salts		R	R	R
Potassium bromate (aqueous)	up to 10%	R	R	R	Soap solutions (can be stress cracking agents)		R	R	R
Potassium bromide (aqueous)	all	R	R	R	Sodium acetate (aqueous)	all	R	R	R
Potassium carbonate (aqueous)	all	R	R	R	Sodium aluminum phosphate		R	—	R
Potassium chlorate (aqueous)	all	R	R	R	Sodium benzoate		R	R	R
Potassium chloride (aqueous)	all	R	R	R	Sodium bicarbonate		R	R	R
Potassium chromate (aqueous)	40%	R	R	R	Sodium bisulphate		R	R	R
Potassium cyanide (aqueous)	all	R	R	R	Sodium bisulphite (aqueous)	all	R	R	R
Potassium dichromate (aqueous)	all	R	R	R	Sodium borate		R	R	R
Potassium ferricynaide (aqueous)	all	R	R	R	Sodium bromide		R	R	R
Potassium ferrocyanide (aqueous)	all	R	R	R	Sodium carbonate (aqueous)	all	R	R	R
Potassium fluoride (aqueous)	all	R	R	R	Sodium chlorate	saturated	R	R	R
Potassium hydroxide (aqueous)	all	R	R	R	Sodium chloride (aqueous)	salt	R	R	R
Potassium hypochlorite		R	R	C	Sodium chlorite		R	R	R
Potassium hydrogen carbonate		R	—	R	Sodium chromate		R	—	R
Potassium hydrogen sulfate	saturated	R	—	R	Sodium cyanide		R	R	R
Potassium hydrogen sulfide		R	—	R	Sodium dichromate		R	—	R
Potassium iodide		R	—	R	Sodium dichromate, acid		R	C	—

TABLE 5
Chemical Resistance of HDPE Pipe Resins, continued

Chemical Name	Concentration	73°F	120°F	140°F
Sodium dodecylbenzenesulphonate		R	—	R
Sodium ferricyanide		R	R	R
Sodium ferrocyanide		R	R	R
Sodium fluoride		R	R	R
Sodium hexacyanoferrate		R	—	R
Sodium hydrogen carbonate		R	—	R
Sodium hydrogen phosphate		R	—	R
Sodium hydrogen sulfite		R	—	R
Sodium hydroxide, aqueous & solid	all	R	R	R
Sodium hypochlorite		R	R	R
Sodium nitrate (aqueous)	all	R	R	R
Sodium orthophosphate		R	—	R
Sodium perborate (aqueous)	all	R	—	R
Sodium perchlorate (aqueous)	R	R	—	R
Sodium peroxide	10%	R	—	R
Sodium peroxide (aqueous)	saturated	C	—	—
Sodium phosphate (aqueous)	saturated	R	—	R
Sodium salts (aqueous)		R	R	—
Sodium silicate		R	—	R
Sodium sulphate		R	R	R
Sodium sulfide		R	R	R
Sodium sulfite		R	—	R
Sodium thiosulphate		R	R	R
Soft soap		R	—	R
Soybean oil		R	—	R
Spindle oil		R to C	—	C
Stain removers		R to C	—	C
Stannic chlorides		R	R	R
Stannous chloride		R	R	R
Starch (aqueous)	up to 100%	R	R	R
Stearic acid		R	R	R to C
Styrene		C	—	N
Stoddard solvent		R	C	—
Succinic acid	50%	R	—	R
Sulfur dioxide, dry		R	R	R
Sulfur dioxide, wet		R	R	R to C
Sulfite liquor		R	R	—
Sulfur		R	R	R
Sulfuric acid	up to 50%	R	R	R
Sulfuric acid	50%-70%	R	R	R
Sulfuric acid	70%-90%	R	C to N	C
Sulfuric acid	90%	C to N	N	N
Sulfuric acid fuming		N	N	N
Sulfurous acid		R	R	R
Sulfuric ether		R to C	—	C
Sulfur trioxide		N	N	N
Sulfuryl chloride		N	—	—
Syrups & sugars		R	R	R
Tall oil		R	R	—
Tallow	pure	R	R	R
Tannic acid		R	R	R
Tanning liquors		R	R	—
Tartaric acid (aqueous)		R	R	R
Tetrabromoethane		N	N	N
Tetrachloroethane		C to N	C to N	N
Tetrachloromethylene		—	—	—
Tetraethyl lead		—	—	—
Tetrahydrofuran		C to N	N	N

Chemical Name	Concentration	73°F	120°F	140°F
Tetrahydronaphthalene		R	—	C to N
Thioglycolic acid		R	—	R
Thionyl chloride		N	N	N
Thiophene		C	—	N
Thread cutting oil		—	—	—
Terpineol		—	—	—
Titanium tetrachloride		R	R	—
Toluene		C to N	C to N	N
Toluene - kerosene	25%-75%	C	—	N
Transformer oil		R	R to C	R to C
Tributyl citrate		C	C	—
Tributyl phosphate		R to C	C	—
Trichloroacetic acid	pure	R	R	C to N
Trichloroacetic acid	50%	R	C	R
Trichloroethylene	pure	C to N	C to N	N
Trichlorobenzene		N	—	N
Tricresyl phosphate		R to C	C	R
Triethanolamine		R	C	R to CD
Triethylene glycol		R	—	R
Triethylamine		R	C	—
Triethyl borate		R	—	C to N
Trimethyl propane		C	C	—
Trimethyol propane (aqueous)		R	—	R
Tri-B-chloroethyl phosphate		R	—	R
Trioctyl phosphate		R	—	C
Trisodium phosphate		R	—	R
Turpentine		C	C to N	N
Tutogen® U		R	—	R
Tween® 20 & 80		R	—	N
Two-stroke engine oil		R	—	C
Urea	up to 33%	R	R	R
Uric acid		R	—	R
Urine		R	R	R
Vaseline		R to C	R	C
Vegetable oils		R	R	R
Vinegar		R	R	R
Vinyl acetate		R	—	R
Walnut oil		R	—	C
Water, distilled, fresh, mine, salt, tap		R	R	R
Wax alcohols		C	·	C
Waxes		R	—	R to C
Whey		R	—	R
Whiskey		R	R	R
Wine		R	R	R
Wood stains		R	—	R to C
Xylene		C to N	C to N	N
Yeast		R	R	R
Zinc carbonate		R	R	R
Zinc chloride		R	—	R
Zinc oxide		R	—	R
Zinc salts (aqueous)	all	R	R	R
Zinc sulfate		R	—	R
Zinc sludge		R	—	R
Zinc stearate		R	—	R

Chemical Resistance Factors

NOTE: This experimental method was developed in Europe but is not used in the United States. It is presented here only as a reference for the interested reader.

Since the experimental method uses compression-molded specimens immersed in the chemical media, the results can only be applied under the same conditions. When plastic parts come into contact with chemical agents, it is important to know how those parts will be affected. When a mechanical stress is superimposed on a chemical one, a plastic may exhibit a completely different behavior. The standard chemical resistance immersion tests that are conducted on compression-molded plaques are suitable only if the plastic part in question will not be under stress. Chemicals that do not normally affect the properties of a stress-free plastic part may cause cracking to occur when a stress is applied. Various chemicals can accelerate crack propagation and therefore cause early failures.

Realizing this need of information for pressure pipes, a task committee within the International Standards Organization (ISO/TC 138 WG 3) began to test HDPE and polypropylene pipe. It was decided to use the creep-rupture internal pressure test, which had been exhaustively investigated using water as the internal medium[14,26].

The creep rupture internal pressure test is usually conducted using water in the inside of the pipe. A graph showing typical failure times versus hoop stresses for HDPE and polypropylene, known as a creep-rupture curve, is shown in Figure 26. The creep rupture curves for pipes under stress with water are the basis for the determination of the chemical resistance factors.

A dimensionless ratio that quantifies the influence of a chemical medium to that of water is called the chemical resistance factor, f_{cr}. By using the chemical as the internal medium, a new creep-rupture curve can be drawn and compared to the standard "water" curve. From these data, two chemical resistance factors are obtained. These indicate the service and the stress life of the pipe with the chemical compound as compared to the pipe with water.

Figure 27 illustrates the basis for determining the two resistance factors:

Chemical Resistance Time Factor, fcrt = tm / tw
Chemical Resistance Stress Factor,

$$fcp\sigma = \frac{\sigma m}{\sigma w}$$

WHERE
tm = service life using chemical medium
tw = service life using water
σm = hoop stress using chemical medium
σw = hoop stress using water

Chapter 3 | 97
Engineering Properties

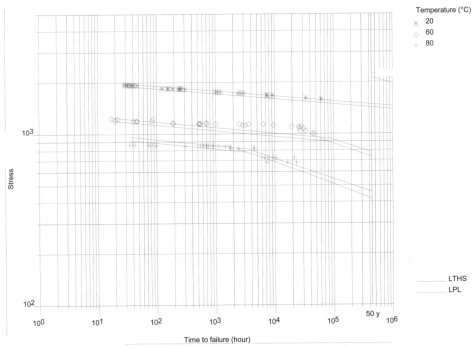

Figure 26 Typical Creep-Rupture Curve for HDPE Pipe

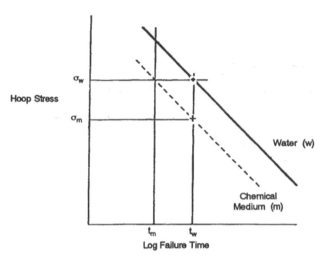

$fcr\sigma = \dfrac{\sigma_m}{\sigma_w}$ If the same failure time for both medium 1 and water is required, the hoop stress time of medium 1 is equal to the water curve hoop stress multiplied by fcrσ.

$fcrt = \dfrac{t_m}{t_w}$ If the same hoop stress for both medium 1 and water is required, the failure time of medium 1 is equal to the water curve failure time multiplied by fcrt.

Figure 27 Creep-Rupture Curve

The Time Factor is determined by dividing the failure time of the pipe with the chemical compound by the failure time using water; the same hoop stress must be used. Conversely, the Stress Factor is determined by dividing the hoop stress with the chemical compound by the hoop stress with water at the same failure time.

The chemical resistance stress factor indicates the level of stress to which the pipe containing the chemical compound can be subjected and still achieve the same service life as water. The chemical resistance time factor (fcrt) indicates what the service life of the pipe containing the chemical compound would be as compared to the pipe containing water at the same hoop stress. Table 6 lists some chemical resistance factors.

TABLE 6
Resistance Factors for Pipes Made from HDPE

Medium	Concentration %	Temp. °C	Time Factor f_{crt}	Stress Factor f_{cr}
Air	100	80	10	1
Mains Water	100	80	1	1
Wetting Agent Solution, Aqueous	2	80	0.25	0.6
Sulphuric Acid	80	80	5	1
Common Salt Solution	25	80	9	1
Caustic Soda Solution	50	80	15	1
Sodium Hypochlorite	20	80	0.02	0.5
Chromic Acid	10	80	0.5	0.8
Chromic Acid	20	80	0.25	0.6
Nitric Acid	65	80	0.01	0.2
Hydrochloric Acid	33	80	0.35	0.7
Methanol	100	60	1	1
Octanol	100	60	0.3	0.8
Acetic Acid	100	60	0.07	0.3
Acetic Acid	60	60	0.4	0.4
Ethyl Acetoacetate	100	80	0.2	0.7
Fuel Oil	100	60	0.2	0.7
Unfractionated Crude Oil	100	60	0.04	0.4
Unfractionated Crude Oil	—	20	>0.5	>0.6
Petrol	100	60	0.01	0.3
Carbon Tetrachloride	100	60	0.025	0.5

Environmental Stress Crack Resistance (ESCR)

Under certain conditions of temperature and stress in the presence of certain chemicals, polyethylene may begin to crack sooner than at the same temperature and stress in the absence of these chemicals. This phenomenon is called environmental stress cracking (ESC). Stress cracking agents for polyethylenes tend to be polar materials such as alcohols, detergents (wetting agents), halogens and aromatics. The property of a material to resist ESC is called environmental stress crack resistance or simply ESCR. The mechanism has been fully researched over the years. Failures from ESC tend to be due to the development of cracks that slowly grow and propagate over time. Stress cracking can be avoided by using stress crack resistant materials and by limiting stresses and strains during pipe installation.

There are over 40 different ESCR test methods used to determine the chemical resistance of various materials. A standard test currently used in the polyethylene industry is the bent-strip test. It is also called the "Bell Test," since it was developed during the 1950's for wire and cable coatings for the telephone industry.

ASTM D1693[3], *Standard Test Method for Environmental Stress-Cracking of Ethylene Plastics*, describes the test method used to determine the ESCR value for polyethylene. Ten small compression-molded specimens are notched and bent and then placed into a holder. The holder is immersed into a tube of a surfactant, typically one such as Igepal C630 (from GAF Corp. NY, NY), at 212°F (100°C) and 100% concentration, and the time to failure is noted. The results are reported using the notation F_{xx}, where xx is the percentage of samples that have failed. For example, the statement F_{20} = 500 hours means that 20% of the samples have failed within 0 to 500 hours. Figure 28 illustrates the bent-strip ESCR test.

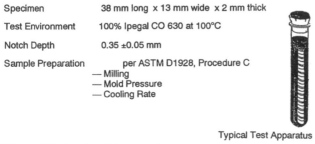

Specimen	38 mm long x 13 mm wide x 2 mm thick
Test Environment	100% Ipegal CO 630 at 100°C
Notch Depth	0.35 ±0.05 mm
Sample Preparation	per ASTM D1928, Procedure C — Milling — Mold Pressure — Cooling Rate

Typical Test Apparatus

Figure 28 Details of the Bent Strip Test

This test was developed when the time to failure was less than 10 hours and the decay of stress did not affect the results. However, the current polyethylene pipe resins generally do not fail this test. This is due to the excellent stress crack resistance of modern resins, but it is also due to the fact that the stress, which is produced by bending the samples, decays over time. Therefore, the intensity of the test diminishes after a few hundred hours. This test is used mainly as a quality assurance test rather than providing definitive ranking of pipe performance.

Compressed Ring ESCR Test

ASTM test method, F1248[5], *Standard Test Method for Determination of Environmental Stress Crack Resistance (ESCR) of Polyethylene Pipe*, determines the polyethylene pipe's resistance to stress cracking in the presence of a stress crack agent at elevated temperatures. A ring specimen of pipe, having a controlled imperfection at one location, is exposed to a stress-cracking agent while compressed between two parallel plates. Figure 29 illustrates the apparatus used to conduct this test. The time to failure is recorded. This test is believed to be more realistic than the bent-strip ESCR test (ASTM D1693) since the test specimens are actual pieces of pipe rather than compression-molded specimens.

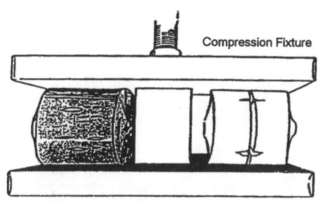

Figure 29 Details of the Rader Compressed Ring ESCR Test

Specimen
Pipe Sample (actual production lot) cut to 1/2 in. length

Test Environment
25% Ipegal CO 630 at 50°C

Notch Depth
20% of minimum wall
(e.g., 0.043 in. for 2-in. SDR 11)

Notch Length
3/4 in.

PENT Test

ASTM test method, F1473[5], *Notch Tensile Test to Measure the Resistance to Slow Crack Growth of PE Pipes and Resins (PENT test)*, is used to measure the slow crack growth properties of a polyethylene material under standard conditions. Specimens cut from either compression-molded plaques or pipe are precisely notched and then exposed to a constant tensile stress at elevated temperature in air. The test is generally performed at 80°C with an applied stress of 2.4 MPa, but may also be carried out at alternate temperatures and stresses. The time to failure is recorded. This test method can be used in place of the standard bent-strip test method for classifying a material's slow crack growth resistance in ASTM D3350.

Specimen Cut from plaque or pipe, 50 mm long, 25mm wide and 10 mm thick.
Test Environment 80°C, 2.4 MPa stress in air.
Notch Depth 3.5 mm.
Side Notches 1.0 mm.

Aging

Weatherability

Ultraviolet (UV) radiation and oxygen-induced degradation in plastics can alter its physical and mechanical properties[17]. The function of UV stabilizers is to inhibit the physical and chemical processes of UV-induced degradation. The prime UV stabilizer used in the polyethylene pipe industry is finely divided carbon black, which is the most effective additive capable of stopping these UV-induced reactions.

The weatherability of plastic pipe used in outdoor applications is not just a function of the material's color. The colorant is only one part of a two-part package that protects the pipe from degradation due to the effects of UV radiation from the sun.

There are two separate issues when dealing with UV protection. The first is weatherability, which is defined as the capability of the resin to resist changes in the physical properties when exposed in an outdoor environment. The other parameter, which is often used to test the durability of a pigment system, is called color fastness. That test measures the time it takes for an article to fade or to change colors.

Since all plastics are susceptible to attack by UV light, the first step is to protect the plastic resin by using a UV stabilizer, of which there are many different types on the market. The type of UV stabilizer is the predominant factor affecting the service life of a part. The next step is to choose a colorant that will not decrease the effectiveness of the UV stabilizer package. Many colorants act in a synergistic manner with the UV stabilizer. The type of pigment may either increase or decrease the physical properties of the article.

It should be noted that the type of pigment selected also has an effect on the lifespan of the product. For example, phthalo-cyanine blue provides better protection, using the same UV stabilizer package in polyethylene, than does ultramarine blue. Therefore, particular attention should be paid to the type of pigment used and not just the general color. Carbon black is considered to be the best color for outdoor articles because of its powerful UV absorptivity. It is also the least expensive pigment for plastics since it is a colorant and, at the same time, one of the best UV stabilizers for outdoor plastic articles. Some general guidelines concerning the weatherability of colorants are shown as follows.

TABLE 7
Relative Weatherability Of UV Stabilized Pigmented Polyethylene

Black	Best Protection
White, Blue	
Red (Inorganic)	
Yellow (Inorganic)	
Green	
Red, Yellow (organic)	
Natural (nonpigmented)	Least Protection

There are also other factors that can affect the life expectancy of a pipe. For further technical information, please refer to PPI Technical Report TR-18[37a], *Weatherability of Thermoplastic Piping*.

Stabilization

Stabilizers are added to the resin to prevent oxidation and the subsequent loss of physical properties. Free radical molecules combine with oxygen and form unstable compounds that continue to react with the polyethylene. This free radical chain process is controlled by using compounds that will react with the free radicals to form stable species incapable of further reaction.

A stabilizer system usually comprises a primary and a secondary antioxidant. It has been found that two different types of antioxidants can work synergistically and provide better protection using lower concentrations and therefore lower costs. The primary antioxidant is used to protect the resin during the extrusion process. Commonly used compounds include BHT, hindered phenols, and secondary amines. Since oxidation attack is a continuous process, secondary antioxidants protect the finished product from long-term oxidation during its service life. Compounds used for this purpose include phosphites and thioethers.

The high temperatures encountered during extrusion facilitate free radical formation. Therefore it is very crucial to protect the polymer during this step. Since there is a finite amount of antioxidant added to the polymer, high processing temperatures combined with a certain time factor could fully deplete the antioxidant ingredient. Once the antioxidant is depleted, the polymer will undergo degradative steps, such as chain scission and/or cross-linking. This degradation process will reduce the physical properties of the polymer so the pipe will not meet industry standards.

There are several tests that can be used to indicate the severity of processing that a polymer has undergone. Two common methods include Differential Scanning Colorimetry (DSC) and carbonyl index tests. DSC induction time or temperature measurements indicate the degree of stabilizer usage. The carbonyl index indicates the degree of oxidative degradation by measuring the type and amount of carbonyl (C=0) functionalities created during UV exposure.

Toxicological Properties

Health Effects

The Food and Drug Administration (FDA) issues requirements for materials that may contact food, either directly or indirectly, under the Code of Federal Regulations (CFR) Title 21, parts 170 to 199. Most natural polyethylene resins do comply with these regulations. Some grades of furnace carbon blacks are certified for compliance to these FDA requirements. If there are any questions concerning FDA compliance for carbon black pigment, contact the resin supplier.

Potable water piping materials, fittings, and pipe are currently tested according to the standards developed by the National Sanitation Foundation (NSF). The most recent standard to be written by the NSF is Standard 61 [30], *Drinking Water System*

Components - Health Effects. It sets forth toxicological standards for all potable water system components including plastics.

Many municipalities and other organizations have adopted the potable water standards that are administered by NSF. Those standards verify that physical, chemical, toxicological, taste and odor requirements have been met by any materials that bear the NSF mark. NSF enforces the standards by conducting unannounced visits to all companies that are listed with them. Non-compliance items are ordered withdrawn from the market place.

Biological Resistance

Biological attack can be described as degradation caused by the action of microorganisms such as bacteria and fungi. Virtually all plastics are resistant to this type of attack. Once installed, polyethylene pipe will not be affected by micro-organisms, such as those found in normal sewer and water systems. Polyethylene is not a nutrient medium for bacteria, fungi, spores, etc.

Research has shown that rodents and gnawing insects are compelled to maintain their teeth in good condition by gnawing on objects. The surface of the pipe serves as a deterrent to gnawing rodents since their teeth slide off the round surface of the pipe. Rodents will gnaw at plastic simply because it is a soft material. Other materials such as wood, copper, lead, and all other plastics would fall prey to this phenomena if installed in rodent-infested areas.

Termites pose no threat to polyethylene pipe. Several studies have been made where polyethylene pipe was exposed to termites. Some slight damage was observed, but this was due to the fact that the plastic was in the way of the termites' traveling pathway. PPI Technical Report, TR-11[37], *Resistance of Thermoplastic Piping Materials to Micro- and Macro-Biological Attack,* has further information on this matter.

Conclusion

The information contained in this chapter should help the reader to understand the fundamental properties of polyethylene. A basic understanding of these properties will aid the engineer or designer in the use of polyethylene pipe and serve to maximize the utility of the service into which it is ultimately installed.

While every effort has been made to present the fundamental properties as thoroughly as possible, it is obvious that this discussion is not all-inclusive. For further information concerning the engineering properties of polyethylene pipe, the reader is referred to a variety of sources including the pipe manufacturers' literature, additional publications of the Plastics Pipe Institute and the References at the end of this chapter.

References

1. *AGA Plastic Pipe Manual for Gas Service*. (1985). Catalog No. XR065. American Gas Association, Arlington, VA, 1985.
2. *ASTM Annual Book*, Volume 08.01 Plastics (I): C177 - D1600, American Society for Testing and Materials, Philadelphia, PA.
3. *ASTM Annual Book*, Volume 08.02 Plastics (II): D1601 - D3099, American Society for Testing and Materials, Philadelphia, PA.
4. *ASTM Annual Book*, Volume 08.03 Plastics (III): D3100 - Latest, American Society for Testing and Materials, Philadelphia, PA.
5. *ASTM Annual Book*, Volume 08.04, Plastic Pipe and Building Products, American Society for Testing and Materials, Philadelphia, PA.
6. *ASTM Annual Book*, Volume 03.01, Metals - Mechanical Testing; Elevated and Low-Temperature Tests; Metallography, American Society for Testing and Materials, Philadelphia, PA.
7. *ASTM Annual Book*, Volume 05.01, Petroleum Products and Lubricants (I), D56 - D1947, American Society for Testing and Materials, Philadelphia, PA.
8. Ayres, R. L. (1981, May 18-20). *Basics of Polyethylene Manufacture, Structure, and Properties*, American Gas Association Distribution Conference, Anaheim, CA.
9. Barker, M. B., J. Bowman, & M. Bevis (1983). The Performance and Causes of Failure of Polyethylene Pipes Subjected to Constant and Fluctuating Internal Pressure Hea*dings, Journal of Materials Science*, 18, 1095-1118.
10. Barker, M. B., & J. Bowman. (1986, December). A Methodology for Describing Creep-Fatigue Interactions in Thermoplastic Component*s, Polymer Engineering and Science*, Vol. 26, No. 22, 1582-1590.
11. Bowman, J. (1989). Can Dynamic Fatigue Loading be a Valuable Tool to Assess MDPE Pipe System Quality, Proceedings of the 11th Plastic Fuel Gas Pipe Symposium, 235-248.
12. Broutman, L. J., D. E. Duvall, & P. K. So. (1990). *Application of Crack Initiation and Growth Data to Plastic Pipe Failure Analysis*, Proceedings of the Society of Plastics Engineers 48th Annual Technical Conference, Vol. 36, 1495-1497.
13. *Designing with Plastic - The Fundamentals*. (1989). Hoechst Celanese Corporation, Engineering Plastics Division Chatham, NJ.
14. Diedrich, G., B. Kempe, & K. Graf. (1979). Zeilstandfestigheit von Rohren aus Polyethln hart (HDPE) und Polypropylen (PP) unter Chemikallenwirkung (Creep Rupture Strength of Polyethylene (HDPE) and Polypropylene (PP) Pipes in the Presence of Chemicals), *Kunstoffe* 69, 470-476.
15. Dieter, G. E. (1966). *Mechanical Metallurgy*, 3rd Edition, McGraw-Hill Book Company, New York, NY.
16. *Driscopipe Engineering Characteristics*. (1981). Phillips Driscopipe, Inc., Richardson, TX.
17. Gaechter, R., & H. Mueller (ed.) (1963). *Plastics Additives*, Hanser Publications, New York, NY.
18. Griffith, A. A., Phil. Trans. (1920). Royal Society of London, Vol. A 221, p. 163.
19. Haag, J., Griffith. (1989, January). Measuring Viscoelastic Behavior, *American Laboratory*, No. 1, 48-58.
20. Harper, C. A. (ed.) (1975). *Handbook of Plastics and Elastomers*, McGraw-Hill Book Company, New York, NY.
21. Heger, F., R. Chambers, & A. Deitz. (1982). *Structural Plastics Design Manual*, American Society of Civil Engineers, New York, NY.
22. Hertzberg, R. W. (1983). *Deformation and Fracture Mechanics of Engineering Materials*, 2nd Edition, J. Wiley & Sons, New York, NY.
23. Hoechst Plastics. (1981). *Hostalen*, Brochure No. HKR IOle-8081, Hoechst AG, Frankfort, Germany.
24. Hoechst Plastics. (1982). *Pipes*, Brochure No. HKR 111e-8122, Hoechst AG, Frankfort, Germany.
25. Hoff, A., & S. Jacobsson. (1981). Thermo-Oxidative Degradation of Low-Density Polyethylene Close to Industrial Processing Conditions, *Journal of Applied Polymer Science*, Vol. 26, 3409-3423.
26. Kemp, G. (1984). Pruefmethoden zur Emitttung des Verhaltens von Polyolefinen bei der Einwirkung von Chemikallen (Methods to Determine the Behavior of Polyolefins in Contact with Chemicals), Zeitschrift fuer, *Werkstofftech*, 15,157-172.
27. Krishnaswamy, P., et al. (1986). A Design Procedure and Test Method to Prevent Rapid Crack Propagation in Polyethylene Gas Pipe, Battelle Columbus Report to the Gas Research Institute.
28. Levenspiel, O. (1982). *Chemical Reaction Engineering*, John Wiley & Sons, New York, NY.
29. Mruk, S. A. (1985). Validating the Hydrostatic Design Basis of PE Piping Materials, Proceedings of the Ninth Plastics Fuel Gas Pipe Symposium, 202-214.
30. *NSF Standard 61: Drinking Water System Components - Health Effects*, National Sanitation Foundation, Ann Arbor, MI.
31. O'Donoghue, P. E., et al. (1989). A Fracture Mechanic's Assessment of the Battelle Slow Crack Growth Test for Polyethylene Pipe Materials, Proceedings of the 11th Plastics Fuel Gas Pipe Symposium, 364-376.
32. Palermo, E. F. (1983). Rate Process Method as a Practical Approach to a Quality Control Method for Polyethylene Pipe, Proceedings of the Eighth Plastics Fuel Gas Pipe Symposium.
33. Palermo, E. F., & I. K. DeBlieu. (1985). Rate Process Concepts Applied to Hydrostatically Rating Polyethylene Pipe, Proceedings of the Ninth Plastics Fuel Gas Pipe Symposium.
34. Plastics Pipe Institute. (1990). Technical Note 11, Suggested Temperature Limits for Thermoplastic Pipe Installation and for Non-Pressure Pipe Operation, Washington, DC.
35. Plastics Pipe Institute. (1992). Technical Report TR-3, Policies and Procedures for Developing Recommended Hydrostatic Design Stresses for Thermoplastic Pipe Materials, Washington, DC.
36. Plastics Pipe Institute. (1992). Technical Report TR-4, Recommended Hydrostatic Strengths and Design Stresses for Thermoplastic Pipe and Fitting Compounds, Washington, DC.

37. Plastics Pipe Institute. (1989). Technical Report TR-11, Resistance of Thermoplastic Piping Materials to Micro- and Macro-Biological Attack, Washington, DC.
37a. Plastics Pipe Institute. (1973). Technical Report TR-18, Weatherability of Thermoplastic Piping, Washington, DC.
38. Plastics Pipe Institute. (1991). Technical Report TR-19, Thermoplastic Piping for the Transport of Chemicals, Washington, DC.
39. Plastics Pipe Institute. (1990). Statement N, Pipe Permeation, Washington, DC.
40. Powell, P. C. (1983). *Engineering with Polymers*, Chapman and Hall, New York, NY.
41. Richards, D., *Abrasion Resistance of Polyethylene Dredge Pipe*, US Army Engineer Waterways Experiment Station, Hydraulics Laboratory, Vicksburg, MS.
42. Rodriguez, F. (1970). *Principles of Polymer Systems*, McGraw-Hill Book Company, New York, NY.
43. Rooke, D. P., & D. J. Cartwright. (1974). *Compendium of Stress Intensity Factors*, Her Majesty's Stationary Office, London.
44. Sih, G. C. (1973). *Handbook of Stress Intensity Factors for Researchers and Engineers*, Lehigh University, Bethlehem, PA.
45. So, P. K., et al. (1987). Crack Initiation Studies in PE Pipe Grade Resins, Proceedings of the 10th Plastics Fuel Gas Pipe Symposium, 240-254.
46. Van Vlack, L. H. (1975). *Elements of Material Science and Engineering*, Addison-Wesley Publishing Co., Inc.
47. The Vinyl Institute. (1987). *Combustion Gases of Various Building Materials*, Wayne, NJ.
48. The Vinyl Institute. (1986). *Combustion Toxicity Testing*, Wayne, NJ.
49. Plastics Pipe Institute. (1992). Technical Note TN-16, Rate Process Method for Evaluating Performance of Polyethylene Pipe, Washington, DC.

Chapter 4

Polyethylene Pipe and Fittings Manufacturing

Introduction

The principles of pipe and fitting production are to melt and convey polyethylene into a particular shape and hold that shape during the cooling process. This is necessary to produce solid wall and profile wall pipe as well as compression and injection molded fittings.

All diameters of solid wall polyethylene pipe are continuously extruded through an annular die. Whereas, for large diameter profile wall pipes, the profile is spirally wound onto a mandrel and heat-fusion sealed along the seams.

Solid wall polyethylene pipe is currently produced in sizes ranging from 1/2 inch to 63 inches in diameter. Spirally wound profile pipe may be made up to 10 feet in diameter or more. There are several specification standards that govern the manufacturing processes for polyethylene pipe, but the main standards for solid wall and profile pipe include:

ASTM D2239 Standard Specification for Polyethylene (PE) Plastic Pipe (SIDR-PR) Based on Controlled Inside Diameter

ASTM D2447 Standard Specification for Polyethylene (PE) Plastic Pipe, Schedules 40 and 80, Based on outside Diameter

ASTM D2513 Standard Specification for Thermoplastic Gas Pressure Pipe, Tubing, and Fittings

ASTM D3035 Standard Specification for Polyethylene (PE) Plastic Pipe (SDR-PR) Based on Controlled Outside Diameter

ASTM F714 Standard Specification for Polyethylene (PE) Plastic Pipe (SDR-PR) Based on Outside Diameter

ASTM F894 Standard Specification for Polyethylene (PE) Large Diameter Profile Wall Sewer and Drain Pipe

AWWA C906 AWWA Standard for Polyethylene (PE) Pressure Pipe and Fittings, 4 in. (100 mm) through 63 in. (1,575 mm) for Water Distribution and Transmission

Generally, thermoplastic fittings are injection or compression molded, fabricated using sections of pipe, or machined from molded plates. Injection molding is used to produce fittings up through 12 inches in diameter, and fittings larger than 12 inches are normally fabricated from sections of pipe. The main ASTM specifications for injection molded fittings include:

ASTM D2683 Standard Specification for Socket-Type Polyethylene Fittings for outside Diameter-Controlled Polyethylene Pipe and Tubing

ASTM D3261 Standard Specification for Butt Heat Fusion Polyethylene (PE) Plastic Fittings for Polyethylene (PE) Plastic Pipe and Tubing

ASTM F1055 Standard Specification for Electrofusion-Type Polyethylene Fittings for Outside Diameter-Controlled Pipe and Tubing

The ASTM specification for fabricated fittings is ASTM F2206 Standard Specification for Fabricated Fittings of Butt-Fused Polyethylene (PE) Plastic Pipe, Fittings, Sheet Stock, Plate Stock, or Block Stock.

All of these standards specify the type and frequency of quality control tests that are required. There are several steps during the manufacturing process that are closely monitored to ensure that the product complies with these rigorous standards. Some of these steps are discussed in the section of this chapter on quality control and assurance.

Pipe Extrusion

The principal aspects of a solid wall polyethylene pipe manufacturing facility are presented in Figures 1 and 2. This section will describe the production of solid wall pipe from raw material handling, extrusion, sizing, cooling, printing, and cutting, through finished product handling. Details concerning profile wall pipe are also discussed in the appropriate sections.

Raw Materials Description

The quality of the starting raw material is closely monitored at the resin manufacturing site. As discussed in the chapter on test methods and codes in this handbook, a battery of tests is used to ensure that the resin is of prime quality. A certification sheet is sent to the pipe and fitting manufacturer documenting important physical properties such as melt index, density, ESCR (environmental strength crack resistance), stabilizer tests and tensile strength. The resin supplier and pipe manufacturer agree upon the specific tests to be conducted.

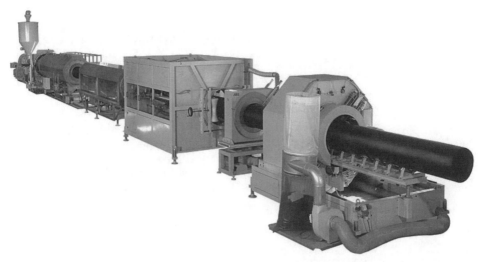

Figure 1 Typical Conventional Extrusion Line

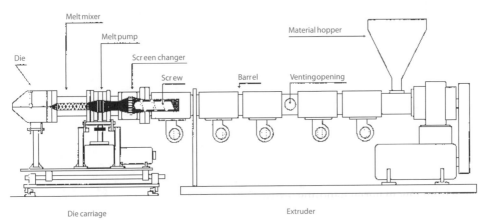

Figure 2 Diagram of Typical Extruder

Extrusion Line

The raw materials used to manufacture polyethylene pipe are generally supplied as pellets. The resin is stabilized as required in ASTM D3350 against thermal oxidation. The resin is supplied in either the natural state which is blended at the extrusion facility to add pecitized color and UV stabilizers or in a pre-colored form. In North America, the most common colors are black and yellow. The choice of color will depend upon the intended application and the requirements of the pipe purchaser. Carbon black is the most common pigment used for water, industrial, sewer and above-ground uses. Other colors are manufactured for telecommunications and other specialty markets.

All ASTM and many other industry standards specify that a PPI-listed compound shall be used to produce pipe and fittings. A compound is defined as the blend of natural resin and color concentrate and the ingredients that make up each of those two materials. The pipe producer must not change any of the ingredients in the listed compound, such as substituting a different color concentrate that could affect the long-term strength performance of the pipe. These stringent requirements ensure that only previously tested and approved compounds are being used.

If the resin is supplied as a natural pellet, the pipe producer will blend a color concentrate with the resin prior to extrusion. In order to obtain a PPI Listing, each manufacturer producing pipe in this manner is required to submit data, according to ASTM 2837, to the PPI Hydrostatic Stress Board. A careful review of the data is made according to PPI Policy TR-3 [5] to assess the long-term strength characteristics of the in-plant blended compound. When those requirements are met, the compound is listed in the PPI Publication TR-4 [6], which lists compounds that have satisfied the requirements of TR-3. Producers of potable water pipe are usually required to have the approval of the NSF International or an equivalent laboratory. NSF conducts un-announced visits during which time they verify that the correct compounds are being used to produce pipe that bears their seal.

Raw Materials Handling

After the material passes the resin manufacturer's quality control tests, it is shipped to the pipe manufacturer's facility in 180,000- to 200,000-pound capacity railcars, 40,000-pound bulk trucks, or 1000- to 1400-pound boxes.

Each pipe producing plant establishes quality control procedures for testing incoming resin against specification requirements. The parameters that are typically tested include: melt index, density, tensile strength, and environmental stress crack resistance (ESCR). Many resin producers utilize statistical process control (SPC) on certain key physical properties to ensure consistency of the product.

Resin is pneumatically conveyed from the bulk transporters to silos at the plant site. The resin is then transferred from the silos to the pipe extruder by a vacuum transfer system. Pre-colored materials can be moved directly into the hopper above the extruder. If a natural material is used, it must first be mixed homogeneously with a color concentrate. The resin may be mixed with the color concentrate in a central blender remote from the extruder or with an individual blender mounted above the extruder hopper. The blender's efficiency is monitored on a regular basis to ensure that the correct amount of color concentrate is added to the raw material.

Color concentrate is important for protecting the pipe from the effects of ultraviolet radiation that can cause degradation. Black concentrate alone is very effective in absorbing ultra-violet radiation whereas non-black concentrates use an ultra-violet stabilizer to provide protection. Refer to the chapter on engineering properties of polyethylene in this Handbook for further details.

Drying

Polyethylene is not hygroscopic but, for ease of processing and to ensure finished product quality, the resin and black concentrate should be dried prior to extrusion. The drying step ensures that the pipe quality will not be affected due to voids caused by water vapor trapped within the pipe wall. The resin manufacturer is the best source of specific recommendations for drying times and temperatures.

Extrusion Principles

The function of the extruder is to heat, melt, mix, and convey the material to the die, where it is shaped into a pipe [8]. The extruder screw design is critical to the performance of the extruder and the quality of the pipe. The mixing sections of the screw are important for producing a homogeneous mix when extruding natural and concentrate blends. A typical extruder is shown in Figure 3.

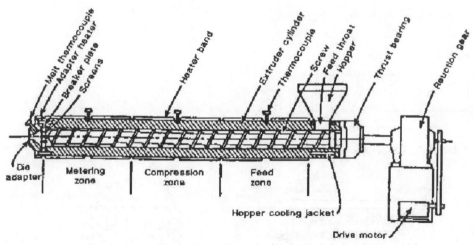

Figure 3 Typical Single-Stage, Single-Screw Extruder (Resin Flow from Right to Left)

There are many different types of screw designs [10], but they all have in common the features shown in Figure 4. Each screw is designed specifically for the type of material being extruded.

The extruder screw operates on the stick/slip principle. The polymer needs to stick to the barrel so that, as the screw rotates, it forces the material in a forward direction. In the course of doing this, the polymer is subjected to heat, pressure and shear (mechanical heating). The extent to which the material is subjected to these three parameters is the function of the screw speed, the barrel temperature settings and the screw design. The design of the screw is important in the production of high quality pipe. The wrong screw design can degrade the resin by overheating and shearing it, which will reduce the physical properties of the pipe.

Figure 4 Typical Extrusion Screw

If a natural resin and concentrate blend is used, the screw will also have to incorporate the colorant into the natural resin. Various mixing devices are used for this purpose as shown in Figure 5. They include mixing rings or pins, fluted or cavity transfer mixers, blister rings, and helix shaped mixers, which are an integral part of the screw. The pipe extrusion line generally consists of the extruder, die, cooling systems, puller, printer, saw and take-off equipment. Each of these items will be addressed in the following section.

Figure 5 Typical Resin Mixing Devices

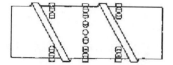

Figure 5.1 Mixing Pins

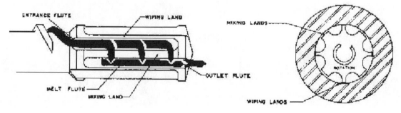

Figure 5.2 Fluted Mixer

Figure 5.3 Helical Mixer

Extruders

The single-screw extruder is generally used to produce polyethylene pipe [3].

An extruder is usually described by its bore and barrel length. Pipe extruders typically have an inside diameter of 2 to 6 inches with barrel lengths of 20 to 32 times the bore diameter. The barrel length divided by the inside diameter is referred to as the L/D ratio. An extruder with an L/D ratio of 24:1 or greater will provide adequate residence time to produce a homogeneous mixture.

The extruder is used to heat the raw material and then force the resulting melted polymer through the pipe extrusion die. The barrel of the machine has a series of four to six heater bands. The temperature of each band is individually controlled by an instrumented thermocouple. During the manufacturing process, the major portion of the heat supplied to the polymer is provided by the motor. This supply of heat can be further controlled by applying cooling or heating to the various barrel zones on the extruder by a series of air or water cooling systems. This is important since the amount of heat that is absorbed by the polymer should be closely monitored. The temperature of the extruder melted polymer is usually between 390°F and 450°F, and under high pressure (2000 to 4000 psi).

Breaker Plate/Screen Pack

The molten polymer leaves the extruder in the form of two ribbons. It then goes through a screen pack which consists of one or more wire mesh screens, positioned against the breaker plate. The breaker plate is a perforated solid steel plate. Screen packs prevent foreign contaminants from entering the pipe wall and assist in the development of a pressure gradient along the screw. This helps to homogenize the polymer. To assist in the changing of dirty screen packs, many extruders are equipped with a screen changer device. It removes the old pack while it inserts the new pack without removing the die head from the extruder.

Die Design

The pipe extrusion die supports and distributes the homogeneous polymer melt around a solid mandrel, which forms it into an annular shape for solid wall pipe [9]. The production of a profile wall pipe involves extruding the molten polymer through a die which has a certain shaped profile.

The die head is mounted directly behind the screen changer unless the extruder splits and serves two offset dies. There are two common types of die designs for solid wall pipe; the spider die design and the basket die design. They are illustrated in Figure 6. These designs refer to the manner in which the melt is broken and distributed into an annular shape and also the means by which the mandrel is supported.

Figure 6 Typical Pipe Dies

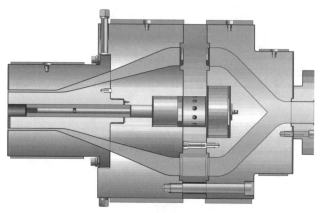

Figure 6.1 Pipe Die with Spider Design

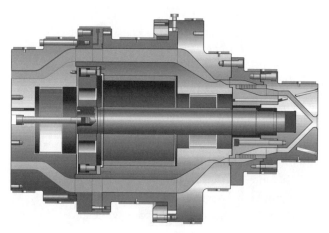

Figure 6.2 Pipe Die with Basket Design

In the spider die (Figure 6.1), the melt stream is distributed around the mandrel by a cone which is supported by a ring of spokes. Since the melt has been split by the spider legs, the flow must be rejoined.

Flow lines caused by mandrel supports should be avoided. This is done by reducing the annular area of the flow channel just after the spider legs to cause a buildup in die pressure and force the melt streams to converge, minimizing weld or spider lines. After the melt is rejoined, the melt moves into the last section of the die, called the land.

The land is the part of the die that has a constant cross-sectional area. It reestablishes a uniform flow and allows the final shaping of the melt and also allows the resin a certain amount of relaxation time. The land can adversely affect the surface finish of the pipe if it is too short in length. Typical land lengths are 15 to 20 times the annular spacing.

The basket design (Figure 6.2) has an advantage over the spider die concerning melt convergence. The molten polymer is forced through a perforated sleeve or plate, which contains hundreds of small holes. Polymer is then rejoined under pressure as a round profile. The perforated sleeve, which is also called a screen basket, eliminates spider leg lines.

Pipe Sizing Operation

The dimensions and tolerances of the pipe are determined and set during the sizing and cooling operation. The sizing operation holds the pipe in its proper dimensions during the cooling of the molten material. For solid wall pipe, the process is accomplished by drawing the hot material from the die through a sizing sleeve and into a cooling tank. Sizing may be accomplished by using either vacuum or pressure techniques. Vacuum sizing is generally the preferred method.

During vacuum sizing, the molten extrudate is drawn through a sizing tube or rings while its surface is cooled enough to maintain proper dimensions and a circular form. The outside surface of the pipe is held against the sizing sleeve by vacuum. After the pipe exits the vacuum sizing tank, it is moved through a second vacuum tank or a series of spray or immersion cooling tanks.

Figure 7 External Sizing Systems

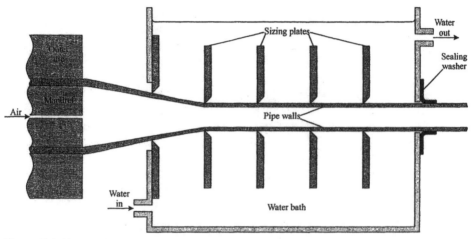

Figure 7.1 Vacuum Tank Sizing[11]

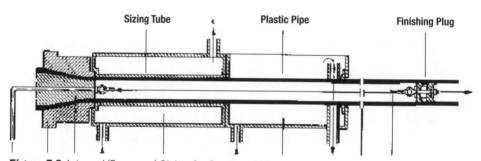

Figure 7.2 Internal (Pressure) Sizing for Small and Medium Pipe Diameters

In the pressure sizing system, a positive pressure is maintained on the inside of the pipe by the use of a plug attached to the die face by a cable or, on very small bore pipe, by closing or pinching off the end of the pipe. The pressure on the outside of the pipe remains at ambient and the melt is forced against the inside of the calibration sleeve with the same results as in the vacuum system.

The production of corrugated pipe is typically done with large external molds to form the outer corrugations. Dual-wall corrugated pipe includes a co-extruded smooth inner layer.

The production of very large diameter profile pipe, up to 10 feet in diameter, uses mandrel sizing. In one form of this process, the extruded profile is wrapped around a mandrel. As the mandrel rotates, the extruded profile is wrapped such that each turn overlaps the previous turn. In some other techniques, the turns are not overlapped. A typical profile wall polyethylene pipe is shown in Figure 8.

Figure 8 Typical Polyethylene Profile Wall from ASTM Standard F894

Figure 8.1 Laying Lengths

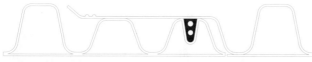

Figure 8.2 Typical Profile Wall Section Showing Bell End (right) and Spigot End (left)

Cooling

For either the vacuum or pressure sizing technique, the pipe should be cool enough to maintain its circularity before exiting the cooling tank. Various methods of cooling are utilized to transfer the heat out of the polyethylene pipe. Depending upon the pipe size, the system may use either total immersion or spray cooling. Spray cooling is usually applied to large diameter pipe where total immersion would be inconvenient. Smaller diameter pipe is usually immersed in a water bath. Cooling water temperatures are typically in the optimum range of 40° to 50°F (4° to 10°C). The total length of the cooling baths must be adequate to cool the pipe below 185°F (85°C) in order to withstand subsequent handling operations.

Stresses within the pipe wall are controlled by providing annealing zones.[4] These zones are spaces between the cooling baths which allow the heat contained within the inner pipe wall to radiate outward and anneal the entire pipe wall. Proper cooling bath spacing is important in controlling pipe wall stresses. Long-term pipe performance is improved when the internal pipe wall stresses are minimized.

Pullers

The puller must provide the necessary force to pull the pipe through the entire cooling operation. It also maintains the proper wall thickness control by providing a constant pulling rate. The rate at which the pipe is pulled, in combination with the extruder screw speed, determines the wall thickness of the finished pipe. Increasing the puller speed at a constant screw speed reduces the wall thickness, while reducing the puller speed at the same screw speed increases the wall thickness.

Standards of ASTM International and other specifications require that the pipe be marked at frequent intervals. The markings include nominal pipe size, type of plastic, SDR and/or pressure rating, and manufacturer's name or trademark and manufacturing code. The marking is usually ink, applied to the pipe surface by an offset roller. Other marking techniques include hot stamp, ink jet and indent printing. If indent printing is used, the mark should not reduce the wall thickness to less than the minimum value for the pipe or tubing, and the long-term strength of the pipe or tubing must not be affected. The mark should also not allow leakage channels when gasket or compression fittings are used to join the pipe or tubing.

Take-off Equipment

Most pipe four inches or smaller can be coiled for handling and shipping convenience. Some manufacturers have coiled pipe as large as 6 inch. Equipment allows the pipe to be coiled in various lengths. Depending upon the pipe diameter, lengths of 10,000 feet are possible. This is advantageous when long uninterrupted lengths of pipe are required - for example, when installing gas and water pipes.

Saw Equipment and Bundling

Pipe four inches or more in diameter is usually cut into specified lengths for storage and shipping. Typical lengths are 40 to 50 feet, which can be shipped easily by rail or truck. The pipe is usually bundled before it is placed on the truck or railcar. Bundling provides ease of handling and safety during loading and unloading.

Fittings Overview

The polyethylene pipe industry has worked diligently to make polyethylene piping systems as comprehensive as possible. As such, various fittings are produced which increase the overall use of the polyethylene piping systems. Some typical fittings are shown in Figure 9.

Polyethylene fittings may be injection molded, fabricated or thermoformed. The following section will briefly describe the operations of each technique.

Injection Molded Fittings

Injection molded polyethylene fittings are manufactured in sizes through 12-inch nominal diameter. Typical molded fittings are tees, 45° and 90° elbows, reducers, couplings, caps, flange adapters and stub ends, branch and service saddles, and self-tapping saddle tees. Very large parts may exceed common injection molding equipment capacities, so these are usually fabricated.

Equipment to mold fittings consists of a mold and an injection molding press, as shown in Figure 10. The mold is a split metal block that is machined to form a part-shaped cavity in the block. Hollows in the part are created by core pins shaped into the part cavity. The molded part is created by filling the cavity in the mold block through a filling port, called a gate. The material volume needed to fill the mold cavity is called a shot.

The injection molding press has two parts; a press to open and close the mold block, and an injection extruder to inject material into the mold block cavity. The injection extruder is similar to a conventional extruder except that, in addition to rotating, the extruder screw also moves lengthwise in the barrel. Injection molding is a cyclical process. The mold block is closed and the extruder barrel is moved into contact with the mold gate. The screw is rotated and then drawn back, filling the barrel ahead of the screw with material. Screw rotation is stopped and the screw is rammed forward, injecting molten material into the mold cavity under high pressure. The part in the mold block is cooled by water circulating through the mold block. When the part has solidified, the extruder barrel and mold core pins are retracted, the mold is opened, and the part is ejected.

Typical quality inspections are for knit line strength, voids, dimensions and pressure tests. A knit line is formed when the shot flows around a core pin and joins together on the other side. Voids can form from material shrinkage during cooling, particularly in heavier sections. Voids can be detected nondestructively by using x-ray scans. If this is not available, samples can be cut into thin sections and inspected visually.

Figure 9 Typical Polyethylene Pipe Fittings

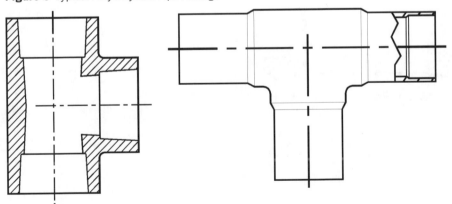

Figure 9.1 Socket Tee   **Figure 9.2** Butt Tee

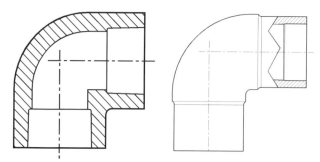

Figure 9.3 90° Socket Elbow   **Figure 9.4** 90° Butt Elbow

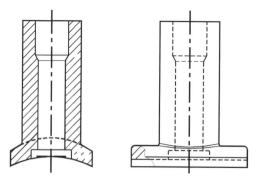

Figure 9.5 Saddle Fusion Fittings

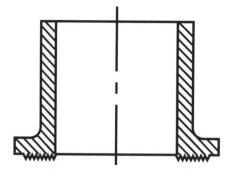

Figure 9.6 Butt Flange Adapter/Stub End

a. injection stage
b. freeze time with follow-up pressure
c. demoulding of finished article

1. locking mechanism
2. moving mounting plate
3. mold cavity plate
4. mold core plate
5. stationery mounting plate
6. plasticating cylinder
7. feed hopper
8. hydralic motor (screw drive)
9. hydralic cylinder of injection unit
10. pressure gauge
11. follw-up pressure limit switch
12. screw stroke adjusment

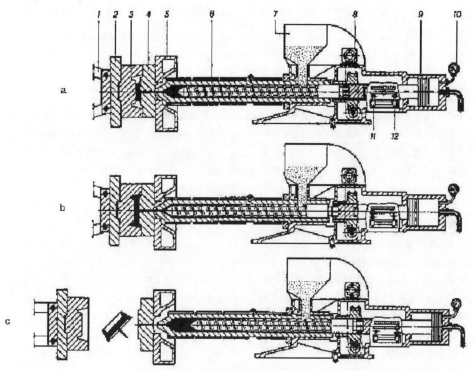

Figure 10 Construction and Mode of Operation of a Reciprocating Screw Injection Unit (Courtesy of Hoechst Celanese Corporation)

Fabricated Fittings

Fully pressure-rated, full bore fabricated fittings are available from select fittings fabricators. Fabricated fittings are constructed by joining sections of pipe, machined blocks, or molded fittings together to produce the desired configuration. Components are joined by heat fusion, hot gas welding or extrusion welding techniques. It is not recommended to use either hot gas or extrusion welding for pressure service fittings since the joint strength is significantly less than that of a heat fusion joint.

Fabricated fittings designed for full pressure service are joined by heat fusion and must be designed with additional material in areas subject to high stress. The common commercial practice is to increase wall thickness in high-stress areas by using heavy-wall pipe sections. The increased wall thickness may be added to the OD, which provides for a full-flow ID; or it may be added to the ID, which slightly restricts ID flow. This is similar to molded fittings that are molded with a larger OD, heavier body wall thickness. If heavy-wall pipe sections are not used, the conventional practice is to reduce the pressure rating of the fitting. The lowest-pressure-rated component determines the operating pressure of the piping system.

Various manufacturers address this reduction process in different manners. Reinforced over-wraps are sometimes used to increase the pressure rating of a fitting. Encasement in concrete, with steel reinforcement or rebar, is also used for the same purpose. Contact the fitting manufacturer for specific recommendations.

Very large diameter fittings require special handling during shipping, unloading, and installation. Precautions should be taken to prevent bending moments that could stress the fitting during these periods. Consult the fittings manufacturer for specifics. These fittings are sometimes wrapped with a reinforcement material, such as fiberglass, for protection.

Thermoformed Fittings

Thermoformed fittings are manufactured by heating a section of pipe and then using a forming tool to reshape the heated area. Examples are sweep elbows, swaged reducers, and forged stub ends. The area to be shaped is immersed in a hot liquid bath and heated to make it pliable. It is removed from the heating bath and reshaped in the forming tool. Then the new shape must be held until the part has cooled.

Electrofusion Couplings

Electrofusion couplings and fittings are manufactured by either molding in a similar manner as that previously described for butt and socket fusion fittings or manufactured from pipe stock. A wide variety of couplings and other associated fittings are available from ½" CTS thru 28" IPS. Fittings are also available for ductile iron sized polyethylene pipe. These couplings are rated as high as FM 200.

Electrofusion fittings are manufactured with a coil-like integral heating element. These fittings are installed utilizing a fusion processor, which provides the proper energy to provide a fusion joint stronger than the joined pipe sections. All electrofusion fittings are manufactured to meet the requirements of ASTM F-1055.

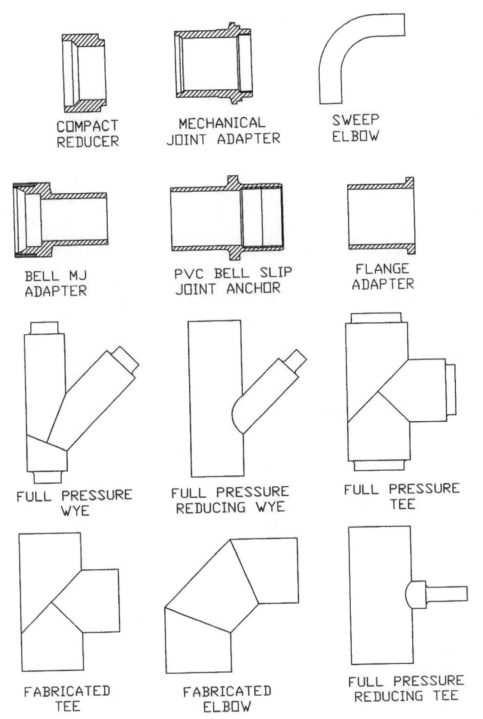

Figure 11 Typical Fabricated Fittings

Quality Control/Quality Assurance Testing

Quality is engineered into the pipe product during the entire manufacturing process. The three phases of quality control for the pipe manufacturer involve the incoming raw material, the pipe or fitting production and the finished product. The combination of all three areas ensures that the final product will fulfill the requirements of the specification to which it was made.

Testing the incoming resin is the first step in the quality control program. It is usually checked for contamination, melt index, density, tensile strength and environmental stress crack resistance (ESCR). Any resin that does not meet the raw material specification is not used for the production of specification-grade pipe.

During the manufacturing step, the pipe producer routinely performs quality assurance tests on samples. This verifies that proper production procedures and controls were implemented during production.

Once the product has been produced, it undergoes a series of quality control tests to ensure that it meets the minimum specifications as required by the appropriate standard. (See Handbook Chapter on Test Methods and Codes.)

The manufacturing specifications for piping products list the tests that are required. There are several quality control tests that are common in most ASTM polyethylene standards. For gas service piping systems, refer to PPI Technical Report TR-32 [7] for a typical quality control program for gas system piping, or to the AGA Plastic Pipe Manual for Gas Service [1]. The typical QC/QA tests found in most standards are described below.

Workmanship, Finish, and Appearance

According to ASTM product specifications, the pipe, tubing, and fittings shall be homogeneous throughout and free of visible cracks, holes, foreign inclusions, blisters, and dents or other injurious defects. The pipe tubing and fittings shall be as uniform as commercially practicable in color, opacity, density and other physical properties.

Dimensions

Pipe diameter, wall thickness, ovality, and length are measured on a regular basis to insure compliance with the prevailing specification. All fittings have to comply with the appropriate specification for proper dimensions and tolerances. All measurements are made in accordance with ASTM D2122, Standard Test Method of Determining Dimensions of Thermoplastic Pipe and Fittings [2].

Pressure Tests

There are three pressure tests that are used to detect defects in the pipe manufacturing process. They are the quick burst, the sustained pressure, and the elevated temperature tests. The details of these test methods are presented in ASTM D1598, Standard Test Method for Time-to-Failure of Plastic Pipe Under Constant Internal Pressure [2], and ASTM D1599, Standard Test Method for Short-Time Hydraulic Failure Pressure of Plastic Pipe, Tubing, and Fittings.

In the quick burst test, the PE pipe sample is pressurized at a temperature of 73.4°F (23°C) and a constant pressure to obtain a burst failure within 60 to 70 seconds. The actual burst pressure measured in the test must exceed the minimum burst pressure requirements in the applicable product specification. This test is used to determine, in a very short time period, if the pipe production has any serious problems.

The sustained pressure test requires that the pipe samples are maintained at a constant pressure for a minimum of 1000 hours at a temperature of 73.4°F (23°C) without failure. Again, this test will indicate if there are any malfunctions in the pipe production.

The elevated temperature test is conducted at 176°F (80°C) using pressurized pipe samples. The failure times must exceed the minimum value as listed in the applicable pipe specification. This test accelerates the failure times due to the higher temperature. It is more sensitive to changes in processing conditions or resin formulation than the other two pressure tests.

Physical Property Tests

Several tests are conducted to ensure that the final pipe product complies to the applicable specification. Depending upon the specification, the type and the frequency of testing will vary. More details about industry standard requirements can be found in the chapter on specifications, test methods and codes in this Handbook.

The following tests, with reference to the applicable ASTM standard [2], are generally required in many product specifications such as natural gas service. The following list of tests was taken from the American Gas Association Manual for Plastic Gas Pipe [1] to serve as an example of typical tests for gas piping systems.

ASTM TESTS

Sustained Pressure	D1598
Burst Pressure	D1599
Apparent Tensile Strength	D2290
Flattening	D2412
Chemical Resistance	D543
Impact Resistance	D2444
Compressed Ring ESCR	F1248

There are other tests that are used that are not ASTM test methods. They are accepted by the industry since they further ensure product reliability. One such test is the Bend-Back Test[1] which is used to indicate inside surface brittleness under highly strained test conditions. In this test, a ring of the pipe is cut and then subjected to a reverse 180-degree bend. Any signs of surface embrittlement, such as cracking or crazing, constitute a failure. The presence of this condition is cause for rejection of the pipe.

Quality Assurance Summary

Through the constant updating of industry standards, the quality performance of the polyethylene pipe and fitting industry is continually evolving. Each year, PPI and ASTM work to improve standards on plastic pipe which include the latest test methods and recommended practices. Resin producers, pipe extruders, and fittings manufacturers incorporate these revisions into their own QA/QC practices to insure compliance with these standards. In this way, the exceptional performance and safety record of the polyethylene pipe industry is sustained.

Summary

This chapter provides an overview of the production methods used in the manufacture of polyethylene pipe and fittings. The purpose of this chapter is to create a familiarity with the processes by which these engineered piping products are made. Through a general understanding of these fundamental processes, the reader should be able to develop an appreciation for the utility and integrity of polyethylene piping systems.

References

1. *AGA Plastic Pipe Manual for Gas Service*. (2001). Catalog No. XR0185, American Gas Association, Arlington, VA.
2. *Annual Book of ASTM Standards*. (2005). Volume 08.04, Plastic Pipe and Building Products, American Society for Testing and Materials, Philadelphia, PA.
3. Gebler, H., H. O. Schiedrum, E. Oswald, & W. Kamp. (1980). The Manufacture of Polypropylene Pipes, *Kunststoffe 70*, pp. 186-192, English Translation.
4. Kamp, W., & H. D. Kurz. (1980). Cooling Sections in Polyolefin Pipe Extrusion, *Kunststoffe 70*, pp. 257-263, English Translation.

5. Policies and Procedures for Developing Hydrostatic Design Basis (HDB), Pressure Design Basis (PDB), Strength Design Basis (SDB), and Minimum Required Strength (MRS) Rating for Thermoplastic Piping Materials or Pipe (2005). Report TR-3, Plastics Pipe Institute, Washington, DC.
6. PPI Listing of Hydrostatic Design Basis (HDB), Strength Design Basis (SDB), Pressure Design Basis (PDB) and Minimum Required Strength (MRS) Ratings for Thermoplastic Piping Materials or Pipe (2005). Report TR-4, Plastics Pipe Institute, Washington, DC.
7. Recommended Minimum In-Plant Quality Control Program for Production of Polyethylene Gas Distribution Piping Systems. (1989). Report TR-32, Plastics Pipe Institute, Washington, DC.
8. Schiedrum, H. O. (1974). The Design of Pipe Dies, *Plastverarbeiter*, No. 10, English Translation.
9. Rauwendaal, C. (1986). *Polymer Extrusion*, MacMillan Publishing Company, Inc., New York, NY.
10. *Screw and Barrel Technology*. (1985). Spirex Corporation, Youngstown, OH.
11. Peacock, Andrew J. (2000). *Handbook of Polyethylene*, Marcel Decker, Inc. New York, NY.

# Chapter 5

Specifications, Test Methods and Codes for Polyethylene Piping Systems

Introduction

The specification, design and use of polyethylene piping systems are governed by a number of standards, methods and codes such as American Society for Testing and Materials (ASTM), American Water Works Association (AWWA) and Canadian Standards Association (CSA), as well as Technical Reports (TR's) and Technical Notes (TN's) published by the Plastics Pipe Institute. This chapter discusses these guidelines with respect to both the polyethylene materials and the finished piping systems used for pressure pipe applications. There are also many standards and guidelines for non-pressure pipe applications, but they are not covered in this chapter. Emphasis is placed on developing an understanding of:

1. Material specifications relating to properties and classifications of polyethylene for piping

2. Test methods and specifications relating to pipe pressure rating, dimensions, fittings and joints

3. Codes, standards and recommended practices governing the application of polyethylene pipe systems in a variety of end uses

Included at the end of this chapter is a current list of some of the major or most frequently used standards and codes for polyethylene piping.

Properties and Classification of Polyethylene Materials

The properties and performance of polyethylene piping systems, to a great extent, are determined by the polyethylene material itself. As its name suggests, polyethylene is made by the polymerization of ethylene, generally with the addition of another alpha-olefin such as propylene, butene or hexene. For pipe applications, the polyethylene resins are generally made from the combination of thousands of these units. A variety of polymerization catalysts and processes exist commercially which are used to control the number of monomer units in the polymer chain, the type, frequency and distribution of the comonomer unit, the amount and type of branching off of the main polymer chain, and the relative uniformity of the polymer chain lengths in the bulk polyethylene resin.

To a greater or lesser extent, each of the above variables can influence the properties of the polyethylene resin and determine its suitability for piping applications. Three basic parameters of polyethylene can be used to give general indications of the resins' properties and its suitability for the piping applications. These are: density, molecular weight, and molecular weight distribution. The engineering properties of polyethylene chapter of this handbook gives further detail on these properties and their interrelationships.

Table 1 provides a generalized indication of the effects of these three important polyethylene characteristics on resin properties [1]. As can be seen from this table, some of the physical properties of polyethylene are primarily determined by only one of the above parameters and are independent of the other two (e.g., hardness or stiffness as functions of density), while some properties are influenced by all three parameters (e.g., low temperature brittleness). Most properties are influenced to a certain degree by a least two of the parameters. Each of these interrelationships of performance properties to molecular characteristics can be complex, subtle or even masked by such overriding influences as thermal history, formulations, sample preparation, etc. As such, it is important when designing or selecting polyethylene materials for piping applications to realize that such relationships can exist.

TABLE 1
Influence of Basic Resin Parameters on Polyethylene Resin Properties*

Property	As Density Increases	As Average Molecular Weight Increases (Melt Index Decreases)	As Molecular Weight Distribution Broadens
Stiffness	Increases	Increases	Decreases
Hardness	Increases	Increases	—
Tensile Strength @ Yield	Increases	Increases	—
Elongation	Decreases	Increases	—
Tensile Strength @ Rupture	Increases	Increases	No Significant Changes
Softening Temperature	Increases	—	Increases
Retention of Strength Under Long-Term Loading	No Significant Change	Increases	No Significant Changes
Resistance to Low Temperature Brittleness	Decreases	Increases	Increases
Permeability	Decreases	—	—
ESCR	—	Increases	Increases
Chemical Resistance	Increases	Increases	—

*Changes in other parameters may alter these effects.

Material Selection and Specification

There are hundreds of types of polyethylene, and it is important to be able to specify the right one for a plastic piping application. In the past, ASTM D 1248 was used to help define the material properties. However, ASTM D 1248 has been modified and now only deals with wire and cable coating grades of polyethylene. Today, the main ASTM standard for aiding in the specification of PE materials for piping applications is D 3350 *"Standard Specification for Polyethylene Plastics Pipe and Fittings Materials."* This ASTM standard defines the most important material properties that need to be considered when choosing a PE material for a pressure pipe application, and defines a classification system to ease the specification process.

ASTM D-3350 *"Standard Specification for Polyethylene Plastics Pipe and Fittings Materials"*

This standard defines basic material requirements, as well as classifies polyethylene piping materials according to a cell classification system consisting of six digits and one letter. Table 2 shows the cell system of ASTM D 3350, with each cell representing a property or characteristic of polyethylene that has been recognized as being significant to processing and/or performance. The properties are divided up into ranges, or cells, so a certain property value can easily be specified.

Another important part of this standard is that not only are the values of the properties specified, but the methodology used to determine the properties is also defined. This way, it is possible to more accurately specify material properties as measured by a certain method, and easier to directly compare different polyethylene materials with a more "apples-to-apples" approach.

Thermal Stability

In addition to the cell classification, in order for a material to meet the requirements of ASTM D 3350, it must also meet the thermal stability requirements. This insures that only those materials that have been adequately stabilized with protective antioxidants and heat stabilizers will be used in long-term piping applications.

TABLE 2
Cell Classification System from ASTM D-3350

PROPERTY	TEST METHOD	0	1	2	3	4	5	6	7	8
Density, g/cm3	D 1505	un-specified	0.925 or lower	>0.925 - 0.940	>0.940 - 0.947	>0.947 - 0.955	>0.955	—	specify value	—
Melt Index	D 1238	un-specified	>1.0	1.0 to 0.4	<0.4 to 0.15	<0.15	A	—	specify value	—
Flexural Modulus, MPa (psi), 2% secant	D 790	un-specified	<138 (<20,000)	138-<276 (20,000 to <40,000)	276-<552 (40,000 to <80,000)	552-<758 (80,000 to <110,000)	758-<1103 (110,000 to <160,000)	>1103 (>160,000)	specify value	—
Tensile strength at yield, MPa (psi)	D638	un-specified	<15 (<2000)	15-<18 (2200-<2600)	18-<21 (2600-<3000)	21-<24 (3000-<3500)	24-<28 (3500-<4000)	>28 (>4000)	specify value	—
Slow Crack Growth Resistance I. ESCR	D1693	un-specified								specify value
a. Test condition			A	B	C	D	—	—	—	
b. Test duration, hours			48	24	192	600	—	—	—	
c. Failure, max. %			50	50	20	20	—	—	—	
Slow Crack Growth Resistance II. PENT (hours) Molded Plaque, 80°C, 2.4MPa, notch depth Table 1	F 1473	un-specified	—	—	—	10	30	100	500	specify value
Hydrostatic Strength Classification I. Hydrostatic design basis, MPa, (psi), (23°C)	D2837	NPR[B]	5.52 (800)	6.89 (1000)	8.62 (1250)	11.03 (1600)	—	—	—	—
Hydrostatic Strength Classification II. Minimum Required Strength, MPa (psi), (20°C)	ISO 12162	—	—	—	—	—	8 (1160)	10 (1450)	—	—

Notes to Table 2: A Refer to 10.1.4.1 (ASTM D 3350) B NPR = Not Pressure Rated

As noted earlier, the ASTM D3350 cell classification consists of a string of six digits and one alpha character. The ending code letter designates the color and UV stabilizer as follows:

Code Letter	Color and UV Stabilizer
A	Natural
B	Colored
C	Black with 2% minimum carbon black
D	Natural with UV stabilizer
E	Colored with UV stabilizer

Note: UV = ultraviolet

Polyethylene Grade - D 3350

The grade designation originally established in ASTM D 1248 has been added to D 3350, and further modified to be consistent with newer PE piping materials. The grade consists of the letters "PE", followed by two numbers. The letters "PE", of course, are to designate the material as being polyethylene. The first number designates the density cell class of the material. The second number designates the cell class for resistance to slow crack growth of a material when tested in accordance with ASTM D 1693 or ASTM F 1473.

As an example, a high-density polyethylene material- **PE**- with a density of 0.945 g/cc - **cell class 3**, and an ESCR, condition C, of greater than 600 hours - **cell class 4**, would be a grade **PE34**.

An example of the application of this system is given in Table 3 for a polyethylene material with a cell class designation of **345464C**, a standard designation for a modern PE 3408 pipe grade resins.

TABLE 3
Properties of a Class PE345464C Material

Cell Property	Class
Density (0.941 - 0.955 g/cc)	3
Melt Index (<0.15)	4
Flexural Modulus (758- <1103 MPa)(110,000 to <160,000 psi)	5
Tensile Strength at yield (21- < 24 Mpa) (3000 < 3500 psi)	4
Slow Crack Growth Resistance, PENT (100 hours)	6
Hydrostatic Design basis at 23°C at 11.03 MPa (1600 psi)	4
Black with 2% minimum carbon black	C

Such a product would be described as a high-density, high-molecular-weight polyethylene having a hydrostatic design basis at 73°F (23°C) of 1600 psi (11.03 MPa). It would be black and contain 2% (minimum) carbon black. The flexural modulus, tensile strength and Slow Crack Growth Resistance would be as defined by their respective cell values.

A specification writer would want to use the cell classification as a minimum requirement. Additional clarifications and specific project requirements should be included in any material specification for piping applications. PPI has published several Model Specifications that are available to use as a guide.

PPI Designations

The use of the cell classifications per ASTM D3350 provides a detailed description of a polyethylene material for piping. The Plastics Pipe Institute has augmented the grade designation from ASTM D 3350 to include the Hydrostatic Design Stress (HDS) by adding two digits on the material's grade. The Hydrostatic Design Stress is the maximum long-term stress the material can be subjected to after applying a design factor of 0.5 to the material's established Hydrostatic Design Basis (HDB). By truncating the standard HDS in hundreds, the PPI has adopted the use of **04** for 400 psi (2.26 MPa) HDS, **06** for 630 psi (4.31 MPa) HDS, and **08** for 800 psi (5.4 MPa) HDS. More information on the relationship between the HDB and HDS is in subsequent sections.

Using this format, the PPI designation for a polyethylene material with a grade of **PE34** and a hydrostatic design stress of 800 psi, is a **PE 3408**. This approach is commonly referred to as the thermoplastic material designation code as defined in ASTM F 412.

Test Methods and Standards for Stress Rating, Dimensioning, Fittings and Joining of Polyethylene Pipe Systems

In order to properly specify polyethylene pipe, it is helpful to understand some of the terminology and nomenclature associated with the stress rating of the PE material and the pressure rating of pipe and fittings made from that material.

Pressure Rating of Polyethylene Pipe

Fundamental to the pressure rating of polyethylene piping systems is the concept of the Long-Term Hydrostatic Strength (LTHS) of the material. ASTM D 1598, "Time-to- Failure of Plastic Pipe Under Constant Internal Pressure," is the standard test method by which polyethylene pipe samples are subjected to constant pressure and their time-to-failure is noted as a function of applied stress. Using the relationship known as the "ISO" equation, it is possible to relate the test pressure and pipe dimensions to the resultant hoop or circumferential stress generated in the wall of the pipe by that internal pressure. See the Engineering Properties chapter of this handbook for further information on the derivation of this equation. The ISO equation can be written for either outside diameter (Eq. 1) or inside diameter (Eq. 2) based pipe dimensions:

(1) $$S = \frac{P(OD - t)}{2t}$$

(2) $$S = \frac{P(ID + t)}{2t}$$

WHERE
S = Hoop Stress (psi or MPa)
P = Internal Pressure (psi or MPa)
ID = Average Inside Diameter (in or mm)
OD = Average Outside Diameter (in or mm)
t = Minimum Wall Thickness (in or mm)

After obtaining a number of stress vs. time-to-failure points it is possible to analyze the data to predict the estimated long-term performance of the piping material by calculating a long-term hydrostatic strength (LTHS) of the polyethylene material, and then categorizing the LTHS into a Hydrostatic Design Basis, or HDB, as shown in Table 4. The HDB then becomes the baseline strength for the material when performing any pressure rating calculations. The procedures for the extrapolation of the data are given in ASTM D 2837, "Obtaining Hydrostatic Design Basis for Thermoplastic Materials." This standard method contains not only the least squares calculations for obtaining the linear log-log regression equation of hoop stress vs. hours-to-failure, but also prescribes the minimum number of failure points, their distribution with respect to time, certain statistical tests for the quality of the data, its fit to the least squares line, and the maximum slope of the regression equation. At least one data point must exceed 10,000 hours. PPI's Technical Report 3, "Policies and Procedures for Developing Hydrostatic Design Basis (HDB), Pressure Design Basis (PDB), Strength Design Basis (SDB), and Minimum Required Strengths (MRS) Ratings for Thermoplastic Piping Materials or Pipe," gives very detailed information on how to properly develop and utilize this design criterion.

TABLE 4
LTHS and HDB Categories from ASTM D2837

Range of Calculated LTHS Values		Hydrostatic Design Basis	
psi	(MPa)	psi	(MPa)
190 to< 240	1.31 to< 1.65	200	1.38
240 to< 300	1.65 to< 2.07	250	1.72
300 to< 380	2.07 to< 2.62	315	2.17
380 to< 480	2.62 to< 3.31	400	2.76
480 to< 600	3.31 to< 4.14	500	3.45
600 to< 760	4.14 to< 5.24	630	4.34
760 to< 960	5.24 to< 6.62	800	5.52
960 to<1200	6.62 to< 8.27	1000	6.89
1200 to< 1530	8.27 to< 10.55	1250	8.62
1530 to< 1920	10.55 to< 13.24	1600	11.03
1920 to< 2400	13.24 to< 16.55	2000	13.79
2400 to< 3020	16.55 to< 20.82	2500	17.24
3020 to< 3830	20.82 to< 26.41	3150	21.72
3830 to< 4800	26.41 to< 33.09	4000	27.58
4800 to< 6040	33.09 to< 41.62	5000	34.47
6040 to< 6810	41.62 to< 46.92	6300	43.41
6810 to< 7920	46.92 to< 54.62	7100	48.92

The HDB categories are based on an R-10 series, wherein each HDB category is 125% of the preceding category. The range of calculated LTHS values allowed within an HDB category are -4% to +20% of the category value.

Figure 1 shows the log-log relationship of hoop stress vs. time-to-failure for a PE 3408 material tested in water at 73°F (23°C) according to ASTM D 1598. The solid line shown represents the least squares analysis of this data according to the techniques of ASTM D 2837. Note that although the actual test data is only approximately 10,000 hours, the line is extrapolated (dashed portion) to 100,000 hours (11.4 years). It is this projected hoop stress at 100,000 hours that is used to establish the Long-Term Hydrostatic Strength (LTHS). Again, referring to ASTM D 2837 (Table 1, therein), it is possible to place this LTHS within a hydrostatic design basis category (HDB).

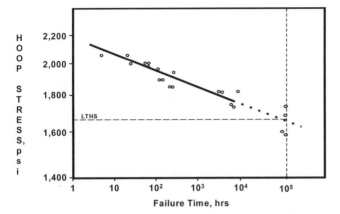

Figure 1 Typical Stress-Rupture Plot for PE 3408 Material

The 100.000-hour stress intercept established in ASTM D 2837 must also be "validated." Certain statistical hydrostatic testing conducted at higher temperatures is performed on specimens from the same lot of pipe material to insure that the curve determined in accordance with the ASTM D 2837 at 73°F remains linear throughout the 100,000 time period. That is to say that the polyethylene material being evaluated remains ductile in character over the course of the 100,000 regression analysis.

Design Factors And Hydrostatic Design Stress
It is necessary in designing plastic piping systems that the HDB be reduced by multiplying it with a design factor (DF) to allow for a greater margin of safety in use and to accommodate potential stresses on the pipe beyond those of internal or line pressure. Other factors to consider are: service and environmental conditions, temperatures higher than 73°F, other fluid mediums, etc. Current industry accepted design factors are 0.5 for water pressure service at 73°F (23°C) and 0.32 for natural gas distribution service. The more demanding and aggressive the application or service conditions, the smaller the design factor may need to be.

The reader should note that the evolution of polyethylene pipe usage has experienced progressive improvements in material science, technical performance and extrusion technology. As a result of this continual advancement in product capability, the design factors are subject to change as material and, hence, pipe performance improvements are recognized. The reader should consult with the pipe manufacturer regarding current recommendations for design factor in light of the anticipated service conditions.

Calculation of the hydrostatic design stress (HDS) is given by equation 3:

(3) $HDS = HDB \times DF$

WHERE
HDS = Hydrostatic Design Stress
HDB = Hydrostatic Design Basis
DF = Design Factor
  = 0.50 for water service at 73°F
  = 0.32 for natural gas distribution at 73°F

A more thorough discussion on design factors is presented in the design chapter of this Handbook, and additional information on Design Factors is given in PPI TR-9 *"Recommended Design Factors for Pressure Applications of Thermoplastic Pipe Materials."*

Dimensioning Systems

The standard dimensions for piping systems are an important part of the design for several reasons. The diameter of the pipe will dictate its ability to carry a needed volume of fluids. The wall thickness will dictate the strength of the pipe and it's ability to handle internal and external pressures as well as affecting the potential flow capacity. As such, the ratio of the diameter and wall thickness - known as the dimension ratio (DR) - becomes an important design factor for pipe. Standardization of these dimensions means it is possible for accessories such as fittings, valves, and installation equipment to be designed for a limited number of sizes, while at the same time, allowing for enough sizes to give the design engineer flexibility to build the system as needed.

ASTM standards include both imperial and metric dimensions for plastic pipe. For simplicity, metric dimensions are not included in this section.

Diameters

The diameter of the pipe can be either outside diameter or inside diameter. Depending on the calculations needing to be done, both numbers are important. Historically, anything called "pipe" has an inside diameter approximately equal to the nominal diameter of the pipe — 4" pipe has an ID of about 4". This general rule will vary depending on wall thickness. For Outside Diameter controlled pipe, the OD stays constant and the wall thickness changes for different pressure ratings. Tubing sizes are based on the outside diameter — 1" tubing has an OD of about 1". This is generally true as wall thickness affects the ID, not the OD of tubing.

Pipe outside diameters are based on one of several sizing systems. The most common on polyethylene pipe is the old Iron Pipe Size, or IPS, system. Since design familiarity was developed with these standard sizes, polyethylene pipe adopted them for continuity. Other sizing systems in use are Copper Tubing Sizes - CTS, and Ductile Iron Pipe Sizes - DIPS. A product standard will have a complete listing of sizes that are applicable to that standard. For ease of reference a complete set of HDPE pipe sizing tables is presented in the Appendix to Design Chapter 4 of this handbook.

Standard Dimension Ratio

The design of any piping systems is made easier by the use of standard dimension systems, based on either inside or outside pipe diameter. As mentioned earlier, the diameter divided by the wall thickness becomes an important design parameter. This result is called the Dimension Ratio - DR. The dimension ratio is very important for pipe design because the diameter to wall thickness relationship dictates the stress carrying capabilities of the pipe.

The dimension ratio, DR may be based on inside diameter (ID) controlled pipe or outside diameter (OD) controlled pipe.

Standard Inside Dimension Ratio (SIDR) is the ratio of the average specified inside diameter to the minimum specified wall thickness (D_i/t) for inside diameter controlled plastic pipe. In this system the inside diameter of the pipe remains constant and the OD changes with wall thickness. The standard intervals for this system are derived by subtracting one from the pertinent number selected from the ANSI Preferred Number Series 10. Some of the more common values are shown in Table 5.

TABLE 5
Standard Dimension Ratios Based on Controlled Inside or Outside Diameter Pipe

ANSI Preferred Number Series 10	SDR = Series 10 + 1	SIDR = Series 10 - 1
5.0	6.0	4.0
6.3	7.3	5.3
8.0	9.0	7.0
10.0	11.0	9.0
12.5	13.5	11.5
16.0	17.0	15.0
20.0	21.0	19.0
25.0	26.0	24.0
31.5	32.5	30.5
40.0	41.0	39.0
50.0	51.0	49.0
63.0	64.0	62.0

The Standard Dimension Ratio (SDR) is the ratio of the average specified outside diameter to the minimum specified wall thickness (D_o/t) for outside diameter-controlled plastic pipe. In this system the outside diameter of the pipe remains constant and the ID changes with wall thickness. The standard intervals for this system are derived by adding one to the pertinent number selected from the ANSI Preferred Number Series 10. Some of the more common SDR values are shown in Table 5.

Where existing system conditions or special local requirements make other diameters or dimension ratios necessary, they are acceptable in engineered products when mutually agreed upon by the customer and manufacturer if (1) the pipe is manufactured from plastic compounds meeting the material requirements of the end use specification, and (2) the strength and design requirements are calculated on the same bases as those used in the end use specification.

The SDR system is of further use in that a table of pressure ratings can be constructed based on SDR regardless of the pipe's diameter. Utilizing SDR = OD / t (equation 3) for HDS and the accepted design factor (DF), the ISO equation (1) can be rewritten to calculate the maximum internal pressure a pipe can sustain over time.

$$(4) \quad P = \frac{2 \times HDS \times (DF)}{(SDR - 1)}$$

WHERE
P = Internal pressure (psi or MPa)
HDS = Hydrostatic Design Stress (psi or MPa)
SDR = Standard Dimension Ratio
DF = Design Factor
   = 0.50 for water at 73°F

Table 6 shows the pressure ratings for water applications of some of the more common SDR's and HDB's encountered in polyethylene piping system design, assuming a standard operating temperature of 73°F.

TABLE 6
Maximum Pressure Ratings of SDR Pipe at 73°F Using A Design Factor of 0.5

SDR	HDB, psi 1600*	1250*	1000*
32.5	50	40	32
26	65	50	40
21	80	62	50
17	100	80	62
13.5	130	100	80
11	160	125	100

* Value shown are psig

Standard Specifications for Fittings and Joinings

One of the best attributes of PE pipe is its ability to be joined by heat fusion (butt, socket and saddle). Butt fusion is performed by heating the ends of the pipe and/or fitting with an electrically heated plate at about 400°F until the ends are molten. The ends are then forced together at a controlled rate and pressure, and held until cooled. Performed properly, this results in a joint that is integral with the pipe itself, is totally leak-proof, and is typically stronger than the pipe itself. Heat fusion joining can be also be used for saddle fusion of service lines from a main line — even while the main line is in service. Another type of heat fusion is electrofusion. The main difference between conventional heat fusion and electrofusion is the method by which heat is supplied.

While heat fusion is a good method for joining PE pipe and fittings, mechanical fittings are another option. Mechanical fittings consist of compression fittings, flanges, or other types of manufactured transition fittings. There are many types and styles of fittings available from which the user may choose. Each offers its particular advantages and limitations for each joining situation the user may encounter.

The chapter on joining polyethylene pipe within this Handbook provides more detailed information on these procedures. It should be noted that, at this time, there are no known adhesives or solvent cements that are suitable for joining polyethylene pipes.

Joining of polyethylene pipe can be done by either mechanical fittings or by heat fusion. All joints and fittings must be designed at the same high level of performance and integrity as the rest of the piping system. For gas distribution systems, the installation of a plastic pipe system must provide that joining techniques comply with Department of Transportation 49 CFR 192 subpart F-Joining of Materials Other Than by Welding. The general requirements for this subpart are:

General

a. The pipeline must be designed and installed so that each joint will sustain the longitudinal pullout or thrust forces caused by contraction or expansion of the piping or by anticipated external or internal loading.

b. Each joint must be made in accordance with written procedures that have been proven by test or experience to produce strong, gas-tight joints.

c. Each joint must be inspected to ensure compliance with this subpart. Within 49 CFR 192 subpart F, 192.281 specifies selected requirements for plastic joints; 192.282 specifies requirements for qualifying joining procedures; 192.285 specifies qualifying persons to make joints; and 192.287 specifies inspection of joints.

Since fittings need to be able to handle the same stresses as the pipe, fusion fittings for polyethylene pipe are produced from the same stress rated materials as are used to make the pipe itself. However, since the geometry of the fittings is different from the pipe, the stress induced by internal pressure is different. Therefore, fittings are designed to handle a specific maximum working pressure and the pressure-to-stress equations based on OD and wall thickness may not apply. Typically, the fitting will be rated to handle the same stress as the pipe to which it is designed to be joined. If there is a question about the pressure rating of the fitting, contact the fitting manufacturer.

Specifications for socket, butt fusion, and electrofusion fittings have been developed by ASTM:

- D 2683 "Standard Specification for Socket-Type Polyethylene Fittings for Outside Diameter-Controlled Polyethylene Pipe and Fittings."
- D 3261 "Standard Specification for Butt Heat Fusion Polyethylene (PE) Plastic Fittings for Polyethylene Plastic Pipe and Tubing."
- F 1055 "Electrofusion Type Polyethylene Fittings for Outside Diameter Controlled Polyethylene Pipe and Tubing."
- D 2657 "Standard Practice for Heat Fusion Joining of Polyolefin Pipe and Fittings."

A generic joining procedure for polyethylene gas pipe has also been published by the PPI: TR-33 "Generic Butt Fusion Joining Procedure for Polyethylene Gas Pipe." In addition to these standards and procedures, each manufacturer will have published joining procedures for their pipe and/or fittings. Some of the relevant standards that pertain to fitting performance or joining practices are listed in the Appendix.

Codes, Standards and Recommended Practices for Polyethylene Piping Systems

There are a large number of codes, standards and practices that govern, or greatly influence the polyethylene piping industry. These standards cover a broad range of applications for polyethylene pipe and fittings. Some standards pertain to the product performance requirements for a specific application, while other standards are guidelines and practices detailing how a certain type of activity is to be performed. Some are test methods that define exactly how a particular test is to be run so that a direct comparison can be made between results. There are several standards writing organizations that deal directly with the manufacture, testing, performance, and use of polyethylene pipe and fittings. Some of the major codes and standards organizations are discussed below. A more inclusive listing can be found in the Appendix of this chapter.

Plastics Pipe Institute (PPI)

The Plastics Pipe Institute is a trade association dedicated to promoting the effective use of plastics piping systems. Prior sections of this chapter reviewed the subjects of PPI designations, pressure ratings and hydrostatic design basis (HDB). The assignment of a recommended hydrostatic design basis for a thermoplastic material falls under the jurisdiction of the Hydrostatic Stress Board - HSB - of the Plastics Pipe Institute. The Hydrostatic Stress Board has the responsibility of developing policies and procedures for the recommendation of the estimated long-term strength for commercial thermoplastic piping materials. The document most widely used for this is Technical Report-3, TR-3 "Policies and Procedures for Developing Hydrostatic Design Bases (HDB), Pressure Design Bases (PDB), Strength Design Bases (SDB), and Minimum Required Strengths (MRS) for Thermoplastic Piping Materials or Pipe." The material stress ratings themselves are published in TR-4, "PPI Listing of Hydrostatic Design Bases (HDB), Strength Design Bases (SDB), Pressure Design Bases (PDB) and Minimum Required Strengths (MRS) Ratings for Thermoplastic Piping Materials or Pipe." There are many other publications pertaining to various aspects of polyethylene pipe available from PPI such as: TN's - Technical Notes, TR's - Technical Reports, Model Specifications, and White Papers on specific positions addressed by the industry. Check the website www.plasticpipe.org for up-to-date publications.

Technical Report 3
TR-3 is a publication that is under the jurisdiction of the Hydrostatic Stress Board and is a compilation of the policies and procedures for recommending the stress or pressure rating of thermoplastic materials such as those used in pressure pipe and fitting or multi-layer pipes intended for use in pressure applications. This recommendation can be in the form of an HDB established according to ASTM D 2837, a PDB for a multi-layer pipe also established according to D 2837, a MRS established according to ISO TR9080, or an SDB established according to ASTM F 2018. In order to better understand the purpose and limitations of this document, it is strongly suggested that the Foreword and Notes to the Reader of TR-3 be read. Further questions should be directed to the Chairman of the HSB - who is the Technical Director of the Plastics Pipe Institute.

Technical Report 4
The recommendations of the Hydrostatic Stress Board are published in TR-4, "PPI Listing of Hydrostatic Design Bases (HDB), Strength Design Bases (SDB), Pressure Design Bases (PDB) and Minimum Required Strengths (MRS) Ratings for Thermoplastic Piping Materials or Pipe." TR-4 lists these thermoplastic piping materials according to the material type - PVC, CPVC, PE, PEX, POM, PVDF, and PA - and the HDB/HDS category for selected temperatures. It also lists the actual

pressure ratings recommended for pipes of multi-layer construction which behave like thermoplastic piping materials during testing and can be evaluated by the same methodology. Again, the Foreword and Notes to the Reader should be studied to better understand how to apply these ratings.

Current printings of both TR-3 and TR-4 can be found at the PPI website at www.plasticpipe.org. It should also be noted that the PPI produces a number of related documents and publications to assist the designer or installer in the use HDPE pipe. These guides, reports, and technical notes are all available for download from the same website.

ASTM

The American Society for Testing and Materials (ASTM) is a consensus standards writing organization, and has published standards for a multitude of industries and applications. Those pertaining to polyethylene pipe are found in Volume 8.04 "Plastic Pipe and Building Products." ASTM employees do not write these standards; rather they are written by interested parties and experts within the industry who are members of ASTM. Most anyone can be a member of ASTM and participate in the standard writing process. Other standards, pertaining to plastics in general are found in other books within Volume 8 - 8.01, 8.02, or 8.03.

ASTM Standards pertaining to PE pipe can be a Standard Specification that defines the product requirements and performance for a specific application. It can also be a Standard Practice, which defines how a particular activity is to be performed, or a Standard Test Method, which defines how a particular test on PE pipe, fittings, or materials is to be done. While ASTM standards are mainly used in North America, many are also ANSI approved for international recognition, or are equivalent to an ISO standard. When a manufacturer prints the ASTM Standard on a product, the manufacturer is certifying that the product meets all of the requirements of that standard.

The typical sections covered in an ASTM Product Standard are:

Scope - what products and applications are covered under this standard.

Referenced Documents - what other standards or specifications are referenced in this standard.

Terminology - lists definitions that are specific to this standard.

Materials - defines material requirements for products that conform to this standard.

Requirements - details the performance requirements that the product must meet. This section will also contain dimensions.

Test Methods - details how the testing is to be performed to determine conformance to the performance requirements.

Marking - details the print that must be on the product. Includes the standard number, manufacturer's name, size, date of manufacture, and possibly the application such as "water." There may be other wording added to the print as the purchaser requires.

This is only a typical example of sections that may be included. While ASTM has defined protocol for product standards, each one may contain sections unique to that standard. Each standard should be reviewed individually for its requirements. A listing of major ASTM standards pertaining to PE pipe and fittings is in the Appendix. Current publications of these standards can be found at the website www.astm.org.

ISO

The International Organization for Standardization (ISO) is a network of national standards institutes from 140 countries working in partnership with international organizations, governments, industry, business and consumer representatives. It serves as a bridge between public and private sectors.

The ISO committee responsible for development of plastics pipe standards is Technical Committee 138. The committee's stated scope is: Standardization of pipes, fittings, valves and auxiliary equipment intended for the transport of fluids and made from all types of plastic materials, including all types of reinforced plastics. Metal fittings used with plastics pipes are also included. The main committee has seven subcommittees devoted to specific issues.

TC 138 has 35 participating countries, including the United States and Canada, and 27 observer countries. For ISO matters the United States is represented by the American National Standards Institute (ANSI). Canadian representation is through the Standards Council of Canada (SCC). The United States representation has been passed through ANSI to the Plastics Pipe Institute.

NSF International

NSF International plays a vital role in the use of polyethylene pipe and fittings for potable water applications. NSF is an independent, not-for-profit organization of scientists, engineers, educators and analysts. It is a trusted neutral agency, serving government, industry and consumers in achieving solutions to problems relating to public health and the environment. NSF standards are developed with the active participation of public health and other regulatory officials, users and industry. The standards specify the requirements for the products, and may include requirements relating to materials, design, construction, and performance. NSF has policies that establish additional requirements that a company must comply with to be able to obtain and maintain certification of products and authorization to use the NSF Mark for potable water applications.

There are two NSF Standards that are of particular importance to the polyethylene pipe and fittings industry: Standard 14, "Plastic Piping components and Related Materials" and Standard 61, "Drinking Water System Components-Health Effects." Standard 14 includes both performance requirements from product standards and provisions for health effects covered in Standard 61. NSF Standard 14 does not contain performance requirements itself, but rather NSF will certify that a product conforms to a certain ASTM, AWWA, etc... product performance standard. In order to be certified for potable water applications under Standard 14, the product must also satisfy the toxicological requirements of Standard 61.

It is also an option to be certified under Standard 61 only, without certifying the performance aspects of the product. In the early 1990's NSF separated the toxicological sections of Standard 14 into a new Standard 61. This was done for several reasons, but mainly to make it easier to bring new, innovative products to market without undue expense and time, while continuing to keep the public safe. This was a great benefit to the industry. Now manufacturers have a choice of staying with Standard 14 or switching to Standard 61. Many manufacturers who have in-house quality programs and the ability to perform the necessary tests switched to this new potable water certification option.

AWWA

The American Water Works Association (AWWA) is a leader in the development of water resource technology. While AWWA prepares and issues standards, they are not specifications. These standards describe minimum requirements and do not contain all of the engineering and administrative information normally contained in specifications. The AWWA standards usually contain options that must be evaluated by the user of the standard. Until each optional feature is specified by the user, the product or service is not fully defined. The use of AWWA standards is entirely voluntary. They are intended to represent a consensus of the water supply industry that the product described will provide satisfactory service.

There are currently two AWWA standards that pertain to polyethylene pipe: AWWA C901, "Polyethylene (PE) Pressure Pipe and Tubing, 1/2 inch through 3 inch, for Water Service" and AWWA C906, "Polyethylene (PE) Pressure Pipe and Fittings, 4 inch through 63 inches, for Water Distribution." Standard C901 addresses polyethylene pressure pipe and tubing for use primarily as potable water service lines in the construction of underground distribution systems. It includes dimensions for pipe and tubing made from PE materials with standard PE designations PE 2406 and PE 3408, in pressure classes of 80 psi, 100 psi, 125 psi, 160 psi and 200 psi. Pipe, ranging in nominal size from 1/2 inch through 3 inch conforms to outside-diameter dimensions of iron pipe sizes (OD based, IPS pipe) or to the inside-diameter dimensions of iron pipe sizes (ID based, IPS pipe). Tubing, ranging in size from 1/2 inch through 2 inch, conforms to the outside-diameter dimensions of copper tubing sizes (CTS). There are also sections on materials, testing and marking requirements; inspection and testing by manufacturer; and in-plant inspection by purchaser.

AWWA Standard C906 addresses larger diameter polyethylene pressure pipe made from materials conforming to standard PE designations PE 2406 and PE 3408. The pipe is primarily intended for use in transporting potable water in either buried or above-ground installations. The standard covers 10 dimension ratios (DR's) for nominal pipe sizes ranging from 4 inch through 63 inch. The available pipe sizes are limited by a maximum wall thickness of 3 inch. Pipe outside diameters (OD's) conform to the outside diameter dimensions of iron pipe sizes (IPS), ductile iron pipe size (DIPS), or those established by the International Organization for Standardization (ISO). Pressure class ratings range from 40 psi to 198 psi for PE 2406 materials, and from 51 psi to 254 psi for PE 3408 materials.

At the time of this writing, another important resource is being developed with the AWWA forum, Manual 55. This soon to be published AWWA manual is a design and installation guide for the use of polyethylene pipe in potable water applications. The manual is intended to supplement C901 and C906 and provide specific design recommendations as it relates to the use of polyethylene pipe in potable water systems. The publication of this important document is anticipated in 2006.

Plumbing Codes

Piping systems used in buildings must meet standards established in the plumbing code adopted by the jurisdiction in which the building is to be constructed. Within the United States there are several "model" codes, any one of which can be used as the basis for a local jurisdiction's code. Most widely used model codes include the International Plumbing Code (IPC), produced by the International Code Council (ICC) and the Uniform Plumbing Code (UPC), produced by the International Association of Plumbing and Mechanical Officials (IAPMO). One of the model codes may be adopted in its entirety or modified by the jurisdiction. Some states adopt a statewide code which municipalities may or may not be allowed to amend based on state law. Both designers and contractors need to be familiar with the code that applies to a particular project with a specific jurisdiction.

Other Codes and Standards

There are several other codes and standards writing organizations which pertain to polyethylene pipe. These groups usually have a type of certification program for products to be used in a certain industry or application, and may or may not write their own performance standards. If they do not write their own standards, they will certify products to an existing standard such as ASTM, AWWA, etc. The certification process will normally consist of an initial application stating what specific products are requesting certification, an on-site inspection of the production facilities, and testing of the product to assure performance to the relevant product specification. This is followed up by annual random inspections and product testing.

The Canadian Standards Association (CSA) provides a good example of the type of compliance certification program that relates to the use of polyethylene pipe in both water (CSA B137.1) and gas distribution (C137.4) applications. CSA's certification of compliance to the standards to which a particular polyethylene pipe is made allows the producer of that product to place the CSA mark on the product. The presence of the mark assures the purchaser that the product has met the requirements of the CSA certification program and insures that the product meets the appropriate product specifications as determined by the audits and inspections conducted by the Canadian Standards Association.

Factory Mutual

Factory Mutual Research (FM), an affiliate of FM Global, is a non-profit organization that specializes in property loss prevention knowledge. The area that pertains to HDPE pipe is the FM Standard "Plastic Pipe and Fittings for Underground Fire Protection Service." Certification to this standard may be required by an insurance company for any PE pipe and fittings being used in a fire water system. FM Global requires an initial inspection and audit of production facilities to be assured that the facility has the proper quality systems in place similar to ISO 9000 requirements. Then testing of the pipe must be witnessed by an FM representative. This testing must pass the requirements set forth in the FM Standard for PE pipe. After initial certification, unannounced audits are performed on at least an annual basis. More information can be found at their website www.fmglobal.com, or by calling at (401) 275-3000.

Conclusion

Polyethylene resins are produced to cover a very broad range of applications. The physical performance properties of these various formulations of polyethylene vary significantly making each grade suitable for a specific range of applications. To that end, the polyethylene pipe industry has worked diligently to establish effective standards and codes which will assist the designer in the selection and specification of piping systems produced from polyethylene materials which lend themselves to the type of service life sought. As such, the discussion which has been presented here should assist the designer and/or installer in his understanding of these standards and their significance relative to the use of these unique plastic piping materials.

Extensive reference has been made throughout the preceding discussion to standards writing or certifying organizations such as ASTM, AWWA, NSF, etc. The standards setting process is dynamic, as is the research and development that continues within the polyethylene pipe industry. As such, new standards and revisions of existing standards are developed on an ongoing basis. For this reason, the reader is encouraged to obtain copies of the most recent standards available from these various standards organizations.

References

1. ASTM Annual Book of Standards, Volume 8.03 Plastics, (III): D 3100 - Latest, American Society for Testing and Materials, West Conshohocken, PA.
2. ASTM Annual Book of Standards, Volume 8.04 Plastic Pipe and Building Products, American Society for Testing and Materials, West Conshohocken, PA.
3. Plastics Pipe Institute, Various Technical Reports, Technical Notes, Model Specifications, Washington, DC.
4. NSF Standard 14, Plastic Piping Components and Related Materials, NSF International, Ann Arbor, MI.
5. NSF Standard 61, Drinking Water System Components - Health Effects, NSF International, Ann Arbor, MI.

Appendix 1- Major Standards, Codes and Practices

General

ASTM

D 3350	Polyethylene Plastics Pipe and Fittings Materials
D 1598	Time-to-Failure of Plastic Pipe Under Constant Internal Pressure
D 1599	Short-Time Hydraulic Failure Pressure of Plastic Pipe, Tubing and Fittings
D 2122	Determining Dimensions of Thermoplastic Pipe and Fittings
D 2837	Obtaining Hydrostatic Design Basis for Thermoplastic Pipe Materials
D 2488	Description and Identification of Soils (Visual-Manual Procedure)
D 2657	Heat-Joining Polyolefin Pipe and Fittings
D 2683	Socket Type Polyethylene Fittings for Outside Diameter Controlled Polyethylene Pipe and Tubing
F 412	Terminology Relating to Plastic Piping Systems
F 480	Thermoplastic Well Casing Pipe and Couplings Made in Standard Dimension Ratios (SDRs), SCH 40, and SCH 80
F 948	Time-to-Failure of Plastic Piping Systems and Components Under Constant Internal Pressure With Flow
F 1055	Electrofusion Type Polyethylene Fittings for Outside Diameter Controlled Polyethylene Pipe and Tubing
F 1248	Test Method for Determination of Environmental Stress Crack Resistance (ESCR) of Polyethylene Pipe
F1290	Electrofusion Joining Polyolefin Pipe and Fittings
F 1473	Notch Tensile Test to Measure the Resistance to Slow Crack Growth of Polyethylene Pipes and Resins
F 1533	Deformed Polyethylene (PE) Liner
F 1901	Polyethylene (PE) Pipe and Fittings for Roof Drain Systems
F 1962	Standard Guide for Use of Maxi-Horizontal Directional Drilling for Placement of Polyethylene Pipe or Conduit Under Obstacles, Including River Crossing
F 2164	Standard Practice for Field Leak Testing of Polyethylene (PE) Pressure Piping Systems Using Hydrostatic Pressure
F 2231	Standard Test Method for Charpy Impact Test on Thin Specimens of Polyethylene Used in Pressurized Pipes
F 2263	Standard Test Method for Evaluating the Oxidative Resistance of Polyethylene (PE) Pipe to Chlorinated Water

Chapter 5
Specifications, Test Methods and Codes for Polyethylene Piping Systems

PPI TECHNICAL REPORTS

TR-3	Policies and Procedures for Developing Hydrostatic Design Bases (HDB), Pressure Design Bases (PDB), Strength Design Bases (SDB), and Minimum Required Strengths (MRS) Ratings for Thermoplastic Piping Materials for Pipe
TR-4	PPI Listing of Hydrostatic Design Bases (HDB), Strength Design Bases (SDB), Pressure Design Bases (PDB) and Minimum Required Strength (MRS) Ratings for Thermoplastic Piping Materials or Pipe
TR-7	Recommended Methods for Calculation of Nominal Weight of Solid Wall Plastic Pipe
TR-9	Recommended Design Factors for Pressure Applications of Thermoplastic Pipe Materials
TR-11	Resistance of Thermoplastic Piping Materials to Micro- and Macro-Biological Attack
TR-14	Water Flow Characteristics of Thermoplastic Pipe
TR-18	Weatherability of Thermoplastic Piping Systems
TR-19	Thermoplastic Piping for the Transport of Chemicals
TR-21	Thermal Expansion and Contraction in Plastics Piping Systems
TR-30	Investigation of Maximum Temperatures Attained by Plastic Fuel Gas Pipe Inside Service Risers
TR-33	Generic Butt Fusion Joining Procedure for Polyethylene Gas Pipe
TR-34	Disinfection of Newly Constructed Polyethylene Water Mains
TR-35	Chemical & Abrasion Resistance of Corrugated Polyethylene Pipe
TR-36	Hydraulic Considerations for Corrugated Polyethylene Pipe
TR-37	CPPA Standard Specification (100-99) for Corrugated Polyethylene (PE) Pipe for Storm Sewer Applications
TR-38	Structural Design Method for Corrugated Polyethylene Pipe
TR-39	Structural Integrity of Non-Pressure Corrugated Polyethylene Pipe
TR-40	Evaluation of Fire Risk Related to Corrugated Polyethylene Pipe

PPI TECHNICAL NOTES

TN-4	Odorants in Plastic Fuel Gas Distribution Systems
TN-5	Equipment used in the Testing of Plastic Piping Components and Materials
TN-6	Polyethylene (PE) Coil Dimensions
TN-7	Nature of Hydrostatic Stress Rupture Curves
TN-11	Suggested Temperature Limits for the Operation and Installation of Thermoplastic Piping in Non-Pressure Applications
TN-13	General Guidelines for Butt, Saddle and Socket Fusion of Unlike Polyethylene Pipes and Fittings
TN-14	Plastic Pipe in Solar Heating Systems
TN-15	Resistance of Solid Wall Polyethylene Pipe to a Sanitary Sewage Environment
TN-16	Rate Process Method for Projecting Performance of Polyethylene Piping Components
TN-17	Cross-linked Polyethylene (PEX) Tubing
TN-18	Long-Term Strength (LTHS) by Temperature Interpolation.
TN-19	Pipe Stiffness for Buried Gravity Flow Pipes
TN-20	Special Precautions for Fusing Saddle Fittings to Live PE Fuel Gas Mains Pressurized on the Basis of a 0.40 Design Factor
TN-21	PPI PENT test investigation
TN-23	Guidelines for Establishing the Pressure Rating for Multilayer and Coextruded Plastic Pipes

Chapter 5
Specifications, Test Methods and Codes for Polyethylene Piping Systems

Gas Pipe, Tubing and Fittings
ASTM

D 2513	Thermoplastic Gas Pressure Pipe, Tubing and Fittings
F 689	Determination of the Temperature of Above-Ground Plastic Gas Pressure Pipe Within Metallic Castings
F 1025	Selection and Use of Full-Encirclement-Type Band Clamps for Reinforcement or Repair of Punctures or Holes in Polyethylene Gas Pressure Pipe
F 1041	Squeeze-Off of Polyolefin Gas Pressure Pipe and Tubing
F 1563	Tools to Squeeze Off Polyethylene (PE) Gas Pipe or Tubing
F 1734	Practice for Qualification of a Combination of Squeeze Tool, Pipe, and Squeeze-Off Procedure to Avoid Long-Term Damage in Polyethylene (PE) Gas Pipe
F 1924	Plastic Mechanical Fittings for Use on Outside Diameter Controlled Polyethylene Gas Distribution Pipe and Tubing
F 1948	Metallic Mechanical Fittings for Use on Outside Diameter Controlled Thermoplastic Gas Distribution Pipe and Tubing
F 1973	Factory Assembled Anodeless Risers and Transition Fittings in Polyethylene (PE) Fuel Gas Distribution Systems
F 2138	Standard Specification for Excess Flow Valves for Natural Gas Service

PPI

TR-22	Polyethylene Plastic Piping Distribution Systems for Components of Liquid Petroleum Gase
MS-2	Model Specification for Polyethylene Plastic Pipe, Tubing and Fittings for Natural Gas Distribution

OTHER STANDARDS FOR GAS PIPING APPLICATIONS

Title 49, CFR part 192	Transportation of Natural Gas and Other Gas by Pipe Line
AGA	AGA Plastic Pipe Manual for Gas Service (American Gas Association)
API	API Spec 15LE Specification for Polyethylene Line Pipe (American Petroleum Institute)

Water Pipe, Tubing and Fittings
ASTM

D 2104	Polyethylene (PE) Plastic Pipe, Schedule 40
D 2239	Polyethylene (PE) Plastic Pipe (SIDR-PR) Based on Controlled Inside Diameter
D 2447	Polyethylene (PE) Plastic Pipe, Schedules 40 to 80, Based on Outside Diameter
D 2609	Plastic Insert Fittings for Polyethylene (PE) Plastic Pipe
D 2683	Socket-Type Polyethylene Fittings for Outside Diameter-Controlled Polyethylene Pipe and Tubing
D 2737	Polyethylene (PE) Plastic Tubing
D 3035	Polyethylene (PE) Plastic Pipe (SDR-PR) Based on Controlled Outside Diameter
D 3261	Butt Heat Fusion Polyethylene (PE) Plastic Fittings for Polyethylene (PE) Plastic Pipe and Tubing
F 405	Corrugated Polyethylene (PE) Tubing and Fittings
F 667	Large Diameter Corrugated Polyethylene (PE) Tubing and Fittings
F 714	Polyethylene (PE) Plastic Pipe (SIDR-PR) Based on Controlled Outside Diameter
F 771	Polyethylene (PE) Thermoplastic High-Pressure Irrigation Pipeline Systems
F 810	Smooth Wall Polyethylene (PE Pipe for Use in Drainage and Waste Disposal Absorption Fields
F 982	Polyethylene (PE) Corrugated Pipe with a Smooth Interior and Fittings
F 894	Polyethylene (PE) Large Diameter Profile Wall Sewer and Drain Pipe
F 905	Qualification of Polyethylene Saddle Fusion Joints
F 1055	Electrofusion Type Polyethylene Fittings for Outside Diameter Controlled Polyethylene Pipe and Tubing
F 1056	Socket Fusion Tools for Use in Socket Fusion Joining Polyethylene Pipe or Tubing and Fittings
F 1759	Standard Practice for Design of High-Density Polyethylene (HDPE) Manholes for Subsurface Applications
F 2206	Standard Specification for Fabricated Fittings of Butt-Fused Polyethylene (PE) Plastic Pipe, Fittings, Sheet Stock, Plate Stock, or Block Stock

PEX

F 876	Cross-linked Polyethylene (PEX) Tubing
F 877	Cross-linked Polyethylene (PEX) Plastic Hot- and Cold-Water Distribution Systems
F 1281	Standard Specification for Cross-linked Polyethylene/Aluminum/ Cross-linked Polyethylene (PEX - AL -PEX) Pressure Pipe
F 1282	Standard Specification Polyethylene/Aluminum/ Polyethylene (PE -Al - PE) Composite Pressure Pipe
F 1807	Metal Insert Fittings Utilizing a Copper Crimp Ring for SDR 9 Cross-linked Polyethylene (PEX) Tubing
F 1865	Mechanical Cold Expansion Insert Fitting With Compression Sleeve for Cross-linked Polyethylene (PEX) Tubing
F 1960	Mechanical Cold Expansion Insert Fittings with PEX Reinforcing Rings for Use with Cross-linked (PEX) Tubing
F 1961	Metal Mechanical Cold Flare Compression Fittings with Disc Spring for Cross-linked Polyethylene (PEX) Tubing
F 1974	Standard Specification for Metal Insert Fittings for Polyethylene/Aluminum/ Polyethylene and Cross-linked Polyethylene/Aluminum/ Crosslinked Polyethylene Composite Pressure Pipe
F 2023	Standard Test Method for Evaluating the Oxidative Resistance of Cross-linked Polyethylene (PEX) Tubing and Systems to Hot Chlorinated Water
F 2080	Cold Expansion Fittings with Metal Compression Sleeves for Cross-linked Polyethylene (PEX) Pipe
F 2098	Stainless Steel Clamps for Securing SDR 9 Cross-linked Polyethylene (PEX) Tubing to Metal Insert Fittings
F 2159	Standard Specification for Plastic Insert Fittings Utilizing a Copper Crimp Ring for SDR9 Cross-linked Polyethylene (PEX) Tubing
F 2262	Standard Specification for Cross-linked Polyethylene/Aluminum/ Crosslinked Polyethylene Tubing OD Controlled SDR9

PPI

MS-3	Model Specification for Polyethylene Plastic Pipe, Tubing and Fittings for Water Mains and Distribution

AWWA

C 901	Polyethylene (PE) Pressure Pipe, Tubing, and Fittings, 1/2 inch through 3 inch for Water Service
C 906	Polyethylene (PE) Pressure Pipe and Fittings, 4 inch through 63 inch for Water Distribution
M 55	AWWA Manual 55: PE Pipe - Design and Installation

CSA

B 137.1	Polyethylene Pipe, Tubing and Fittings for Cold Water Pressure Services
B137.4	Polyethylene Piping Systems for Gas Services (Canadian Standards Association)

Installation
ASTM

D 2321	Underground Installation of Flexible Thermoplastic Sewer Pipe
D 2774	Underground Installation of Thermoplastic Pressure Piping
F 449	Subsurface Installation of Corrugated Thermoplastic Tubing for Agricultural Drainage or Water Table Control
F 481	Installation of Thermoplastic Pipe and Corrugated Tubing in Septic Tank Leach Fields
F 585	Insertion of Flexible Polyethylene Pipe into Existing Sewers
F 645	Selection, Design and Installation of Thermoplastic Water Pressure Pipe System
F 690	Underground Installation of Thermoplastic Pressure Piping Irrigation Systems
F 1176	Design and Installation of Thermoplastic Irrigation Systems with Maximum Working Pressure of 63 psi
F 1417	Test Method for Installation Acceptance of Plastic Gravity Sewer Lines Using Low-Pressure Air
F 1606	Standard Practice for Rehabilitation of Existing Sewers and Conduits with Deformed Polyethylene (PE) Liner
F 1668	Guide for Construction Procedures for Buried Plastic Pipe
F 1759	Standard Practice for Design of High-Density Polyethylene (HDPE) Manholes for Subsurface Applications
F 1743	Qualification of a Combination of Squeeze Tool, Pipe, and Squeeze-Off Procedures to Avoid Long-Term Damage in Polyethylene (PE) Gas Pipe
F 1804	Determine Allowable Tensile Load For Polyethylene (PE) Gas Pipe During Pull-in Installation
F 1962	Guide for Use of Maxi-Horizontal Directional Drilling for Placement of Polyethylene Pipe of Conduit Under Obstacles, Including River Crossings
F 2164	Standard Practice for Field Leak Testing of Polyethylene (PE) Pressure Piping Systems Using Hydrostatic Pressure

CONDUIT

F 2160	Standard Specification for Solid Wall High Density Polyethylene (HDPE) Conduit Based on Controlled Outside Diameter (OD)
F 2176	Standard Specification for Mechanical Couplings Used on Polyethylene Conduit, Duct, and Innerduct

ISO STANDARDS

9080	Thermoplastics pipes and fittings for the transport of fluid - Methods of extrapolation of hydrostatic stress rupture data to determine the long-term hydrostatic strength of thermoplastics pipe materials
4427	Polyethylene (PE) pipes for water supply
4437	Buried polyethylene (PE) pipes for the supply of gaseous fuels - Metric series - Specifications
12162	Thermoplastics materials for pipes and fittings for pressure applications - Classification and designation - Overall service (design) coefficient

AASHTO STANDARDS

M 252	Plastic and Corrugated Drainage Tubing
M 294	Corrugated Polyethylene Pipe, 12 to 24 inch Diameter
F 2136	Standard Test Method for Notched, Constant Ligament-Stress (NCLS) Test to Determine Slow-Crack Growth Resistance of HDPE Resins or HDPE Corrugated Pipe

AMERICAN SOCIETY OF AGRICULTURAL ENGINEERS

| S376.1 | Design, Installation and Performance of Underground, Thermoplastic Irrigation Pipelines |

Chapter 6

Design of Polyethylene Piping Systems

Introduction

Design of a polyethylene piping system is generally no different than the design undertaken with any ductile and flexible piping material. The design equations and relationships are well-established in the literature, and they can be employed in concert with the distinct performance properties of this material to create a piping system which will provide years of service for the intended application.

In the pages which follow, we will explore the basic design methods for using polyethylene pipe in a very broad range of applications. The material is divided into three distinct sections. The first will deal with hydraulic design of a polyethylene piping system. In this section we will present the design equations and examples for determining fluid flow in both pressurized and gravity flow.

Section 2 will focus on burial design and flexible pipeline design theory. From this discussion, the designer will develop a clear understanding of the nature of pipe/soil interaction and the relative importance of trench design as it relates to the use of a flexible piping material.

Finally, Section 3 will deal with the response of polyethylene pipe to temperature change. As with any construction material, polyethylene expands and contracts in response to changes in temperature. Specific design methodologies will be presented in this section to address this very important aspect of pipeline design as it relates to the use of HDPE pipe.

This chapter will conclude with a fairly extensive appendix which details the physical characteristics and dimensions of polyethylene pipe produced in accordance with various industry standards.

Section 1 **Design for Flow Capacity**

Piping systems generally transport a fluid, liquid, slurry or gas, from one location to another. This section provides design information for determining the polyethylene pipe diameter required for point-to-point application flow requirements. This section addresses polyethylene pipe diameter and pressure rating, general fluid flows in pipes and fittings, liquid (water and water slurry) flow under pressure, non-pressure (gravity) liquid flow, and compressible gas flow under pressure. Network flow analysis and design is not addressed.[1,2]

The procedure for piping system design is frequently an iterative process. For pressure liquid flows, combinations of sustained internal pressure, surge pressure, and head loss pressure can affect pipe selection. For non-pressure and gravity flow systems, piping design typically requires selecting a pipe size that provides adequate reserve flow capacity and a wall thickness or profile design that sufficiently addresses anticipated static and dynamic earthloads. After a trial pipe size and pressuring or external load rating is selected, it is evaluated to determine if it is appropriate for the design requirements of the application. Evaluation may show that a different size, pressure rating or external load capacity may be required and, if so, a different pipe is selected, and the new profile is evaluated. The appendix to this chapter provides design and engineering information for polyethylene pipes made to selected industry standards discussed in this chapter and throughout this handbook.

Pipe ID for Flow Calculations

Polyethylene pipes are manufactured under industry standards that control either outside diameter or inside diameter.

Thermoplastic pipes are generally produced in accordance with a dimension ratio (DR) system. The dimension ratio, DR or IDR, is the ratio of the pipe diameter to the respective minimum wall thickness, either OD or ID, respectively. As the diameter changes, the pressure rating remains constant for the same material, dimension ratio and application. The exception to this practice is production of thermoplastic pipe in accordance with the industry established SCH 40 and SCH 80 dimensions such as referenced in ASTM D 2447.

Pipe Diameter for OD Controlled Pipe

OD-controlled pipe is dimensioned by outside diameter and wall thickness. Several sizing systems are used including IPS, which is the same OD as IPS steel pipe; DIPS, which is the same OD as ductile iron pipe; and CTS, which is the same OD as copper tubing. For flow calculations, inside diameter is calculated by deducting twice the wall thickness from the outside diameter. OD-controlled pipe standards include ASTM D2513, ASTM D2737, ASTM D2447, ASTM D3035, ASTM F714, AWWA C901, AWWA C906 and API 15LE.[3,4,5,6,7,8,9,10] The appendix provides specific dimensional information for outside diameter controlled polyethylene pipe and tubing made in accordance with selected ASTM and AWWA standards.

Equation 1-1 may be used to determine an average inside diameter for OD-controlled polyethylene pipe made to dimension ratio (DR) specifications in accordance with the previously referenced standards. In these standards, pipe dimensions are specified as average outside diameter and, typically, wall thickness is specified as a minimum dimension, and a +12% tolerance is applied. Therefore, an average ID for flow calculation purposes may be determined by deducting twice the average wall thickness (minimum wall thickness plus half the wall tolerance or 6%) from the average outside diameter.

(1-1)
$$D_A = D_O - 2.12\left(\frac{D_O}{DR}\right)$$

WHERE
D_A = pipe average inside diameter, in
D_O = pipe outside diameter, in
DR = dimension ratio

(1-2)
$$DR = \frac{D_O}{t}$$

t = pipe minimum wall thickness, in

Pipe Diameter for ID Controlled Pipe
Standards for inside diameter controlled pipes provide average dimensions for the pipe inside diameter that are used for flow calculations. ID-controlled pipe standards include ASTM D2104, ASTM D2239, ASTM F894 and AWWA C901. [11,12,13]

The terms "DR" and "IDR" are used with outside diameter controlled and inside diameter controlled pipe respectively. Certain dimension ratios that meet an ASTM-specified number series are "standardized dimension ratios," that is SDR or SIDR. Standardized dimension ratios are: 41, 32.5, 26, 21, 17, 13.5, 11, 9, and 7.3. From one SDR or SIDR to the next, there is about a 25% difference in minimum wall thickness.

Pressure Rating for Pressure Rated Pipes
Conventionally extruded (solid wall) polyethylene pipes have a simple cylindrical shape, and are produced to industry standards that specify outside diameter and wall thickness, or inside diameter and wall thickness. OD controlled pressure pipes are pressure rated using Equation 1-3. ID controlled pressure pipes are pressure rated using Equation 1-4.

Equations 1-3 and 1-4 utilize the Hydrostatic Design Basis, HDB at 73°F (23°C) to establish the performance capability of the pipe profile at that temperature. HDB's for various polyethylene pipe materials are published in PPI TR-4, "PPI Listing of Hydrostatic Design Basis (HDB), Strength Design Basis (SDB), Pressure Design Basis (PDB) and Minimum Required Strength (MRS) Ratings for Thermoplastic Piping Materials or Pipe (2005)". Materials that are suitable for use at higher temperatures above 100°F (38°C) will also have elevated temperature HDB's which are published in PPI TR-4. Two design factors, DF and F_T, are used to relate environmental conditions and service temperature conditions to the product. See Tables 1-2 and 1-3. If the HDB at an elevated temperature is known, that HDB value should be used in Equation 1-3 or 1-4, and the service temperature design factor, F_T, would then be 1. If the elevated HDB is not known, then F_T should be used, but this will generally result in a lower or more conservative pressure rating.

(1-3) $$P = \frac{2\,HDB \times DF \times F_T}{(DR-1)}$$

(1-4) $$P = \frac{2\,HDB \times DF \times F_T}{(IDR+1)}$$

WHERE
P = Pressure rating, psi
HDB = Hydrostatic Design Basis, psi
DF = Design Factor, from Table 1-2
F_T = Service Temperature Design Factor, from Table 1-3, 1.0 if the elevated temperature HDB is not used.
DR = OD -Controlled Pipe Dimension Ratio

(1-5) $$DR = \frac{D_O}{t}$$

D_O = OD-Controlled Pipe Outside Diameter, in.
t = Pipe Minimum Wall Thickness, in.
IDR = ID -Controlled Pipe Dimension Ratio

(1-6) $$IDR = \frac{D_I}{t}$$

D_I = ID-Controlled Pipe Inside Diameter, in.

Chapter 6
Design of Polyethylene Piping Systems

TABLE 1-1
Hydrostatic Design Basis Ratings and Service Temperatures

Property	ASTM Standard	PE 3408	PE 2406
HDB at 73°F (23°C)	D 2837	1600 psi (11.04 MPa)	1250 psi (8.62 MPa)
Maximum recommended temperature for Pressure Service	–	140°F (60°C)*	140°F (60°C)
Maximum Recommended Temperature for Non-Pressure Service	–	180°F (82°C)	180°F (82°C)

* Some polyethylene piping materials are stress rated at temperatures as high as 180° F.
For more information regarding these materials and their use, the reader is referred to PPI, TR-4

The long-term strength of thermoplastic pipe is based on regression analysis of stress-rupture data obtained in accordance with ASTM D2837. Analysis of the data obtained in this procedure is utilized to establish a stress intercept for the material under evaluation at 100,000 hours. This intercept when obtained at 73°F is called the long-term hydrostatic strength or LTHS. The LTHS typically falls within one of several categorized ranges that are detailed in ASTM D2837. This categorization of the LTHS for a given pipe material establishes its hydrostatic design basis or HDB. The HDB is then utilized in either equation 1-3 or 1-4 to establish the pressure rating for a particular pipe profile by the application of a design factor (DF). The DF for water service is 0.50, as indicated in Table 1-2. Additional information regarding the determination of the LTHS and the D2837 protocol is presented in the Engineering Properties chapter of this handbook.

TABLE 1-2
PE Pipe Design Factors (DF)

Pipe Environment	Design Factor (DF) at 73°F (23°C)
Water; Aqueous solutions of salts, acids and bases; Sewage; Wastewater; Alcohols; Glycols (anti-freeze solutions);	0.50
Nitrogen; Carbon dioxide; Methane; Hydrogen sulfide; non-federally regulated applications involving dry natural gas other non-reactive gases	0.50
LPG vapors (propane; propylene; butane) †	0.40
Natural Gas Distribution (Federally regulated under CFR Tile 49, Part 192)*	0.32 **
Fluids such as solvating/permeating chemicals in pipe or soil (typically hydrocarbons) in 2% or greater concentrations, natural or other fuel-gas liquid condensates, crude oil, fuel oil, gasoline, diesel, kerosene, hydrocarbon fuels	0.25

* An overall design factor of 0.32 is mandated by the US Code of Federal Regulations, Title 49, Part 192. In addition, DOT limits maximum service pressure of PE pipe made after July 2004 to 125 psi.
** Design factors (Service Fluid Factors) in Canada are governed by CSA Z622.99. There are small differences between the values required by the US Code of Federal Regulations, Title 49, Part 192 and CSA Z622.99
†See paragraphs which follow regarding limitations on use of PE pipe in LPG service.

As indicated in Table 1-2, polyethylene pipe which meets the requirements of ASTM D2513 may be used for the transport of liquefied petroleum gas (LPG). NFPA 58 recommends a maximum operating pressure of 30 psig for LPG gas applications involving polyethylene pipe. This design limit is established in recognition of the higher condensation temperature for LPG as compared to that of natural gas and, thus, the maximum operating pressure is recommended to ensure that plastic pipe is not subjected to excessive exposure to LPG condensates. For further information the reader is referred to PPI's TR-22, Polyethylene Piping Distribution Systems for Components of Liquid Petroleum Gases.[14]

TABLE 1-3
Service Temperature Design Factors, F_T

Maximum Continuously Applied Service Temp., °F(°C)	Temperature Compensation Factor, FT, for PE3408
≤ 80 (26)	1.00
≤ 90 (32)	0.90
≤ 100 (38)	0.78
≤ 110 (43)	0.75
≤ 120 (49)	0.63
≤ 130 (54)	0.60
≤ 140 (60)	0.50

Fluid Flow in Polyethylene Piping

Head Loss in Pipes – Darcy-Weisbach/Fanning/Colebrook/Moody

Viscous shear stresses within the liquid and friction along the pipe walls create resistance to flow within a pipe. This resistance within a pipe results in a pressure drop, or loss of head in the piping system.

The Darcy-Weisbach or Fanning formula, Equation 1-7, and the Colebrook formula, Equation 1-10, are generally accepted methods for calculating friction losses due to liquids flowing in full pipes.[15,16] These formulas recognize dependence on pipe bore and pipe surface characteristics, liquid viscosity and flow velocity.

The Darcy-Weisbach formula is:

(1-7)
$$h_f = f \frac{L V^2}{d' 2g}$$

WHERE
h_f = friction (head) loss, ft. of liquid
L = pipeline length, ft.
d' = pipe inside diameter, ft.
V = flow velocity, ft/sec.

(1-8)
$$V = \frac{0.4085 \, Q}{D_1^2}$$

g = constant of gravitational acceleration (32.2ft/sec²)
Q = flow rate, gpm
D_I = pipe inside diameter, in
f = friction factor (dimensionless, but dependent upon pipe surface roughness and Reynolds number)

Liquid flow in pipes will assume one of three flow regimes. The flow regime may be laminar, turbulent or in transition between laminar and turbulent. In laminar flow (Reynolds number, Re, below 2000), the pipe's surface roughness has no effect and is considered negligible. As such, the friction factor, f, is calculated using Equation 1-9.

(1-9)
$$f = \frac{64}{Re}$$

For turbulent flow (Reynolds number, Re, above 4000), the friction factor, f, is dependent on two factors, the Reynolds number and pipe surface roughness. The friction factor may be determined from Figure 1-1, the Moody Diagram, which can be used for various pipe materials and sizes.[17] In the Moody Diagram, relative roughness, ε/d' (see Table 1-4 for ε) is used. The friction factor may then be determined using the Colebrook formula. The friction factor can also be read from the Moody diagram with enough accuracy for calculation.

The Colebrook formula is:

(1-10)
$$\frac{1}{\sqrt{f}} = -2 \log_{10}\left\{\frac{\varepsilon}{3.7d'} + \frac{2.51}{Re\sqrt{f}}\right\}$$

For Formulas 1-9 and 1-10, terms are as previously defined, and:
ε = absolute roughness, ft.
Re = Reynolds number, dimensionless

(1-11)
$$Re = \frac{Vd'}{v} = \frac{Vd'\rho}{\mu g}$$

(1-12)
$$\text{Re} = \frac{3126Q}{D_I k}$$

v = kinematic viscosity, ft2/sec

(1-13)
$$v = \frac{\Gamma g}{\rho}$$

ρ = fluid density, lb/ft3
Γ = dynamic viscosity, lb-sec/ft2
k = kinematic viscosity, centistokes

(1-14)
$$k = \frac{z}{s}$$

z = dynamic viscosity, centipoises
s = liquid density, gm/cm3

When the friction loss through one size pipe is known, the friction loss through another pipe of different diameter may be found by:

(1-15)
$$h_{f1} = h_{f2}\left(\frac{d'_1}{d'_2}\right)^5$$

The subscripts 1 and 2 refer to the known and unknown pipes. Both pipes must have the same surface roughness, and the fluid must be the same viscosity and have the same flow rate.

TABLE 1-4
Surface Roughness for Various New Pipes

Type of Pipe	'S' Absolute Roughness of Surface, ft		
	Values for New Pipe Reported by Reference [18]	Values for New Pipe and Recommended Design Values Reported by Reference [19]	
		Mean Value	Recommended Design Value
Riveted steel	0.03 - 0.003	–	–
Concrete	0.01 - 0.001	–	–
Wood stave	0.0003 - 0.0006	–	–
Cast Iron – Uncoated	0.00085	0.00074	0.00083
Cast Iron – Coated	–	0.00033	0.00042
Galvanized Iron	0.00050	0.00033	0.00042
Cast Iron – Asphalt Dipped	0.0004	–	–
Commercial Steel or Wrought Iron	0.00015	–	–
Drawn Tubing	0.000005 corresponds to "smooth pipe"	–	–
Uncoated Stee	–	0.00009	0.00013
Coated Steel	–	0.00018	0.00018
Uncoated Asbestos – Cement	–		
Cement Mortar Relined Pipes (Tate Process)	–	0.00167	0.00167
Smooth Pipes (PE and other thermoplastics, Brass, Glass and Lead)	–	"smooth pipe" (≤ 0.00001) (See Note)	"smooth pipe" (≤ 0.00001) (See Note)

Note: Pipes that have absolute roughness equal to or less than 0.00001inch are considered to exhibit "smooth pipe" characteristics.

Pipe Deflection Effects

Pipe flow formulas generally assume round pipe. Because of its flexibility, buried PE pipe may deform slightly under earth and other loads to assume somewhat of an elliptical shape having a slightly increased lateral diameter and a correspondingly reduced vertical diameter. Elliptical deformation slightly reduces the pipe's flow area. Practically speaking, this phenomenon can be considered negligible as it relates to pipe flow capacity. Calculations reveal that a deformation of about 7% in polyethylene pipe results in a flow reduction of approximately 1%.

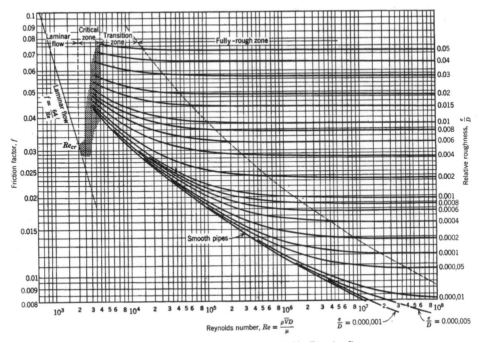

Note for the Moody Diagram: D = pipe inside diameter, ft

Figure 1-1 The Moody Diagram

Head Loss in Fittings

Fluids flowing through a fitting or valve will experience a friction loss that can be directly expressed using a resistance coefficient, K', for the particular fitting.[20] As shown in the discussion that follows, head loss through a fitting can be conveniently added into system flow calculations as an equivalent length of straight pipe having the same diameter as system piping. Table 1-5 presents K' factors for various fittings.

Where a pipeline contains a large number of fittings in close proximity to each other, this simplified method of predicting flow loss may not be adequate due to the cumulative systems effect. Where this is a design consideration, the designer should consider an additional frictional loss allowance, or a more thorough treatment of the fluid mechanics.

The equivalent length of pipe to be used to estimate the friction loss due to fittings may be obtained by Eq. 1-18 where LEFF = Effective Pipeline length, ft; D is pipe bore diameter in ft.; and K' is obtained from Table 1-5.

(1-16) $LEFF = K'D$

TABLE 1-5
Representative Fittings Factor, K', to Determine Equivalent Length of Pipe

Piping Component	K'
90° Molded Elbow	40
45° Molded Elbow	21
15° Molded Elbow	6
90° Fabricated Elbow	32
75° Fabricated Elbow	27
60° Fabricated Elbow	21
45° Fabricated Elbow	16
30° Fabricated Elbow	11
15° Fabricated Elbow	5
45° Fabricated Wye	60
Equal Outlet Tee, Run/Branch	60
Equal Outlet Tee, Run/Run	20
Globe Valve, Conventional, Fully Open	340
Angle Valve, Conventional, Fully Open	145
Butterfly Valve, ≥ 8-in, Fully Open	40
Check Valve, Conventional Swing	135

Head Loss Due to Elevation Change

Line pressure may be lost or gained from a change in elevation. For liquids, the pressure for a given elevation change is given by:

(1-17) $$h_E = h_2 - h_1$$

WHERE
h_E = Elevation head, ft of liquid
h_1 = Pipeline elevation at point 1, ft
h_2 = Pipeline elevation at point 2, ft

If a pipeline is subject to a uniform elevation rise or fall along its length, the two points would be the elevations at each end of the line. However, some pipelines may have several elevation changes as they traverse rolling or mountainous terrain. These pipelines may be evaluated by choosing appropriate points where the pipeline slope changes, then summing the individual elevation heads for an overall pipeline elevation head.

In a pipeline conveying liquids and running full, pressure in the pipe due to elevation exists whether or not liquid is flowing. At any low point in the line, internal pressure will be equal to the height of the liquid above the point multiplied by the specific weight of the liquid. If liquid is flowing in the line, elevation head and head loss due to liquid flow in the pipe are added to determine the pressure in the pipe at a given point in the pipeline.

Pressure Flow of Water – Hazen-Williams

The Darcy-Weisbach method of flow resistance calculation may be applied to liquid and gases, but its solution can be complex. For many applications, empirical formulas are available and, when used within their limitations, reliable results are obtained with greater convenience. For example, Hazen and Williams developed an empirical formula for the flow of water in pipes at 60° F.

The Hazen-Williams formula for water at 60° F (16°C) can be applied to water and other liquids having the same kinematic viscosity of 1.130 centistokes (0.00001211 ft²/sec), or 31.5 SSU. The viscosity of water varies with temperature, so some error can occur at temperatures other than 60°F (16°C).

Hazen-Williams formula for friction (head) loss in feet:

(1-18)
$$h_f = \frac{0.002083\, L}{D_I^{4.8655}} \left(\frac{100\,Q}{C}\right)^{1.85}$$

Hazen-Williams formula for friction (head) loss in psi:

(1-19)
$$p_f = \frac{0.00090150\,L}{D_I^{4.8655}} \left(\frac{100\,Q}{C}\right)^{1.85}$$

Terms are as previously defined, and:

D_I = pipe inside diameter, in

C = Hazen-Williams Friction Factor, dimensionless c = 150_155 for PE , (not related to Darcy-Weisbach friction factor, f)

Q = flow rate, gpm

Other forms of these equations are prevalent throughout the literature.[21] The reader is referred to the references at the end of this chapter.

TABLE 1-6
Properties of Water

Temperature, °F/°C	Specific Weight, lb/ft³	Kinematic Viscosity, Centistokes
32 / 0	62.41	1.79
60 / 15.6	62.37	1.13
75 / 23.9	62.27	0.90
100 / 37.8	62.00	0.69
120 / 48.9	61.71	0.57
140 / 60	61.38	0.47

Water flow through pipes of different materials and diameters may be compared using the following formula.

(1-20)
$$\% \ flow = 100 \frac{D_{I2}}{D_{I1}} \left(\frac{C_2}{C_1} \right)^{0.3806}$$

Where the subscripts 1 and 2 refer to the designated properties for two separate pipe profiles, in this case, the pipe inside diameter (D_1 in inches) of the one pipe (1) versus that of the second pipe (2) and the Hazen-Williams factor for each respective profile.

Pipe Flow Design Example

A polyethylene pipeline conveying water at 60°F is 15,000 feet long and is laid on a uniform grade that rises 150 feet. What is the friction head loss in 4" IPS DR 17 PE 3408 pipe for a 50 gpm flow? What is the elevation head? What is the internal pressure at the bottom of the pipe when water is flowing uphill? When flowing downhill? When full but not flowing?

Using equation 1-21 and C = 150

$$P_f = \frac{0.0009015(15000)}{3.938^{4.8655}} \left(\frac{100(50)}{150} \right)^{1.85} = 11.3 \ psi$$

To determine the elevation head, assume point 1 is at the bottom of the elevation, and point 2 is at the top. Using Equation 1-17,

$$h_E = 150 - 0 = 150 \ ft \ of \ water$$

The specific weight of water at 60°F is 62.37 lb/ft³, which is a pressure of 62.37 lb over a 1 ft square area, or a pressure of 62.37/144 = 0.43 lb/in². Therefore,

$$h_E = (150 - 0)0.43 = 64.5 \ psi$$

When water is flowing, elevation head and the friction head are added. The maximum friction head acts at the source point, and the maximum elevation head at the lowest point. Therefore, when flowing uphill, the pressure, P, at the bottom is elevation head plus the friction head because the flow is from the bottom to the top.

$$P = h_E + p_f = 64.5 + 11.3 = 75.8 \ psi$$

When flowing downhill, water flows from the top to the bottom. Friction head applies from the source point at the top, so the pressure developed from the downhill flow is applied in the opposite direction as the elevation head. Therefore,

$$P = h_E - p_f = 64.5 - 11.3 = 53.2 \ psi$$

When the pipe is full, but water is not flowing, no friction head develops.

$$P = h_E + p_f = 64.5 + 0 = 64.5 \; psi$$

Surge Considerations

A piping system must be designed for continuous operating pressure and for transient (surge) pressures imposed by the particular application. Surge allowance and temperature effects vary from pipe material to pipe material, and erroneous conclusions may be drawn when comparing the Pressure Class (PC) of different pipe materials.

The ability to handle temporary pressure surges is a major advantage of polyethylene. Due to the viscoelastic nature of polyethylene, a piping system can safely withstand momentarily applied maximum pressures that are significantly above the pipe's PC. The strain from an occasional, limited load of short duration is met with an elastic response, which is relieved upon the removal of the load. This temporary elastic strain causes no damage to the polyethylene material and has no adverse effect on the pipe's long-term strength.[22,23,24]

In order to determine the appropriate DR required for a pressure polyethylene pipe system, the designer must calculate both the continuous working pressure, potential pressure surges and the Working Pressure Rating (WPR) for the pipe.

Surge Pressure

Transient pressure increases (water hammer) are the result of sudden changes in velocity of the flowing fluid. For design purposes, the designer should consider two types of surges:

1. **Recurring Surge Pressure (P_{RS}).** Recurring surge pressures occur frequently and are inherent to the design and operation of the system. Recurring surge pressures would include normal pump start up or shutdown, normal valve opening and closing, and/or "background" pressure fluctuation associated with normal pipeline operation.

2. **Occasional Surge Pressure (P_{OS}).** Occasional surge pressures are caused by emergency operations. Occasional surge pressures are usually the result of a malfunction, such as power failure or system component failure, which includes pump seize-up, valve stem failure and pressure relief valve failure.

To determine the WPR for a selected DR, the pressure surge must be calculated. The following equations may be used to estimate the pressure surge created in pressure water piping systems.

An abrupt change in the velocity of a flowing liquid generates a pressure wave. The velocity of the wave may be determined using Equation 1-21.

(1-21)
$$a = \frac{4660}{\sqrt{1 + \frac{K_{BULK}}{E_d}(DR - 2)}}$$

WHERE
a = Wave velocity (celerity), ft/sec
K_{BULK} = Bulk modulus of fluid at working temperature (typically 300,000 psi for water at 73°F)
E_d = Dynamic instantaneous effective modulus of pipe material (typically 150,000 psi for PE pipe)
DR = Pipe dimension ratio

The resultant transient surge pressure, Ps, may be calculated from the wave velocity, a, and the change in fluid velocity, Δv.

(1-22)
$$P_s = a\left(\frac{\Delta V}{2.31g}\right)$$

WHERE
P_s = Transient surge pressure, psig
a = Wave velocity (celerity), ft/sec
ΔV = Sudden velocity change, ft/sec
g = Constant of gravitational acceleration, 32.2 ft/sec²

Figure 1-2 represents the pressure surge curves for PE3408 as calculated using Equations 1-21 and 1-22 for standard Dimension Ratios (DR's).

Chapter 6
Design of Polyethylene Piping Systems

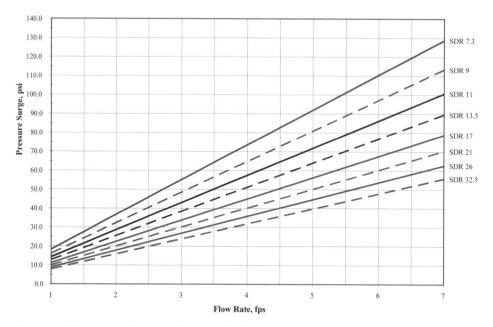

* A value of 150,000 psi and 300,000 psi were used for E_d and K, respectively.
** Calculated surge pressure values applicable to water at temperatures not exceeding 80°F (27°C).

Figure 1-2 Sudden Velocity Change vs. Pressure Surge for PE3408

For Δv, the velocity change must be abrupt, such as a rapid valve operation or a pump startup or shutdown. The critical time for sudden velocity change is calculated using Equation 1-23.

(1-23)
$$T_c = \frac{2 \cdot L}{a}$$

WHERE
T_C = Critical time, seconds
L = Pipeline length, ft
a = Wave velocity (celerity), ft/sec

Pressure Class

The Pressure Class (PC) is used to define the pressure capacity under a pre-defined set of operating conditions. For polyethylene, the PC denotes the maximum allowable working pressure for water with a predefined allowance for pressure surges and a maximum operating temperature of 80°F(27°C).

The predefined allowances for pressure surges as given in Table 1-7 are determined as follows:

1. For recurrent surges, the allowance is 50% of the PC.
2. For occasional surges, the allowance is 100% of the PC.

This pressure allowance for surge is applied exclusively to pressure that occurs during a surge event and never to the sustained operating pressure.

Table 1-7 shows the PC ratings (Eq. 1-3), surge allowance (Eq. 1-22) and corresponding allowable sudden velocity change in flow velocity for standard Dimension Ratios (DR's).

TABLE 1-7
Pressure Class (PC) Ratings*, Surge Allowance and Corresponding Sudden Velocity Change for PE3408 Pipe

DR	PC*, psig	Recurring Surge, P_{RS}		Occasional Surge, P_{OS}	
		Surge Allowance, P_{RS}, psig	Sudden Velocity Change, ΔV, fps	Surge Allowance, P_{OS}, psig	Sudden Velocity Change, ΔV, fps
7.3	254	127	6.9	254	13.8
9	200	100	6.2	200	12.4
9.3	193	96	6.1	193	12.2
11	160	80	5.6	160	11.1
13.5	128	64	5.0	128	10.0
15.5	110	55	4.7	110	9.3
17	100	50	4.4	100	8.9
21	80	40	4.0	80	8.0
26	64	32	3.6	64	7.2
32.5	51	25	3.2	51	6.4

Pressure Class ratings are for water not exceeding 80°F (27°C). Pressure Class ratings can vary for other fluids (Eq. 1-3 & Table 1-2) and service temperatures (Table 1-8).

When polyethylene pipe is operated at temperatures above 80°F (27°C), a temperature compensation factor must be applied to the PC. Temperature Compensation Factors, FT, are given in Table 1-8 for operating temperatures up to 140°F (60°C).

Working Pressure Rating

A pipeline containing flowing liquid is periodically subjected to two modes of hydrostatic stress: sustained stress from Working Pressure and transient stress from sudden water velocity changes. The pipe must be designed to handle both. This is verified by calculating the Working Pressure Rating (WPR).

The PC rating is the Working Pressure Rating (WPR) if the following conditions are valid:

1. The water temperature does not exceed 80°F (27°C), and
2. The expected sudden velocity changes do not exceed the values in Table 1-7. If the temperature exceeds 80°F then

(1-24) $\quad WPR = PC \times F_T$

In the event that these conditions are not valid, then the following conditions must be evaluated, with the WPR of the pipe being the lesser value of the PC or PCxF$_T$ and the WPR as computed by either equation 1-25 or 1-26. One and a half times the pipe's nominal PC, less the maximum pressure resulting from recurring surge pressures.

(1-25) $\quad WPR = (1.5 \times PC \times F_T) - P_{RS}$

Note: $P_{RS} = P_S$ as defined in Eq. 1-22.

and, two times the pipe's nominal PC, less the maximum pressure resulting from occasional surge pressures.

(1-26) $\quad WPR = (2 \times PC \times F_T) - P_{OS}$

Note: $P_{OS} = P_S$ as defined in Eq. 1-22.

The lowest calculated WPR from equations 1-24 and 1-25 or 1-24 and 1-26 must be compared to the working pressure.

(1-27) $\quad WPR \geq WP$

If a pipe is operating at a working pressure below the pipe's nominal PC, its surge pressure capacity is generally greater in accordance with the following equations:

1. Recurring Surge Pressure (P_{RS})

(1-28) $\quad WP + P_{RS} \leq 1.5 PC \times F_T$

2. Occasional Surge Pressure (P_{OS})

(1-29) $\quad WP + P_{OS} \leq 2.0 PC \times F_T$

In all cases, the following condition must be satisfied.

(1-30) $\quad WP \leq PC \times F_T$

Water Pressure Pipe Design Example

A water utility is considering PC80 (DR21 per Table 1-7) PE3408 pipe for buried water main service for which the sustained working pressure will be 75psig, and the maximum water service temperature is lower than 80°F (27°C). However, the designer expects the pipe to endure operating conditions in which the recurring surge pressures result from a sudden velocity change of up to 5 fps. Determine if DR21 is acceptable for this application.

Referring to Figure 1-2 for DR21 and a sudden velocity change of 5 fps, the Pressure Surge is 50 psig. Alternatively, Eq. 1-22 may be used to calculate the Pressure Surge.

Reviewing the criteria in Table 1-7, WPR does not equal PC since the sudden velocity change for Recurring Surge exceeds the maximum of 4 fps for DR21. Since Eq. 1-24 is not valid, determine the WPR using Eq. 1-25 (recurring surge.)

$$WPR = (1.5 \times 80 \times 1.0) - 50 = 70 \text{psig}$$

Since the WPR is less than the WP of 75 psig for the system, DR21 is not acceptable for these anticipated conditions. DR21 would only be acceptable for this application if the designer can operate the water main at a working pressure of 70 psig or less, or if the the occurrence of the surges was only occasional, in which case Table 1-7 gives a maximum flow velocity of 8 fps for DR21 for occasional surge. Otherwise, DR17 should be evaluated using the same three-step procedure as outlined above.

Referring to Figure 1-2 for DR17 and a sudden velocity change of 5fps, the Pressure Surge is 56 psig. Alternatively, Eq. 1-22 may be used to calculate the Pressure Surge.

Reviewing the criteria in Table 1-7, WPR is less than PC since the sudden velocity change for Recurring Surge exceeds the maximum of 4.4 fps for DR17. Determine the WPR using Eq. 1-25.

From Table 1-7, the PC for DR17 is 100 psig.

$$WPR = (1.5 \times 100 \times 1.0) - 56 = 94 \text{psig}$$

The WPR is the lesser of the WPR calculated by Eq. 1-25 and the PC, or 94 psig (94 psig ≥ 75 psig). Although for DR 17 the surge capacity was exceeded, the sustained working pressure capacity was underutilized (See Eq. 1-28). Therefore, DR17 would be considered acceptable for this application.

Figure 1-3 is a graphical representation of the WPR for DR21 and DR17 HDPE for recurring surge pressures. By referring to Figure 1-2, the Pressure Surge may be determined for each of the velocity changes from 1 - 7 ft/sec (the Pressure Surge may also be determined using Eq. 1-22). Using these values in Eq. 1-25, the WPR can be determined.

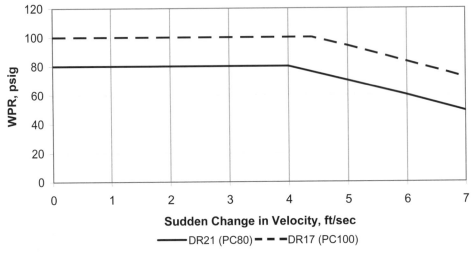

Figure 1-3 HDPE WPR vs. Sudden Change in Velocity for Recurring Surge Pressures for DR21 (PC80) and DR17 (PC100) @ 80°F

Controlling Surge Effects

Reducing the rate at which a change in flow velocity occurs can help control water hammer effects. That is to say that transient surge pressure may be minimized or avoided altogether when the velocity change is controlled in such a way that it occurs over a time that exceeds the critical time, t_c, as presented in Equation 1-25.

In hilly regions, a liquid flow may separate at high points, and cause surge pressures when the flow rejoins. Reducing the downhill, downstream pipeline bore may help keep the pipeline full by reducing the flow rate. Flow separation is more likely to occur with oversize pipelines. Vacuum breakers, air relief valves and flow control valves can also be effective.

Pressure Flow of Liquid Slurries

Liquid slurry piping systems transport solid particles entrained in a liquid carrier. Water is typically used as a liquid carrier, and solid particles are commonly granular materials such as sand, fly-ash or coal. Key design considerations involve the solid material, particle size and the carrier liquid.

Turbulent flow is preferred to ensure that particles are suspended in the liquid. Turbulent flow also reduces pipeline wear because particles suspended in the carrier liquid will bounce off the pipe inside surface. Polyethylene pipe has elastic properties that combine with high molecular weight toughness to provide service life that can significantly exceed many metal piping materials. Flow velocity that is too low to maintain fully turbulent flow for a given particle size can allow solids to drift to the bottom of the pipe and slide along the surface. However, compared to metals, polyethylene is a softer material. Under sliding bed and direct impingement conditions, polyethylene may wear appreciably. Polyethylene directional fittings are generally unsuitable for slurry applications because the change of flow direction in the fitting results in direct impingement. Directional fittings in liquid slurry applications should employ hard materials that are resistant to wear from direct impingement.

Particle Size

As a general recommendation, particle size should not exceed about 0.2 in (5 mm), but larger particles are occasionally acceptable if they are a small percentage of the solids in the slurry. With larger particle slurries such as fine sand and coarser particles, the viscosity of the slurry mixture will be approximately that of the carrying liquid. However, if particle size is very small, about 15 microns or less, the slurry viscosity will increase above that of the carrying liquid alone. The rheology of fine particle slurries should be analyzed for viscosity and specific gravity before determining flow friction losses. Inaccurate assumptions of a fluid's rheological properties can lead to significant errors in flow resistance analysis. Examples of fine particle slurries are water slurries of fine silt, clay and kaolin clay.

Slurries frequently do not have uniform particle size, and some particle size non-uniformity can aid in transporting larger particles. In slurries having a large proportion of smaller particles, the fine particle mixture acts as a more viscous carrying fluid that helps suspend larger particles. Flow analysis of non-uniform particle size slurries should include a rheological characterization of the fine particle mixture.

Solids Concentration and Specific Gravity

Equations 1-31 through 1-34 are useful in determining solids concentrations and mixture specific gravity.

(1-31)
$$C_V = \frac{S_M - S_L}{S_S - S_L}$$

(1-32) $$C_W = \frac{C_V S_S}{S_M}$$

(1-33) $$S_M = C_V (S_S - S_L) + S_L$$

(1-34) $$S_M = \frac{S_L}{1 - \frac{C_W (S_S - S_L)}{S_S}}$$

WHERE
S_L = carrier liquid specific gravity
S_S = solids specific gravity
S_M = slurry mixture specific gravity
C_V = percent solids concentration by volume
C_W = percent solids concentration by weight

Critical Velocity

As pointed out above, turbulent flow is preferred to maintain particle suspension. A turbulent flow regime avoids the formation of a sliding bed of solids, excessive pipeline wear and possible clogging. Reynolds numbers above 4000 will generally insure turbulent flow.

Maintaining the flow velocity of a slurry at about 30% above the critical settlement velocity is a good practice. This insures that the particles will remain in suspension thereby avoiding the potential for excessive pipeline wear. For horizontal pipes, critical velocity may be estimated using Equation 1-35.

Individual experience with this equation varies. Other relationships are offered in the literature. See Thompson and Aude (26). A test section may be installed to verify applicability of this equation for specific projects.

(1-35) $$V_C = F_L \sqrt{2gd' (S_S - 1)}$$

Where terms are previously defined and
V_C = critical settlement velocity, ft/sec
F_L = velocity coefficient (Tables 1-11 and 1-12)
d' = pipe inside diameter, ft

An approximate minimum velocity for fine particle slurries (below 50 microns, 0.05 mm) is 4 to 7 ft/sec, provided turbulent flow is maintained. A guideline minimum velocity for larger particle slurries (over 150 microns, 0.15 mm) is provided by Equation 1-36.

(1- 36) $$V_{min} = 14\sqrt{d'}$$

WHERE
V_{min} = approximate minimum velocity, ft/sec

Critical settlement velocity and minimum velocity for turbulent flow increases with increasing pipe bore. The relationship in Equation 1-37 is derived from the Darcy-Weisbach equation.

(1- 37) $$V_2 = \frac{\sqrt{d'_2}}{\sqrt{d'_1}} V_1$$

The subscripts 1 and 2 are for the two pipe diameters.

TABLE 1-8
Scale of Particle Sizes

Tyler Screen Mesh	U.S. Standard Mesh	Inches	Microns	Class
–	–	1.3 – 2.5	33,000 – 63,500	Very coarse gravel
–	–	0.6 – 1.3	15,200 – 32,000	Coarse gravel
2.5	–	0.321	8,000	Medium gravel
5	5	0.157	4,000	Fine gravel
9	10	0.079	2,000	Very fine gravel
16	18	0.039	1,000	Very coarse sand
32	35	0.0197	500	Coarse sand
60	60	0.0098	250	Medium sand
115	120	0.0049	125	Fine sand
250	230	0.0024	62	Very fine sand
400	–	0.0015	37	Coarse silt
–	–	0.0006 – 0.0012	16 – 31	Medium silt
–	–	–	8 – 13	Fine silt
–	–	–	4 – 8	Very fine silt
–	–	–	2 – 4	Coarse clay
–	–	–	1 – 2	Medium clay
–	–	–	0.5 - 1	Fine clay

TABLE 1-9
Typical Specific Gravity and Slurry Solids Concentration

Material	Specific Gravity	Typical Solids Concentration	
		% by Weight	% by Volume
Gilsonite	1.05	40–45	39–44
Coal	1.40	45–55	37–47
Sand	2.65	43–43	23–30
Limestone	2.70	60–65	36–41
Copper Concentrate	4.30	60–65	26–30
Iron Ore	4.90	–	–
Iron Sands	1.90	–	–
Magnetite	4.90	60 - 65	23 - 27

TABLE 1-10
Water-Base Slurry Specific Gravities

C_W	Solid Specific Gravity, S_S								
	1.4	1.8	2.2	2.6	3.0	3.4	3.8	4.2	4.6	5.0
5	1.01	1.02	1.03	1.03	1.03	1.04	1.04	1.04	1.04	1.04
10	1.03	1.05	1.06	1.07	1.07	1.08	1.08	1.08	1.08	1.09
15	1.04	1.07	1.09	1.10	1.11	1.12	1.12	1.13	1.13	1.14
20	1.05	1.10	1.12	1.14	1.15	1.16	1.17	1.18	1.19	1.19
25	1.08	1.13	1.16	1.18	1.20	1.21	1.23	1.24	1.24	1.25
30	1.09	1.15	1.20	1.23	1.25	1.27	1.28	1.30	1.31	1.32
35	1.11	1.18	1.24	1.27	1.30	1.33	1.35	1.36	1.38	1.39
40	1.13	1.22	1.28	1.33	1.36	1.39	1.42	1.44	1.46	1.47
45	1.15	1.25	1.33	1.38	1.43	1.47	1.50	1.52	1.54	1.56
50	1.17	1.29	1.38	1.44	1.50	1.55	1.58	1.62	1.64	1.67
55	1.19	1.32	1.43	1.51	1.58	1.63	1.69	1.72	1.76	1.79
60	1.21	1.36	1.49	1.59	1.67	1.73	1.79	1.84	1.89	1.92
65	1.23	1.41	1.55	1.67	1.76	1.85	1.92	1.98	2.04	2.08
70	1.25	1.45	1.62	1.76	1.88	1.98	2.07	2.14	2.21	2.27

TABLE 1-11
Velocity Coefficient, F_L (Uniform Particle Size)

Particle Size, mm	Velocity Coefficient, F_L			
	$C_V = 2\%$	$C_V = 5\%$	$C_V = 10\%$	$C_V = 15\%$
0.1	.76	0.92	0.94	0.96
0.2	0.94	1.08	1.20	1.28
0.4	1.08	1.26	1.41	1.46
0.6	1.15	1.35	1.46	1.50
0.8	1.21	1.39	1.45	1.48
1.0	1.24	1.04	1.42	1.44
1.2	1.27	1.38	1.40	1.40
1.4	1.29	1.36	1.67	1.37
1.6	1.30	1.35	1.35	1.35
1.8	1.32	1.34	1.34	1.34
2.0	1.33	1.34	1.34	1.34
2.2	1.34	1.34	1.34	1.34
2.4	1.34	1.34	1.34	1.34
2.6	1.35	1.35	1.35	1.35
2.8	1.36	1.36	1.36	1.36
≥ 3.0	1.36	1.36	1.36	1.36

TABLE 1-12
Velocity Coefficient, F_L (50% Passing Particle Size)

Particle Size, mm	Velocity Coefficient, F_L			
	$C_V = 5\%$	$C_V = 10\%$	$C_V = 20\%$	$C_V = 30\%$
0.01	0.48	0.48	0.48	0.48
0.02	0.58	0.59	0.60	0.61
0.04	0.70	0.72	0.74	0.76
0.06	0.77	0.79	0.81	0.83
0.08	0.83	0.86	0.86	0.91
0.10	0.85	0.88	0.92	0.95
0.20	0.97	1.00	1.05	1.08
0.40	1.09	1.13	1.18	1.23
0.60	1.15	1.21	1.26	1.30
0.80	1.21	1.25	1.31	1.33
1.0	1.24	1.29	1.33	1.35
2.0	1.33	1.36	1.38	1.40
3.0	1.36	1.38	1.39	1.40

Equation 1-7, Darcy-Weisbach, and Equations 1-18 and 1-19, Hazen-Williams, may be used to determine friction head loss for pressure slurry flows provided the viscosity limitations of the equations are taken into account. Elevation head loss is increased by the specific gravity of the slurry mixture.

(1-38) $h_E = S_M (h_2 - h_1)$

Compressible Gas Flow

Flow equations for smooth pipe may be used to estimate compressible gas flow through polyethylene pipe.

Empirical Equations for High Pressure Gas Flow

Equations 1-39 through 1-42 are empirical equations used in industry for pressure greater than 1 psig.[26] Calculated results may vary due to the assumptions inherent in the derivation of the equation.

Mueller Equation

(1-39)
$$Q_h = \frac{2826 D_I^{2.725}}{S_g^{0.425}} \left(\frac{p_1^2 - p_2^2}{L} \right)^{0.575}$$

Weymouth Equation

(1-40)
$$Q_h = \frac{2034 D_I^{2.667}}{S_g^{0.5}} \left(\frac{p_1^2 - p_2^2}{L} \right)^{0.5}$$

IGT Distribution Equation

(1-41)
$$Q_h = \frac{2679 D_I^{2.667}}{S_g^{0.444}} \left(\frac{p_1^2 - p_2^2}{L} \right)^{0.555}$$

Spitzglass Equation

(1-42)
$$Q_h = \frac{3410}{S_g^{0.5}} \left(\frac{p_1^2 - p_2^2}{L} \right)^{0.5} \left(\frac{D_I^5}{1 + \frac{3.6}{D_I} + 0.03 D_I} \right)^{0.5}$$

WHERE
Q_h = flow, standard ft³/hour
S_g = gas specific gravity
p_1 = inlet pressure, lb/in² absolute
p_2 = outlet pressure, lb/in² absolute
L = length, ft
D_I = pipe inside diameter, in

Empirical Equations for Low Pressure Gas Flow

For applications where internal pressures are less than 1 psig, such as landfill gas gathering or wastewater odor control, Equations 1-43 or 1-44 may be used.

Mueller Equation

(1-43)
$$Q_h = \frac{2971 D_I^{2.725}}{S_g^{0.425}} \left(\frac{h_1 - h_2}{L}\right)^{0.575}$$

Spitzglass Equation

(1-44)
$$Q_h = \frac{3350}{S_g^{0.5}} \left(\frac{h_1 - h_2}{L}\right)^{0.5} \left(\frac{D_I^5}{1 + \frac{3.6}{D_I} + 0.03 D_I}\right)^{0.5}$$

Where terms are previously defined, and
h_1 = inlet pressure, in H_2O
h_2 = outlet pressure, in H_2O

Gas Permeation

Long distance pipelines carrying compressed gasses may deliver slightly less gas due to gas permeation through the pipe wall. Permeation losses are small, but it may be necessary to distinguish between permeation losses and possible leakage. Equation 1-45 may be used to determine the volume of a gas that will permeate through polyethylene pipe of a given wall thickness:

(1-45)
$$q_P = \frac{K_P A_s \Theta P_A}{t'}$$

WHERE
q_P = volume of gas permeated, cm³ (gas at standard temperature and pressure)
K_P = permeability constant (Table 1-13)
A_s = surface area of the outside wall of the pipe, 100 in²
P_A = pipe internal pressure, atmospheres (1 atmosphere = 14.7 lb/in²)
Θ = elapsed time, days
t' = wall thickness, mils

TABLE 1-13
Permeability Constants (27)

Gas	K_P
Methane	85
Carbon Monoxide	80
Hydrogen	425

TABLE 1-14
Physical Properties of Gases (Approx. Values at 14.7 psi & 68°F)

Gas	Chemical Formula	Molecular Weight	Weight Density, lb/ft^3, σ	Specific Gravity, S_g
Acetylene (ethylene)	C_2H_2	26.0	0.0682	0.907
Air	–	29.0	0.0752	1.000
Ammonia	NH_3	17.0	0.0448	0.596
Argon	A	39.9	0.1037	1.379
Butane	C_4H_{10}	58.1	0.1554	2.067
Carbon Dioxide	CO_2	44.0	0.1150	1.529
Carbon Monoxide	CO	28.0	0.0727	0.967
Ethane	C_2H_6	30.0	0.0789	1.049
Ethylene	C_2H_4	28.0	0.0733	0.975
Helium	H_e	4.0	0.0104	0.138
Hydrogen Chloride	HCl	36.5	0.0954	1.286
Hydrogen	H	2.0	0.0052	0.070
Hydrogen Sulphide	H_2S	34.1	0.0895	1.190
Methane	CH_4	16.0	0.0417	0.554
Methyl Chloride	CH_3Cl	50.5	0.1342	1.785
Natural Gas	–	19.5	0.0502	0.667
Nitric Oxide	NO	30.0	0.0708	1.037
Nitrogen	N_2	28.0	0.0727	0.967
Nitrous Oxide	N_2O	44.0	0.1151	1.530
Oxygen	O_2	32.0	0.0831	1.105
Propane	C_3H_8	44.1	0.1175	1.562
Propene (Propylene)	C_3H_6	42.1	0.1091	1.451
Sulfur Dioxide	SO_2	64.1	0.1703	2.264
Landfill Gas (approx. value)	–	–	–	1.00
Carbureted Water Gas	–	–	–	0.63
Coal Gas	–	–	–	0.42
Coke-Oven Gas	–	–	–	0.44
Refinery Oil Gas	–	–	–	0.99
Oil Gas (Pacific Coast)	–	–	–	0.47
"Wet" Gas (approximate value)	–	–	–	0.75

Gravity Flow of Liquids

In a pressure pipeline, a pump of some sort, generally provides the energy required to move the fluid through the pipeline. Such pipelines can transport fluids across a level surface, uphill or downhill. Gravity flow lines, on the other hand, utilize the energy associated with the placement of the pipeline discharge below the inlet. Like pressure flow pipelines, friction loss in a gravity flow pipeline depends on viscous shear stresses within the liquid and friction along the wetted surface of the pipe bore.

Some gravity flow piping systems may become very complex, especially if the pipeline grade varies, because friction loss will vary along with the varying grade. Sections of the pipeline may develop internal pressure, or vacuum, and may have varying liquid levels in the pipe bore.

Manning

For open channel water flow under conditions of constant grade, and uniform channel cross section, the Manning equation may be used.[28,29,30] Open channel flow exists in a pipe when it runs partially full. Like the Hazen-Williams formula, the Manning equation is limited to water or liquids with a kinematic viscosity equal to water.

Manning Equation
(1- 46)
$$V = \frac{1.486}{n} r_H^{2/3} S^{1/2}$$

WHERE
V = flow velocity, ft/sec
n = roughness coefficient, dimensionless
r_H = hydraulic radius, ft

(1- 47)
$$r_H = \frac{A_C}{P_W}$$

A_C = cross-sectional area of pipe bore, ft²
P_W = perimeter wetted by flow, ft
S_H = hydraulic slope, ft/ft

(1-48)
$$S_H = \frac{h_U - h_D}{L} = \frac{h_f}{L}$$

h_U = upstream pipe elevation, ft
h_D = downstream pipe elevation, ft
h_f = friction (head) loss, ft of liquid

It is convenient to combine the Manning equation with
(1-49) $$Q = A_c V$$

To obtain
(1-50) $$Q = \frac{1.486 A_C}{n} r_H^{2/3} S_H^{1/2}$$

Where terms are as defined above, and
Q = flow, ft³/sec

When a circular pipe is running full or half-full,
(1-51) $$r_H = \frac{d'}{4} = \frac{D_I}{48}$$

WHERE
d' = pipe inside diameter, ft
D_I = pipe inside diameter, in

Full pipe flow in ft³ per second may be estimated using:
(1-52) $$Q_{FPS} = (6.136 \times 10^{-4}) \frac{D_I^{8/3} S^{1/2}}{n}$$

Full pipe flow in gallons per minute may be estimated using:
(1-53) $$Q' = 0.275 \frac{D_I^{8/3} S^{1/2}}{n}$$

Nearly full circular pipes will carry more liquid than a completely full pipe. When slightly less than full, the hydraulic radius is significantly reduced, but the actual flow area is only slightly lessened. Maximum flow is achieved at about 93% of full pipe flow, and maximum velocity at about 78% of full pipe flow.

TABLE 1-15
Values of n for Use with Manning Equation

Surface	n, typical design
Polyethylene pipe	0.009
Uncoated cast or ductile iron pipe	0.013
Corrugated steel pipe	0.024
Concrete pipe	0.013
Vitrified clay pipe	0.013
Brick and cement mortar sewers	0.015
Wood stave	0.011
Rubble masonry	0.021

Note: The n-value of 0.009 for polyethylene pipe is for clear water applications. An n-value of 0.010 is typically utilized for applications such as sanitary sewer, etc.

Comparative Flows for Slipliners

Deteriorated gravity flow pipes may be rehabilitated by sliplining with polyethylene pipe. This process involves the installation of a polyethylene liner inside of the deteriorated original pipe as described in subsequent chapters within this manual. For conventional sliplining, clearance between the liner outside diameter and the existing pipe bore is required to install the liner; thus after rehabilitation, the flow channel is smaller than the original pipe. However, it is often possible to rehabilitate with a polyethylene slipliner, and regain all or most of the original flow capacity due to the extremely smooth inside surface of the polyethylene pipe and its resistance to deposition or build-up. Comparative flow capacities of circular pipes may be determined by the following:

(Eq. 1-54)

$$\% \; flow = 100 \frac{Q_1}{Q_2} = 100 \frac{\left(\dfrac{D_{I1}^{8/3}}{n_1} \right)}{\left(\dfrac{D_{I2}^{8/3}}{n_2} \right)}$$

Table 1-16 was developed using Equation 1-54 where D_{I1} = the inside diameter (ID) of the liner, and D_{I2} = the original inside diameter of the deteriorated host pipe.

TABLE 1-16
Comparative Flows for Slipliners

Existing Sewer ID, in	Liner OD, in.	Liner DR 32.5			Liner DR 26			Liner DR 21			Liner DR 17		
		Liner ID, in.†	% flow vs. concrete	% flow vs. clay	Liner ID, in.†	% flow vs. concrete	% flow vs. clay	Liner ID, in.†	% flow vs. concrete	% flow vs. clay	Liner ID, in.†	% flow vs. concrete	% flow vs. clay
4	3.500	3.272	97.5%	84.5%	3.215	93.0%	80.6%	3.147	87.9%	76.2%	3.064	81.8%	70.9%
6	4.500	4.206	64.6%	56.0%	4.133	61.7%	53.5%	4.046	58.3%	50.5%	3.939	54.3%	47.0%
6	5.375	5.024	103.8%	90.0%	4.937	99.1%	85.9%	4.832	93.6%	81.1%	4.705	87.1%	75.5%
8	6.625	6.193	84.2%	73.0%	6.085	80.3%	69.6%	5.956	75.9%	65.8%	5.799	70.7%	61.2%
8	7.125	6.660	102.2%	88.6%	6.544	97.5%	84.5%	6.406	92.1%	79.9%	6.236	85.8%	74.4%
10	8.625	8.062	93.8%	81.3%	7.922	89.5%	77.6%	7.754	84.6%	73.3%	7.549	78.8%	68.3%
12	10.750	10.049	103.8%	90.0%	9.873	99.1%	85.9%	9.665	93.6%	81.1%	9.409	87.1%	75.5%
15	12.750	11.918	90.3%	78.2%	11.710	86.1%	74.6%	11.463	81.4%	70.5%	11.160	75.7%	65.6%
15	13.375	12.503	102.5%	88.9%	12.284	97.8%	84.8%	12.025	92.4%	80.1%	11.707	86.1%	74.6%
16	14.000	13.087	97.5%	84.5%	2.858	93.0%	80.6%	12.587	87.9%	76.2%	12.254	81.8%	70.9%
18	16.000	14.956	101.7%	88.1%	14.695	97.0%	84.1%	14.385	91.7%	79.4%	14.005	85.3%	74.0%
21	18.000	16.826	92.3%	80.0%	16.532	88.1%	76.3%	16.183	83.2%	72.1%	15.755	77.5%	67.1%
24	20.000	18.695	85.6%	74.2%	18.369	81.7%	70.8%	17.981	77.2%	66.9%	17.506	71.9%	62.3%
24	22.000	20.565	110.4%	95.7%	20.206	105.3%	91.3%	19.779	99.5%	86.2%	19.256	92.6%	80.3%
27	24.000	22.434	101.7%	88.1%	22.043	97.0%	84.1%	21.577	91.7%	79.4%	21.007	85.3%	74.0%
30	28.000	26.174	115.8%	100.4%	25.717	110.5%	95.8%	25.173	104.4%	90.5%	24.508	97.2%	84.2%
33	30.000	28.043	108.0%	93.6%	27.554	103.0%	89.3%	26.971	97.3%	84.3%	26.259	90.6%	78.5%
36	32.000	29.913	101.7%	88.1%	29.391	97.0%	84.1%	28.770	91.7%	79.4%	28.009	85.3%	74.0%
36	34.000	31.782	119.5%	103.6%	31.228	114.1%	98.9%	30.568	107.7%	93.4%	29.760	100.3%	86.9%
42	36.000	33.652	92.3%	80.0%	33.065	88.1%	76.3%	32.366	83.2%	72.1%	31.511	77.5%	67.1%
48	42.000	39.260	97.5%	84.5%	38.575	93.0%	80.6%	37.760	87.9%	76.2%	36.762	81.8%	70.9%
54	48.000	44.869	101.7%	88.1%	44.086	97.0%	84.1%	43.154	91.7%	79.4%	42.014	85.3%	74.0%
60	54.000	50.478	105.1%	91.1%	49.597	100.3%	86.9%	48.549	94.8%	82.1%	47.266	88.2%	76.5%

† Liner ID calculated per Equation 1-1.

Flow Velocity

Acceptable flow velocities in polyethylene pipe depend on the specific details of the system. For water systems operating at rated pressures, velocities may be limited by surge allowance requirements. See Tables 1-7 and 1-8. Where surge effects are reduced, higher velocities are acceptable, and if surge is not a consideration, water flow velocities exceeding 25 feet per second may be acceptable.

Liquid flow velocity may be limited by the capabilities of pumps or elevation head to overcome friction (head) loss and deliver the flow and pressure required for the application. Polyethylene pipe is not eroded by water flow. Liquid slurry pipelines may be subject to critical minimum velocities that ensure turbulent flow and maintain particle suspension in the slurry.

Gravity liquid flows of 2 fps (0.6 m/s) and higher can help prevent or reduce solids deposition in sewer lines. When running full, gravity flow pipelines are subject to the same velocity considerations as pressure pipelines.

Flow velocity in compressible gas lines tends to be self-limiting. Compressible gas flows in polyethylene pipes are typically laminar or transitional. Fully turbulent flows are possible in short pipelines, but difficult to achieve in longer transmission and distribution lines because the pressure ratings for polyethylene pipe automatically limit flow capacity and, therefore, flow velocity.

Pipe Surface Condition, Aging

Aging acts to increase pipe surface roughness in most piping systems. This in turn increases flow resistance. Polyethylene pipe resists typical aging effects because polyethylene does not rust, rot, corrode, tuberculate, or support biological growth, and it resists the adherence of scale and deposits. In some cases, moderate flow velocities are sufficient to prevent deposition, and where low velocities predominate, occasional high velocity flows will help to remove sediment and deposits. As a result, the design capabilities for pressure and gravity flow pipelines are retained as the pipeline ages.

Where cleaning is needed to remove depositions in low flow rate gravity flow pipelines, water-jet cleaning or forcing a "soft" (plastic foam) pig through the pipeline are effective cleaning methods. Bucket, wire and scraper-type cleaning methods will damage polyethylene pipe and must not be used.

Section 2 **Buried PE Pipe Design**

Introduction

Buried PE Pipe Design covers basic engineering information for calculating earth and live-load pressures on PE pipe, for finding the pipe's response to these pressures taking into account the interaction between the pipe and its surrounding soil, and for judging that an adequate safety factor exists for a given application.

Soil pressure results from the combination of soil weight and surface loads. As backfill is placed around and over a PE pipe, the soil pressure increases and the pipe deflects vertically and expands laterally into the surrounding soil. The lateral expansion mobilizes passive resistance in the soil which, in combination with the pipe's inherent stiffness, resists further lateral expansion and consequently further vertical deflection.

During backfilling, ring (or hoop) stress develops within the pipe wall. Ring bending stresses (tensile and compressive) occur as a consequence of deflection, and ring compressive stress occurs as a consequence of the compressive thrust created by soil compression around the pipe's circumference. Except for shallow pipe subject to live load, the combined ring stress from bending and compression results in a net compressive stress.

The magnitude of the deflection and the stress depends not only on the pipe's properties but also on the soil's properties. The magnitude of deflection and stress must be kept safely within PE pipe's performance limits. Excessive deflection may cause loss of stability and flow restriction, while excessive compressive stress may cause wall crushing or ring buckling. Performance limits for PE pipe are given in Watkins, Szpak, and Allman[1] and illustrated in Figure 2-1.

Calculations

Section 2 describes how to calculate the soil pressure acting on PE pipe due to soil weight and surface loads, how to determine the resulting deflection based on pipe and soil properties, and how to calculate the allowable (safe) soil pressure for wall compression (crushing) and ring buckling for PE pipe.

Detailed calculations are not always necessary to determine the suitability of a particular PE pipe for an application. Pipes that fall within the AWWA C906 Committee Report "Design Window"[2] regarding pipe DR, installation, and burial depth meet specified deflection limits for PE pipe, have a safety factor of at least 2 against buckling, and do not exceed the allowable material compressive stress for PE. Thus, the designer need not perform extensive calculations for pipes that are sized and installed in accordance with the Design Window. The Design Window is more fully explained in the sub-section on Standard Installation later in this section. Many applications meet the requirements of the Design Window and therefore it is usually worth checking before performing any calculations.

Installation Categories

For the purpose of calculation, buried installations of PE pipe can be separated into four categories depending on the depth of cover, surface loading, groundwater level and pipe diameter. Each category involves slightly different equations for determining the load on the pipe and the pipe's response to the load. The boundaries between the categories are not definite, and engineering judgment is required to select the most appropriate category for a specific installation. The categories are:

1. Standard Installation-Trench or Embankment installation with a maximum cover of 50 ft with or without traffic, rail, or surcharge loading. To be in this category, where live loads are present the pipe must have a minimum cover of at least one diameter or 18" whichever is greater. Earth pressure applied to the pipe is found using the prism load (geostatic soil stress). The Modified Iowa Formula is used for calculating deflection. Crush and buckling are performance limits as well. The Standard Installation section also presents the AWWA "Design Window."

2. Shallow Cover Vehicular Loading Installation applies to pipes buried at a depth of at least 18" but less than one pipe diameter. This installation category uses the same equations as the Standard Installation but with an additional equation relating wheel load to the pipe's bending resistance and the soil's supporting strength.

3. Deep Fill Installation applies to embankments with depths exceeding 50 ft. The soil pressure calculation may be used for profile pipe in trenches less than 50 ft. The Deep Fill Installation equations differ from the Standard Installation equations by considering soil pressure based on arching, calculating deflection from the Watkins-Gaube Graph, and calculating buckling with the Moore-Selig Equation.

4. Shallow Cover Flotation Effects applies to applications where insufficient cover is available to either prevent flotation or hydrostatic collapse. Hydrostatic buckling is introduced in this chapter because of its use in subsurface design.

Section 2 of the Design Chapter is limited to the design of polyethylene pipes buried in trenches or embankments. The load and pipe reaction calculations presented may not apply to pipes installed using trenchless technologies such as pipe bursting and directional drilling. These pipes may not develop the same soil support as pipe installed in a trench. The purveyor of the trenchless technology should be consulted for piping design information. See the Chapter on "Polyethylene Pipe for Horizontal

Directional Drilling" and ASTM F1962, *Use of Maxi-Horizontal Directional Drilling (HDD) for Placement of Polyethylene Pipe or Conduit Under Obstacles, Including River Crossings* for additional information on design of piping installed using directional drilling.

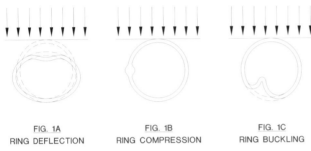

FIG. 1A RING DEFLECTION

FIG. 1B RING COMPRESSION

FIG. 1C RING BUCKLING

Figure 2-1 Performance Limits for Buried PE Pipe

Design Process

The interaction between pipe and soil, the variety of field-site soil conditions, and the range of available pipe Dimension Ratios make the design of buried pipe seem challenging. This section of the Design Chapter has been written with the intent of easing the designer's task. While some very sophisticated design approaches for buried pipe systems may be justified in certain applications, the simpler, empirical methodologies presented herein have been proven by experience to provide reliable results for virtually all PE pipe installations.

The design process consists of the following steps:

1. Determine the **vertical soil** pressure acting at the crown of the pipe due to earth, live, and surcharge loads.

2. Select a **trial pipe**, which means selecting a **trial dimension ratio (DR)** or, in the case of profile pipe, a **trial profile** as well.

3. Select an embedment material and degree of compaction. As will be described later, soil type and compaction are relatable to a specific **modulus of soil reaction value (E')**. (As deflection is proportional to the combination of pipe and soil stiffness, pipe properties and embedment stiffness can be traded off to obtain an optimum design.)

4. For the trial pipe and trial modulus of soil reaction, **calculate** the deflection due to the vertical soil pressure. **Compare** the pipe deflection to the deflection limit. If deflection exceeds the limit, it is generally best to look at increasing the modulus of soil reaction rather than reducing the DR or changing to a heavier profile. Repeat step 4 for the new E' and/or new trial pipe.

5. For the trial pipe and trial modulus of soil reaction, **calculate** the allowable soil pressure for wall crushing and for wall buckling. **Compare** the allowable soil pressure to the applied vertical pressure. If the allowable pressure is equal to or higher than the applied vertical pressure, the design is complete. If not, select a different pipe DR or heavier profile or different E', and repeat step 5.

Since design begins with calculating vertical soil pressure, it seems appropriate to discuss the different methods for finding the vertical soil pressure on a buried pipe before discussing the pipe's response to load within the four installation categories.

Vertical Soil Pressure

The weight of the earth, as well as surface loads above the pipe, produce soil pressure on the pipe. The weight of the earth or "earth load" is often considered to be a "dead-load" whereas surface loads are referred to as "surcharge loads" and may be temporary or permanent. When surcharge loads are of short duration they are usually referred to as "live loads." The most common live load is vehicular load. Other common surcharge loads include light structures, equipment, and piles of stored materials or debris. This section gives formulas for calculating the vertical soil pressure due to both earth and surcharge loads. The soil pressures are normally calculated at the depth of the pipe crown. The soil pressures for earth load and each surcharge load are added together to obtain the total vertical soil pressure which is then used for calculating deflection and for comparison with wall crush and wall buckling performance limits.

Earth Load

In a uniform, homogeneous soil mass, the soil load acting on a horizontal plane within the mass is equal to the weight of the soil directly above the plane. If the mass contains areas of varying stiffness, the weight of the mass will redistribute itself toward the stiffer areas due to internal shear resistance, and arching will occur. Arching results in a reduction in load on the less stiff areas. Flexible pipes including PE pipes are normally not as stiff as the surrounding soil, so the resulting earth pressure acting on PE pipe is reduced by arching and is less than the weight of soil above the pipe. (One minor exception to this is shallow cover pipe under dynamic loads.) For simplicity, engineers often ignore arching and assume that the earth load on the pipe is equal to the weight of soil above the pipe, which is referred to as the "prism load" or "geostatic stress." Practically speaking, the prism load is a conservative loading for PE pipes. It may be safely used in virtually all designs. Equation 2-1 gives the vertical soil pressure due to the prism load. The depth of cover is the depth from the ground surface to the pipe crown.

(2-1) $$P_E = wH$$

WHERE
P_E = vertical soil pressure due to earth load, psf
w = unit weight of soil, pcf
H = depth of cover, ft

UNITS CONVENTION: To facilitate calculations for PE pipes, the convention used with rigid pipes for taking the load on the pipe as a line load along the longitudinal axis in units of **lbs/lineal-ft** of pipe length is not used here. Rather, the load is treated as a soil pressure acting on a horizontal plane at the pipe crown and is given in units of **lbs/ft² or psf.**

Soil weight can vary substantially from site to site and within a site depending on composition, density and load history. Soil weights are often found in the construction site geotechnical report. The saturated unit weight of the soil is used when the pipe is below the groundwater level. For design purposes, the unit weight of soil is commonly assumed to be 120 pcf, when site-specific information is not available.

Generally, the soil pressure on profile pipe and on DR pipe in deep fills is significantly less than the prism load due to arching. For these applications, soil pressure is best calculated using the calculations that account for arching in the "Deep Fill Installation" section.

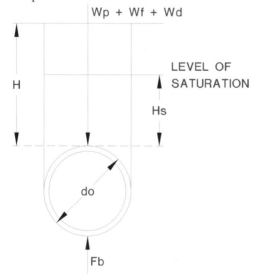

Figure 2-2 Prism Load

Live Load

Even though wheel loadings from cars and other light vehicles may be frequent, these loads generally have little impact on subsurface piping compared to the less frequent but significantly heavier loads from trucks, trains, or other heavy vehicles. For design of pipes under streets and highways, only the loadings from these heavier vehicles are considered. The pressure transmitted to a pipe by a vehicle depends on the pipe's depth, the vehicle's weight, the tire pressure and size, vehicle speed, surface smoothness, the amount and type of paving, the soil, and the distance from the pipe to the point of loading. For the more common cases, such as H20 (HS20) truck traffic on paved roads and E-80 rail loading, this information has been simplified and put into Tables 2-2, 2-3, and 2-4 to aid the designer. For special cases, such as mine trucks, cranes, or off-road vehicles, Equations 2-2 and 2-4 may be used.

The maximum load under a wheel occurs at the surface and diminishes with depth. Polyethylene pipes should be installed a minimum of one diameter or 18", whichever is greater, beneath the road surface. At this depth, the pipe is far enough below the wheel load to significantly reduce soil pressure and the pipe can fully utilize the embedment soil for load resistance. Where design considerations do not permit installation with at least one diameter of cover, additional calculations are required and are given in the section discussing "Shallow Cover Vehicular Loading Installation." State highway departments often regulate minimum cover depth and may require 2.5 ft to 5 ft of cover depending on the particular roadway.

During construction, both permanent and temporary underground pipelines may be subjected to heavy vehicle loading from construction equipment. It may be advisable to provide a designated vehicle crossing with special measures such as temporary pavement or concrete encasement, as well as vehicle speed controls to limit impact loads.

The following information on AASHTO Loading and Impact Factor is not needed to use Tables 2-2 and 2-4. It is included to give the designer an understanding of the surface loads encountered and typical impact factors. If the designer decides to use Equations 2-2 or 2-4 rather than the tables, the information will be useful.

AASHTO Vehicular Loading

Vehicular loads are typically based on The American Association of State Highway and Transportation Officials (AASHTO) standard truck loadings. For calculating the soil pressure on flexible pipe, the loading is normally assumed to be an H20 (HS20) truck. A standard H20 truck has a total weight of 40,000 lbs (20 tons). The weight is distributed with 8,000 lbs on the front axle and 32,000 lbs on the rear axle. The HS20 truck is a tractor and trailer unit having the same axle loadings as the H20 truck but with two rear axles. See Figure 2-3. For these trucks, the maximum wheel load is found at the rear axle(s) and equals 40 percent of the total weight of the truck.

The maximum wheel load may be used to represent the static load applied by either a single axle or tandem axles. Some states permit heavier loads. The heaviest tandem axle loads normally encountered on highways are around 40,000 lbs (20,000 lbs per wheel). Occasionally, vehicles may be permitted with loads up to 50 percent higher.

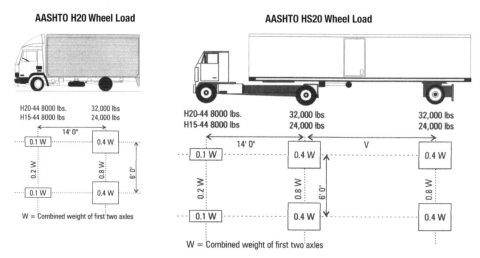

Figure 2-3 AASHTO H20 & HS20 Vehicle Loads

Impact Factor

Road surfaces are rarely smooth or perfectly even. When vehicles strike bumps in the road, the impact causes an instantaneous increase in wheel loading. Impact load may be found by multiplying the static wheel load by an impact factor. The factor varies with depth. Table 2-1 gives impact factors for vehicles on paved roads. For unpaved roads, impact factors of 2.0 or higher may occur, depending on the road surface.

TABLE 2-1
Typical Impact Factors for Paved Roads

Cover Depth, ft	Impact Factor, I_f
1	1.35
2	1.30
3	1.25
4	1.20
6	1.10
8	1.00

Derived from Illinois DOT dynamic load formula (1996).

Vehicle Loading through Highway Pavement (Rigid)

Pavement reduces the live load pressure reaching a pipe. A stiff, rigid pavement spreads load out over a large subgrade area thus significantly reducing the vertical soil pressure. Table 2-2 gives the vertical soil pressure underneath an H20 (HS20) truck traveling on a paved highway (12-inch thick concrete). An impact factor is incorporated. For use with heavier trucks, the pressures in Table 2-2 can be adjusted proportionally to the increased weight as long as the truck has the same tire area as an HS20 truck.

TABLE 2-2
Soil Pressure under H20 Load (12" Thick Pavement)

Depth of cover, ft.	Soil Pressure, lb/ft^2
1	1800
1.5	1400
2	800
3	600
4	400
5	250
6	200
7	175
8	100
Over 8	Neglect

Note: For reference see ASTM F7906. Based on axle load equally distributed over two 18 by 20 inch areas, spaced 72 inches apart. Impact factor included.

Vehicle Loading through Flexible Pavement or Unpaved Surface

Flexible pavements (or unpaved surfaces) do not have the bridging ability of rigid pavement and thus transmit more pressure through the soil to the pipe than given by Table 2-2. In some cases, the wheel loads from two vehicles passing combine to create a higher soil pressure than a single dual-tire wheel load. The maximum pressure may occur directly under the wheels of one vehicle or somewhere in between the wheels of the two vehicles depending on the cover depth. Table 2-3 gives the largest of the maximum pressure for two passing H20 trucks on an unpaved surface. No impact factor is included. The loading in Table 2-3 is conservative and about 10% higher than loads found by the method given in AASHTO Section 3, LRFD Bridge Specifications Manual based on assuming a single dual-tire contact area of 20 x 10 inches and using the equivalent area method of load distribution.

TABLE 2-3
Soil Pressure Under H20 Load (Unpaved or Flexible Pavement)

Depth of cover, ft.	Soil Pressure, lb/ft²
1.5	2000
2.0	1340
2.5	1000
3.0	710
3.5	560
4.0	500
6.0	310
8.0	200
10.0	140

Note: Based on integrating the Boussinesq equation for two H20 loads spaced 4 feet apart or one H20 load centered over pipe. No pavement effects or impact factor included.

Off-Highway Vehicles

Off-highway vehicles such as mine trucks and construction equipment may be considerably heavier than H20 trucks. These vehicles frequently operate on unpaved construction or mine roads which may have very uneven surfaces. Thus, except for slow traffic, an impact factor of 2.0 to 3.0 should be considered. For off-highway vehicles, it is generally necessary to calculate live load pressure from information supplied by the vehicle manufacturer regarding the vehicle weight or wheel load, tire footprint (contact area) and wheel spacing.

The location of the vehicle's wheels relative to the pipe is also an important factor in determining how much load is transmitted to the pipe. Soil pressure under a point load at the surface is dispersed through the soil in both depth and expanse. Wheel loads not located directly above a pipe may apply pressure to the pipe, and this pressure can be significant. The load from two wheels straddling a pipe may produce a higher pressure on a pipe than from a single wheel directly above it.

For pipe installed within a few feet of the surface, the maximum soil pressure will occur when a single wheel (single or dual tire) is directly over the pipe. For deeper pipes, the maximum case often occurs when vehicles traveling above the pipe pass within a few feet of each other while straddling the pipe, or in the case of off-highway vehicles when they have closely space axles. The minimum spacing between the centerlines of the wheel loads of passing vehicles is assumed to be four feet. At this spacing for H20 loading, the pressure on a pipe centered midway between the two passing vehicles is greater than a single wheel load on a pipe at or below a depth of about four feet.

For design, the soil pressure on the pipe is calculated based on the vehicle location (wheel load locations) relative to the pipe that produces the maximum pressure. This generally involves comparing the pressure under a single wheel with that occurring with two wheels straddling the pipe. The Timoshenko Equation can be used to find the pressure directly under a single wheel load, whereas the Boussinesq Equation can be used to find the pressure from wheels not directly above the pipe.

Timoshenko's Equation

The Timoshenko Equation gives the soil pressure at a point directly under a distributed surface load, neglecting any pavement.

(2-2)
$$P_L = \frac{I_f W_w}{a_C}\left(1 - \frac{H^3}{(r_T^2 + H^2)^{1.5}}\right)$$

WHERE
P_L = vertical soil pressure due to live load, lb/ft²
I_f = impact factor
W_w = wheel load, lb
a_C = contact area, ft²
r_T = equivalent radius, ft
H = depth of cover, ft

The equivalent radius is given by:
(2-3)
$$r_T = \sqrt{\frac{a_C}{\pi}}$$

For standard H2O and HS20 highway vehicle loading, the contact area is normally taken for dual wheels, that is, 16,000 lb over 10 in. by 20 in. area.

Timoshenko Example Calculation

Find the vertical pressure on a 24" polyethylene pipe buried 3 ft beneath an unpaved road when an R-50 truck is over the pipe. The manufacturer lists the truck with a gross weight of 183,540 lbs on 21X35 E3 tires, each having a 30,590 lb load over an imprint area of 370 in².

SOLUTION: Use Equations 2-2 and 2-3. Since the vehicle is operating on an unpaved road, an impact factor of 2.0 is appropriate.

$$r_T = \sqrt{\frac{370/144}{\pi}} = 0.90 ft \qquad P_L = \frac{(2.0)(30,590)}{\frac{370}{144}}\left(1 - \frac{3^3}{(0.90^2 + 3^2)^{1.5}}\right)$$

$$P_L = 2890 lb/ft^2$$

Chapter 6
Design of Polyethylene Piping Systems

Boussinesq Equation

The Boussinesq Equation gives the pressure at any point in a soil mass under a concentrated surface load. The Boussinesq Equation may be used to find the pressure transmitted from a wheel load to a point that is not along the line of action of the load. Pavement effects are neglected.

(2-4)
$$P_L = \frac{3 I_f W_w H^3}{2\pi r^5}$$

WHERE
PL = vertical soil pressure due to live load lb/ft²
Ww = wheel load, lb
H = vertical depth to pipe crown, ft
I_f = impact factor
r = distance from the point of load application to pipe crown, ft

(2-5)
$$r = \sqrt{X^2 + H^2}$$

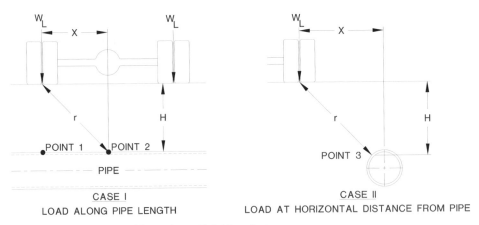

Figure 2-4 Illustration of Boussinesq Point Loading

Example Using Boussinesq Point Loading Technique

Determine the vertical soil pressure applied to a 12" pipe located 4 ft deep under a dirt road when two vehicles traveling over the pipe and in opposite lanes pass each other. Assume center lines of wheel loads are at a distance of 4 feet. Assume a wheel load of 16,000 lb.

SOLUTION: Use Equation 2-4, and since the wheels are traveling, a 2.0 impact factor is applied. The maximum load will be at the center between the two wheels, so X = 2.0 ft. Determine r from Equation 2-5.

$$r = \sqrt{4^2 + 2.0^2} = 4.47 \, ft$$

Then solve Equation 2-4 for PL, the load due to a single wheel.

$$P_L = \frac{3(2.0)(16,000)(4)^3}{2\pi(4.47)^5}$$

$$P_L = 548 \, lb/ft^2$$

The load on the pipe crown is from both wheels, so

$$2 P_L = 2(548) = 1096 \, lb/ft^2$$

The load calculated in this example is higher than that given in Table 2-3 for a comparable depth even after correcting for the impact factor. Both the Timoshenko and Boussinesq Equations give the pressure applied at a point in the soil. In solving for pipe reactions it is assumed that this point pressure is applied across the entire surface of a unit length of pipe, whereas the actual applied pressure decreases away from the line of action of the wheel load. Methods that integrate this pressure over the pipe surface such as used in deriving Table 2-3 give more accurate loading values. However, the error in the point pressure equations is slight and conservative, so they are still effective equations for design.

TABLE 2-4
Live Load Pressure for E-80 Railroad Loading

Depth of cover, ft.	Soil Pressure, lb/ft²
2.0	3800
5.0	2400
8.0	1600
10.0	1100
12.0	800
15.0	600
20.0	300
30.0	100
Over 30.0	Neglect

For reference see ASTM A796.

Railroad Loads

The live loading configuration used for pipes under railroads is the Cooper E-80 loading, which is an 80,000 lb load that is uniformly applied over three 2 ft by 8 ft areas on 5 ft centers. The area represents the 8 ft width of standard railroad ties and the standard spacing between locomotive drive wheels. Live loads are based on the axle weight exerted on the track by two locomotives and their tenders coupled together in doubleheader fashion. See Table 2-4. Commercial railroads frequently require casings for pressure pipes if they are within 25 feet of the tracks, primarily for safety reasons in the event of a washout. Based upon design and permitting requirements, the designer should determine whether or not a casing is required.

Surcharge Load

Surcharge loads may be distributed loads, such as a footing, foundation, or an ash pile, or may be concentrated loads, such as vehicle wheels. The load will be dispersed through the soil such that there is a reduction in pressure with an increase in depth or horizontal distance from the surcharged area. Surcharge loads not directly over the pipe may exert pressure on the pipe as well. The pressure at a point beneath a surcharge load depends on the load magnitude and the surface area over which the surcharge is applied. Methods for calculating vertical pressure on a pipe either located directly beneath a surcharge or located near a surcharge are given below.

Pipe Directly Beneath a Surcharge Load

This design method is for finding the vertical soil pressure under a rectangular area with a uniformly distributed surcharge load. This may be used in place of Tables 2-2 to 2-4 and Equations 2-2 and 2-4 to calculate vertical soil pressure due to wheel loads. This requires knowledge of the tire imprint area and impact factor.

The point pressure on the pipe at depth, H, is found by dividing the rectangular surcharge area (ABCD) into four sub-area rectangles (a, b, c, and d) which have a common corner, E, in the surcharge area, and over the pipe. The surcharge pressure, P_L, at a point directly under E is the sum of the pressure due to each of the four sub-area loads. Refer to Figure 2-5 A.

The pressure due to each sub-area is calculated by multiplying the surcharge pressure at the surface by an Influence Value, I_V. Influence Values are proportionality constants that measure what portion of a surface load reaches the subsurface point in question. They were derived using the Boussinesq Equation and are given in Table 2-5.

(2-6) $$P_L = p_a + p_b + p_c + p_d$$

WHERE
P_L = vertical soil pressure due to surcharge pressure, lb/ft²
p_a = pressure due to sub-area a, lb/ft²
p_b = pressure due to sub-area b, lb/ft²
p_c = pressure due to sub-area c, lb/ft²
p_d = pressure due to sub-area d, lb/ft²

Pressure due to the surcharge applied to the i-th sub-area equals:

(2-7) $\quad p_i = I_V \, w_S$

WHERE
I_V = Influence Value from Table 2-5
w_S = distributed pressure of surcharge load at ground surface, lb/ft²

If the four sub-areas are equivalent, then Equation 12 may be simplified to:

(2-8) $\quad P_L = 4 I_V \, w_S$

The load calculated in this example is higher than that given in Table 2-3 for a comparable depth even after correcting for the impact factor (1096 psf versus 1000 psf). The influence value is dependent upon the dimensions of the rectangular area and upon the depth to the pipe crown, H. Table 2-5 Influence Value terms depicted in Figure 2-5, are defined as:

H = depth of cover, ft
M = horizontal distance, normal to the pipe centerline, from the center of the load to the load edge, ft
N = horizontal distance, parallel to the pipe centerline, from the center of the load to the load edge, ft

Interpolation may be used to find values not given in Table 2-5. The influence value gives the portion (or influence) of the load that reaches a given depth beneath the corner of the loaded area.

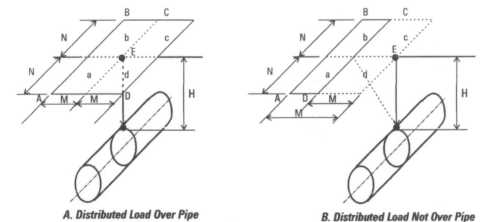

A. Distributed Load Over Pipe   B. Distributed Load Not Over Pipe

Figure 2-5 Illustration of Distributed Loads

TABLE 2-5
Influence Values, I_V for Distributed Loads

M/H	N/H													
	0.1	0.2	0.3	0.4	0.5	0.6	0.7	0.8	0.9	1.0	1.2	1.5	2.0	∞
0.1	0.005	0.009	0.013	0.017	0.020	0.022	0.024	0.026	0.027	0.028	0.029	0.030	0.031	0.032
0.2	0.009	0.018	0.026	0.033	0.039	0.043	0.047	0.050	0.053	0.055	0.057	0.060	0.061	0.062
0.3	0.013	0.026	0.037	0.047	0.056	0.063	0.069	0.073	0.077	0.079	0.083	0.086	0.089	0.090
0.4	0.017	0.033	0.047	0.060	0.071	0.080	0.087	0.093	0.098	0.101	0.106	0.110	0.113	0.115
0.5	0.020	0.039	0.056	0.071	0.084	0.095	0.103	0.110	0.116	0.120	0.126	0.131	0.135	0.137
0.6	0.022	0.043	0.063	0.080	0.095	0.107	0.117	0.125	0.131	0.136	0.143	0.149	0.153	0.156
0.7	0.024	0.047	0.069	0.087	0.103	0.117	0.128	0.137	0.144	0.149	0.157	0.164	0.169	0.172
0.8	0.026	0.050	0.073	0.093	0.110	0.125	0.137	0.146	0.154	0.160	0.168	0.176	0.181	0.185
0.9	0.027	0.053	0.077	0.098	0.116	0.131	0.144	0.154	0.162	0.168	0.178	0.186	0.192	0.196
1.0	0.028	0.055	0.079	0.101	0.120	0.136	0.149	0.160	0.168	0.175	0.185	0.194	0.200	0.205
1.2	0.029	0.057	0.083	0.106	0.126	0.143	0.157	0.168	0.178	0.185	0.196	0.205	0.209	0.212
1.5	0.030	0.060	0.086	0.110	0.131	0.149	0.164	0.176	0.186	0.194	0.205	0.211	0.216	0.223
2.0	0.031	0.061	0.088	0.113	0.135	0.153	0.169	0.181	0.192	0.200	0.209	0.216	0.232	0.240
∞	0.032	0.062	0.089	0.116	0.137	0.156	0.172	0.185	0.196	0.205	0.212	0.223	0.240	0.250

Vertical Surcharge Example # 1

Find the vertical surcharge load for the 4' x 6', 2000 lb/ft² footing shown below.

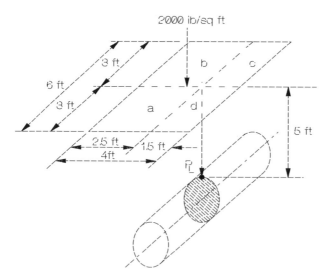

SOLUTION: Use equations 2-6 and 2-7, Table 2-5, and Figure 2-5. The 4 ft x 6 ft footing is divided into four sub-areas, such that the common corner of the sub-areas is directly over the pipe. Since the pipe is not centered under the load, sub-areas a and b have dimensions of 3 ft x 2.5 ft, and sub-areas c and d have dimensions of 3 ft x 1.5 ft.

Determine sub-area dimensions for M, N, and H, then calculate M/H and N/H. Find the Influence Value from Table 2-5, then solve for each sub area, p_a, p_b, p_c, p_d, and sum for P_L.

	Sub-area			
	a	b	c	d
M	2.5	2.5	1.5	1.5
N	3.0	3.0	3.0	3.0
M/H	0.5	0.5	0.3	0.3
N/H	0.6	0.6	0.6	0.6
I_v	0.095	0.095	0.063	0.063
p_i	190	190	126	126

Therefore: P_L = 632 lbs/ft²

Pipe Adjacent to, but Not Directly Beneath, a Surcharge Load

This design method may be used to find the surcharge load on buried pipes near, but not directly below, uniformly distributed loads such as concrete slabs, footings and floors, or other rectangular area loads, including wheel loads that are not directly over the pipe.

The vertical pressure is found by first adding an imaginary loaded area that covers the pipe, then determining the surcharge pressure due to the overall load (actual and imaginary) based on the previous section, and finally by deducting the pressure due to the imaginary load from that due to the overall load.

Refer to Figure 2-5 B. Since there is no surcharge directly above the pipe centerline, an imaginary surcharge load, having the same pressure per unit area as the actual load, is applied to sub-areas c and d. The surcharge pressure for sub-areas a+d and b+c are determined, then the surcharge loads from the imaginary areas c and d are deducted to determine the surcharge pressure on the pipe.

(2-9) $$P_L = P_{a+d} + P_{b+c} - P_d - P_c$$

Where terms are as previously defined above, and

p_{a+d} = surcharge load of combined sub-areas a and d, lb/ft²
p_{b+c} = surcharge load of combined sub-areas b and c, lb/ft²

Vertical Surcharge Example # 2

Find the vertical surcharge pressure for the 6' x 10', 2000 lb/ft2 slab shown below.

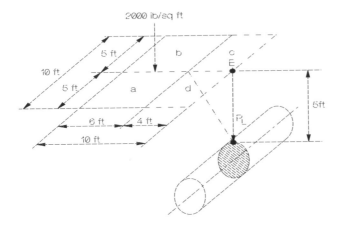

SOLUTION: Use Equations 2-7 and 2-9, Table 2-5, and Figure 2-5 B. The surcharge area is divided into two sub-areas, a and b. The area between the surcharge and the line of the pipe crown is divided into two sub-areas, c and d, as well. The imaginary load is applied to sub-areas c and d. Next, the four sub-areas are treated as a single surcharge area. Unlike the previous example, the pipe is located under the edge of the surcharge area rather than the center. So, the surcharge pressures for the combined sub-areas a+d and b+c are determined, and then for the sub-areas c and d. The surcharge pressure is the sum of the surcharge pressure due to the surcharge acting on sub-areas a+d and b+c, less the imaginary pressure due to the imaginary surcharge acting on sub-areas c and d.

	Sub-area			
	a + d	b + c	c	d
M	10	10	4	4
N	5	5	5	5
M/H	2.0	2.0	0.8	0.8
N/H	1.0	1.0	1.0	1.0
I_v	0.200	0.200	0.160	0.160
P_i	400	400	(320)	(320)

Therefore P_L = 160 lb/ft2

Installation Category #1: **Standard Installation-Trench or Embankment**

After calculating the soil pressure acting on the pipe, the next step in design is to select a trial pipe and embedment modulus of soil reaction, E'. Using the trial values, deflection is calculated and compared to the deflection limits, the compressive wall stress is calculated and compared to the allowable material stress to ensure a safety factor against wall crushing, and the allowable buckling pressure is calculated and compared to the applied vertical pressure.

The Standard Installation category applies to pipes that are installed between 1.5 and 50 feet of cover. Where surcharge, traffic, or rail load may occur, the pipe must have at least one full diameter of cover. If such cover is not available, then the application design must also consider limitations under the Shallow Cover Vehicular Loading Installation category. Where ground water occurs above the pipe's invert and the pipe has less than two diameters of cover, the potential for the occurrence of flotation or upward movement of the pipe may exist. See Shallow Cover Flotation Effects.

While the Standard Installation is suitable for up to 50 feet of cover, it may be used for more cover. The 50 feet limit is based on A. Howard's [3] recommended limit for use of E' values. Above 50 feet, the E' values given in Table 2-6 are generally thought to be overly conservative as they are not corrected for the increase in embedment stiffness that occurs with depth as a result of the higher confinement pressure within the soil mass. In addition, significant arching occurs at depths greater than 50 feet.

The Standard Installation, as well as the other design categories for buried PE pipe, looks at a ring or circumferential cross-section of pipe and neglects longitudinal loading, which is normally insignificant. They also ignore the re-rounding effect of internal pressurization. Since re-rounding reduces deflection and stress in the pipe, ignoring it is conservative.

Ring Deflection
Ring deflection is the normal response of flexible pipes to soil pressure. It is also a beneficial response in that it leads to the redistribution of soil stress and the initiation of arching. Ring deflection can be controlled within acceptable limits by the selection of appropriate pipe embedment materials, compaction levels, trench width and, in some cases, the pipe itself.

The magnitude of ring deflection is inversely proportional to the combined stiffness of the pipe and the embedment soil. M. Spangler [4] characterized this relationship in the Iowa Formula in 1941. R. Watkins [5] modified this equation to allow a simpler approach for soil characterization, thus developing the Modified Iowa Formula. In 1964, Burns and Richards [6] published a closed-form solution for ring deflection and pipe stress based on classical linear elasticity. In 1976 M. Katona et. al. [7] developed a finite element program called CANDE (Culvert Analysis and Design) which is now available in a PC version and can be used to predict pipe deflection and stresses.

The more recent solutions may make better predictions than the Iowa Formula, but they require detailed information on soil and pipe properties, e.g. more soil lab testing. Often the improvement in precision is all but lost in construction variability.

Therefore, the Modified Iowa Formula remains the most frequently used method of determining ring deflection.

Spangler's Modified Iowa Formula can be written for use with solid wall HDPE pipe as:

(2-10)
$$\frac{\Delta X}{D_M} = \frac{1}{144} \left(\frac{K_{BED} L_{DL} P_E + K_{BED} P_L}{\frac{2E}{3}\left(\frac{1}{DR-1}\right)^3 + 0.061 F_s E'} \right)$$

and for use with ASTM F894 profile wall pipe as:

(2-11)
$$\frac{\Delta X}{D_I} = \frac{P}{144} \left(\frac{K_{BED} L_{DL}}{\frac{1.24(RSC)}{D_M} + 0.061 F_s E'} \right)$$

WHERE
ΔX = Horizontal deflection, in
K_{BED} = Bedding factor, typically 0.1
L_{DL} = Deflection lag factor
P_E = Vertical soil pressure due to earth load, psf
P_L = Vertical soil pressure due to live load, psf
E = Apparent modulus of elasticity of pipe material, lb/in²
E' = Modulus of Soil reaction, psi
F_S = Soil Support Factor
RSC = Ring Stiffness Constant, lb/ft
DR = Dimension Ratio, OD/t
D_M = Mean diameter (D_I+2z or D_O-t), in
z = Centroid of wall section, in
t = Minimum wall thickness, in
D_I = pipe inside diameter, in
D_O = pipe outside diameter, in

Deflection is reported as a percent of the diameter which can be found by multiplying 100 times $\Delta X/D_M$ or $\Delta X/D_I$. (When using RSC, the units of conversion are accounted for in Equation 2-11.)

Apparent Modulus of Elasticity for Pipe Material, E

The modulus of PE is dependent on load-rate and temperature. Table 2-6 gives values for the apparent modulus of elasticity for use in Spangler's Iowa Formula. It has long been an industry practice to use the short-term modulus in the Iowa Formula for thermoplastic pipe. This is based on the idea that, in granular embedment soil, deformation is a series of instantaneous deformations consisting of rearrangement and fracturing of grains while the bending stress in the pipe wall is decreasing due to stress relaxation. Use of the short-term modulus has proven effective and reliable for corrugated and profile wall pipes. These pipes typically have pipe stiffness values of 46 psi or less when measured per ASTM D2412. Conventional DR pipes starting with DR17 or lower have significantly higher stiffness and therefore they may carry a greater proportion of the earth and live load than corrugated or profile pipe; so it is conservative to use the 50-year modulus for DR pipes when determining deflection due to earth load.

Vehicle loads are generally met with a higher modulus than earth loads, as load duration may be nearly instantaneous for moving vehicles. The deflection due to a combination of vehicle or temporary loads and earth load may be found by separately calculating the deflection due to each load using the modulus appropriate for the expected load duration, then adding the resulting deflections together to get the total deflection. When doing the deflection calculation for vehicle load, the Lag Factor will be one. An alternate, but conservative, method for finding deflection for combined vehicle and earth load is to do one calculation using the 50-year modulus, but separate the vertical soil pressure into an earth load component and a live load component and apply the Lag Factor only to the earth load component.

TABLE 2-6
Design Values for Apparent Modulus of Elasticity, E @ 73°F

Load Duration	Short-Term	10 hours	100 hours	1000 hours	1 year	10 years	50 years
HDPE Modulus of Elasticity, psi	110,000	57,500	51,200	43,700	38,000	31,600	28,200
MDPE Modulus of Elasticity, psi	88,000	46,000	41,000	35,000	30,400	25,300	22,600

Ring Stiffness Constant, RSC

Profile wall pipes manufactured to ASTM F894, "Standard Specification for Polyethylene (PE) Large Diameter Profile Wall Sewer and Drain Pipe," are classified on the basis of their Ring Stiffness Constant (RSC). Equation 2-12 gives the RSC.

(2-12) $$RSC = \frac{6.44\ EI}{D_M^2}$$

WHERE
E = Apparent modulus of elasticity of pipe material (Short-term value Table 2-6) @73°F
I = Pipe wall moment of inertia, in^4/in (t^3/12, if solid wall construction)
z = Pipe wall centroid in
D_I = Pipe inside diameter in
D_M = Mean diameter (DI + 2z or D_O-t), in
t = Minimum wass thickness, in

Modulus of Soil Reaction, E'

The soil reaction modulus is proportional to the embedment soil's resistance to the lateral expansion of the pipe. There are no convenient laboratory tests to determine the soil reaction modulus for a given soil. A. Howard [8] determined E' values empirically from numerous field deflection measurements by substituting site parameters (i.e. depth of cover, soil weight) into Spangler's equation and "back-calculating" E'. Howard developed a table for the Bureau of Reclamation relating E' values to soil types and compaction efforts. See Table 2-7. In back-calculating E', Howard assumed the prism load was applied to the pipe. Therefore, Table 2-7 E' values indirectly include load reduction due to arching and are suitable for use only with the prism load.

Due to differences in construction procedures, soil texture and density, pipe placement, and insitu soil characteristics, pipe deflection varies along the length of a pipeline. Petroff [9] has shown that deflection measurements along a pipeline typically fit the Normal Distribution curve. To determine the anticipated maximum deflection using Eq. 2-10 or 2-11, variability may be accommodated by reducing the Table 2-7 E' value by 25%, or by adding to the calculated deflection percentage the 'correction for accuracy' percentage given in Table 2-7.

In shallow installations, the full value of the E' given in Table 2-7 may not develop. This is due to the lack of "soil confining pressure" to hold individual soil grains tightly together and stiffen the embedment. Increased weight or equivalently, depth, increases the confining pressure and, thus, the E'. J. Hartley and J. Duncan [10] published recommended E' values based on depth of cover. See Table 2-8. These are particularly useful for shallow installations.

TABLE 2-7
Values of E' for Pipe Embedment (See Howard [8])

Soil Type-pipe Embedment Material (Unified Classification System)[1]	E' for Degree of Embedment Compaction, lb/in²			
	Dumped	Slight, <85% Proctor, <40% Relative Density	Moderate, 85%-95% Proctor, 40%-70% Relative Density	High, >95% Proctor, >70% Relative Density
Fine-grained Soils (LL > 50)[2] Soils with medium to high plasticity; CH, MH, CH-MH	No data available: consult a competent soils engineer, otherwise, use E' = 0.			
Fine-grained Soils (LL < 50) Soils with medium to no plasticity, CL, ML, ML-CL, with less than 25% coarse grained particles.	50	200	400	1000
Fine-grained Soils (LL < 50) Soils with medium to no plasticity, CL, ML, ML-CL, with more than 25% coarse grained particles; Coarse-grained Soils with Fines, GM, GC, SM, SC[3] containing more than 12% fines.	100	400	1000	2000
Coarse-grained soils with Little or No Fines GW, GP, SW, SP[3] containing less than 12% fines	200	1000	2000	3000
Crushed Rock	1000	3000	3000	3000
Accuracy in Terms of Percentage Deflection[4]	±2%	±2%	±1%	±0.5%

[1] ASTM D-2487, USBR Designation E-3
[2] LL = Liquid Limit
[3] Or any borderline soil beginning with one of these symbols (i.e., GM-GC, GC-SC).
[4] For ±1% accuracy and predicted deflection of 3%, actual deflection would be between 2% and 4%.

Note: Values applicable only for fills less than 50 ft (15 m). Table does not include any safety factor. For use in predicting initial deflections only; appropriate Deflection Lag Factor must be applied for long-term deflections. If embedment falls on the borderline between two compaction categories, select lower E' value, or average the two values. Percentage Proctor based on laboratory maximum dry density from test standards using 12,500 ft-lb/cu ft (598,000 J/m²) (ASTM D-698, AASHTO T-99, USBR Designation E-11). 1 psi = 6.9 KPa.

TABLE 2-8
Values of E' for Pipe Embedment (See Duncan and Hartley[10])

Type of Soil	Depth of Cover, ft	E' for Standard AASHTO Relative Compaction, lb/in²			
		85%	90%	95%	100%
Fine-grained soils with less than 25% sand content (CL, ML, CL-ML)	0-5	500	700	1000	1500
	5-10	600	1000	1400	2000
	10-15	700	1200	1600	2300
	15-20	800	1300	1800	2600
Coarse-grained soils with fines (SM, SC)	0-5	600	1000	1200	1900
	5-10	900	1400	1800	2700
	10-15	1000	1500	2100	3200
	15-20	1100	1600	2400	3700
Coarse-grained soils with little or no fines (SP, SW, GP, GW)	0-5	700	1000	1600	2500
	5-10	1000	1500	2200	3300
	10-15	1050	1600	2400	3600
	15-20	1100	1700	2500	3800

Soil Support Factor, Fs

Ring deflection and the accompanying horizontal diameter expansion create lateral earth pressure which is transmitted through the embedment soil and into the trench sidewall. This may cause the sidewall soil to compress. If the compression is significant, the embedment can move laterally, resulting in an increase in pipe deflection. Sidewall soil compression is of particular concern when the insitu soil is loose, soft, or highly compressible, such as marsh clay, peat, saturated organic soil, etc. The net effect of sidewall compressibility is a reduction in the soil-pipe system's stiffness. The reverse case may occur as well if the insitu soil is stiffer than the embedment soil; e.g. the insitu soil may enhance the embedment giving it more resistance to deflection. The Soil Support Factor, F_s, is a factor that may be applied to E' to correct for the difference in stiffness between the insitu and embedment soils. Where the insitu soil is less stiff than the embedment, F_s is a reduction factor. Where it is stiffer, F_s is an enhancement factor, i.e. greater than one.

The Soil Support Factor, Fs, may be obtained from Tables 2-9 and 2-10 as follows:

- Determine the ratio B_d/D_O, where B_d equals the trench width at the pipe springline (inches), and D_O equals the pipe outside diameter (inches).
- Based on the native insitu soil properties, find the soil reaction modulus for the in situ soil, E'_N in Table 2-9.
- Determine the ratio E'_N/E'.
- Enter Table 2-10 with the ratios B_d/D_O and E'_N/E' and find Fs.

TABLE 2-9
Values of E'_N, Native Soil Modulus of Soil Reaction, Howard [3]

Native In Situ Soils					E'_N (psi)
Granular		Cohesive			
Std. Pentration ASTM D1586 Blows/ft	Description	Unconfined Compressive Strength (TSF)	Description		
> 0 - 1	very, very loose	> 0 - 0.125	very, very soft		50
1 - 2	very loose	0.125 - 0.25	very soft		200
2 - 4	very loose	0.25 - 0.50	soft		700
4 - 8	loose	0.50 - 1.00	medium		1,500
8 - 15	slightly compact	1.00 - 2.00	stiff		3,000
15 - 30	compact	2.00 - 4.00	very stiff		5,000
30 - 50	dense	4.00 - 6.00	hard		10,000
> 50	very dense	> 6.00	very hard		20,000
Rock	–	–	–		50,000

TABLE 2-10
Soil Support Factor, F_S

E'_N/E'	B_d/D_0 1.5	B_d/D_0 2.0	B_d/D_0 2.5	B_d/D_0 3.0	B_d/D_0 4.0	B_d/D_0 5.0
0.1	0.15	0.30	0.60	0.80	0.90	1.00
0.2	0.30	0.45	0.70	0.85	0.92	1.00
0.4	0.50	0.60	0.80	0.90	0.95	1.00
0.6	0.70	0.80	0.90	0.95	1.00	1.00
0.8	0.85	0.90	0.95	0.98	1.00	1.00
1.0	1.00	1.00	1.00	1.00	1.00	1.00
1.5	1.30	1.15	1.10	1.05	1.00	1.00
2.0	1.50	1.30	1.15	1.10	1.05	1.00
3.0	1.75	1.45	1.30	1.20	1.08	1.00
5.0	2.00	1.60	1.40	1.25	1.10	1.00

Lag Factor and Long-Term Deflection

Spangler observed an increase in ring deflection with time. Settlement of the backfill and consolidation of the embedment under the lateral pressure from the pipe continue to occur after initial installation. To account for this, he recommended applying a lag factor to the Iowa Formula in the range of from 1.25 to 1.5. Lag occurs in installations of both plastic and metal pipes. Howard [3, 11] has shown that the lag factor varies with the type of embedment and the degree of compaction. Many plastic pipe designers use a Lag Factor of 1.0 when using the prism load as it accounts for backfill settlement. This makes even more sense when the Soil Support Factor is included in the calculation.

Vertical Deflection Example

Estimate the vertical deflection of a 24" diameter HDPE DR 26 pipe installed under 18 feet of cover. The embedment material is a well-graded sandy gravel, compacted to a minimum 90 percent of Standard Proctor density, and the native ground is a saturated, soft clayey soil. The anticipated trench width is 42".

SOLUTION: Use the prism load, Equation 2-1, Tables 2-7, 2-9, and 2-10, and Equation 2-10. Table 2-7 gives an E' for a compacted sandy gravel or GW-SW soil as 2000 lb/in². To estimate maximum deflection due to variability, this value will be reduced by 25%, or to 1500 lb/in². Table 2-9 gives an E'_N of 700 psi for soft clay. Since B_d/D equals 1.75 and E'_N/E' equals 0.47, F_S is obtained by interpolation and equal 0.60.

The prism load on the pipe is equal to:

$$P_E = (120)(18) = 2160 \, lb/ft^2$$

Substituting these values into Equation 2-10 gives:

$$\frac{\Delta X}{D_M} = \frac{2160}{144} \left(\frac{(0.1)(1.0)}{\frac{2(110,000)}{3}(\frac{1}{26-1})^3 + (0.061)(0.60)(1500)} \right)$$

$$\frac{\Delta X}{D_M} = 0.025 = 2.5\%$$

Deflection Limits

The designer limits ring deflection in order to control geometric stability of the pipe, wall bending strain, pipeline hydraulic capacity and compatibility with cleaning equipment, and, for bell-and-spigot jointed pipe, its sealing capability. Only the limits for geometric stability and bending strain will be discussed here. Hydraulic capacity is not impaired at deflections less than 7.5%, and bell and spigot deflection limits are established per ASTM D3212.

Geometric stability is lost when the pipe crown flattens and loses its ability to support earth load. Crown flattening occurs with excessive deflection as the increase in horizontal diameter reduces crown curvature. At 25% to 30% deflection, the crown may completely reverse its curvature inward and collapse. See Figure 2-1A. A deflection limit of 7.5% provides at least a 3 to 1 safety factor against reverse curvature.

Bending strain occurs in the pipe wall as a result of ring deflection—outer-fiber tensile strain at the pipe springline and outer-fiber compressive strain at the crown and invert. While strain limits of 5% have been proposed, Jansen [12] reported that, on tests of PE pipe manufactured from pressure-rated resins and subjected to soil pressure only, "no upper limit from a practical design point of view seems to exist for the bending strain." In other words, as deflection increases, the pipe's performance limit will not be overstraining but reverse curvature collapse.

Thus, for non-pressure applications, a 7.5 percent deflection limit provides a large safety factor against instability and strain and is considered a safe design deflection. Some engineers will design profile wall pipe and other non-pressure pipe applications to a 5% deflection limit, but allow spot deflections up to 7.5% during field inspection.

The deflection limits for pressurized pipe are generally lower than for non-pressurized pipe. This is primarily due to strain considerations. Hoop strain from pressurization adds to the outer-fiber tensile strain. But the internal pressure acts to reround the pipe and, therefore, Eq. 2-10 overpredicts the actual long-term deflection for pressurized pipe. Safe allowable deflections for pressurized pipe are given in Table 2-11. Spangler and Handy [13] give equations for correcting deflection to account for rerounding.

TABLE 2-11
Safe Deflection Limits for Pressurized Pipe

DR or SDR	Safe Deflection as % of Diameter
32.5	7.5
26	7.5
21	7.5
17	6.0
13.5	6.0
11	5.0
9	4.0
7.3	3.0

*Based on Long-Term Design Deflection of Buried Pressurized Pipe given in ASTM F1962.

Compressive Ring Thrust

Earth pressure exerts a radial-directed force around the circumference of a pipe that results in a compressive ring thrust in the pipe wall. (This thrust is exactly opposite to the tensile hoop thrust induced when a pipe is pressurized.) See Figure 2-1b. Excessive ring compressive thrust may lead to two different performance limits:

crushing of the material or buckling (loss of stability) of the pipe wall. See Figure 2-1c. This section will discuss crushing, and the next section will discuss buckling.

As is often the case, the radial soil pressure causing the stress is not uniform around the pipe's circumference. However, for calculation purposes it is assumed uniform and equal to the vertical soil pressure at the pipe crown.

Pressure pipes often have internal pressure higher than the radial pressure applied by the soil. As long as there is pressure in the pipe that exceeds the external pressure, the net thrust in the pipe wall is tensile rather than compressive, and wall crush or buckling checks are not necessary. Whether one needs to check this or not can be quickly determined by simply comparing the internal pressure with the vertical soil pressure.

Crushing occurs when the compressive stress in the wall exceeds the yield stress of the pipe material. Equations 2-13 and 2-14 give the compressive stress resulting from earth and live load pressure for conventional extruded DR pipe and for ASTM F894 profile wall PE Pipe:

(2-13)
$$S = \frac{(P_E + P_L)\,DR}{288}$$

(2-14)
$$S = \frac{(P_E + P_L)\,D_O}{288A}$$

WHERE
P_E = vertical soil pressure due to earth load, psf
P_L = vertical soil pressure due to live-load, psf
S = pipe wall compressive stress, lb/in2
DR = Dimension Ratio, D_O/t
D_O = pipe outside diameter (for profile pipe $D_O = D_I + 2H_P$), in
D_I = pipe inside diameter, in
H_P = profile wall height, in
A = profile wall average cross-sectional area, in2/in

(Note: These equations contain a factor of 144 in the denominator for correct units conversions.)

Equation 2-14 may overstate the wall stress in profile pipe. Ring deflection in profile wall pipe induces arching. The "Deep Fill Installation" section of this chapter discusses arching and gives equations for calculating the earth pressure resulting from arching, P_{RD}. P_{RD} is given by Equation 2-23 and may be substituted for PE to determine the wall compressive stress when arching occurs.

The compressive stress in the pipe wall can be compared to the pipe material allowable compressive stress. If the calculated compressive stress exceeds the allowable stress, then a lower DR (heavier wall thickness) or heavier profile wall is required.

Allowable Compressive Stress

Table 2-12 gives allowable long-term compressive stress values for PE 3408 and PE 2406 material.

TABLE 2-12
Long-Term Compressive Stress at 73°F (23°C)

Material	Long-Term Compressive Stress, lb/in²
PE 3408	1000
PE 2406	800

The long-term compressive stress value should be reduced for elevated temperature pipeline operation. Temperature design factors used for hydrostatic pressure may be used, i.e. 0.5 @ 140°F. Additional temperature design factors may be obtained by reference to Table 1-11 in Section 1 of this chapter.

Ring Compression Example

Find the pipe wall compressive ring stress in a DR 32.5 HDPE pipe buried under 46 ft of cover. The ground water level is at the surface, the saturated weight of the insitu silty-clay soil is 120 lbs/ft³.

SOLUTION: Find the vertical earth pressure acting on the pipe. Use Equation 2-1.

Although the net soil pressure is equal to the buoyant weight of the soil, the water pressure is also acting on the pipe. Therefore the total pressure (water and earth load) can be found using the saturated unit weight of the soil.

Next, solve for the compressive stress.

$$P_E = (120 \ pcf)(46 \ ft) = 5520 \ psf$$

$$S = \frac{(5520 \ lb/ft^2)(32.5)}{288} = 623 \ lb/inch^2$$

The compressive stress is within the 1000 lb/in² allowable stress for HDPE given in Table 2-12.

Constrained (Buried) Pipe Wall Buckling

Excessive compressive stress (or thrust) may cause the pipe wall to become unstable and buckle. Buckling from ring compressive stress initiates locally as a large "dimple," and then grows to reverse curvature followed by structural collapse. Resistance to buckling is proportional to the wall thickness divided by the diameter raised to a power. Therefore the lower the DR, the higher the resistance. Buried pipe has an added resistance due to support (or constraint) from the surrounding soil.

Non-pressurized pipes or gravity flow pipes are most likely to have a net compressive stress in the pipe wall and, therefore, the allowable buckling pressure should be calculated and compared to the total (soil and ground water) pressure. For most pressure pipe applications, the fluid pressure in the pipe exceeds the external pressure, and the net stress in the pipe wall is tensile. Buckling needs only be considered for that time the pipe is not under pressure, such as during and immediately after construction and during system shut-downs.

This chapter gives two equations for calculating buckling. The modified Luscher Equation is for buried pipes that are beneath the ground water level, subject to vacuum pressure, or under live load with a shallow cover. These forces act to increase even the slightest eccentricity in the pipe wall by following deformation inward. While soil pressure alone can create instability, soil is less likely to follow deformation inward, particularly if it is granular. So, dry ground buckling is only considered for deep applications and is given by the Moore-Selig Equation found in the section, "Buckling of Pipes in Deep, Dry Fills."

Luscher Equation for Constrained Buckling Below Ground Water Level

For pipes below the ground water level, operating under a full or partial vacuum, or subject to live load, Luscher's equation may be used to determine the allowable constrained buckling pressure. Equation 2-15 and 2-16 are for DR and profile pipe respectively.

(2-15)
$$P_{WC} = \frac{5.65}{N}\sqrt{RB'E'\frac{E}{12(DR-1)^3}}$$

(2-16)
$$P_{WC} = \frac{5.65}{N}\sqrt{RB'E'\frac{EI}{D_M^3}}$$

WHERE
P_{WC} = allowable constrained buckling pressure, lb/in²
N = safety factor

(2-17) $$R = 1 - 0.33\frac{H_{GW}}{H}$$

WHERE
R = buoyancy reduction factor
H_{GW} = height of ground water above pipe, ft
H = depth of cover, ft

(2-18) $$B' = \frac{1}{1 + 4e^{(-0.065H)}}$$

WHERE
e = natural log base number, 2.71828
E' = soil reaction modulus, psi
E = apparent modulus of elasticity, psi
DR = Dimension Ratio
I = pipe wall moment of inertia, in^4/in (t^3/12, if solid wall construction)
D_M = Mean diameter (D_I + 2z or D_O − t), in

Although buckling occurs rapidly, long-term external pressure can gradually deform the pipe to the point of instability. This behavior is considered viscoelastic and can be accounted for in Equations 2-15 and 2-16 by using the apparent modulus of elasticity value for the appropriate time and temperature of the loading. For instance, a vacuum event is resisted by the short-term value of the modulus whereas continuous ground water pressure would be resisted by the 50 year value. For modulus values see Table 2-6.

For pipes buried with less than 4 ft or a full diameter of cover, Equations 2-15 and 2-16 may have limited applicability. In this case the designer may want to use Equations 2-39 and 2-40.

The designer should apply a safety factor commensurate with the application. A safety factor of 2.0 has been used for thermoplastic pipe.

The allowable constrained buckling pressure should be compared to the total vertical stress acting on the pipe crown from the combined load of soil, and ground water or floodwater. It is prudent to check buckling resistance against a ground water level for a 100-year-flood. In this calculation the total vertical stress is typically taken as the prism load pressure for saturated soil, plus the fluid pressure of any floodwater above the ground surface.

For DR pipes operating under a vacuum, it is customary to use Equation 2-15 to check the combined pressure from soil, ground water, and vacuum, and then to use the unconstrained buckling equation, Equation 2-39, to verify that the pipe can operate with the vacuum independent of any soil support or soil load, in case construction does not develop the full soil support. Where vacuum load is short-term, such as during water hammer events two calculations with Equation 2-14 are necessary. First determine if the pipe is sufficient for the ground water and soil pressure using a long-term modulus; then determine if the pipe is sufficient for the combined ground water, soil pressure and vacuum loading using the short-term modulus.

Constrained Buckling Example

Does a 36" SDR 26 HDPE pipe have satisfactory resistance to constrained buckling when installed with 18 ft of cover in a compacted soil embedment? Assume ground water to the surface and an E' of 1500 lb/in².

SOLUTION: Solve Equation 2-15. Since this is a long-term loading condition, the stress relaxation modulus can be assumed to be 28,200 psi. Soil cover, H, and ground water height, H_{GW}, are both 18 feet. Therefore, the soil support factor, B', is found as follows;

$$B' = \frac{1}{1+4e^{-(0.065)(18)}} = 0.446$$

and the bouyancy reduction factor, R, is found as follows:

$$R = 1 - 0.33\frac{18}{18} = 0.67$$

Solve Equation 2-15 for the allowable long-term constrained buckling pressure:

$$P_{WC} = \frac{5.65}{2}\sqrt{\frac{0.67(0.446)1500(28,200)}{12(26-1)^3}}$$

$P_{WC} = 23.2 \; psi = 3340 \; psf$

The earth pressure and ground water pressure applied to the pipe is found using Equation 2-1 (prism load) with a saturated soil weight. The saturated soil weight being the net weight of both soil and water.

$$P_E = (120)(18) = 2160 \frac{lb}{ft^2}$$

Compare this with the constrained buckling pressure. Since P_{WC} exceeds P_E, DR 26 has satisfactory resistance to constrained pipe buckling.

AWWA Design Window

The AWWA Committee Report, "Design and Installation of Polyethylene (PE) Pipe Made in Accordance with AWWA C906" describes a Design Window. Applications that fall within this window require no calculations other than constrained buckling per Equation 2-15. It turns out that if pipe is limited to DR 21 or lower as in Table 2-13, the constrained buckling calculation has a safety factor of at least 2, and no calculations are required.

The design protocol under these circumstances (those that fall within the AWWA Design Window) is thereby greatly simplified. The designer may choose to proceed with detailed analysis of the burial design and utilize the AWWA Design Window guidelines as a means of validation for his design calculations and commensurate safety factors. Alternatively, he may proceed with confidence that the burial design for these circumstances (those outlined within the AWWA Design Window) has already been analyzed in accordance with the guidelines presented in this chapter.

The Design Window specifications are:

- Pipe made from pressure-rated high-density PE material.
- Essentially no dead surface load imposed over the pipe, no ground water above the surface, and provisions for preventing flotation of shallow cover pipe have been provided.
- The embedment materials are coarse-grained, compacted to at least 85% Standard Proctor Density and have an E' of at least 1000 psi. The native soil must be stable; in other words the native soil must have an E' of at least 1000 psi.
- The unit weight of the native soil does not exceed 120 pcf.
- The pipe is installed in accordance with manufacturer's recommendations for controlling shear and bending loads and minimum bending radius, and installed in accordance with ASTM D2774 for pressure pipes or ASTM D2321 for non-pressure pipes.
- Minimum depth of cover is 2 ft (0.61 m); except when subject to AASHTO H20 truck loadings, in which case the minimum depth of cover is the greater of 3 ft (0.9 m) or one pipe diameter.
- Maximum depth of cover is 25 ft (7.62 m).

TABLE 2-13
Design Window Maximum & Minimum Depth of Cover *

DR	Min. Depth of Cover With H20 Load	Min. Depth of Cover Without H20 Load	Maximum Depth of Cover
7.3	3 ft	2 ft	25 ft
9	3 ft	2 ft	25 ft
11	3 ft	2 ft	25 ft
13.5	3 ft	2 ft	25 ft
17	3 ft	2 ft	25 ft
21	3 ft	2 ft	25 ft

* Limiting depths where no calculations are required. Pipes are suitable for deeper depth provided a sufficient E' is accomplished during installations. Calculations would be required for depth greater than 25 ft.

Installation Category #2: **Shallow Cover Vehicular Loading**

The Standard Installation methodology assumes that the pipe behaves primarily as a "membrane" structure, that is, the pipe is almost perfectly flexible with little ability to resist bending. At shallow cover depths, especially those less than one pipe diameter, membrane action may not fully develop, and surcharge or live loads place a bending load on the pipe crown. In this case the pipe's flexural stiffness carries part of the load and prevents the pipe crown from dimpling inward under the load. Equation 2-19, published by Watkins [14] gives the soil pressure that can be supported at the pipe crown by the combination of the pipe's flexural stiffness (bending resistance) and the soil's internal resistance against heaving upward. In addition to checking Watkins' formula, the designer should check deflection using Equations 2-10 or 2-11, pipe wall compressive stress using Equations 2-13 or 2-14, and pipe wall buckling using Equations 2-15 or 2-16.

Watkins' equation is recommended only where the depth of cover is greater than one-half of the pipe diameter and the pipe is installed at least 18 inches below the road surface. In other words, it is recommended that the pipe regardless of diameter always be at least 18" beneath the road surface where there are live loads present; more may be required depending on the properties of the pipe and installation. In some cases, lesser cover depths may be sufficient where there is a reinforced concrete cap or a reinforced concrete pavement slab over the pipe. Equation 2-19 may be used for both DR pipe and profile pipe. See definition of "A" below.

$$(2\text{-}19) \quad P_{WAT} = \frac{12w(KH)^2}{N_S D_O} + \frac{7387(I)}{N_S D_O^2 c}\left(S_{MAT} - \frac{w D_O H}{288A}\right)$$

WHERE
P_{WAT} = allowable live load pressure at pipe crown for pipes with one diameter or less of cover, psf
w = unit weight of soil, lb/ft³
D_O = pipe outside diameter, in
H = depth of cover, ft
I = pipe wall moment of inertia (t³/12 for DR pipe), in⁴/in
A = profile wall average cross-sectional area, in²/in, for profile pipe or wall thickness (in) for DR pipe
c = outer fiber to wall centroid, in
c = H_P – z for profile pipe and c = 0.5t for DR pipe, in
H_P = profile wall height, in
z = pipe wall centroid, in
S_{MAT} = material yield strength, lb/in², Use 3000 PSI for PE3408
N_S = safety factor
K = passive earth pressure coefficient

(2-20)
$$K = \frac{1 + SIN(\phi)}{1 - SIN(\phi)}$$

Ø = angle of internal friction, deg

Equation 2-19 is for a point load applied to the pipe crown. Wheel loads should be determined using a point load method such as given by Equations 2-2 (Timoshenko) or 2-4 (Boussinesq).

When a pipe is installed with shallow cover below an unpaved surface, rutting can occur which will not only reduce cover depth, but also increase the impact factor.

Shallow Cover Example
Determine the safety factor against flexural failure of the pipe accompanied by soil heave, for a 36″ RSC 100 F894 profile pipe 3.0 feet beneath an H20 wheel load. Assume an asphalt surface with granular embedment.

SOLUTION: The live load pressure acting at the crown of the pipe can be found using Equation 2-4, the Boussinesq point load equation. At 3.0 feet of cover the highest live load pressure occurs directly under a single wheel and equals:

$$PL = \frac{(3)(2.0)(16000)(3.0)^3}{2\pi(3.0)^5} = 1697 \ psf$$

WHERE
$I_f = 2.0$
$W = 16,000$ lbs
$H = 3.0$ ft
$w = 120$ pcf

The live load pressure is to be compared with the value in Equation 2-19. To solve Equation 2-19, the following parameters are required:

$I = 0.171$ in^4/in
$A = 0.470$ in^2/in
$H_P = 2.02$ in (Profile Wall Height)
$DO = D_I + 2*h = 36.00 + 2*2.02 = 40.04$ in
$Z = 0.58$ in
$C = h - z = 1.44$ in
$S = 3000$ psi
$\phi = 30$ deg.

Determine the earth pressure coefficient:

$$K = \frac{1+\sin(30)}{1-\sin(30)} = \frac{1+0.5}{1-0.5} = 3.0$$

The live load pressure incipient to failure equals:

$$P_{WAT} = \frac{(12)120(3.0*3.0)^2}{40.04} + \frac{7387*0.171}{40.04^2(1.44)}(3000 - \frac{120(40.04)3.0}{288*0.470})$$

$$P_{WAT} = 2904 + 1584 = 4498 \text{ psf}$$

The resulting safety factor equals:

$$N = \frac{P_{WAT}}{P_L} = \frac{4498}{1697} = 2.65$$

Installation Category #3: **Deep Fill Installation**

The performance limits for pipes in a deep fill are the same as for any buried pipe. They include:

1. Compressive ring thrust stress
2. Ring deflection
3. Constrained pipe wall buckling

The suggested calculation method for pipe in deep fill applications involves the introduction of design routines for each performance limit that are different than those previously given.

Compressive ring thrust is calculated using soil arching. The arching calculation may also be used for profile pipe designs in standard trench applications. Profile pipes are relatively low stiffness pipes where significant arching may occur at relatively shallow depths of cover.

At a depth of around 50 feet or so it becomes impractical to use Spangler's equation as published in this chapter because it neglects the significant load reduction due to arching and the inherent stiffening of the embedment and consequential increase in E' due to the increased lateral earth pressure applied to the embedment. This section gives an alternate deflection equation for use with polyethylene pipes. It was first introduced by Watkins et al. [1] for metal pipes, but later Gaube extended its use to include polyethylene pipes. [15]

Where deep fill applications are in dry soil, Luscher's equation (Eq. 2-15 or 2-16) may often be too conservative for design as it considers a radial driving force from ground water or vacuum. Moore and Selig[17] developed a constrained pipe wall buckling equation suitable for pipes in dry soils, which is given in a following section.

Considerable care should be taken in the design of deeply buried pipes whose failure may cause slope failure in earthen structures, or refuse piles or whose failure may have severe environmental or economical impact. These cases normally justify the use of methods beyond those given in this Chapter, including finite element analysis and field testing, along with considerable professional design review.

Compressive Ring Thrust and the Vertical Arching Factor
The combined horizontal and vertical earth load acting on a buried pipe creates a radially-directed compressive load acting around the pipe's circumference. When a PE pipe is subjected to ring compression, thrust stress develops around the pipe hoop, and the pipe's circumference will ever so slightly shorten. The shortening permits "thrust arching," that is, the pipe hoop thrust stiffness is less than the soil hoop thrust stiffness and, as the pipe deforms, less load follows the pipe. This occurs much like the vertical arching described by Marston.[18] Viscoelasticity enhances this effect. McGrath[19] has shown thrust arching to be the predominant form of arching with PE pipes.

Burns and Richard[6] have published equations that give the resulting stress occurring in a pipe due to arching. As discussed above, the arching is usually considered when calculating the ring compressive stress in profile pipes. For deeply buried pipes McGrath [19] has simplified the Burns and Richard's equations to derive a vertical arching factor as given by Equation 2-21.

$$\text{(2-21)} \quad VAF = 0.88 - 0.71 \frac{S_A - 1}{S_A + 2.5}$$

WHERE
VAF = Vertical Arching Factor
S_A = Hoop Thrust Stiffness Ratio

$$\text{(2-22)} \quad S_A = \frac{1.43 \, M_S \, r_{CENT}}{EA}$$

WHERE
r_{CENT} = radius to centroidal axis of pipe, in
M_S = one-dimensional modulus of soil, psi
E = apparent modulus of elasticity of pipe material, psi
A = profile wall average cross-sectional area, in²/in, or wall thickness (in) for DR pipe

One-dimensional modulus values for soil can be obtained from soil testing, geotechnical texts, or Table 2-14 which gives typical values. The typical values in Table 2-14 were obtained by converting values from McGrath[20].

TABLE 2-14
Typical Values of M_S, One-Dimensional Modulus of Soil

Vertical Soil Stress¹ (psi)	Gravelly Sand/Gravels 95% Std. Proctor (psi)	Gravelly Sand/Gravels 90% Std. Proctor (psi)	Gravelly Sand/Gravels 85% Std. Proctor (psi)
10	3000	1600	550
20	3500	1800	650
40	4200	2100	800
60	5000	2500	1000
80	6000	2900	1300
100	6500	3200	1450

* Adapted and extended from values given by McGrath[20]. For depths not shown in McGrath[20], the MS values were approximated using the hyperbolic soil model with appropriate values for K and n where n=0.4 and K=200, K=100, and K=45 for 95% Proctor, 90% Proctor, and 85% Proctor, respectively.
¹ Vertical Soil Stress (psi) = [soil depth (ft) x soil density (pcf)]/144

The radial directed earth pressure can be found by multiplying the prism load (pressure) by the vertical arching factor as shown in Eq. 2-23.

$$\text{(2-23)} \quad P_{RD} = (VAF)wH$$

WHERE
PRD = radial directed earth pressure, lb/ft²
w = unit weight of soil, pcf
H = depth of cover, ft

The ring compressive stress in the pipe wall can be found by substituting P_{RD} from Equation 2-23 for P_E in Equation 2-13 for DR pipe and Equation 2-14 for profile wall pipe.

Radial Earth Pressure Example
Determine the radial earth pressure acting on a 36" RSC 100 profile wall pipe buried 30 feet deep. The pipe's cross-sectional area, A, equals 0.470 inches²/inch, its radius to the centroidal axis is 18.00 inches plus 0.58 inches, and its modulus is 28,250 psi. Its wall height is 2.02 in and its D_O equals 36 in +2 (2.02 in) or 40.04 in. Assume the pipe is installed in a clean granular soil compacted to 90% Standard Proctor (Ms = 1875 psi), the insitu soil is as stiff as the embedment, and the backfill weighs 120 pcf. (Where the excavation is in a stable trench, the stiffness of the insitu soil can generally be ignored in this calculation.)

$$S_A = \frac{1.43(1875\frac{lbs}{inch^2})(18.58\,inch)}{(28250\frac{lbs}{inch^2})(0.470\frac{inch^2}{inch})} = 3.75$$

$$VAF = 0.88 - 0.71\frac{3.75-1}{3.75+2.5} = 0.57$$

$$P_{RD} = 0.57(120\,pcf)(30\,ft) = 2052\frac{lb}{ft^2}$$

$$S = \frac{P_{RD}D_O}{288A} = \frac{2052\,psf(40.04\,in)}{288\,(0.470\,in^2/in)} = 607\,psi \leq 1000\,psi$$

Ring Deflection of Pipes Using Watkins-Gaube Graph
R. Watkins[1] developed an extremely straight-forward approach to calculating pipe deflection in a fill that does not rely on E'. It is based on the concept that the deflection of a pipe embedded in a layer of soil is proportional to the compression or settlement of the soil layer and that the constant of proportionality is a function of the relative stiffness between the pipe and soil. Watkins used laboratory testing to

establish and graph proportionality constants, called Deformation Factors, D_F, for the stiffness ranges of metal pipes. Gaube[15, 16] extended Watkins' work by testing to include PE pipes. In order to predict deflection, the designer first determines the amount of compression in the layer of soil in which the pipe is installed using conventional geotechnical equations. Then, deflection equals the soil compression multiplied by the D_F factor. This bypasses some of the inherent problems associated with using E' values. The designer using the Watkins-Gaube Graph should select conservative soil modulus values to accommodate variance due to installation. Two other factors to consider when using this method is that it assumes a constant Deformation Factor independent of depth of cover and it does not address the effect of the presence of ground water on the Deformation Factor.

To use the Watkins-Gaube Graph, the designer first determines the relative stiffness between pipe and soil, which is given by the Rigidity Factor, R_F. Equation 2-24 and 2-25 are for DR pipe and for profile pipe respectively:

(2-24)
$$R_F = \frac{12 E_S (DR-1)^3}{E}$$

(2-25)
$$R_F = \frac{E_S D_m^3}{EI}$$

WHERE
DR = Dimension Ratio
E_S = Secant modulus of the soil, psi
E = Apparent modulus of elasticity of pipe material, psi
I = Pipe wall moment of inertia of pipe, in^4/in
D_m = Mean diameter (D_I + 2z or D_O – t), in

The secant modulus of the soil may be obtained from testing or from a geotechnical engineer's evaluation. In lieu of a precise determination, the soil modulus may be related to the one-dimensional modulus, M_S, from Table 2-14 by the following equation where μ is the soil's Poisson ratio.

(2-26)
$$E_S = M_S \frac{(1+\mu)(1-2\mu)}{(1-\mu)}$$

TABLE 2-15
Typical range of Poisson's Ratio for Soil (Bowles[21])

Soil Type	Poisson Ratio, μ
Saturated Clay	0.4-0.5
Unsaturated Clay	0.1-0.3
Sandy Clay	0.2-0.3
Silt	0.3-0.35
Sand (Dense)	0.2-0.4
Coarse Sand (Void Ratio 0.4-0.7)	0.15
Fine-grained Sand (Void Ratio 0.4-0.7)	0.25

Next, the designer determines the Deformation Factor, D_F, by entering the Watkins-Gaube Graph with the Rigidity Factor. See Fig. 2-6. The Deformation Factor is the proportionality constant between vertical deflection (compression) of the soil layer containing the pipe and the deflection of the pipe. Thus, pipe deflection can be obtained by multiplying the proportionality constant D_F times the soil settlement. If D_F is less than 1.0 in Fig. 2-6, use 1.0.

The soil layer surrounding the pipe bears the entire load of the overburden above it without arching. Therefore, settlement (compression) of the soil layer is proportional to the prism load and not the radial directed earth pressure. Soil strain, ε_S, may be determined from geotechnical analysis or from the following equation:

(2-27)
$$\varepsilon_S = \frac{wH}{144 E_S}$$

WHERE
w = unit weight of soil, pcf
H = depth of cover (height of fill above pipe crown), ft
E_S = secant modulus of the soil, psi

The designer can find the pipe deflection as a percent of the diameter by multiplying the soil strain, in percent, by the deformation factor:

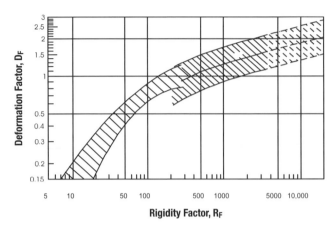

Figure 2-6 Watkins-Gaube Graph

$$(2\text{-}28) \quad \frac{\Delta X}{D_M}(100) = D_F \varepsilon_S$$

WHERE

$\Delta X / D_M$ multiplied by 100 gives percent deflection.

Watkins – Gaube Calculation Technique

Find the deflection of a 6″ SDR 11 pipe under 140 ft of fill with granular embedment containing 12% or less fines, compacted at 90% of standard proctor. The fill weighs 75 pcf.

SOLUTION: First, calculate the vertical soil pressure equation, Eq. 2-1.

Eq. 2-1: $P_E = wH$

$P_E = (75 \text{lb/ft}^3)(140 \text{ ft})$

$P_E = 10,500 \text{ lb/ft}^2$ or 72.9 psi

The M_S is obtained by interpolation from Table 2-14 and equals 2700. The secant modulus can be found assuming a Poisson Ratio of 0.30

$$E_S = \frac{2700 \, psi \, (1+0.30)(1-2(0.30))}{(1-0.30)} = 2005 \, psi$$

The rigidity factor is obtained from Equation 2-24.

$$R_F = \frac{12(2005)(11-1)^3}{28250} = 852$$

Using Figure 2-6, the deformation factor is found to be 1.2. The soil strain is calculated by Equation 2-27.

$$\varepsilon_S = \frac{75pcf * 140ft}{144 * 2005\frac{lbs}{inch^2}} \bullet 100 = 3.6\%$$

The deflection is found by multiplying the soil strain by the deformation factor:

$$\frac{\Delta X}{D_M}(100) = 1.2 * 3.6 = 4.4\%$$

Moore-Selig Equation for Constrained Buckling in Dry Ground

As discussed previously, a compressive thrust stress exists in buried pipe. When this thrust stress approaches a critical value, the pipe can experience a local instability or large deformation and collapse. In an earlier section of this chapter, Luscher's equation was given for constrained buckling under ground water. Moore and Selig[17] have used an alternate approach called the continuum theory to develop design equations for contrained buckling due to soil pressure (buckling of embedded pipes). The particular version of their equations given below is more appropriate for dry applications than Luscher's equation. Where ground water is present, Luscher's equation should be used.

The Moore-Selig Equation for critical buckling pressure follows: (Critical buckling pressure is the pressure at which buckling will occur. A safety factor should be provided.)

(2-29)
$$P_{CR} = \frac{2.4 \, \varphi \, R_H}{D_M}(EI)^{\frac{1}{3}}(E_S^*)^{\frac{2}{3}}$$

WHERE
P_{CR} = Critical constrained buckling pressure, psi
φ = Calibration Factor, 0.55 for granular soils
R_H = Geometry Factor
E = Apparent modulus of elasticity of pipe material, psi
I = Pipe wall moment of Inertia, in^4/in (t^3/12, if solid wall construction)
E_S^* = ES/(1-μ)
E_S = Secant modulus of the soil, psi
μ_s = Poisson's Ratio of Soil

The geometry factor is dependent on the depth of burial and the relative stiffness between the embedment soil and the insitu soil. Moore has shown that for deep burials in uniform fills, R_H equals 1.0.

Critical Buckling Example

Determine the critical buckling pressure and safety factor against buckling for the 6" SDR 11 pipe in the previous example.

SOLUTION:

$$E_S^* = \frac{2000}{(1-0.3)} = 2860 \frac{lbs}{inch^2}$$

$$P_{CR} = \frac{2.4*0.55*1.0}{5.987}(28250*0.018)^{\frac{1}{3}}(2860)^{\frac{2}{3}} = 354\frac{lbs}{inch^2}$$

Determine the S.F. against buckling:

$$S.F. = \frac{P_{CR}}{P_E} = \frac{354*144}{140*75} = 4.9$$

Installation Category #4: **Shallow Cover Flotation Effects**

Shallow cover presents some special considerations for flexible pipes. As already discussed, full soil structure interaction (membrane effect) may not occur, and live loads are carried in part by the bending stiffness of the pipe. Even if the pipe has sufficient strength to carry live load, the cover depth may not be sufficient to prevent the pipe from floating upward or buckling if the ground becomes saturated with ground water. This section addresses:

- Minimum soil cover requirements to prevent flotation
- Hydrostatic buckling (unconstrained)

Design Considerations for Ground Water Flotation

High ground water can float buried pipe, causing upward movement off-grade as well as catastrophic upheaval. This is not an issue for plastic pipes alone. Flotation of metal or concrete pipes may occur at shallow cover when the pipes are empty.

Flotation occurs when the ground water surrounding the pipe produces a buoyant force greater than the sum of the downward forces provided by the soil weight, soil friction, the weight of the pipe, and the weight of its contents. In addition to the disruption occurring due to off-grade movements, flotation may also cause significant reduction of soil support around the pipe and allow the pipe to buckle from the external hydrostatic pressure.

Flotation is generally not a design consideration for buried pipe where the pipeline runs full or nearly full of liquid or where ground water is always below the pipe invert. Where these conditions are not met, a quick "rule of thumb" is that pipe buried in soil having a saturated unit weight of at least 120 lb/ft³ with at least 1½ pipe diameters of cover will not float. However, if burial is in lighter weight soils or with lesser cover, ground water flotation should be checked.

Mathematically the relationship between the buoyant force and the downward forces is given in Equation 2-30. Refer to Figure 2-7. For an empty pipe, flotation will occur if:

(2-30) $\quad F_B > W_P + W_S + W_D + W_L$

WHERE
F_B = buoyant force, lb/ft of pipe
W_P = pipe weight, lb/ft of pipe
W_S = weight of saturated soil above pipe, lb/ft of pipe
W_D = weight of dry soil above pipe, lb/ft of pipe
W_L = weight of liquid contents, lb/ft of pipe

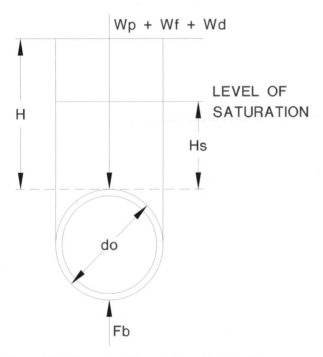

Figure 2-7 Schematic of Ground Water Flotation Forces

For a 1 ft length of pipe totally submerged, the upward buoyant force is:

(2-31) $\quad F_B = \omega_G \dfrac{\pi}{4} d_O^{\,2}$

WHERE
d_O = pipe outside diameter, ft
ω_G = specific weight of ground water
(fresh water = 62.4 lb/ft³)
(sea water = 64.0 lb/ft³)

The average pipe weight, W_P in lbs/ft may be obtained from manufacturers' literature or from Equation 2-32.

(2-32)
$$W_P = \pi d_O^2 \frac{(1.06 \cdot DR - 1.12)}{DR^2} 59.6$$

Equation 2-33 gives the weight of soil per lineal foot of pipe.

(2-33) $$W_D = \omega_d (H - H_S) d_O$$

WHERE
ω_d = unit weight of dry soil, pcf (See Table 2-16 for typical values.)
H = depth of cover, ft
H_S = level of ground water saturation above pipe, ft

TABLE 2-16
Saturated and Dry Soil Unit Weight

Soil Type	Unit Weight, lb/ft³	
	Saturated	Dry
Sands & Gravel	118-150	93-144
Silts & Clays	87-131	37-112
Glacial Till	131-150	106-144
Crushed Rock	119-137	94-125
Organic Silts & Clay	81-112	31-94

(2-34)
$$W_S = (\omega_S - \omega_G)\left(\frac{d_O^2(4-\pi)}{8} + d_O H_S\right)$$

WHERE
ω_S = saturated unit weight of soil, pcf

When an area is submerged, the soil particles are buoyed by their immersion in the ground water. The effective weight of submerged soil, (W_S – W_G), is the soil's saturated unit weight less the density of the ground water. For example, a soil of

120 pcf saturated unit weight has an effective weight of 57.6 pcf when completely immersed in water (120 - 62.4 = 57.6 pcf).

Equation 2-35 gives the weight per lineal foot of the liquid in a full pipe.

(2-35)
$$W_L = \omega_L \frac{\pi d'^2}{4}$$

WHERE
W_L = weight of the liquid in the pipe, lb/ft
ω_L = unit weight of liquid in the pipe, pcf

and if half-full, the liquid weight is

(2-36)
$$W_L = \omega_L \frac{\pi d'^2}{8}$$

WHERE
ω_L = unit weight of the liquid in the pipe, lb/ft³
d' = pipe inside diameter, ft

For liquid levels between empty and half-full (0% to 50%), or between half-full and full (50% to 100%), the following formulas provide an approximate liquid weight with an accuracy of about ±10%. Please refer to Figure 2-8.

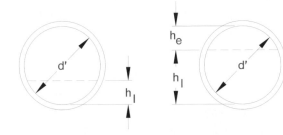

LIQUID LEVEL
BETWEEN 0% & 50%

LIQUID LEVEL
BETWEEN 50% & 100%

Figure 2-8 Flotation and Internal Liquid Levels

For a liquid level between empty and half-full, the weight of the liquid in the pipe is approximately

(2-37)
$$W_L = \omega_L \frac{4 h_l^3}{3} \sqrt{\frac{d' - h_l}{h_l}} + 0.392$$

WHERE
h_l = liquid level in pipe, ft

For a liquid level between half-full and full, the weight of the liquid in the pipe is approximately

(2-38)
$$W_L = \omega_L \left(\frac{\pi d'^2}{4} - 1.573\, h_e \right)$$

WHERE $h_e = d' - h_l$

Unconstrained Pipe Wall Buckling (Hydrostatic Buckling)

The equation for buckling given in this section is here to provide assistance when designing shallow cover applications. However, it may be used to calculate the buckling resistance of above grade pipes subject to external air pressure due to an internal vacuum, for submerged pipes in lakes or ponds, and for pipes placed in casings without grout encasement.

Unconstrained pipe are pipes that are not constrained by soil embedment or concrete encasement. Above ground pipes are unconstrained, as are pipes placed in a casing prior to grouting. Buried pipe may be considered essentially unconstrained where the surrounding soil does not significantly increase its buckling resistance beyond its unconstrained strength. This can happen where the depth of cover is insufficient to prevent the pipe from floating slightly upward and breaking contact with the embedment below its springline. Ground water, flooding, or vacuum can cause buckling of unconstrained pipe.

A special case of unconstrained buckling referred to as "upward" buckling may happen for shallow buried pipe. Upward buckling occurs when lateral pressure due to ground water or vacuum pushes the sides of the pipe inward while forcing the pipe crown and the soil above it upward. (Collapse looks like pipe deflection rotated 90 degrees.) A pipe is susceptible to upward buckling where the cover depth is insufficient to restrain upward crown movement. It has been suggested that a minimum cover of four feet is required before soil support contributes to averting upward buckling; however, larger diameter pipe may require as much as a diameter and a half to develop full support.

A conservative design for shallow cover buckling is to assume no soil support, and design the pipe using the unconstrained pipe wall buckling equation. In lieu of this, a concrete cap, sufficient to resist upward deflection, may also be placed over the pipe and then the pipe may be designed using Luscher's equation for constrained buckling.

Equations 2-39 and 2-40 give the allowable unconstrained pipe wall buckling pressure for DR pipe and profile pipe, respectively.

(2-39)
$$P_{WU} = \frac{f_O}{N_S} \frac{2E}{(1-\mu^2)} \left(\frac{1}{DR-1}\right)^3$$

(2-40)
$$P_{WU} = \frac{f_O}{N_S} \frac{24EI}{(1-\mu^2)D_M^3}$$

WHERE
P_{WU} = allowable unconstrained pipe wall buckling pressure, psi
DR = Dimension Ratio
E = apparent modulus of elasticity of pipe material, psi
f_O = Ovality Correction Factor, Figure 2-9
N_S = safety factor
I = Pipe wall moment of inertia, in4/in
μ = Poisson's ratio
D_M = Mean diameter, (DI + 2z or DO -t), in
D_I = pipe inside diameter, in
z = wall-section centroidal distance from inner fiber of pipe, in

Although buckling occurs rapidly, long-term external pressure can gradually deform the pipe to the point of instability. This behavior is considered viscoelastic and can be accounted for in Equations 2-39 and 2-40 by using the apparent modulus of elasticity value for the appropriate time and temperature of the specific application as given in Table 2-6. For Poisson's ratio, designers typically use a value of 0.45 for long-term loading on polyethylene pipe, and 0.35 for short-term loading.

Ovality or deflection of the pipe diameter increases the local radius of curvature of the pipe wall and thus reduces buckling resistance. Ovality is typically reported as the percentage reduction in pipe diameter or:

(2-41)
$$\%DEFLECTION = 100\left(\frac{D_I - D_{MIN}}{D_I}\right)$$

WHERE
D_I = pipe inside diameter, in
D_{MIN} = pipe minimum inside diameter, in

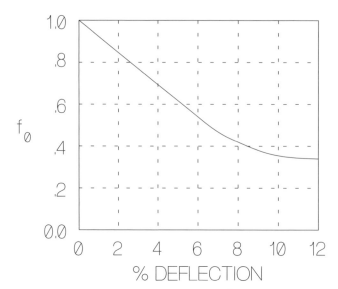

Figure 2-9 Ovality Compensation Factor, $f_Ø$

The designer should compare the critical buckling pressure with the actual anticipated pressure, and apply a safety factor commensurate with their assessment of the application. A safety factor of 2.5 is common, but specific circumstances may warrant a higher or lower safety factor. For large-diameter submerged pipe, the anticipated pressure may be conservatively calculated by determining the height of water from the pipe invert rather than from the pipe crown.

Ground Water Flotation Example
Find the allowable flood water level above a 10" DR 26 HDPE pipe installed with only 2 ft of cover. Assume the pipe has 3 percent ovality due to shipping, handling, and installation loads.

SOLUTION: Use Equation 2-39. The pipe wall buckling pressure depends upon the duration of the water level above the pipe. If the water level is constant, then a long-term value of the stress relaxation modulus should be used, but if the water level rises only occasionally, a shorter term elastic modulus may be applied.

Case (a): For the constant water above the pipe, the stress relaxation modulus at 50 year, 73°F is approximately 28,200 lb/in² for a typical P3408 material. Assuming 3% ovality (f_O equals 0.76) and a 2.5 to 1 safety factor, the allowable long-term pressure, P_{WU} is given by:

$$P_{WU} = \frac{(0.76)}{2.5} \frac{2(28,200)}{(1-0.45^2)} \left(\frac{1}{26-1}\right)^3 = 1.4 \; psi = 3.2 \; ft - Hd$$

Case (b): Flooding conditions are occasional happenings, usually lasting a few days to a week or so. However, ground water elevations may remain high for several weeks following a flood. The 1000 hour (41.6 days) elastic modulus value has been used to approximate the expected flood duration.

$$P_{WU} = \frac{(0.76)}{2.5} \frac{2(43,700)}{(1-0.45^2)} \left(\frac{1}{26-1}\right)^3 = 2.1 \, psi = 4.9 \, ft - Hd$$

Section 3 **Thermal Design Considerations**

Introduction

Like most materials, polyethylene is affected by changing temperature. Unrestrained, polyethylene will experience greater expansion and contraction than many other materials due to increasing or decreasing (respectively) temperatures. However, its low elastic modulus eases the challenge of arresting this movement, and very often end restraints may be employed to eliminate the effects of temperature changes.

Polyethylene pipe can be installed and operated in sub-freezing conditions. Ice in the pipe will restrict or stop flow, but not cause pipe breakage. In sub-freezing conditions, polyethylene is not as impact resistant as it is at room temperature. In all cases, one should follow the unloading guidelines in the handling and storage section of the PPI Engineering Handbook chapter on inspections, tests, and safety considerations that calls for use of lifting devices to safely unload polyethylene piping products.

Unrestrained Thermal Effects

The theoretical change in length for an unrestrained pipe placed on a frictionless surface can be determined from Equation 3-1.

(3-1) $$\Delta L = L \alpha \Delta T$$

WHERE
ΔL = pipeline length change, in
L = pipe length, ft
α = thermal expansion coefficient, in/in/°F
ΔT = temperature change, °F

The coefficient of thermal expansion for polyethylene pipe material is approximately 1×10^{-4} in/in/°F. As a "rule of thumb," temperature change for *unrestrained* PE pipe is about "1/10/100," that is, 1 inch for each 10°F temperature change for each 100 foot of pipe. A temperature rise results in a length increase while a temperature drop results in a length decrease.

End Restrained Thermal Effects

A length of pipe that is restrained or anchored on both ends and one placed on a frictionless surface will exhibit a substantially different reaction to temperature change than the unrestrained pipe discussed above. If the pipe is restrained in a straight line between two points and the temperature decreases, the pipe will attempt to decrease in length. Because the ends are restrained or anchored, length change cannot occur, so a longitudinal tensile stress is created along the pipe. The magnitude of this stress can be determined using Equation 3-2.

(3-2) $\sigma = E \alpha \Delta T$

Where terms are as defined above, and
σ = longitudinal stress in pipe, psi
E = apparent modulus elasticity of pipe material, psi

The value of the apparent modulus of elasticity of the pipe material has a large impact on the calculated stress. As with all thermoplastic materials, polyethylene's modulus, and therefore its stiffness, is dependent on temperature and the duration of the applied load. Therefore, the appropriate elastic modulus should be selected based on these two variables. When determining the appropriate time interval, it is important to consider that heat transfer occurs at relatively slow rates through the wall of polyethylene pipe; therefore temperature changes do not occur rapidly. Because the temperature change does not happen rapidly, the average temperature is often chosen for the modulus selection.

TABLE 3-1
Apparent Modulus Elasticity for HDPE Pipe Material at Various Temperatures

Load Duration	PE 3408 Apparent Elastic Modulus†, 1000 psi (MPa), at Temperature, °F (°C)							
	-20 (-29)	0 (-18)	40 (4)	60 (16)	73 (23)	100 (38)	120 (49)	140 (60)
Short-Term	300.0	260.0	170.0	130.0	110.0	100.0	65.0	50.0
	(2069)	(1793)	(1172)	(896)	(758)	(690)	(448)	(345)
10 h	140.8	122.0	79.8	61.0	57.5	46.9	30.5	23.5
	(971)	(841)	(550)	(421)	(396)	(323)	(210)	(162)
100 h	125.4	108.7	71.0	54.3	51.2	41.8	27.2	20.9
	(865)	(749)	(490)	(374)	(353)	(288)	(188)	(144)
1000 h	107.0	92.8	60.7	46.4	43.7	35.7	23.2	17.8
	(738)	(640)	(419)	(320)	(301)	(246)	(160)	(123)
1 y	93.0	80.6	52.7	40.3	38.0	31.0	20.2	15.5
	(641)	(556)	(363)	(278)	(262)	(214)	(139)	(107)
10 y	77.4	67.1	43.9	33.5	31.6	25.8	16.8	12.9
	(534)	(463)	(303)	(231)	(218)	(178)	(116)	(89)
50 y	69.1	59.9	39.1	29.9	28.2	23.0	15.0	11.5
	(476)	(413)	(270)	(206)	(194)	(159)	(103)	(79)

† Typical values based on ASTM D 638 testing of molded plaque material specimens. An elastic modulus for PE 2406 may be estimated by multiplying the PE 3408 modulus value by 0.875.

As longitudinal stress builds in the pipe wall, a thrust load is created on the end structures. The thrust load is determined by Equation 3-3.

(3-3) $\quad F = \sigma A_p$

Where terms are as defined above, and
F = end thrust, lb
A_P = area of pipe cross section,(/4)(DO² – Di²) in²

Equations 3-2 and 3-3 can also be used to determine the compressive stress and thrust (respectively) from a temperature increase.

Although the length change of polyethylene pipe during temperature changes is greater than many other materials, the amount of force required to restrain the movement is less because of its lower modulus of elasticity.

As pipeline temperature decreases from weather or operating conditions, a longitudinal tensile stress develops along the pipe that can be determined using Equation 3-2. The allowable tensile stress for pipe operating at its pressure rating is determined using Equation 3-4.

(3-4) $\quad \sigma_{allow} = HDB \times DF$

WHERE
σ_{allow} = Allowable tensile stress at 73°F, lb/in²
HDB = Hydrostatic Design Basis, psi (Table 1-1)*
DF = Design Factor, from Table 1-2

* The manufacturer should be consulted for HDB values for temperatures other than 73°F.

Equation 3-3 is used to determine the thrust load applied to structural anchoring devices.

During temperature increase, the pipeline attempts to increase in length, but is restrained by mechanical guides that direct longitudinal compressive thrust to structural anchors that prevent length increase. This in turn creates a longitudinal compressive stress in the pipe and a thrust load against the structural anchors. The compressive stress that develops in the pipe and is resisted by the structural anchors is determined using Equation 3-2. Compressive stress should not exceed the allowable compressive stress per Table 2-12 in Section 2 of this chapter.

Above Ground Piping Systems

The design considerations for polyethylene piping systems installed above ground are extensive and, therefore, are addressed separately in the Handbook chapter on above ground applications for PE pipe.

Buried Piping Systems

A buried pipe is generally well restrained by soil loads and will experience very little lateral movement. However, longitudinal end loads may result that need to be addressed.

Transitions to other pipe materials that use the bell and spigot assembly technique will need to be calculated using the thrust load as delivered by the pressure plus the potential of the load due to temperature changes. Merely fixing the end of the HDPE to the mating material may result in up stream joints pulling apart unless those connections are restrained. The number of joints that need to be restrained to prevent bell and spigot pull out may be calculated using techniques as recommended by the manufacturer of the alternate piping material. Equation 3-3 may be used to calculate the total thrust load due to the temperature changes.

Low thrust capacity connections to manholes or other piping systems as will be present in many no pressure gravity flow systems may be addressed via a longitudinal thrust anchor such as shown in Fig. 3-1. The size of the thrust block will vary depending on soil conditions and the thrust load as calculated via Equation 3-3.

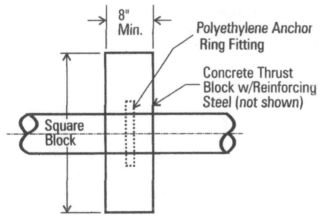

Figure 3-1 Longitudinal Thrust Anchor

Conclusion

The durability and visco-elastic nature of modern polyethylene piping materials makes these products ideally suited for a broad array of piping applications such as: potable water mains and service lines, natural gas distribution, oil and gas gathering, force main sewers, gravity flow lines, industrial and various mining piping. To this end, fundamental design considerations such as fluid flow, burial design and thermal response were presented within this chapter in an effort to provide guidance to the piping system designer on the use of these tough piping materials in the full array of potential piping applications.

For the benefit of the pipeline designer, a considerable amount of background information and/or theory has been provided within this chapter. However, the designer should also keep in mind that the majority of pipeline installations fall within the criteria for the AWWA Design Window approach presented in Section 2 of this chapter. Pipeline installations that fall within the guidelines for the AWWA Window, may be greatly simplified in matters relating to the design and use of flexible polyethylene piping systems.

While every effort has been made to be as thorough as possible in this discussion, it also should be recognized that these guidelines should be considered in light of specific project, installation and/or service needs. For this reason, this chapter on pipeline design should be utilized in conjunction with the other chapters of this Handbook to provide a more thorough understanding of the design considerations that may be specific to a particular project or application using polyethylene piping systems. The reader is also referred to the extensive list of references for this chapter as additional resources for project and or system analysis and design.

References for Section 1

1. Jeppson, Roland W., Analysis of Flow in Pipe Networks, Ann Arbor Science, Ann Arbor, MI.
2. Distribution Network Analysis, *AWWA Manual M32*, American Water Works Association, Denver, CO.
3. ASTM D 2513, Standard Specification for Thermoplastic Gas Pipe, Tubing and Fittings, American Society for Testing and Materials, West Conshohocken, PA.
4. ASTM D 2737, Standard Specification for Polyethylene (PE) Tubing, American Society for Testing and Materials, West Conshohocken, PA.
5. ASTM D 2447, Standard Specification for Polyethylene (PE) Plastic Pipe, Schedules 40 and 80, Based on Outside Diameter, American Society for Testing and Materials, West Conshohocken, PA.
6. ASTM D 3035, Standard Specification for Polyethylene (PE) Plastic Pipe (DR-PR) Based on Controlled Outside Diameter, American Society for Testing and Materials, West Conshohocken, PA.
7. ASTM F 714, Standard Specification for Polyethylene (PE) Plastic Pipe (SDR-PR) Based on Controlled Outside Diameter, American Society for Testing and Materials, West Conshohocken, PA.
8. ANSI/AWWA C901, AWWA Standard for Polyethylene (PE) Pressure Pipe and Tubing, 1/2 In.(13 mm) Through 3 In. (76 mm) for Water Service, American Water Works Association, Denver, CO
9. ANSI/AWWA C906, AWWA Standard for Polyethylene (PE) Pressure Pipe and Fittings, 4 In. Through 63 In. for Water Distribution, American Water Works Association, Denver, CO.
10. API Specification 15LE, Specification for Polyethylene Line Pipe (PE), American Petroleum Institute, Washington DC.
11. ASTM D 2104, Standard Specification for Polyethylene (PE) Plastic Pipe, Schedule 40, American Society for Testing and Materials, West Conshohocken, PA.
12. ASTM D 2239, Standard Specification for Polyethylene (PE) Plastic Pipe (SIDR-PR) Based on Controlled Inside Diameter, American Society for Testing and Materials, West Conshohocken, PA.
13. ASTM F 894, Standard Specification for Polyethylene (PE) Large Diameter Profile Wall Sewer and Drain Pipe, American Society for Testing and Materials, West Conshohocken, PA.
14. PPI TR-22, Polyethylene Piping Distribution Systems for Components of Liquid Petroleum Gases, Plastics Pipe Institute, Washington DC.
15. Nayyar, Mohinder L. (1992). *Piping Handbook*, 6th Edition, McGraw-Hill, New York, NY.
16. Iolelchick, I.E., Malyavskaya O.G., & Fried, E. (1986). *Handbook of Hydraulic Resistance*, Hemisphere Publishing Corporation.
17. Moody, L.F. (1944). *Transactions*, Volume 6, American Society of Mechanical Engineers (ASME), New York, NY.
18. Swierzawski, Tadeusz J. (2000). Flow of Fluids, Chapter B8, *Piping Handbook*, 7th edition, Mohinder L. Nayyar, McGraw- Hill, New York, NY.
19. Lamont, Peter A. (1981, May). Common Pipe Flow Formulas Compared with the Theory of Roughness, *Journal of the American Water Works Association*, Denver, CO.
20. Flow of Fluids through Valves, Fittings and Pipe. (1957). Crane Technical Paper No 410, the Crane Company, Chicago, IL.
21. Chen, W.F., & J.Y. Richard Liew. (2002). *The Civil Engineering Handbook*, 2nd edition, CRC Press, Boca Raton, FL.
22. Bowman, J.A. (1990). The Fatigue Response of Polyvinyl Chloride and Polyethylene Pipe Systems, Buried Plastics Pipe Technology, ASTM STP 1093, American Society for Testing and Materials, Philadelphia.
23. Marshall, GP, S. Brogden, & M.A. Shepherd, Evaluation of the Surge and Fatigue Resistance of PVC and PE Pipeline Materials for use in the UK Water Industry, Proceedings of Plastics Pipes X, Goteborg, Sweden.
24. Fedossof, F.A., & Szpak, E. (1978, Sept 20-22). Cyclic Pressure Effects on High Density Polyethylene Pipe, Paper presented at the Western Canada Sewage Conference, Regian, Saskatoon, Canada.
25. Parmakian, John. (1963). *Waterhammer Analysis*, Dover Publications, New York, NY.
26. Thompson, T.L., & Aude, T.C. (1980). Slurry Pipelines, *Journal of Pipelines*, Elsevier Scientific Publishing Company, Amsterdam.
27. *Handbook of Natural Gas Engineering*. (1959). McGraw-Hill, New York, NY.
28. *AGA Plastic Pipe Manual for Gas Service*. (2001). American Gas Association, Washington DC.
29. ASCE Manuals and Reports on Engineering Practice No. 60. (1982). Gravity Sewer Design and Construction, American Society of Civil Engineers, New York, NY.
30. Hicks, Tyler G. (1999). *Handbook of Civil Engineering Calculations*, McGraw-Hill, New York, NY.
31. PPI TR-14, Water Flow Characteristics of Thermoplastic Pipe, Plastics Pipe Institute, Washington DC.

References for Section 2

1. Watkins, R.K., Szpak, E., & Allman, W.B. (1974). Structural Design of PE Pipes Subjected to External Loads, Engr. Experiment Station, Utah State Univ., Logan.
2. AWWA Committee Report. (1998, October). Design and Installation of Polyethylene (PE) Pipe Made in Accordance with AWWA C906, American Water Works Association, Denver, CO.
3. Howard, A.K. (1996). *Pipeline Installation*, Relativity Printing, Lakewood, Colorado, ISBN 0-9651002-0-0.
4. Spangler, M.G. (1941). The Structural Design of Flexible Pipe Culverts, Bulletin 153, Iowa Engineering Experiment Station, Ames, IA.
5. Watkins, R.K., & Spangler, M.G. (1958). Some Characteristics of the Modulus of Passive Resistance of Soil—A Study in Similitude, Highway Research Board Proceedings 37:576-583, Washington.
6. Burns, J.Q., & Richard, R.M. (1964). Attenuation of Stresses for Buried Cylinders, Proceedings of the Symposium on Soil Structure Interaction, pp.378-392, University of Arizona, Tucson.

7. Katona, J.G., Forrest, F.J., Odello, & Allgood, J.R. (1976). CANDE—A Modern Approach for the Structural Design and Analysis of Buried Culverts, Report FHWA-RD-77-5, FHWA, US Department of Transportation.
8. Howard, A.K. (1977, January). Modulus of Soil Reaction Values for Buried Flexible Pipe, *Journal of the Geotechnical Engineering Division*, ASCE, Vol. 103, No GT 1.
9. Petroff, L.J. (1995). Installation Technique and Field Performance of HDPE, Profile Pipe, Proceedings 2nd Intl. Conference on the Advances in Underground Pipeline Engineering, ASCE, Seattle.
10. Duncan, J.M., & Hartley, J.D. (1982). Evaluation of the Modulus of Soil Reaction, E', and Its Variation with Depth, Report No. UCB/GT/82-02, University of California, Berkeley.
11. Howard, A.K. (1981). The USBR Equation for Predicting Flexible Pipe Deflection, Proceedings Intl. Conf. On Underground Plastic Pipe, ASCE, New Orleans, LA.
12. Janson, L.E. (1991). Long-Term Studies of PVC and PE Pipes Subjected to Forced Constant Deflection, Report No. 3, KP-Council, Stockholm, Sweden.
13. Spangler, M.G., & Handy, R.L. (1982). *Soil Engineering*, 4th ed., Harper & Row, New York.
14. Watkins, R.K. (1977). Minimum Soil Cover Required Over Buried Flexible Cylinders, Interim Report, Utah State University, Logan, UT.
15. Gaube, E. (1977, June). Stress Analysis Calculations on Rigid Polyethylene and PVC Sewage Pipes, *Kunstoffe*, Vol.67, pp. 353-356, Germany.
16. Gaube, E., & Muller, W. (1982, July). Measurement of the long-term deformation of HDPE pipes laid underground, *Kunstoffe*, Vol. 72, pp. 420-423, Germany.
17. Moore, I. D., & Selig, E. T. (1990). Use of Continuum Buckling Theory for Evaluation of Buried Plastic Pipe Stability, Buried Plastic Pipe Technology, ASTM STP 1093, Philadelphia.
18. Marston, A. (1930). Iowa Engineering Experiment Station, Bulletin No. 96.
19. McGrath, T. (1994). *Analysis of Burns & Richard Solution for Thrust in Buried Pipe*, Simpson Gumpertz & Heger, Inc, Cambridge, Mass.
20. McGrath, T.J. (1998). Replacing E' with the Constrained Modulus in Flexible Pipe Design, proceedings Pipeline Div. Conf. Pipelines in the Constructed Environment, ASCE, San Diego, CA.
21. Bowles, J.E. (1982). *Foundation Analysis and Design*, 3rd ed., McGraw-Hill Book Company, New York.

Appendix
List of Design Chapter Variables

υ	=	kinematic viscosity, ft^2/sec
ρ	=	fluid density, lb/ft^3
μ	=	dynamic viscosity, lb-sec/ft^2
Δv	=	Sudden velocity change, ft/sec
a	=	Wave velocity (celerity), ft/sec
A_C	=	Cross-sectional area of pipe bore, ft^2
a_c	=	contact area, ft^2
A	=	profile wall average cross-sectional area, in^2/in, for profile pipe or wall thickness (in) for DR pipe
A_S	=	Area of pipe cross-section or (/4) ($D_O^2 - D_I^2$), in^2
A_P	=	area of the outside wall of the pipe, 100 in^2
C	=	Hazen-Williams Friction Factor, dimensionless, see table 1-7.
c	=	outer fiber to wall centroid, in
C_V	=	percent solids concentration by volume
C_W	=	percent solids concentration by weight
D_A	=	pipe average inside diameter, in
DF	=	Design Factor, from Table 1-2
d'	=	Pipe inside diameter, ft
D_I	=	Pipe inside diameter, in
D_M	=	Mean diameter (DI+2z or D_O-t), in
D_{MIN}	=	pipe minimum inside diameter, in
D_O	=	pipe outside diameter, in
d_O	=	pipe outside diameter, ft
DR	=	Dimension Ratio, D_O/t
E	=	Apparent modulus of elasticity for pipe material, psi
e	=	natural log base number, 2.71828
E'	=	Modulus of soil reaction, psi
E_d	=	Dynamic instantaneous effective modulus of pipe material (typically 150,000 psi for PE pipe)
E_N	=	Native soil modulus of soil reaction, psi
E_S	=	Secant modulus of the soil, psi
E_S^*	=	$E_S/(1-\mu)$
f	=	friction factor (dimensionless, but dependent upon pipe surface roughness and Reynolds number)
F	=	end thrust, lb
F_B	=	buoyant force, lb/ft
F_L	=	velocity coefficient (Tables 1-14 and 1-15)
f_O	=	Ovality Correction Factor, Figure 2-9
F_S	=	Soil Support Factor
F_T	=	Service Temperature Design Factor, from Table 1-11
g	=	Constant gravitational acceleration, 32.2 ft/sec^2
H_P	=	profile wall height, in
H	=	height of cover, ft
h_l	=	liquid level in the pipe, ft
H_{GW}	=	ground water height above pipe, ft
h_1	=	pipeline elevation at point 1, ft

h_1	=	inlet pressure, in H_2O
h_U	=	upstream pipe elevation, ft
h_2	=	pipeline elevation at point 2, ft
h_2	=	outlet pressure, in H_2O
d_D	=	downstream pipe elevation, ft
hD	=	Hydrostatic Design Basis, psi
h_E	=	Elevation head, ft of liquid
h_f	=	friction (head) loss, ft. of liquid
H_S	=	level of ground water saturation above pipe, ft
I_V	=	Influence Value from Table 2-5
I	=	Pipe wall moment of inertia, in^4/in
IDR	=	ID -Controlled Pipe Dimension Ratio
I_f		impact factor
k	=	kinematic viscosity, centistokes
K_{BULK}	=	Bulk modulus of fluid at working temperature
K_{BED}	=	Bedding factor, typically 0.1
K	=	passive earth pressure coefficient
K'	=	Fittings Factor, Table 1-5
K_P	=	permeability constant (Table 1-13)
L_{EFF}	=	Effective Pipeline length, ft.
L	=	Pipeline length, ft
L_{DL}	=	Deflection lag factor
ΔL	=	pipeline length change, in
M	=	horizontal distance, normal to the pipe centerline, from the center of the load to the load edge, ft
M_s	=	one-dimensional modulus of soil, psi
n	=	roughness coefficient, dimensionless
N	=	horizontal distance, parallel to the pipe centerline, from the center of the load to the load edge, ft
N_S	=	safety factor
P	=	Internal Pressure, psi
P_W	=	perimeter wetted by flow, ft
p_1	=	inlet pressure, lb/in^2 absolute
p_2	=	outlet pressure, lb/in^2 absolute
P_A	=	pipe internal pressure, atmospheres (1 atmosphere = 14.7 lb/in^2)
PC	=	Pressure Class
P_{CR}	=	Critical constrained buckling pressure, psi
P_E	=	vertical soil pressure due to earth load, psf
P_f	=	friction (head) loss, psi
P_L	=	vertical soil pressure due to live load, psf
P_{OS}	=	Occasional Surge Pressure
P_{RD}	=	radial directed earth pressure, lb/ft^2
P_{RS}	=	Recurring Surge Pressure
P_s	=	Transient surge pressure, psig
P_{WAT}	=	Allowable live load pressure at pipe crown for pipes with one diameter or less of cover, psf
P_{WC}	=	allowable constrained buckling pressure, lb/in^2
P_{WU}	=	allowable unconstrained pipe wall buckling pressure, psi

p_i	=	Pressure due to sub-area i lb/ft^2
Q	=	flow rate, gpm
Q_{FPS}	=	flow, ft^3/sec
Q_h	=	flow, standard ft^3/hour
q_P	=	volume of gas permeated, cm^3 (gas at standard temperature and pressure)
r_H	=	hydraulic radius, ft
r	=	distance from the point of load application to pipe crown, ft
R	=	buoyancy reduction factor
r_{CENT}	=	radius to centroidal axis of pipe, in
Re	=	Reynolds number, dimensionless
R_H	=	Geometry Factor
RSC	=	Ring Stiffness Constant, lb/ft
r_T	=	equivalent radius, ft
RF	=	Rigidity factor, dimensions
s	=	liquid density, gm/cm^3
S_H	=	hydraulic slope, ft/ft
S	=	pipe wall compressive stress, lb/in^2
S_{MAT}	=	material yield strength, lb/in^2
S_A	=	Hoop Thrust Stiffness Ratio
S_g	=	gas specific gravity
S_L	=	carrier liquid specific gravity
S_M	=	slurry mixture specific gravity
S_S	=	solids specific gravity
t	=	minimum wall thickness, in
t'	=	wall thickness, mils
T_c	=	Critical time, seconds
V	=	flow velocity, ft/sec
VAF	=	Vertical Arching Factor
V_C	=	critical settlement velocity, ft/sec
ν	=	kinematic viscosity. ft^2/sec
V_{Min}	=	approximate minimum velocity, ft/sec
w	=	unit weight of soil, pcf
w	=	unit weight of soil, lb/ft^3
W_D	=	weight of dry soil above pipe, lb/ft of pipe
W_w	=	wheel load, lb
W_L	=	weight of liquid contents, lb/ft of pipe
W_L	=	weight of the liquid in contacts, lb/ft of pipe
WP	=	Working Pressure, psi
W_P	=	pipe weight, lb/ft of pipe
WPR	=	Working Pressure Rating, psi
w_S	=	distributed surcharge pressure acting over ground surface, lb/ft^2
W_S	=	weight of saturated soil above pipe, lb/ft of pipe
ζ	=	dynamic viscosity, centipoises
Z	=	Centroid of wall section, in
Z	=	Pipe wall centroid, in
Z_i	=	wall-section centroidal distance from inner fiber of pipe, in
α	=	thermal expansion coefficient, in/in/°F

ΔL	=	length change, in
ΔT	=	temperature change, °F
ΔX	=	Horizontal deflection, in
ΔV	=	Sudden velocity change., ft/sec
ε	=	absolute roughness, ft.
ε_s	=	Soil strain
Θ	=	elapsed time, days
μ_s	=	Poisson's Ratio of Soil
μ	=	Poisson's ratio
σ	=	longitudinal stress in pipe, psi
σ_{allow}	=	Allowable tensile stress at 73°F, lb/in
φ	=	Calibration Factor, 0.55 for granular soils change in psi
ω_D	=	unit weight of dry soil, lb/ft³ (See Table 2-16 for typical values.)
ω_G	=	unit weight of groundwater lb/ft³
ω_L	=	unit weight of liquid in the pipe, lb/ft³
ω_S	=	unit weight of saturated soil, pcf lb/ft³
ϕ	=	angle of internal friction, deg
Γ	=	Dynamic viscosity, lb-sec/ft²

PIPE WEIGHTS AND DIMENSIONS (IPS)
PE3408 (BLACK)

Nominal in.	OD Actual in.	SDR	Pipe inside diameter (d) in.	Minimum Wall Thickness (t) in.	Weight (w) lb. per foot
1/2	0.840	7	0.59	0.120	0.118
		7.3	0.60	0.115	0.114
		9	0.64	0.093	0.095
		9.3	0.65	0.090	0.093
		11	0.68	0.076	0.080
		11.5	0.69	0.073	0.077
3/4	1.050	7	0.73	0.150	0.184
		7.3	0.75	0.144	0.178
		9	0.80	0.117	0.149
		9.3	0.81	0.113	0.145
		11	0.85	0.095	0.125
		11.5	0.86	0.091	0.120
1	1.315	7	0.92	0.188	0.289
		7.3	0.93	0.180	0.279
		9	1.01	0.146	0.234
		9.3	1.02	0.141	0.227
		11	1.06	0.120	0.196
		11.5	1.07	0.114	0.188
1 1/4	1.660	7	1.16	0.237	0.461
		7.3	1.18	0.227	0.445
		9	1.27	0.184	0.372
		9.3	1.28	0.178	0.362
		11	1.34	0.151	0.312
		11.5	1.35	0.144	0.300
		13.5	1.40	0.123	0.259
1 1/2	1.900	7	1.32	0.271	0.603
		7.3	1.35	0.260	0.583
		9	1.45	0.211	0.488
		9.3	1.47	0.204	0.474
		11	1.53	0.173	0.409
		11.5	1.55	0.165	0.393
		13.5	1.60	0.141	0.340
		15.5	1.64	0.123	0.299

Chapter 6
Design of Polyethylene Piping Systems

OD			Pipe inside diameter (d)	Minimum Wall Thickness (t)	Weight (w)
Nominal in.	Actual in.	SDR	in.	in.	lb. per foot
2	2.375	7	1.66	0.339	0.943
		7.3	1.69	0.325	0.911
		9	1.82	0.264	0.762
		9.3	1.83	0.255	0.741
		11	1.92	0.216	0.639
		11.5	1.94	0.207	0.614
		13.5	2.00	0.176	0.531
		15.5	2.05	0.153	0.467
		17	2.08	0.140	0.429
3	3.500	7	2.44	0.500	2.047
		7.3	2.48	0.479	1.978
		9	2.68	0.389	1.656
		9.3	2.70	0.376	1.609
		11	2.83	0.318	1.387
		11.5	2.85	0.304	1.333
		13.5	2.95	0.259	1.153
		15.5	3.02	0.226	1.015
		17	3.06	0.206	0.932
		21	3.15	0.167	0.764
		26	3.21	0.135	0.623
4	4.500	7	3.14	0.643	3.384
		7.3	3.19	0.616	3.269
		9	3.44	0.500	2.737
		9.3	3.47	0.484	2.660
		11	3.63	0.409	2.294
		11.5	3.67	0.391	2.204
		13.5	3.79	0.333	1.906
		15.5	3.88	0.290	1.678
		17	3.94	0.265	1.540
		21	4.05	0.214	1.262
		26	4.13	0.173	1.030
		32.5	4.21	0.138	0.831

Chapter 6 | 253
Design of Polyethylene Piping Systems

OD			Pipe inside diameter (d)	Minimum Wall Thickness (t)	Weight (w)
Nominal in.	Actual in.	SDR	in.	in.	lb. per foot
5	5.563	7	3.88	0.795	5.172
		7.3	3.95	0.762	4.996
		9	4.25	0.618	4.182
		9.3	4.29	0.598	4.065
		11	4.49	0.506	3.505
		11.5	4.54	0.484	3.368
		13.5	4.69	0.412	2.912
		15.5	4.80	0.359	2.564
		17	4.87	0.327	2.353
		21	5.00	0.265	1.929
		26	5.11	0.214	1.574
		32.5	5.20	0.171	1.270
6	6.625	7	4.62	0.946	7.336
		7.3	4.70	0.908	7.086
		9	5.06	0.736	5.932
		9.3	5.11	0.712	5.765
		11	5.35	0.602	4.971
		11.5	5.40	0.576	4.777
		13.5	5.58	0.491	4.130
		15.5	5.72	0.427	3.637
		17	5.80	0.390	3.338
		21	5.96	0.315	2.736
		26	6.08	0.255	2.233
		32.5	6.19	0.204	1.801
8	8.625	7	6.01	1.232	12.433
		7.3	6.12	1.182	12.010
		9	6.59	0.958	10.054
		9.3	6.66	0.927	9.771
		11	6.96	0.784	8.425
		11.5	7.04	0.750	8.096
		13.5	7.27	0.639	7.001
		15.5	7.45	0.556	6.164
		17	7.55	0.507	5.657
		21	7.75	0.411	4.637
		26	7.92	0.332	3.784

OD			Pipe inside diameter (d)	Minimum Wall Thickness (t)	Weight (w)
Nominal in.	Actual in.	SDR	in.	in.	lb. per foot
		7	7.49	1.536	19.314
		7.3	7.63	1.473	18.656
		9	8.22	1.194	15.618
		9.3	8.30	1.156	15.179
		11	8.68	0.977	13.089
10	10.750	11.5	8.77	0.935	12.578
		13.5	9.06	0.796	10.875
		15.5	9.28	0.694	9.576
		17	9.41	0.632	8.788
		21	9.66	0.512	7.204
		26	9.87	0.413	5.878
		32.5	10.05	0.331	4.742
		7	8.89	1.821	27.170
		7.3	9.05	1.747	26.244
		9	9.75	1.417	21.970
		9.3	9.84	1.371	21.353
		11	10.29	1.159	18.412
12	12.750	11.5	10.40	1.109	17.693
		13.5	10.75	0.944	15.298
		15.5	11.01	0.823	13.471
		17	11.16	0.750	12.362
		21	11.46	0.607	10.134
		26	11.71	0.490	8.269
		32.5	11.92	0.392	6.671
		7	9.76	2.000	32.758
		7.3	9.93	1.918	31.642
		9	10.70	1.556	26.489
		9.3	10.81	1.505	25.745
		11	11.30	1.273	22.199
14	14.000	11.5	11.42	1.217	21.332
		13.5	11.80	1.037	18.445
		15.5	12.09	0.903	16.242
		17	12.25	0.824	14.905
		21	12.59	0.667	12.218
		26	12.86	0.538	9.970
		32.5	13.09	0.431	8.044

Chapter 6
Design of Polyethylene Piping Systems

OD			Pipe inside diameter (d)	Minimum Wall Thickness (t)	Weight (w)
Nominal in.	Actual in.	SDR	in.	in.	lb. per foot
16	16.000	7	11.15	2.286	42.786
		7.3	11.35	2.192	41.329
		9	12.23	1.778	34.598
		9.3	12.35	1.720	33.626
		11	12.92	1.455	28.994
		11.5	13.05	1.391	27.862
		13.5	13.49	1.185	24.092
		15.5	13.81	1.032	21.214
		17	14.00	0.941	19.467
		21	14.38	0.762	15.959
		26	14.70	0.615	13.022
18	18.000	7	12.55	2.571	54.151
		7.3	12.77	2.466	52.307
		9	13.76	2.000	43.788
		9.3	13.90	1.935	42.558
		11	14.53	1.636	36.696
		11.5	14.68	1.565	35.263
		13.5	15.17	1.333	30.491
		15.5	15.54	1.161	26.849
		17	15.76	1.059	24.638
		21	16.18	0.857	20.198
		26	16.53	0.692	16.480
		32.5	16.83	0.554	13.296
20	20.000	7	13.94	2.857	66.853
		7.3	14.19	2.740	64.576
		9	15.29	2.222	54.059
		9.3	15.44	2.151	52.541
		11	16.15	1.818	45.304
		11.5	16.31	1.739	43.535
		13.5	16.86	1.481	37.643
		15.5	17.26	1.290	33.146
		17	17.51	1.176	30.418
		21	17.98	0.952	24.936
		26	18.37	0.769	20.346
		32.5	18.70	0.615	16.415

OD			Pipe inside diameter (d)	Minimum Wall Thickness (t)	Weight (w)
Nominal in.	Actual in.	SDR	in.	in.	lb. per foot
22	22.000	9	16.82	2.444	65.412
		9.3	16.98	2.366	63.574
		11	17.76	2.000	54.818
		11.5	17.94	1.913	52.677
		13.5	18.55	1.630	45.548
		15.5	18.99	1.419	40.107
		17	19.26	1.294	36.805
		21	19.78	1.048	30.172
		26	20.21	0.846	24.619
		32.5	20.56	0.677	19.863
24	24.000	9	18.35	2.667	77.845
		9.3	18.53	2.581	75.658
		11	19.37	2.182	65.237
		11.5	19.58	2.087	62.690
		13.5	20.23	1.778	54.206
		15.5	20.72	1.548	47.731
		17	21.01	1.412	43.801
		21	21.58	1.143	35.907
		26	22.04	0.923	29.299
		32.5	22.43	0.738	23.638
28	28.000	11	22.60	2.545	88.795
		11.5	22.84	2.435	85.329
		13.5	23.60	2.074	73.781
		15.5	24.17	1.806	64.967
		17	24.51	1.647	59.618
		21	25.17	1.333	48.874
		26	25.72	1.077	39.879
		32.5	26.17	0.862	32.174
30	30.000	11	24.22	2.727	101.934
		11.5	24.47	2.609	97.954
		13.5	25.29	2.222	84.697
		15.5	25.90	1.935	74.580
		17	26.26	1.765	68.439
		21	26.97	1.429	56.105
		26	27.55	1.154	45.779
		32.5	28.04	0.923	36.934

OD			Pipe inside diameter (d)	Minimum Wall Thickness (t)	Weight (w)
Nominal in.	Actual in.	SDR	in.	in.	lb. per foot
32	32.000	13.5	26.97	2.370	96.367
		15.5	27.62	2.065	84.855
		17	28.01	1.882	77.869
		21	28.77	1.524	63.835
		26	29.39	1.231	52.086
		32.5	29.91	0.985	42.023
36	36.000	15.5	31.08	2.323	107.395
		17	31.51	2.118	98.553
		21	32.37	1.714	80.791
		26	33.06	1.385	65.922
		32.5	33.65	1.108	53.186
42	42.000	15.5	36.26	2.710	146.176
		17	36.76	2.471	134.141
		21	37.76	2.000	109.966
		26	38.58	1.615	89.727
		32.5	39.26	1.292	72.392
48	48.000	17	42.01	2.824	175.205
		21	43.15	2.286	143.629
		26	44.09	1.846	117.194
		32.5	44.87	1.477	94.552
54	54.000	21	48.55	2.571	181.781
		26	49.60	2.077	148.324
		32.5	50.48	1.662	119.668

PIPE WEIGHTS AND DIMENSIONS (DIPS)
PE3408 (Black)

Nominal in.	Actual in.	DR*	Pipe inside diameter (d) in.	Minimum Wall Thickness (t) in.	Weight (w) lb. per foot
		7	2.76	0.566	2.621
		9	3.03	0.440	2.119
		11	3.20	0.360	1.776
		13.5	3.34	0.293	1.476
3	3.960	15.5	3.42	0.255	1.299
		17	3.47	0.233	1.192
		21	3.56	0.189	0.978
		26	3.64	0.152	0.798
		32.5	3.70	0.122	0.644
		7	3.35	0.686	3.851
		9	3.67	0.533	3.114
		11	3.87	0.436	2.609
		13.5	4.05	0.356	2.168
4	4.800	15.5	4.14	0.310	1.909
		17	4.20	0.282	1.752
		21	4.32	0.229	1.436
		26	4.41	0.185	1.172
		32.5	4.49	0.148	0.946
		7	4.81	0.986	7.957
		9	5.27	0.767	6.434
		11	5.57	0.627	5.392
		13.5	5.82	0.511	4.480
6	6.900	15.5	5.96	0.445	3.945
		17	6.04	0.406	3.620
		21	6.20	0.329	2.968
		26	6.34	0.265	2.422
		32.5	6.45	0.212	1.954
		7	6.31	1.293	13.689
		9	6.92	1.006	11.069
		11	7.31	0.823	9.276
		13.5	7.63	0.670	7.708
8	9.050	15.5	7.81	0.584	6.787
		17	7.92	0.532	6.228
		21	8.14	0.431	5.106
		26	8.31	0.348	4.166
		32.5	8.46	0.278	3.361

OD			Pipe inside diameter (d)	Minimum Wall Thickness (t)	Weight (w)
Nominal in.	Actual in.	DR*	in.	in.	lb. per foot
		7	7.74	1.586	20.593
		9	8.49	1.233	16.652
		11	8.96	1.009	13.955
		13.5	9.36	0.822	11.595
10	11.100	15.5	9.58	0.716	10.210
		17	9.72	0.653	9.369
		21	9.98	0.529	7.681
		26	10.19	0.427	6.267
		32.5	10.38	0.342	5.056
		7	9.20	1.886	29.121
		9	10.09	1.467	23.548
		11	10.66	1.200	19.734
		13.5	11.13	0.978	16.397
12	13.200	15.5	11.39	0.852	14.439
		17	11.55	0.776	13.250
		21	11.87	0.629	10.862
		26	12.12	0.508	8.863
		32.5	12.34	0.406	7.151
		7	10.67	2.186	39.124
		9	11.70	1.700	31.637
		11	12.35	1.391	26.513
		13.5	12.90	1.133	22.030
14	15.300	15.5	13.21	0.987	19.398
		17	13.39	0.900	17.801
		21	13.76	0.729	14.593
		26	14.05	0.588	11.907
		32.5	14.30	0.471	9.607
		7	12.13	2.486	50.601
		9	13.30	1.933	40.917
		11	14.05	1.582	34.290
		13.5	14.67	1.289	28.492
16	17.400	15.5	15.02	1.123	25.089
		17	15.23	1.024	23.023
		21	15.64	0.829	18.874
		26	15.98	0.669	15.400
		32.5	16.26	0.535	12.425

Chapter 6
Design of Polyethylene Piping Systems

OD			Pipe inside diameter (d)	Minimum Wall Thickness (t)	Weight (w)
Nominal in.	Actual in.	DR*	in.	in.	lb. per foot
18	19.500	7	13.59	2.786	63.553
		9	14.91	2.167	51.390
		11	15.74	1.773	43.067
		13.5	16.44	1.444	35.785
		15.5	16.83	1.258	31.510
		17	17.07	1.147	28.916
		21	17.53	0.929	23.704
		26	17.91	0.750	19.342
		32.5	18.23	0.600	15.605
20	21.600	7	15.06	3.086	77.978
		9	16.51	2.400	63.055
		11	17.44	1.964	52.842
		13.5	18.21	1.600	43.907
		15.5	18.65	1.394	38.662
		17	18.91	1.271	35.479
		21	19.42	1.029	29.085
		26	19.84	0.831	23.732
		32.5	20.19	0.665	19.147
24	25.800	11	20.83	2.345	75.390
		13.5	21.75	1.911	62.642
		15.5	22.27	1.665	55.159
		17	22.58	1.518	50.618
		21	23.20	1.229	41.495
		26	23.70	0.992	33.858
		32.5	24.12	0.794	27.317
30	32.000	13.5	26.97	2.370	96.367
		15.5	27.62	2.065	84.855
		17	28.01	1.882	77.869
		21	28.77	1.524	63.835
		26	29.39	1.231	52.086
		32.5	29.91	0.985	42.023

* These DRs (7.3, 9, 11, 13.5, 17, 21, 26, 32.5) are from the standard dimension ratio (SDR) series established by ASTM F 412.51

Chapter 7

Underground Installation of Polyethylene Piping

Introduction

Piping systems are prevalent throughout our everyday world. Most of us think of piping systems as underground structures used to convey liquids of one sort or another. To the novice, the concept of pipeline installation underground sounds relatively straight forward: a) dig a trench, b) lay the pipe in the trench, and c) fill the trench back in.

While this simplified perspective of pipeline construction may be appealing, it does not begin to address the engineering concepts involved in the underground installation of a pipeline. This chapter is written to assist in the development of a comprehensive understanding of the engineering principles utilized in the underground installation of polyethylene pipe.

In the pages which follow, the reader will be introduced to the concept of a pipe soil system and the importance that the soil and the design and preparation of the back-fill materials play in the long-term performance of a buried pipe structure. Specific terminology and design concepts relating to the underground installation of polyethylene pipe will be fully discussed. This will include fundamental guidelines regarding trench design and the placement and subsequent backfill of the polyethylene pipe.

This chapter is intended to assist the pipeline designer in the underground installation of polyethylene piping materials. This chapter is not intended as a substitute for the judgement of a professional engineer. Rather, it is felt that a comprehensive presentation of these design and installation principles may assist the engineer or designer in utilizing polyethylene pipe in a range of applications that require that it be buried beneath the earth.

Those individuals who are installing 24" diameter or smaller pressure pipe less than 16 feet deep are advised to see Appendix 1 which includes simplified installation guidelines. These guidelines were written to be used without review of this entire chapter; however, the installer of smaller pipes would benefit from a familiarity of the principles of pipe installation given in this chapter.

Appendix 2 contains a specification for large diameter (greater than 18") gravity flow pipes. For these sizes it is advised that the installer read this chapter.

Flexible Pipe Installation Theory

Most PE piping is considered flexible. Flexible pipes include corrugated metal pipes, plastic pipes, and some steel and ductile pipes. These pipes deflect under load, such as the overburden load encountered when installed underground. The designer and installer of underground, flexible piping must utilize the soil to construct an envelope of supporting material around the pipe so that the deflection is maintained at an acceptable level. The extent to which the pipe depends on this enveloping soil for support is a function of the depth of cover, surface loading, and the SDR or Ring Stiffness of the pipe.

In general, the supporting envelope is built by surrounding the pipe with firm, stable material. This envelope is often referred to as the "embedment" (see figure 1). The amount of support provided by the embedment is directly proportional to its stiffness. For this reason, often the embedment material is compacted. The stiffness of the material placed above the pipe may also affect the pipe's performance. Considerable load reduction may occur due to arching, that is the redistribution of stresses in the soil above and around the pipe which result in a shifting of load away from the pipe. The stiffer the backfill above the pipe, the more arching occurs.

The designer or installer will consider not only the embedment material but also the undisturbed in-situ soil surrounding the embedment and the ground water. The movement of the in-situ soil in the trench wall can reflect through the embedment and affect the pipe's performance. Therefore, careful consideration must be paid to this material when planning and during installation. This will most likely affect the equipment involved in doing the installation and the installation procedure itself more so than the pipe design. Of particular significance are soft clays and wet, loose silts or sands. When these materials are encountered, unstable trenching can occur with considerable sloughing and loosening of the trench walls during excavation. Flexible pipe can be installed in such ground with limited deflection as long as attention is paid to the proper handling of these conditions.

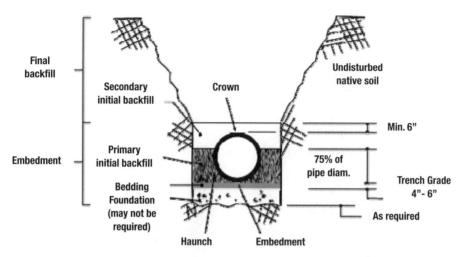

Figure 1 Pipe Trench

Note: When groundwater levels are expected to reach above pipe, the secondary initial backfill should be a continuation of the primary initial backfill in order to provide optimum pipe support. Minimum trench width will depend on site conditions and embedment materials.

Deflection Control

As described previously, the load carrying capability of a PE pipe can be greatly increased by the soil in which it is embedded if, as the pipe is loaded, the load is transferred from the pipe to the soil by a horizontal outward movement of the pipe wall. This enhances contact between pipe and soil and mobilizes the passive resistance of the soil. This resistance aids in preventing further pipe deformation and contributes to the support for the vertical loads. The amount of resistance found in the embedment soil is a direct consequence of the installation procedure. The stiffer the embedment materials are, the less deflection occurs. Because of this, the combination of embedment and pipe is often referred to as a pipe-soil system (see figure 2).

The key objective of a PE pipe installation is to limit or control deflection. (In this chapter the term "deflection" will mean a change in vertical diameter of the pipe, unless otherwise stated.) The deflection of a PE pipe is the sum total of two major components: the "installation deflection," which reflects the technique and care by which the pipe is handled and installed; and the "service deflection," which reflects the accommodation of the constructed pipe-soil system to the subsequent earth loading and other loadings.

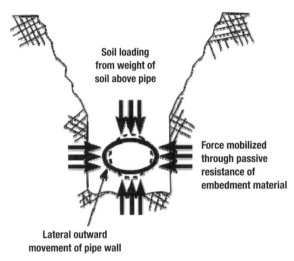

Figure 2 Mobilization of Enveloping Soil through Pipe Deformation

The "service deflection," which is usually a decrease in vertical pipe diameter, may be predicted by a number of reasonably well documented relationships, including those of Watkins and Spangler[1,2], or by use of a finite element analysis such as CANDE[1,2].

The "installation deflection" may be either an increase or decrease in vertical pipe diameter. An increase in vertical pipe diameter is referred to as "rise" and is usually a result of the forces acting on the pipe during compaction of the embedment beside it. Up to a point this may be beneficial in offsetting service deflection. Installation deflection is not predictable by any mathematical formula, although there are empirical methods for accounting for it[3].

Installation deflection is subject to control by the care used in the placement and compaction of the pipe embedment material in relation to the pipe's ring stiffness. For instance, compaction forces from hand operated air or gasoline tampers normally cause little rise, even when obtaining densities of 95 percent, but driving heavy loading equipment or driven compactors on the embedment while it is being placed beside the pipe may cause severe rise.

Commonly, deflection varies along the length of the pipeline due to variations in construction technique, soil type and loading. Field measurements illustrating this variability have been made by the U. S. Bureau of Reclamation and have been published by Howard[3]. Typically, this variation runs around ±2 percent.

Acceptance Deflection

To evaluate and control the quality of a flexible pipe installation, many designers impose an "acceptance deflection" requirement. This is particularly important for gravity flow lines with high SDR pipe. Commonly, pressure pipes are not checked for deflection. The "acceptance deflection" is the maximum vertical pipe deflection permitted following installation. Typically, measurements are made only after most of the initial soil consolidation occurs, usually at least 30 days after installation. The design engineer sets the "acceptance deflection" based on the particular application and type of joints. Commonly, a deflection limit of 5 percent is used, although PE pipe in gravity applications can usually withstand much larger deflections without impairment. When deflection is measured past 30 days, it is common to allow for a higher percentage. (See section on "Inspection.")

Pipe Design Considerations

While control of pipe deflection is a key objective in the installation of PE pipe, deflection is not ordinarily the criterion that determines the selection of a specific pipe SDR or ring stiffness. For typical burial conditions most PE pipes, when properly installed, are adequately stiff to preclude excessive deflection. However, for gravity flow applications, the design may often be controlled by the pipe's buckling resistance or the compressive thrust load in the pipe's wall. The selection of pipe design and construction requirements should be based on anticipated site conditions including trench depth, hydrostatic pressure due to ground water, superimposed static or traffic load, and depth of cover.

Pipe Embedment Materials

The embedment is the material immediately surrounding the pipe. This material may be imported, such as a crushed stone, or it may be the material excavated from the trench to make room for the pipe. In this case, it is referred to as native soil.

The embedment material should provide adequate strength, stiffness, uniformity of contact and stability to minimize deformation of the pipe due to earth pressures. The earth pressure acting on the pipe varies around the pipe's circumference (see figure 3). The pressure on the crown or top will typically be less than the free field stress as is the pressure at the invert or bottom of the pipe. Often, the highest pressure may be acting horizontally at the springline of the pipe, due to mobilization of passive pressure and arching.

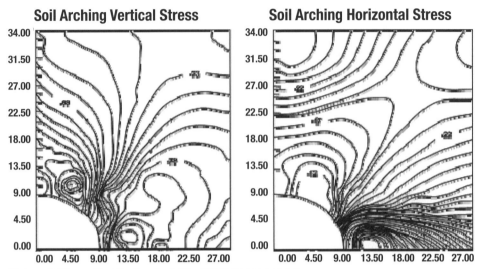

Figure 3 Stress Distribution for 18 in. HDPE pipe in soil box.

Because the earth pressure is acting around the circumference, it is important to completely envelop the pipe in embedment. (This may vary to a greater or lesser extent depending on the earth pressure, burial depth, and SDR.) To ensure that the embedment function should always be carried out under the anticipated job conditions, the design engineer will specify the permissible pipe embedment materials and their minimum acceptable density (compaction).

The properties of the in-situ (or native) soil into which the pipe is placed need not be as demanding as those for the embedment materials (unless it is used as the embedment material). The native soil may experience additional compression and deformation due to the horizontal pressure exerted by the pipe and transferred through the embedment material. This is usually a minor effect, but in some cases it can result in additional pipe deflection. This is most likely to occur where native soils are wet and loose, soft, or where native soil sloughs into the trench during excavation and is not removed. This effect is attenuated as the trench width (or width of embedment material) increases. Therefore, consideration must be given to the in-situ soil to ensure that it has adequate strength to permanently contain the embedment system. This is also discussed in a following section.

Terminology of Pipe Embedment Materials

The materials enveloping a buried pipe are generally identified, as shown by their function or location (see figure 1).

Foundation - A foundation is required only when the native trench bottom does not provide a firm working platform for placement of the pipe bedding material.

Bedding - In addition to bringing the trench bottom to required grade, the bedding levels out any irregularities and ensures uniform support along the length of the pipe.

Haunching - The backfill under the lower half of the pipe (haunches) distributes the superimposed loadings. The nature of the haunching material and the quality of its placement are one of the most important factors in limiting the deformation of PE pipe.

Primary Initial Backfill - This zone of backfill provides the primary support against lateral pipe deformation. To ensure such support is available, this zone should extend from trench grade up to at least 75 percent of the pipe diameter. Under some conditions, such as when the pipe will be permanently below the ground water table, the primary initial backfill should extend to at least 6 inches over the pipe.

Secondary Initial Backfill - The basic function of the material in this zone is to distribute overhead loads and to isolate the pipe from any adverse effects of the placement of the final backfill.

Final Backfill - As the final backfill is not an embedment material, its nature and quality of compaction has a lesser effect on the flexible pipe. However, arching and thus a load reduction on the pipe is promoted by a stiff backfill. To preclude the possibility of impact or concentrated loadings on the pipe, both during and after backfilling, the final backfill should be free of large rocks, organic material, and debris. The material and compaction requirements for the final backfill should reflect sound construction practices and satisfy local ordinances and sidewalk, road building, or other applicable regulations.

Classification and Supporting Strength of Pipe Embedment Materials

The burial of HDPE pipe for gravity flow applications is covered by ASTM D2321 *"Standard Practice for Underground Installation of Thermoplastic Pipe for Sewer and Other Gravity-Flow Applications."* ASTM 2774, *"Standard Practice for Underground Installation of Thermoplastic Pressure Piping,"* covers water pipe and force mains.

Strength of Embedment Soil

When selecting embedment material, consideration should be given to how the grain size, shape, and distribution will affect its supporting strength. The following will help guide the designer or installer in making a choice. In general, soils with large grains such as gravel have the highest stiffness and thus provide the most supporting strengths. Rounded grains tend to roll easier than angular, or sharp grains, which tend to interlock, and resist shear better. Well graded mixtures of soils (GW, SW), which contain a good representation of grains from a wide range of sizes, tend to offer more resistance than uniform graded soils (GP, SP). (See Table 1 for symbol definitions.)

Aside from the grain characteristics, the density has the greatest effect on the embedment's stiffness. For instance, in a dense soil there is considerable interlocking of grains and a high degree of grain-to-grain contact. Movement within the soil mass is restricted as the volume of the soil along the surface of sliding must expand for the grains to displace. This requires a high degree of energy. In a loose soil, movement causes the grains to roll or to slide, which requires far less energy. Thus, loose soil has a lower resistance to movement. Loose soil will permit more deflection of pipe for a given load than a dense soil.

When a pipe deflects, two beneficial effects occur.

1. The pipe pushes into the embedment soil and forces the soil to start moving. As this occurs, a resistance develops within the soil which acts to restrain further deflection.

2. Vertical deflection results in a reduction in earth load transmitted to the pipe due to the mobilization of arching.

Embedment Classification Per ASTM D-2321

Pipe embedment materials have been grouped by ASTM D-2321, "*Underground Installation of Flexible Thermoplastic Sewer Pipe,*" into five embedment classes according to their suitability for that use.

TABLE 1
Embedment Classes per ASTM D-2321

Class	Soil Description	Soil Group Symbol	Average Value of E'			
				Degree of Compaction of Embedment Mateial[1] (Standard Proctor)		
			Dumped	Slight 85%	Moderate 90%	Heavy <95%
IA	Manufactured aggregate, angular open-graded and clean. Includes crushed stone, crushed shells.	None	500	1000	2000	3000
IB	Processed aggregate, angular dense-graded and clean. Includes Class IA material mixed with sand and gravel to minimize migration	None	200	1000	2000	3000
II	Coarse-grained soils, clean. Includes gravels, gravel-sand mixtures, and well and poorly graded sands. Contains little to no fines (less than 5% passing #200).	GW, GP, SW, SP	200	1000	2000	3000
II	Coarse-grained soils, borderline clean to "with fines." Contains 5% to 12% fines (passing #200).	GW- GC SP-SM	200	1000	5000	3000
III	Coarse-grained soils containing 12% to 50% fines. Includes clayey gravel, silty sands, and clayer sands.	GM, GC, SM, SC	100	200	1000	2000
IVA	Fine-grained soils (inorganic). Includes inorganic silts, rock flour, silty-fine sands, clays of low to medium plasticity, and silty or sandy clays.	ML, CL	50	200	400	1000
IVB	Fine-grained soils (inorganic). Includes diatomaceous silts, elastic silts, fat clays.	MH, CH	No data available: consult a competent soils engineer. Otherwise use E' equals zero.			
V	Organic soils. Includes organic silts or clays and peat.	OL, OH, PT	No data avaliable: consult a competent soils engineer. Otherwise use E' equals zero.			

[1] E' values taken from Bureau of Reclamation table of average values and modified slightly herein to make the values more conservative.

The lower numbered classes have larger grain sizes and thus are more suitable as pipe embedment. The materials included in each class are identified in Table 1 and grouped in accordance with their classification per ASTM D-2487, *Standard Unified Soil Classifcation System* (USCS). (See Appendix III for a general discussion of soil classification.) A visual manual procedure for the field identification of soils is offered by ASTM D-2488.

The supporting strength of embedment materials roughly coincides with their embedment class. The supporting strength or stiffness of a soil material is represented by the modulus of soil reaction, E'. Based on extensive evaluation of field and laboratory performance of flexible pipes, the Bureau of Reclamation of the U. S. Department of the Interior has issued a table of Modulus of Soil Reaction Values. This tabulation of E' values, which is reproduced with some modifications based on recent experience in Table 1, provides a measure of available soil support depending on the soil embedment class and its degree of compaction. As discussed above, generally speaking, the classes with finer soils offer less supporting strength. However, within any embedment class, increased compaction greatly improves a soil's supporting strength, hence the critical role of proper compaction, particularly with the finer-grained soils. To ensure that the pipe is always adequately supported, it is the general practice to use materials and degrees of compaction resulting in an E' equal to or greater than 750 psi, although some applications may require higher or lower E' values.

Use of Embedment Materials

The determination of requirements for embedment materials and their placement should take into consideration not only their relative supporting strength but also their stability under end use conditions, ease of placement and compaction, and cost and availability.

Class I and Class II

Class I and Class II soils are granular and tend to provide the maximum embedment support as illustrated by the high E' values that can be achieved with them. Class I material is generally manufactured aggregate, such as crushed stone. Class II materials consist of clean sands and gravels and are more likely to be naturally occurring soils such as river deposits. Class I and Class II materials can be blended together to obtain materials that resist migration of finer soils into the embedment zone (as will be explained below.) In addition, Class I and II materials can be placed and compacted over a wide range of moisture content more easily than can other materials. This tends to minimize pipe deflection during installation. The high permeability of open-graded Class I and II materials aids in de-watering trenches, making these materials desirable in situations such as rock cuts where water problems may be encountered. This favorable combination of characteristics leads many designers to select these materials over others when they are readily and economically available.

Maximum aggregate size of Class I and Class II materials when used next to the pipe (i. e. , bedding, haunching and initial backfill) should not be larger than those given in Table 2 below. (Larger stones up to 1½ inches have been successfully used, but they are difficult to shovel slice and compact.) The smaller the rock size, the easier it is to place in the haunches. Maximum size for the foundation material is not restricted except that it should be graded to prevent the bedding stone from migrating into it.

TABLE 2
Maximum Particle Size vs. Pipe Size

Nominal Pipe Size (in.)	Maximum Particle Size (in.)
2 to 4	½
6 to 8	¾
10 to 15	1
16 and larger	1 ½

Migration

When the pipe is located beneath the ground water level, consideration must be given to the possibility of loss of side support through soil migration (the conveying by ground water of finer particle soils into void spaces of coarser soils). Generally, migration can occur where the void spaces in the embedment material are sufficiently large to allow the intrusion of eroded fines from the trench side walls.

For migration to occur, the in-situ soil must be erodible. Normally, erodible soils are fine sand and silts and special clays known as dispersive clays. (Most clays have good resistance to dispersion.) This situation is exacerbated where a significant gradient exists in the ground water from outside of the trench toward the inside of the trench; i. e. , the trench must act as a drain. (Seasonal fluctuations of the ground water level normally do not create this condition.)

For such anticipated conditions, it is desirable when using granular materials (Class I and II) to specify that they be angular and graded to minimize migration. Rounded particles have a tendency to flow when a considerable amount of water exists and material with a high void content provides "room" for migrating particles. The Army Corps of Engineers developed the following particle size requirements for properly grading adjacent materials to minimize migration:

(1) $D_{15}^E < 5 D_{85}^A$

(2) $D_{50}^E \geq 25 D_{85}^A$

Where the D_{15}, D_{50} and D_{85} are the particle sizes from a particle size distribution plot at 15%, 50% and 85%, respectively, finer by weight and where D^E is the embedment soil and D^A is the adjacent in-situ soil.

Another approach to preventing migration is to use geotextile separation fabrics. The fabric is sized to allow water to flow but to hold embedment materials around the pipe. Figure 4 shows a typical installation.

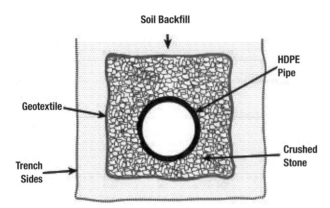

Figure 4 Installation of Geotextile Separation Fabrics

Cement Stabilized Sand

One special case of Class II material is Cement Stabilized Sand. Cement Stabilized Sand, once cured, is generally considered to give the same or better supporting strength as compacted Class I material. Cement Stabilized Sand consists of sand mixed with 3-5 percent cement. To achieve proper density, the material is placed with compaction rather than poured as with concrete. The material must be placed moist (at or near optimum moisture content) and then compacted in lifts as a Class II material. (The optimum moisture content is that moisture content at which a material can achieve its highest density for a given level of compaction.) If desired, deflection can be reduced if the cement sand embedment material is allowed to cure overnight before placement of backfill to grade. If the trench is backfilled immediately, cement sand will give the same support as a Class II material, but the lag factor will be reduced. Cement sand is usually placed in both the primary initial and secondary initial backfill zones (see figure 1).

Class III and Class IVA

Class III and Class IVA materials provide less supporting stiffness than Class I or II materials for a given density or compaction level, in part because of the increased clay content. In addition, they require greater compactive effort to attain specified densities and their moisture content must be closely controlled within the optimum limit. (The optimum moisture content is that moisture content at which a material can achieve its highest density for a given level of compaction.) Placement and compaction of Class IVA materials are especially sensitive to moisture content. If the Class IVA material is too wet, compaction equipment may sink into the material; if the soil is too dry, compaction may appear normal, but subsequent saturation with ground water may cause a collapse of the structure and lead to a loss of support.

Typically, Class IVA material is limited to applications with pressure pipe at shallow cover.

Class IVB and Class V

Class IVB and Class V materials offer hardly any support for a buried pipe and are often difficult to properly place and compact. These materials are normally not recommended for use as pipe embedment unless the pipe has a low SDR (or high ring stiffness), there are no traffic loads, and the depth of cover is only a few feet. In many cases the pipe will float in this type of soil if the material becomes saturated.

Compaction of Embedment Materials

Compaction criteria for embedment materials are a normal requirement in flexible pipe construction. Compaction reduces the void space between individual grains and increases the embedment density, thereby greatly improving pipe load carrying ability while reducing deflection, settlement, and water infiltration problems. Compaction of the embedment often will increase the stiffness of the in-situ soil and provide a sort of pre-stressing for the embedment and in-situ soils. Because of these benefits compaction should be considered on all projects.

Density Requirements

The required degree of compaction for an installation will be set by the designer in consideration of height of cover, extent of live loading, water table elevation and soil properties. Generally, the "moderate" compaction requirements listed in Table 1 are quite satisfactory. When compacting to this "moderate" level, it is suggested that the minimum target values for field measured densities be set as 90 percent Standard Proctor Density. This field density requirement will ensure that the actual densities will always be within the "moderate" range presented in Table 1. The applicable method for measuring density, ASTM D-2029, *Test for Relative Density of Cohesionless Soils*, or ASTM D-698, *Tests For Moisture-Density Relations of Soils and Soil-Aggregate Mixtures*, will be determined by the nature of the embedment material. Generally, the density of granular soils is determined using either test, whereas that of fine grained materials is determined by ASTM D-98. See Appendix 3 for a discussion of the difference between density and compaction and a discussion of the various test methods.

Compaction Techniques

Compaction of the embedment material should be performed by the most economical method available, consistent with providing uniform compaction and attaining the minimum specified density. Typical equipment used for compaction are hand held tamping bars (see figure 5), gasoline driven impact tampers ("whackers"), vibratory plates, and air driven impact tampers ("pogo sticks"). With crushed stone, some degree of densification can be achieved by the technique of shovel slicing, which consists of cutting the soil with a shovel.

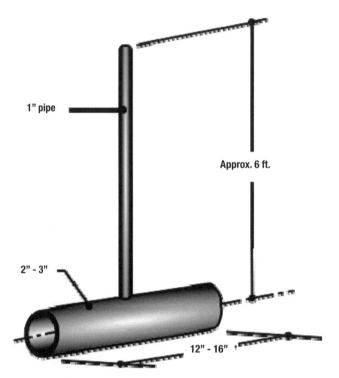

Figure 5 Tamping Tool

Compaction of the haunching material can best be accomplished by hand with tampers or suitable power compactors, taking particular care in the latter case not to disturb the pipe from its line and grade. In 36" and larger pipe, hand tampers are often used to reach under the haunches; they are then followed up with power compaction alongside the pipe.

When compacting the embedment near the pipe with impact-type tampers, caution should be taken to not allow direct contact of the equipment with the pipe. Avoid use of impact tampers directly above the pipe until sufficient backfill (usually 12") has been placed to ensure no local deformation of the pipe. Compaction of the embedment material alongside the pipe should not cause pipe to lift off of grade, but if upward movement occurs, reduce the compaction level below the springline or move the compactor away from the pipe toward the side of the trench.

Compaction of primary initial backfill should be conducted at, or near, the material's optimum moisture content. The backfill should be placed in layers, or lifts, that are brought up evenly on both sides of the pipe, otherwise the pipe could be moved off alignment. Each lift should be thoroughly compacted prior to placement of the next layer. The maximum lift height that will allow development of uniform density will vary depending on the material, its moisture content, and compactive effort. In general, maximum lifts of approximately 12 inches for Class I, 8 inches for Class II, and 6 inches for all others are adequate.

Compaction of Class I and II Materials

Compaction by vibration is most effective with granular (Class I and II) materials. Compaction of stone does not deform the stone but it does move it into a more compact or dense arrangement. In cases where the engineer specifies a minimum soil density of 90 percent of Standard Proctor or higher, as for installations under deep cover, mechanical compaction of Class I materials will be required. Impact tampers will also increase the density of Class I and II materials, primarily due to vibration. Impact tamping also acts to drive the embedment into the in-situ soil, which stiffens the trench wall interface. For this reason, impact compaction of Class I material should be considered for any application where the pipe will be below the ground water table or where the stability of the in-situ soil is in question.

An alternate method of achieving compaction with Class I materials is shovel slicing. Materials having been shovel sliced thoroughly will generally yield a modulus of around 1000 psi. The effectiveness of this method depends on the frequency of slicing along the length of the pipe. This technique should be limited to dry or firm (or better) in-situ soils. Where Class I materials are dumped around the pipe without any compactive effort (or shovel slicing), E's may be considerably lower than those given in the Bureau of Reclamation table. This is especially the case in wet or loose ground. A few passes with a vibratory compactor will increase the density and modulus of soil reaction.

Mechanical compaction of Class II materials can be aided by slight wetting. When so doing, care must be taken not to saturate the material or flood the trench, particularly when the native trench material does not drain freely. Flooding can result in flotation of the pipe.

Compaction by saturation, also called flooding or water tamping, is sometimes used to compact Class II materials. This method of compaction rarely yields Proctor densities greater than 75 percent, and therefore it will generally not give an E' of 750 psi or higher. Flooding is only suited for those applications where the pipe has sufficient internal supporting strength for the design load and does not depend on the soil for side support. (When considering this method for embedment that must provide side support, a geotechnical engineer should be consulted.) Compaction by saturation is limited to applications where both the embedment soil and in-situ soil are free draining. Compaction should be done in lifts not exceeding the radius of the pipe or 24 inches, whichever is smaller. Only enough water should be placed to saturate the material. It should be determined through proper monitoring that the desired level of compaction is being attained in each lift. Compaction by saturation should not be used in freezing weather. Water jetting, or the introduction of water under pressure to the embedment material, should not be used with plastic pipe.

Compaction of Class III and IV Materials

Compaction by impact is usually most effective with Class III and Class IVa materials. The use of mechanical impact tampers is most practical and effective. Depending on the embedment material, its moisture content, and lift height, several compaction passes may be required. A maximum lift height of 6 inches should be used when compacting by impact. Embedment density should be suitably monitored to ensure that specification requirements are met.

Density Checks

It is prudent to routinely check density of the embedment material. Typically, several checks are made during start-up of the project to ensure that the compaction procedure is achieving the desired density. Random checks are subsequently made to verify that the materials or procedures have not changed. Checks should be made at different elevations of the embedment material to assure that the desired compaction is being achieved throughout the embedment zone.

Trench Construction

Trenches should be excavated to line and grade as indicated by contract documents and in accordance with applicable safety standards. Excavation should precede upgrade. Excessive runs of open trench should be avoided to minimize such problems as trench flooding, caving of trench walls and the freezing of trench bottom and backfill material, and to minimize hazards to workmen and traffic. This can be accomplished by closely coordinating excavation with pipe installation and backfilling.

Principal considerations in trench construction are trench width, stability of the native soil supporting and containing the pipe and its embedment soil, stability of trench walls, and water accumulation in the trench. When encountering unstable soils or wet conditions, they should be controlled by providing an alternate foundation, sloping or bracing the trench walls, de-watering the trench bottom, or some other such measure.

Trench Width

Since flexible pipe has to support, at most, only the weight of the "prism" or vertical column of soil directly over the pipe, the precaution of keeping the trench as narrow as possible is not the concern that it is for a rigid pipe, which can be subjected to the weight of the soil beside the prism as well as the prism itself. With PE pipe, widening the trench will generally not cause a loading greater than the prism load on the pipe. Trench width in firm, stable ground is determined by the practical consideration of allowing sufficient room for the proper preparation of the trench bottom and placement and compaction of the pipe embedment materials, and the economic consideration of the costs of excavation and of imported embedment materials. Trench width in firm, stable ground will generally be determined by the pipe size and the compacting equipment used. The following table gives minimum trench width values.

The trench width may need to be increased over the values in Table 3 to allow for sufficient clearance between the trench sidewalls and the pipe for compaction equipment. Typically for large diameter pipe (18″ and larger), this required clearance will vary from 12 to 18 inches. If two or more pipes are laid in the same trench, sufficient space must be provided between the pipes so that embedment material can be compacted.

TABLE 3
Minimum Trench Width in Stable Ground

Nominal Pipe Size (in.)	Minimum Trench Width (in.)
3 to 16	Pipe O. D. + 12
18 to 42	Pipe O. D. + 18
48 and larger	Pipe O. D. + 24

Trench Length

Table 4 lists the recommended lengths of trench openings for each placement of continuous lengths of fused pipe, assembled above the trench. When the trench sidewalls are significantly sloped, somewhat shorter trench openings may be used. When space or ground conditions do not permit these suggested trench openings, the pipe lengths may be joined within the trench, using a joining machine or flanged couplings. When bell-and-spigot jointed pipe or flange-end pipe is used, the trench opening needs to be only long enough to accommodate placement and assembly of a single pipe length.

TABLE 4
Suggested Length of Minimum Trench Opening (Feet) for Installation of Joined Lengths of Polyethylene Pipe

Nominal Pipe Size (in.)	Depth of Trench (Feet)					
	3	5	7	9	11	13
½ to 3	15	20	25	30	35	40
4 to 8	25	30	35	40	45	50
10 to 14	35	40	45	50	55	60
16 to 22	45	50	55	60	65	70
24 to 42	-	60	65	70	75	80
48	-	-	80	90	100	110

Stability of the Trench

Although the native soil in which PE pipe is installed need not be as strong and stiff as the pipe embedment materials, it should provide adequate support and stable containment of the embedment material so that the density of the embedment material does not diminish. If the trenching conditions present construction problems such as trench sidewalls that readily slough or a soft trench floor that will not support workers or compaction, it is termed unstable. The instability is usually a condition of the trench and not the soil. Most often the primary cause of the instability is high groundwater, not the soil. Even soft or loose soils can provide good support for the pipe if they are confined. The problem with unstable conditions generally occurs during the installation. When the trench is opened where groundwater is present, most soils, except firm, cohesive soils (firm clays) or cemented soils, tend to slough off the trench wall. This results in a trench that keeps widening, with loose material falling into the trench floor.

Soil formations that commonly lead to unstable trenching conditions include materials with fine grain soils (silts or clays) saturated with water and uncemented sands saturated with water. In some cases, where the soil has an extremely high water content, such as with peat or with clay (or silt) having a water content beyond the liquid limit, the soil behaves "hydraulically", that is, the water in the soil controls the soil's behavior. Here, the backfill must be designed to sustain all the pressure from the pipe without support from the in-situ soil. These conditions may occur in saturated fine grained soils where the unconfined compressive strength of the soil is less than 500 psf, or in saturated, sandy soils where the standard penetration value, N, is less than 6 blows per ft. In this case, an engineering evaluation should be made to determine the necessity for special procedures such as a "wide" trench or permanent trench sheeting of the trench width.

As mentioned above, most trench stability problems occur in trenches that are excavated below the groundwater level. (However, the designer and the contractor should keep in mind that all trenches pose the risk of collapse and therefore workers should not be in trenches that are not adequately braced or sloped.) Stability can be improved by lowering the water table through deep wells, well-points, or other such means. In some ground the permeability is such that the only option is to remove the water after it has seeped out of the trench walls. Here the contractor will use underdrains or sumps on the trench floor. De-watering should continue throughout the pipe laying operation until sufficient cover is placed over the pipe so that it will not float.

Stability of Trench Floor

Trench floor stability is influenced by the soils beneath the trench. The floor must be stable in order to support the bedding material. A stable bedding minimizes bending of the pipe along its horizontal axis and supports the embedment enveloping the pipe. Generally, if the trench floor can be walked on without showing foot prints it is considered stable.

In many cases the floor can be stabilized by simply dewatering. Where dewatering is not possible or where it is not effective, stabilization of the trench floor may be accomplished by various cost-effective methods which can be suited to overcome all but the most difficult soil conditions. Included among these are the use of alternate trench foundations such as wood pile or sheathing capped by a concrete mat, or wood sheathing with keyed-in plank foundation; stabilization of the soil by the use of special grout or chemicals; geofabric migration barriers; or ballasting (undercutting). A cushion of bedding material must be provided between any special foundation and the pipe. Permanently buried timber should be suitably treated.

Stabilization by ballasting (undercutting) is the removal of a sufficient quantity of undesirable material. This technique is frequently employed to stabilize randomly encountered short sections of unstable soil. The extent of required over-excavation and details of accompanying construction requirements will be determined by the engineer in consideration of the qualities of the unstable soil and the specific design requirements. The following are general guidelines:

The trench bottom should be over-excavated over the full trench width from 18 to 36 inches below the pipe grade (depending on the soil strength and pipe diameter) and then brought back to grade with a foundation of ballast material topped with Class I material. An appropriate bedding should then be placed on the foundation. The grading of the foundation material should be selected so that it acts as an impervious mat into which neither the bedding, other embedment material, nor the surrounding native soil will migrate.

These guidelines are suitable for most situations except for extremely weak soils (such as quicksands, organic silts, and peats) which may call for further overexcavation, or other special treatment.

Stability of Trench Walls

In order to control deflection, the embedment material must be placed from undisturbed trench sidewall to undisturbed trench sidewall. Where trench walls are unstable, it may be necessary to use trench shields, bracing, or permanent sheeting to achieve a stable sidewall while installing the pipe. Where material sloughs into the trench it should be removed. This technique often leads to widening the trench.

Walls of trenches below the elevation of the crown of the pipe should be maintained as vertical as possible. The shape of the trench above the pipe will be determined by the stability of the trench walls, excavation depth, surface loadings near the trench, proximity of existing underground structures, presence of groundwater or runoff water, safety and practical considerations. These will determine if the trench walls may be vertical, excavated with slope or benched sides, or shored. When trench walls are shored or otherwise stabilized, the construction scheme must allow for the proper placement and compaction of pipe embedment materials. Some suggested trench construction schemes follow. The final procedure must be in compliance with all applicable safety regulations.

Sloping of trench walls in granular and cohesionless soils should be provided whenever the walls are more than about four feet in depth or otherwise required by state, local or federal regulations. For safety, if the walls are not sloped, they should be stabilized by alternate means such as shoring or bracing. The slope should be no greater than the angle of repose of the materials being excavated and should be approved by the engineer.

Shoring or bracing will frequently be required in wet fine grained cohesive type soils and clays. Bracing or sheathing that is constructed of treated timber, steel or other acceptable material may be used to stabilize trench walls either permanently or temporarily. Wherever possible, sheathing and bracing should be installed so that its bottom extends no lower than about one-quarter of the pipe diameter below the pipe crown. When so installed, pulling the sheathing will minimally disturb the embedment material and the side support it provides. Sheathing that is installed to project below the pipe springline should be left in place unless, as with some thinner sheathing, it is designed to be pulled and removed without disturbing the embedment next to the pipe. In this case, the trench width should be increased by 12 to 24 inches depending on the pipe diameter to allow for minor disturbance to the embedment near the sheathing. Vibratory placement or extraction of sheeting is not advised. This method can cause severe disturbance to the bedding and liquefaction of the surrounding soils. Where steel sheet piling is used as sheathing and is to be removed or pulled, to minimize disturbance to the pipe embedment, it should be installed so that it is not closer than one pipe diameter or 18 inches, whichever is larger, from either side of the pipe. The void left by removal of the sheathing should be filled with embedment material.

Portable Trench Shield

Portable trench shields or boxes which provide a moveable safe working area for installing pipe can be used with flexible pipe. However, the installation technique of flexible pipe with the shield is not the same as it is for rigid pipe. In order to use the shield with PE pipe, all excavation of the trench below the pipe crown elevation should be done from inside of the shield. That is, the backhoe operator should dig inside of the shield and force the shield down as soil is removed. (The technique of digging out a large hole to pipe invert grade then sliding the shield into it will result in excess deflection of PE pipe.) After placing the pipe in the trench, embedment material should be placed in lifts and the shield vertically raised after each lift is placed so that workers can shovel embedment material under the shield to fill the void created by the shield wall. Figure 6 illustrates the steps used with a Portable Trench Shield.

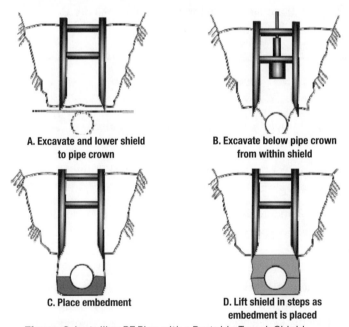

Figure 6 Installing PE Pipe with a Portable Trench Shield

If trench soil quality and applicable safety regulations permit, it is best to use shields that are placed with no portion of their sides extending lower than one-quarter of a pipe diameter below the pipe crown. This minimizes the amount of lifting required and precludes the possibility for disturbing embedment materials. If the sides of the trench box or shield do project below this point, then the box should be lifted vertically as described above, before moving along the trench.

The minimum inside clear width of the box, or shield, should allow for the minimum trench width requirements for the pipe to be satisfied plus an additional 12 to 24 inches depending on the pipe diameter.

Installation Procedure Guidelines

The following guidelines for the installation of PE pipe are based on the discussions of the previous sections. The reader is advised to see PPI Technical Report TR31 for more specific installation recommendations for solid wall SDR pipe. If the reader is interested in installing 24" or smaller pipes at 16 feet or less, see Appendix 1.

The installation procedure discussed in this section consists of trench floor preparation, providing a sufficiently stable working platform, and meeting the design grade requirements. Following pipe placement, backfill material which has been selected with regards to potential material migration, required density, depth of cover, weight of soil and surcharge loads is installed as follows:

1. Bedding material is placed and leveled.
2. Haunching is placed and, if required, compacted so as not to disturb the pipe from its line and grade.
3. The remainder of the primary initial backfill is placed and, if required, compacted in lifts.
4. Secondary backfill is used to protect the pipe during the final backfilling operation and also to provide support for the top portion of the pipe.
5. The final backfill may consist of any qualifying material that satisfies road construction or other requirements and, when required, must be compacted.

Trench Floor Preparation

The trench floor must have sufficient stability and load-bearing capacity to present a firm working platform during construction to maintain the pipe at its required alignment and grade and sustain the weight of the fill materials placed around and over the pipe. The trench bottom should be smooth and free from sloughed sidewall material, large stones, large dirt clods, frozen material, hard or soft spots due to rocks or low-bearing-strength soils, and any other condition that could lead to non-uniform or unstable support of the pipe. The trench bottom must be kept dry during installation of the pipe and the embedment materials. All foundation and bedding materials must be placed and compacted according to the design requirements. Such materials should be selected to provide the necessary migration control when required.

Over-excavation of the trench floor by more than 6 inches beyond grade requires that the over-excavation be filled with acceptable embedment material that is compacted to a density equal to that of the embedment material. If the over excavation exceeds 12 inches, it should be brought to proper grade with a suitably graded Class I or II material that is compacted to the same density as that of the native soil but not less than the density requirements for the embedment materials.

In stable soils the trench floor should be undercut by machine and then brought up to proper grade by use of a well-leveled bedding consisting of a 4 to 6-inch layer of embedment material. This material should be compacted by mechanical means to at least 90 percent Standard Proctor Density. Class I material may be shovel sliced where the depth of cover permits.

In unstable soils that may be too soft, of low load-bearing capacity or otherwise inadequate, the trench bottom must first be stabilized by soil modification, by providing an alternate foundation, or by the removal of the undesirable material and replacement with stable foundation material. A cushion of at least 4 inches of compacted bedding should be provided between any special foundation and the pipe. Adequacy of trench bottom stability is difficult to evaluate by visual observation and is therefore best determined by soil tests or at the site during installation. However, a warning of a potentially unstable soil condition is given by a trench bottom that cannot support the weight of workmen.

Uneven soil support conditions, where the grade line traverses both soft and hard spots, requires special consideration. Ballasting is the most frequently employed technique to deal with randomly encountered short sections of soft soils.

When differential conditions of pipe support might occur, such as in transitions from manholes to trench or from hard to soft soils, a transition support region should be provided to ensure uniform pipe support and preclude the development of shear, or other concentrated loading on the pipe. The following procedure may be used:

The soil next to the more rigid support is over-excavated to a depth of not less than 12 inches over a distance of 2 pipe diameters along the pipe line; over the next 2 diameters away from the rigid support, the depth of over-excavation is gradually decreased until it meets the normal trench depth. See Figures 7 and 8. Pipe grade is then restored by the addition of granular material that is compacted. In the case of connections to manholes and buildings, the distance of over-excavation along the pipe length should be no less than required to reach undisturbed soil.

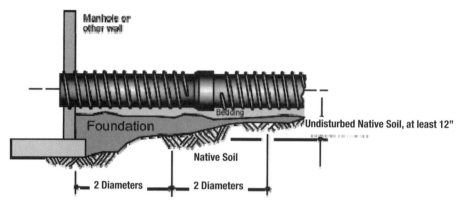

Figure 7 Pipe Support in Transition from Rigid Support to Normal Trench Support

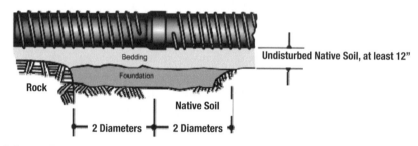

Figure 8 Proper Transition from Rock Trench Bottom to Normal Trench Support

Backfilling and Compaction

Backfilling should follow pipe placement and assembly as closely as possible. Such practice prevents the pipe from being shifted out of line by cave-ins, protects the pipe from external damage, eliminates pipe lifting due to flooding of open trench and in very cold weather lessens the possibility of backfill material becoming frozen. The quality of the backfill materials and their placement and compaction will largely determine the pipe's ultimate deformation and alignment. Backfill material should be selected with consideration of potential material migration to, or from, the trench wall and other layers of embedment material. Under most circumstances, compaction will be required for all material placed in the trench from 6 inches beneath the pipe to at least 6 inches above the pipe.

The required density of the bedding, haunching and the primary and secondary initial backfill material will depend on several considerations such as depth of cover, weight of soil, and surcharge loads. The minimum density for these materials should be equal to 85 percent Standard Proctor Density for Class I and II materials or 90 percent Standard Proctor Density for Class III or IVa materials. For Class II,III, and IVa materials, compaction will always be required to obtain these densities.

Class I material placed by shovel slicing will generally have a minimum density of 85 percent Standard Proctor; however, its E' may not be greater than 750 psi. Just dumping Class I material into the trench may produce densities near 85 percent. However, except in shallow cover without live loads, this method will normally not provide adequate support to the pipe as voids may exist under the pipe haunches or elsewhere in the material.

Backfill Placement

Bedding performs a most important function in that it levels out any irregularities in the trench bottom, assuring uniform support and load distribution along the barrel of each pipe section and supports the haunching material. A mat of at least 6 inches of compacted embedment material will provide satisfactory bedding.

Haunching material must be carefully placed and compacted so as not to disturb the pipe from its line and grade while ensuring that it is in firm and intimate contact with the entire bottom surface of the pipe. Usually a vibratory compactor has less tendency to disturb the pipe than an impact tamper.

Primary initial backfill should be placed and compacted in lifts evenly placed on each side of the pipe. The lifts should not be greater than 12 inches for Class 1, 8 inches for Class II, and 6 inches for Class III and IVa materials. The primary initial backfill should extend up to at least three-quarters of the pipe diameter to perform its function of pipe side support as shown in figure 2. If the construction does not call for the use of a secondary initial backfill, then the primary layer should extend to not less than 6 inches above the pipe crown. In any location where the pipe may be covered by existing or future groundwater, the primary initial backfill should extend up to at least 6 inches over the pipe crown for pipe up to 27-inch diameter and to at least 12 inches over the pipe for larger pipe.

Secondary initial backfill serves to protect the pipe during the final backfilling operation and to provide support to the top portion of the pipe. Secondary initial backfill should extend to 6 inches above pipe for pipe up to 24 inches and to 12 inches for larger pipe. These depths can be modified slightly depending on the depth of burial, groundwater level, and type of native soil. Compaction of this layer should be to the same extent as that specified for the primary initial backfill. If the final backfill material contains large rock (boulder or cobble size) or clumps, then 18 inches of cushion material should be provided in the secondary initial backfill. Secondary initial backfill may consist of a different material than the primary initial backfill; however, in most cases, it should be a material that will produce an E' of at least 750 psi.

The final backfill may consist of any material that satisfies road construction or other requirements. The material must be free of large stones or other dense hard objects which could damage the pipe when dropped into the trench or create concentrated pipe loading. The final backfill may be placed in the trench by machines.

There should be at least one foot of cover over the pipe before compaction of the final backfill by the use of self-powered compactors. Construction vehicles should not be driven over the pipe until a three foot cover of properly compacted material is placed over the pile.

When backfilling on slopes, the final backfill should be well compacted if there is any risk of the newly backfilled trench becoming a "french drain." Greater compaction may be achieved by tamping the final backfill in 4 inch layers all the way from the top of the initial backfill to the ground or surface line of the trench. To prevent water from undercutting the underside of the pipe, concrete collars keyed into the trench sides and foundation may be poured around the pipe or a polyethylene waterstop can be fabricated onto the pipe.

Proper Burial of HDPE Fabricated Fittings

A common question is "Does the installation of heat fused polyethylene solid wall pipe and fittings need thrust blocks?" The simple answer to this question is that heat fused HDPE pipe and fittings are a monolithic structure which does not require thrust blocks to restrain the longitudinal loads resulting from pipe pressurization.

Since fittings are part of the monolithic structure no thrust blocks are needed to keep the fittings from separating from the HDPE pipe. Bell and spigot piping systems must have thrust blocks or restrained joints to prevent separation of pipe from fittings when there is a change of direction.

Pipe movement due to elastic deformation, thermal expansion/contraction, etc. is not detrimental to HDPE pipe, but pipe movement or the attachment of valves or other appurtenances used with HDPE pipe systems can cause excessive loads. Proper backfill prevents excessive loads in most situations.

Common fittings, elbows and equal tees normally require the same backfill as specified for the pipe. When service connections are made from HDPE water mains, no special compaction is required. When service connections are made under an active roadway, 95% Standard Proctor density is normally required around the pipe and the service connection.

In water systems and fire protection piping systems, reducing tees are frequently used to connect from the main to valves and hydrants. The attached drawing shows the use of concrete support pads, thrust blocks on hydrants, self restrained HDPE MJ adapters and sand stabilized with cement around the reducing tee. While no true thrust blocks are on the HDPE pipe or fittings in this arrangement, the sand stabilized with cement provides proper support for the reducing tee. Compaction of the soil around these fittings is difficult and the use of sand stabilized with cement or flowable fill is usually easy.

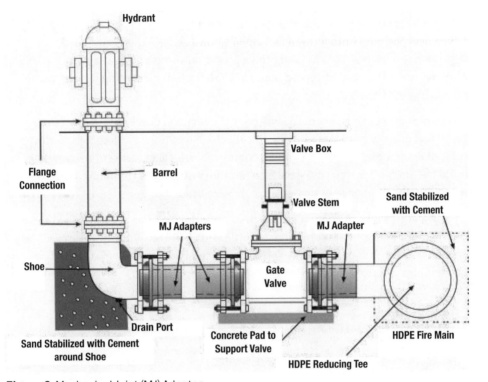

Figure 9 Mechanical Joint (MJ) Adaptor

As with all piping systems, proper compaction of the soil around pipe and fittings is important. In water and/or fire protection systems, when in-situ embedment materials can be compacted to a Standard Proctor density of 85% for installation outside of roadways or 95% Standard Proctor density in roadways, these materials should be used. When in-situ materials do not provide proper support, then sand stabilized with cement or flowable fill should be used.

Figure 9 shows an HDPE self-restrained mechanical joint (MJ) adapter being used to connect to the valve. When large reducing tees or equal tees are used, MJ adapters, flanges or electrofusion couplings should be fused to the reducing tees before it is placed in the trench. The direct connection of long pipe sections or valves can create bending loads on the leg of the reducing tee. The use of MJ's, flanges or electrofusion couplings on the reducing leg of the tee makes installation of reducing tees easier and safer while preventing stresses on the tee.

Inspection

One principal function of the inspector is to insure that the pipe meets the acceptance deflection specified by the engineer. Besides seeing that the installation practice of the contractor meets the specification, the inspector should periodically make deflection measurements of the pipe. Where the pipe can be accessed, inspection can be as simple as going through the pipe and taking diameter measurements. For smaller pipe, a mandrel or deflection measuring device can be pulled through the pipe.

Good installation practice consists of frequent deflection checks at the beginning of the project or anywhere there is a significant change in the installation procedure, soil formation, or materials. A prudent contractor will check deflection every 100 or 200 feet under these circumstances. After the contractor is confident in the procedure, the frequency of inspection can be relaxed.

Typically, acceptance deflection is measured after the pipe has been installed for at least 30 days. This gives the soil time to settle and stabilize. Where pipe exceeds its acceptance limit, it should be uncovered and the embedment material should be replaced and compacted.

References

1. Watkins, R. K. (1975). *Buried Structures*, Foundation Engineering Handbook (edited by H. F. Winterkom and H. Y. Fang), Van Nostrand Reinhold Co., New York, NY.
2. Spangler, M. G. (1951). *Soil Engineering*, International Textbook Co., Scranton, PA.
3. Howard, A. K. (1981). *The USBR Equation for predicting Flexible Pipe Deflection*, Proc. Int. Conf. on Underground Plastic Pipe, ASCE, New Orleans, LA.
4. Howard, A. K. (1972, January). *Modulus of Soil reaction Values for buned Flexible Pipe*, Joumal of the Geotechnical Engineering Division, ASCE, Vol. 103, No. GT1, pp. 3346.

Appendix 1

Simplified Installation Guidelines for Pressure Pipe

(Small diameter pressure pipes usually have adequate stiffness and are usually installed in such shallow depths that it is unnecessary to make an internal inspection of the pipe for deflection.)

A quality job can be achieved for most installations following the simple steps that are listed below. These guidelines apply where the following conditions are met:

1. Pipe Diameter of 24-inch or less
2. SDR equal to or less than 26
3. Depth of Cover between 2. 5 feet and 16 feet
4. Groundwater elevation never higher than 2 feet below the surface
5. The route of the pipeline is through stable soil

Stable soil is an arbitrary definition referring to soil that can be cut vertically or nearly vertically without significant sloughing, or soil that is granular but dry (or de-watered) that can stand vertical to at least the height of the pipe. These soils must also possess good bearing strength. (Quantitatively, good bearing capacity is defined as a minimum unconfined compressive strength of 1000 psf for cohesive soils or a minimum standard penetration resistance of 10 blows per ft for coarse grained soils.) Examples of soils that normally do not possess adequate stability for this method are mucky, organic, or loose and wet soils.

Where the above conditions are met, the specifier can write installation specifications from the following steps. The specifier should insure that all OSHA, state and local safety regulations are met.

The following are general guidelines for the installation of PE pipe. Other satisfactory methods or specifications may be available. This information should not be substituted for the judgment of a professional engineer in achieving specific requirements.

Simplfied Step-by-Step Installation

Trenching

Trench collapses can occur in any soil and account for a large number of worker deaths each year. In unbraced or unsupported excavations, proper attention should be paid to sloping the trench wall to a safe angle. Consult the local codes. All trench shoring and bracing must be kept above the pipe. (If this is not possible, consult the more detailed installation recommendations.) The length of open trench required for fused pipe sections should be such that bending and lowering the pipe into the ditch does not exceed the manufacturer's minimum recommended bend radius and result in kinking. The trench width at pipe grade should be equal to the pipe outer diameter (O. D.) plus 12 inches.

De-watering

For safe and proper construction the groundwater level in the trench should be kept below the pipe invert. This can be accomplished by deep wells, well points or sump pumps placed in the trench.

Bedding

Where the trench bottom soil can be cut and graded without difficulty, pressure pipe may be installed directly on the prepared trench bottom. For pressure pipe, the trench bottom may undulate, but must support the pipe smoothly and be free of ridges, hollows, and lumps. In other situations, and for gravity drain or sewer pipe, bedding may be prepared from the excavated material if it is rock free and well broken up during excavation. For gravity flow systems, the trench bottom

should be graded evenly. The trench bottom should be relatively smooth and free of rock. When rocks, boulders, or large stones are encountered which may cause point loading on the pipe, they should be removed and the trench bottom padded with 4 to 6 inches of tamped bedding material. Bedding should consist of free-flowing material such as gravel, sand, silty sand, or clayey sand that is free of stones or hard particles larger than
one-half inch.

Pipe Embedment

Figure 1 shows trench construction and terminology. Haunching and initial backfill are considered trench embedment materials. The embedment material should be a coarse grained soil, such as gravel or sand, or a coarse grained soil containing fines, such as a silty sand or clayey sand. The particle size should not exceed one-half inch for 2 to 4-inch pipe, three-quarter inch for 6 to 8-inch pipe and one inch for all other sizes. Where the embedment is angular, crushed stone may be placed around the pipe by dumping and slicing with a shovel. Where the embedment is naturally occurring gravels, sands and mixtures with fines, the embedment should be placed in lifts, not exceeding 6 inches in thickness, and then tamped. Tamping should be accomplished by using a mechanical tamper. Compact to at least 85 percent Standard Proctor density as defined in ASTM D-698. Under streets and roads, increase compaction to 95 percent Standard Proctor density.

Pressure Testing

If a pressure test is required, it should be conducted after the embedment material is placed.

Trench Backfill

The final backfill may consist of the excavated material, provided it is free from unsuitable matter such as large lumps of clay, organic material, boulders or stones larger than 8 inches, or construction debris. Where the pipe is located beneath a road, place the final backfill in lifts as mentioned earlier and compact to 95 percent Standard Proctor Density.

Appendix 2

Guidelines for Preparing an Installation Specification General Requirements

General Requirements

Subsurface conditions should be adequately investigated and defined prior to establishing final project specifications. Subsurface investigations are necessary to determine types of soil that are likely to be encountered during construction, existence of rock, thickness of strata layers, relative quality of strata layers, presence of other utilities, and presence of ground water. These findings are useful both in specifying the proper pipe for an application and in planning construction procedures.

Prior to start of construction the on-site surface conditions, including water run-off, traffic and other problems should be appraised by on-site inspections of the proposed pipeline location. Addtionally, all the construction documents, including plans, subsoil information and project specifications should be reviewed. All required permits should be obtained and arrangements made to insure compliance with all applicable federal, state, and local safety regulations.

The installation should be checked throughout the construction period by an inspector who is thoroughly familiar with the contract specifications, materials and installation procedures. The inspection should ensure that significant factors such as trench depth, width, grade, pipe foundation, quality and compaction of embedment and backfill comply with contract requirements.

The following specification may be used for most gravity drain projects in stable or de-watered trenches. Where special methods of stabilization are required as discussed previously, a more detailed specification may be required.

Guide Specification High Density Polyethylene (Hdpe) Gravity-drain Pipe (F-894 Pipe)
Various construction techniques can be used for installing PE pipe. The techniques described below are considered satisfactory, but there may be other techniques which will work equally as well. The information below is considered reliable, but the author makes no warranty, expressed or implied, as to the content, and disclaims all liability therefor. This information should not be substituted for the judgment of a professional engineer in achieving specific requirements.

General

Scope
The work covered by this section includes furnishing all labor, equipment, and materials required to supply, install, and test high-density polyethylene (HDPE) pipe, including accessories, as shown on the drawings and/or specified herein.

Quality Assurance
- The Contractor shall submit to the Engineer in writing that the pipe furnished under this specification is in conformance with the material and mechanical requirements specified herein.
- Each HDPE pipe length shall be clearly marked with the following:

1. Manufacturer's Name
2. Pipe Size
3. SDR or Ring Stiffness Constant Classification
4. Production Code Designating Plant Location, Machine, and Date of Manufacture

Shop Drawings

Complete shop drawings on all piping and accessories shall be submitted to the Engineer.

Storage

All pipe and accessories shall be stored on flat, level ground with no rocks or other objects under the pipe.

The maximum recommended stacking height for HDPE pipe is given in Table A2. 1

TABLE A2. 1
Allowable Stacking Heights for F-894 Pipe

Nominal Pipe Size	Number of Rows High
18	4
21	4
24	3
27	3
30	3
33	2
36	2
42	2
48	1
54	1
<54	1

Products

General

Apart from the structural voids and hollows associated with some profile wall designs, the pipe and fittings shall be homogenous throughout and free from visible cracks, holes, foreign inclusions or other injurious defects. The pipe shall be as uniform as commercially practical in color, opacity, density, and other physical properties.

High-Density Polyethylene (HDPE) Pipe

HDPE Profile wall pipe and fittings shall be manufactured in accordance with the requirements of (Engineer: specify appropriate ASTM designation here. For profile pipe designate ASTM F 894-85. For solid wall DR pipe designate ASTM F 714.)

HDPE profile wall pipe shall be made from a plastic compound meeting the requirements of Type III, Class C, Category 5, Grade P34 as defined in ASTM D 1248 and with an established hydrostatic design basis (HDB) of not less than 1250 psi for water at 73.4°F, determined in accordance with method ASTM D 2837. Materials meeting the requirements of cell classification PE 345464C or higher cell classification in accordance with ASTM D-3350 are also suitable. (Engineer: specify appropriate HDB and Cell class as underlined above.)

Material other than those specified above may be used as part of the profile wall construction, for example, as a core tube to support the shape of the profile during the processing, provided that these materials are compatible with the PE material, are completely encapsulated in the finished product, and in no way compromise the performance of the PE pipe product in the intended use.

Execution

Pipe Laying

A. Before the sewer pipe is placed in position in the trench, the bottom and sides of the trench shall be carefully prepared, the required bedding placed, and bracing and sheeting installed where required. The trench shall be excavated to the dimensions shown on the Engineer's drawings. Each pipe shall be accurately placed to the line and grade called for on the drawings. Grade shall be controlled by a laser beam or batter boards and a Mason's line. All equipment for maintaining grade shall be furnished by the Contractor.

B. All pipe and fittings shall be inspected before they are installed.

C. Pipe laying shall proceed upgrade, starting at the lower end of the grade with the bells uphill.

D. If the trench bottom does not provide a firm and stable working platform, sufficient material shall be removed and replaced with approved compacted materials to provide a firm foundation for the pipe.

E. Pipe trenches shall be kept free from water during pipe laying, jointing and until sufficient backfill has been placed to prevent flotation of the pipe. The minimum height of backfill to prevent flotation may be obtained from the Engineer. The Contractor may use sump pumps, well points, or other devices to remove water from the trench bottom. Small puddles that are no closer than 4" from the bottom of the pipe are acceptable. The contractor shall provide ample means and devices to promptly remove and dispose of all water from any source entering the trench.

F. No connection shall be made where joint surfaces and joint materials have been soiled by dirt in handling until such surfaces are thoroughly cleaned by washing and wiping.

G. As the work progresses, the interior of all pipes shall be kept clean. After each line of pipe has been laid, it shall be carefully inspected and all soil, trash, rags, and other foreign matter removed from the interior.

H. Backfilling of trenches shall be started immediately after the pipe is placed in the trench.

I. If the Engineer determines that no groundwater will be encountered or that the maximum height of the groundwater level (from seepage or other groundwater movement through the existing soil formation or the pipe trench) will not exceed the springline of the pipe during the service life of the line, the pipe shall be backfilled according to detail drawing Figure A1 titled "Dry Installation Bedding Requirements. "

J. If the Engineer determines that groundwater will be encountered or that the ground water level is anticipated to exceed the springline of the pipe during the service life of the line, backfill pipe according to the detail drawing Figure A2, titled "Wet Installation Bedding Requirements. "

K. Shoring, sheeting, or trench shields shall be utilized in such a manner as to minimize disturbance of the backfill material beneath the pipe crown. Trench sheeting that extends below the crown should either be left permanently in place or consist of adequately supported steel sheets 1" (one inch) thick or less which can be extracted with minimal disturbance to the pipe embedment. Where moveable trench shields are used, the following steps shall be followed unless an alternate technique that does not disturb the pipe embedment can be demonstrated:

   1. Excavation of the trench below the elevation of the pipe crown shall be done from inside of the trench shield to prevent the accumulation of loose or sloughed material along the outside of the shield. Excavation of the trench ahead of the shield at an elevation below the pipe crown is not permitted unless approved by the Engineer.

   2. After laying the pipe in the trench, bedding and pipe embedment shall be placed in lifts and the shield must be lifted in steps. As the shield is lifted, embedment material shall be shoveled under the shield so as to fill all voids left by the removal of the shield.

L. Bedding Material. Bedding material to be selected by Engineer. (Note to Engineer: Bedding material to be selected by evaluating depth of cover and E' required to control deflection and buckling.)

When E' = 1,000 is required:
Specify Class I material shovel sliced to a minimum density of 85 percent Standard Proctor, or Class II material with mechanical compaction to a minimum density of 90 percent Standard Proctor, or Class III material with mechanical compaction to a minimum density of 90 percent Standard Proctor. (Embedment Classes are defined in ASTM D-2321.)

When E' = 2,000 is required:
Specify Class I or Class II material with mechanical compaction to a minimum density of 90 percent Standard Proctor.

When E' = 3,000 is required:
Specify Class I material with mechanical compaction to a minimum density of 90 percent Standard Proctor.

M. Backfill material placed under the pipe haunches shall be thoroughly shovel sliced along the length of the pipe.

N. Where compaction of backfill materials is required, compact by mechanical means. Suitable mechanical means includes vibratory sleds, gasoline driven impact tampers, and air driven impact tampers or other approved means. Compact to a minimum of 90 percent Standard Proctor or as required by the Engineer.

O. Pipe embedment soil shall be placed in lifts as follows:
- Lift thickness for Class I material shall not exceed 12 inches.
- Lift thickness for Class II material shall not exceed 8 inches.
- Lift thickness for Class III material shall not exceed 6 inches.

P. After completing backfill in the pipe zone, the trench shall be backfilled to grade with native soil. Where pipe is located beneath streets, compact backfill to a minimum of 95 percent Standard Proctor or otherwise as directed by the Engineer. HDPE profile pipe shall not be subject to a roller or wheel loads until a minimum of one pipe diameter or 36" (whichever is larger) of backfill has been placed over the top of the pipe and a hydrohammer shall not be used until a minimum depth of one pipe diameter or 48" (whichever is larger) of backfill has been placed over the top of the pipe.

Connections

A. Connections to existing lines shall be made by coupling a piece of smooth O. D. HDPE pipe to the existing line. The coupling shall be a flexible elastomeric boot with stainless steel clamps. The coupling is to be encased in cement-stabilized sand, grout, or concrete.

B. Connections to concrete manholes shall be made using smooth O. D. pipe and water stops, profile pipe cast into the concrete, or via elastomeric sleeves or gaskets precast in the manhole. Since the particular technique used is highly dependent on the construction method, these connections cannot be guaranteed by the manufacturer to be leak free.

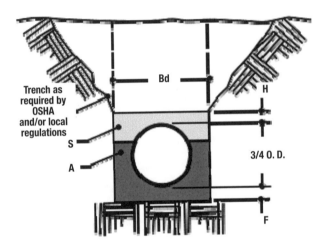

A = ¼" to 1" Class I, II, or III Material
If cover < 18 ft – shovel slice
If cover > 18 ft – compact to at least 90% Standard Proctor per ASTM D-698
If cover > 24 ft – use wet installation bedding requrirements (Figure A2).
S = Selected earth backfill compacted to at least 90% Standard Proctor per ASTM D-698.
H = 6 inches for 18" to 36" pipe
 = 12 inches for 42" to 84" pipe
 = 18 inches for 96" to 120" pipe
F = 4 inches for 18" to 36" pipe
 = 6 inches for 42" to 84" pipe
 = 8 inches for 96" to 120" pipe
Bd = OD + 18"– see Table 3
 = OD + 18 inches for 18" to 36" pipe
 = OD + 24 for 36" to 60" pipe
 = OD + 36 for 66" to 84" pipe
 = OD + 48 for 96" to 120" pipe

Figure A1 Dry[1] Installation Bedding Requirements for HDPE Profile Wall Pipe
([1]Pipe springline elevation permanently above groundwater)

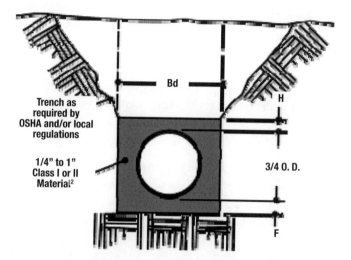

Compact Class I or Class II Material[2] to a minimum of 90% Standard Proctor per ASTM D-698.

H = 6 inches for 18" to 36" pipe
  = 12 inches for 42" to 84" pipe
  = 18 inches for 96" to 120" pipe
F = 4 inches for 18" to 36" pipe
  = 6 inches for 42" to 84" pipe
  = 8 inches for 96" to 120" pipe
Bd = OD + 18" – see Table 3
  = OD + 18 inches for 18" to 36" pipe
  = OD + 24 for 36" to 60" pipe
  = OD + 36 for 66" to 84" pipe
  = OD + 48 for 96" to 120" pipe
See section in this appendix titled "Pipe Laying"

Figure A2 Wet[1] Installation Bedding Requirements for HDPE Profile Wall Pipe
([1]Pipe springline elevation permanently above groundwater)
([2]Selection of bedding material to be made by Engineer based on pipe design requirements)
See section in this appendix titled "Pipe Laying."

C. Connections to HDPE manholes shall be made using closure pieces with shoulder gaskets.

Pipe Tunnels and Casing

A. The annular space between HDPE pipe and the casing pipe shall be filled with concrete grout. (Engineer: Grout is required where the pipe's allowable hydrostatic buckling resistance is less than the water pressure created by groundwater entering the casing.) The Contractor's procedure for placing the grout shall be approved by the Engineer. After installation of pipe in casing, the casing shall be kept dewatered until grouting is completed. Grout shall be placed by gravity flow only. Do not pressure grout PE pipe in a casing.

Inspection and Testing

A. After completion of any section of sewer, the grades, joints, and alignment shall be true to line and grade. There shall be no visual leakage, and the sewer shall be completely free from any cracks and from protruding joint materials, deposits of sand, mortar or other materials on the inside to the satisfaction of the Engineer.

B. At the Engineer's request, a deflection test shall be performed by the Contractor. The deflection can be measured mechanically by a mandrel or manually using an extension ruler. The final deflection test shall not be made on a section of sewer until all the backfill on that section has been in place for 30 days. However, the Contractor shall perform the deflection test on the first 300 – 400 feet of pipe after it has been backfilled in order to verify that the installation procedures are adequate to meet the requirements of the contract. No additional pipe shall be laid until this test has been successfully completed. Pipe deflection may be determined by direct vertical measurement of no less than 4 equally spaced points in each pipe section or by pulling a mandrel.

For solid wall PE pipe, deflection shall not exceed 5 percent of the I. D. For profile pipe, deflection shall not exceed 5 percent of the base I. D. as indicated in Table A2. 2.

All excess deflections shall be corrected. The Contractor shall correct the deficiency and retest the pipe.

TABLE A2. 2
Minimum Acceptable Diameter

Nominal Pipe Size (in.)	Base I. D. (in.)	Minimum Acceptable Diameter 5% of Base I. D. (in.)
18	17.34	16.47
21	20. 26	19.25
24	2319	22.03
27	26.11	24.80
30	29.02	27.57
33	31.94	30.34
36	34.86	33.12
42	40.67	38.64
48	46.48	44.16
54	52.29	49.68
60	58.10	55.20
66	63.91	60.71
72	69.72	66.23
84	81.34	77.27
96	92.96	88.31

C. The Contractor shall conduct either an infiltration test or a water test for leakage as determined by the Engineer. Testing shall be conducted in accordance with all applicable safety standards.

1. **Infiltration Test**

   Infiltration shall not exceed 50 gallons per 24 hours per inch of diameter per mile of sewer. Contractor shall furnish all supplies, materials, labor, service, etc. , needed to make infiltration or exfiltration tests including water. No separate payment will be made for equipment, supplies, material, water, or services.

   Any leakage, including active seepage, shall be corrected where such leakage exists until the pipeline meets the requirements of the allowable leakage specifications.

   Infiltration tests shall be made when groundwater level is 18 inches or more above the top of the outside of the pipe.

2. **Water Test**

   When normal groundwater does not stand at a level outside the pipe so as to enable infiltration tests to be made to the satisfaction of the Engineer, the Contractor shall make exfiltration tests by filling the pipe or sections thereof with water to a head of not less than 2 ft. above the top of the outside of the pipe and observing the amount of water required to maintain this level.

Cleanup
A. After completing each section of the sewer line, the Contractor shall remove all debris, construction materials, and equipment from the site of the work, then grade and smooth over the surface on both sides of the line and leave the entire right-of-way in a clean, neat, and serviceable condition.

Appendix 3

Basic Soil Concepts For Flexible Pipe Installation

Soil Classification
The embedment soil surrounding a flexible pipe prevents pipe from deflecting through its shear strength and stiffness. Shear strength enables the soil to resist distortion much like a solid body. Shear strength, or shear resistance as it is often called, arises from the structure of the soil's fabric. Soil is an assemblage of (1) mineral particles such as silica or aluminum silicates, (2) water, and (3) air. Mineral particles can range in size from the large, such as boulders, to the microscopic, such as the colloidal particles making up clay. The size of the individual soil particles or grains has a significant effect on the soil's behavior. Embedment soil is classified as either "fine" grained or "coarse" grained.

Fine Grain Soil (Clay and Silt)
Very small (colloidal) size soil particles are capable of absorbing large quantities of water, as much as 10 times their own weight. These particles attract each other to produce a mass which sticks together. This property is called cohesion or plasticity. Soils containing such particles are referred to as "cohesive" and include clayey soils. Cohesion gives clayey soils resistance to shear. The strength of clayey soils is dependent on the amount of water within the soil. As the content of water increases, the shear resistance decreases. Therefore, when using clays as pipe embedment beneath the ground water level, one must examine its sensitivity to water. Fat clays (CH), which are highly expansive, usually make poor embedment materials. (CH is the USCF soil classification symbol for fat clay.) Lean clays (CL), or other clays having relative low sensitivity to water, sometimes can be used for embedment.

While silts possess little to no cohesion, they are composed of very fine grains, which makes them behave somewhat like clay in that they can contain a high percentage of water. It is also common for silt and clay to occur together. Therefore, the general classification schemes for pipe embedment usually treat silts and clays similarly. (USCF symbols for inorganic silts are ML and MH, and for organic silts OL and OH.)

Coarse Grain Soils

Assemblages of larger-sized particles such as sands (S) and gravels (G) do not exhibit plasticity. Water has less effect on these materials. These soils are called "cohesionless" or "granular. " Normally, cohesionless soils have high shear resistances. When a mass of cohesionless soil is sheared, individual grains either roll, slide, fracture, or distort along the surface of sliding. Likewise, many cohesive soils contain grains of sand, so they can exhibit significant shear resistance. These materials make excellent embedment in wet or dry conditions.

Density and Compaction

When discussing the installation of embedment material, two terms are use extensively. They are compaction and density. These terms are defined, herein, to assist the reader.

Density refers to the weight of a particular volume of soil As discussed above, soil consists of three materials or phases: a mineral phase, water, and air. As the soil is compacted, the mineral phase may undergo some change, but typically the air and water are expelled from the soil and the overall volume is reduced. The weight of the mineral phase stays the same. Thus, a given weight of mineral phase occupies a smaller volume after compaction. Typically, when densities are given, they are based on the dry unit weight of the soil (which is the weight of the mineral phase only) occupying a given volume, say a cubic foot.

Compaction, on the other hand, refers to the amount of energy imparted into the soil to reduce its volume. Typically, more energy, often called compactive effort, is required to increase the density of a fine grain soil than a coarse grain soil. One reason for this is that the fine grain soil has cohesion which must be overcome in order for the mineral phase particles to be pushed closer together. Another reason is that it is harder to force the water out of a fine grain material because of its low permeability.

Methods of Measuring Density

There are two general categories of density measures. One method involves imparting a standard amount of energy into the soil, say a fixed number of blows with a specified weight. The Standard and Modified Proctor density tests are such methods. The other measure involves comparing the in-place density with the most dense and least dense arrangement that can be achieve with that soil. An example of this method is the Relative Density test.

The Proctor Density is the most common method used with pipe embedment and will be discussed in somewhat more detail. Typically, a soil sample is taken from the embedment material and tested in the laboratory, where a precisely defined amount of compaction energy is applied to it, which compacts the sample to its

Proctor density. (This amount of energy is defined by the particular Proctor test, whether it is the Standard Proctor defined in ASTM D-698 or the Modified Proctor defined in ASTM D-1599. It is essentially independent of the soil type.) The sample is then dried and its density measured. This density is the standard for this material and is considered to be 100 percent of the Proctor density. The technician then makes measurement of the density (dry unit weight per cubic ft.) of the compacted embedment in the field using, say, the nuclear density gauge. That density can then be compared with the density obtained in the laboratory. The comparison is usually expressed in percent. Typically, the field density must be at least 90 percent of the laboratory density. In this case, we would say the minimum density is 90 percent of the Proctor.

For pipe installation, the important factor is soil stiffness. If two soils are compacted to the same Proctor density, that does not mean that the two soils provide equal supporting stiffness for the pipe. A crushed stone at 90 percent Proctor will be much stiffer than a clay compacted to 90 percent Proctor. This fact is illustrated by the different E' values assigned to these materials at these densities. In the case of the crushed stone its E' equals 3000 psi, whereas the clay has an E' of only a 1000 psi. Methods used to measure soil stiffness such as the California Bearing Ratio test are not convenient for field testing of pipe Therefore, it is common to measure and monitor density.

Comparison of Installation of Rigid and Flexible Pipe

The underground installation of PE piping is similar to the installation of other piping materials. The performance of the pipe will depend on the quality of the installation. Most PE piping is considered flexible, which means that the pipe will depend to some extent on the support of the embedment soil. Often the installation of flexible pipe is contrasted with the installation of rigid pipe, but general requirements for both types of pipe are similar. A narrow trench keeps loads on both types of pipe at a minimum. Both pipes require firm, stable bedding and uniform support under the haunches. The major difference between the two types of pipes is that the flexible pipe requires side support, whereas the rigid pipe does not. Side support comes from the placement of firm, stable material beside the pipe. Often this is the same material used beside the rigid pipe with the exception that the material must be compacted. Sufficient space alongside the pipe must be provided for compacting the embedment material. The trench backfill placed above the pipe can be treated in the same manner for both flexible and rigid pipe. The denser the material above the pipe, the smaller the load applied to the pipe.

PE pipe interacts advantageously with the embedment soil. The viscoelastic properties of PE and most soils are similar. As the pipe deflects, much of the earth is transmitted by arching action to the soil around the pipe. Thus the need for stable soil beside the pipe. Rigid pipe is typically manufactured from materials that are not compliant with soil deformation. As the soil settles, load accumulates on the rigid pipe. If this load exceeds the pipe materials' yield strength, the pipe will fail by a sudden rupture or crack. PE is a ductile material that can yield. Under excessive loads, PE pipe will deform without cracking. The deformation is often sufficient to relieve the accumulated stresses, so performance is not interrupted.

Deflection is usually the main criterion for judging the performance of a flexible pipe. Pipes that deflect have two advantages over rigid pipe: (1) the deflection permits the release of accumulated stresses which promotes arching and causes a more uniform distribution of earth pressure around the pipe and (2) the deflection affords a convenient method of inspecting the quality of the installation - generally the less deflection the better the installation.

Chapter 8

Above-Ground Applications for Polyethylene Pipe

Introduction

A significant portion of applications require that a pipe be laid out, or "strung out," across the prevailing terrain. It may simply be placed on the surface, or it may be suspended or "cradled" in support structures. These types of installations may be warranted by any one of several factors. One is the economic considerations of a temporary piping system. Another is the ease of inspection and maintenance. Still another is simply that prevailing local conditions prevent burial of the pipe.

Polyethylene pipe provides unique joint integrity, toughness, flexibility, and low weight. These factors combine to make its use practical for many "above-ground" applications. This resilient material has been used for temporary water lines, various types of bypass lines, dredge lines, mine tailings, and fines-disposal piping. Polyethylene pipe is used for slurry transport in many industries such as those that work with kaolins and phosphates. The ease of installation and exceptional toughness of polyethylene pipe often make it practical for oil and gas collection. The economics and continued successful performance of this unique piping material is evident despite the extreme climatic conditions that exist in these diverse applications.

This chapter presents design criteria and prevailing engineering methods that are used for above-ground installation of polyethylene pipe. The effects of temperature extremes, chemical exposure, ultraviolet radiation, and mechanical impact are discussed in detail. Engineering design methodology for both "on-grade" and suspended or cradled polyethylene pipe installations are presented and illustrated with typical sample calculations. All equations in the design methodology were obtained from published design references. These references are listed so the designer can verify the applicability of the methodology to his particular project. Additional installation considerations are also discussed.

Design Criteria

The design criteria that can influence the behavior of polyethylene pipe installed above ground include:

- Temperature
- Chemical exposure
- Ultraviolet radiation
- Potential mechanical impact or loading

Figure 1 Above-Ground Installation of Polyethylene Pipe in a Wyoming Mining Operation

Temperature

The diversity of applications for which polyethylene pipes are used in above-ground applications reflects the usable temperature range for this material. Above-grade installations are exposed to demanding fluctuations in temperature extremes as contrasted to a buried installation where system temperatures can be relatively stable. Irradiation by sunlight, seasonal changes, and day-to-night transitions can impose a significant effect on any piping material installed above the ground.

As a general rule, polyethylene pipe can be used safely at temperatures as low as -75°F (-60°C) and as high as 150°F (65°C). However, polyethylene is a thermoplastic material and, as such, these extremes impact the engineering properties of the piping. Additional information in this regard is available within the engineering properties chapter of this handbook.

Pressure Capability

The pressure capability of a polyethylene pipe is predicated on the long-term hydrostatic strength (LTHS) of the polymer used in its manufacture. This strength is then classified into one of a series of hydrostatic design bases (HDB's) in accordance with ASTM D2837.[1] This information is, in turn, used by the Hydrostatic Stress

Board of the Plastics Pipe Institute to establish a recommended hydrostatic design stress (HDS) for pipe made from a specific material.[2] The hydrostatic design basis of a PE3408 piping material is 1600 psi at 73°F (23°C). This yields a hydrostatic design stress (HDS) of 800 psi at the same temperature for water-related applications. The HDS is used in Equation 1 to determine the pressure capability for a specific wall thickness or pipe series used to transport a specific medium, such as water or natural gas.[3]

(1) Pressure Rating, P

$$P = \frac{2(HDS)}{(DR-1)} = \frac{2(HDS)}{((OD/T)-1)}$$

WHERE
P = Pressure rating in psi
HDS = Recommended hydrostatic design stress in psi
    = HDB x DF
HDB = 1600 psi for PE3408
    = 1250 psi for PE2406
DF = Service design factor
    = 0.50 for water
    = 0.32 for natural gas
DR = Ratio of OD to wall thickness (OD/t)
OD = Outside diameter of pipe in inches
t = Minimum wall thickness in inches

As the temperature to which the polyethylene pipe is exposed increases above the reference temperature of 73°F (23°C), the LTHS decreases. Correspondingly, the HDS and the pressure rating of a specific DR is reduced as the service temperature increases. On the other hand, if the service temperature is lowered below 73°F (23°C), the LTHS and HDS increase. In other words, the pressure rating of a polyethylene pipe with the same DR increases as the service temperature decreases. Temperature has a similar effect on material stiffness. As the temperature increases, the modulus of elasticity decreases.

In consideration of the effect of temperature change on material properties, the pressure rating relationship (Eq.1) is often re-written with the inclusion of Temperature Design Factor, F_T. The revised relationship is presented in the form of Equation 2, below.

(2)

$$P = \frac{2(HDB)(DF)(F_T)}{(DR-1)}$$

WHERE

P = Pressure rating in psi
HDB = 1600 psi for PE3408 at 73° F (23° C)
= 1250 psi for PE2406 at 73° F (23° C)
DF = Service design factor
= 0.50 for water
= 0.32 for gas
F_T = Temperature Design Factor at a specific operating temperature from Table 1
DR = Ratio of OD to wall thickness (OD/t)
OD = Outside diameter of pipe in inches
t = Minimum wall thickness in inches

TABLE 1
Service Temperature vs. Modulus of Elasticity and Temperature Design Factors (F_T) for HDPE pipe

Service Temperature °F (°C)	Apparent Modulus of Elasticity (E_S) psi	Apparent Long – Term Modulus of Elasticity (E_L) psi	Temperature Design Factor, F_T
140 (60)	50,000	12,000	0.50
130 (55)	57,000	13,000	0.60
120 (49)	65,000	15,000	0.63
110 (44)	80,000	18,000	0.75
100 (38)	100,000	23,000	0.78
90 (32)	103,000	24,000	0.90
80 (27)	108,000	25,000	1.00
73 (23)	130,000	30,000	1.00
60 (16)	150,000	35,000	1.15
50 (10)	165,000	38,000	1.30
40 (4)	170,000	39,000	1.40
30 (-1)	200,000	46,000	1.60

The temperature effects on elasticity and pressure ratings for polyethylene pipe are illustrated in Table 1. This table lists design factors that are applied to the standard pressure ratings at 73°F (23°C) to derive an estimation of the true long-term pressure capability of a polyethylene pipe at a specific service temperature. The manner by which these factors are utilized will be discussed within the design methodology section of this chapter.

The values listed in Table 1 represent a generalization of the temperature effect on the HDB of all PE3408 piping materials. Information regarding the temperature-responsive nature of a specific polyethylene pipe is available from the respective pipe manufacturer.

Low Temperature Extremes

Generally speaking, the limitation for extremely low environmental service temperature is the potential for embrittlement of the material. Note, however, that most polyethylene piping materials are tested at extremely low temperatures with no indication of embrittlement.

The effect of low temperature on polyethylene pipe is unique. As shown in Table 1, the modulus of elasticity increases as temperatures are lowered. In effect, the pipe becomes stiffer but retains its ductile qualities. The actual low temperature embrittlement of most polyethylene is below -180°F (-118°C). In actual practice, polyethylene pipe has been used in temperatures as low as -75°F (-60° C).[1.5] Obviously, service conditions at these extremes may warrant insulation to prevent heat loss and freezing of the material being conveyed.

It should be noted that in extreme service applications operating at high pressure and increasingly lower temperature that the ability of some polyethylene piping materials to absorb and dissipate energy such as that associated with sudden impact may be compromised. In these situations, it is feasible that, with the addition of a sustaining or driving force, a through-wall crack can form which is capable of traveling for significant distances along the longitudinal axis of the pipe. This phenomenon is generally referred to as rapid crack propagation or RCP, and can occur in any pressure piping or pressure vessel design regardless of the material of manufacture.

This type of phenomenon is generally not experienced in polyethylene in fluid transport applications as the energy dissipation associated with the sudden release of fluid from the pipe mediates the driving force required to sustain the crack. Gas or compressed air handling applications do not provide for the dissipation of energy and, as such, a driving or sustaining force is a potential possibility. For these reasons, the operation of polyethylene pipe above ground in extremely cold environments (<32°F) should be carefully researched in light of the potential application and prevailing service conditions. The reader is referred to the pipe manufacturer for additional information regarding RCP and specific design measurers for above ground, cold weather installations.

Expansion and Contraction

The coefficient of linear expansion for unrestrained polyethylene pipe is approximately ten times that of metal or concrete. The end result is that large changes in the length of unrestrained polyethylene piping may occur due to temperature fluctuations. While the potential for expansion (or contraction) is large when compared with that of metal, concrete, or vitrified clay pipe, note that the modulus of elasticity for polyethylene is substantially lower than that of these alternative piping materials. This implies that the degree of potential movement associated with a specific temperature change may be higher for the polyethylene, but the stress associated with restraint of this movement is significantly less. The end result is that the means of restraint required to control this movement potential is often less elaborate or expensive. The stresses imposed by contraction or expansion of a polyethylene piping system are usually on an order of 5% to 10% of those encountered with rigid piping materials.

Chemical Resistance

Unlike many piping materials, polyethylene pipe will not rust, rot, pit, or corrode as a result of chemical, electrolytic, or galvanic action. The primary chemical environments that pose potentially serious problems for polyethylene pipe are strong oxidizing agents or certain hydrocarbons. Concentrated sulphuric and nitric acids are strong oxidizers, while diesel and fuel oils typify the hydrocarbons.

Environments that contain these harsh chemicals may affect the performance characteristics of an above-ground system made from polyethylene pipe. The continued exposure of polyethylene to strong oxidizing agents may lead to crack formation or crazing of the pipe surface. Occasional or intermittent exposure to these agents will not, however, significantly affect the long-term performance of a polyethylene pipe.

Hydrocarbon exposures normally cause only temporary effects on polyethylene. The result of the exposure is, for the most part, evident only as long as the exposure is maintained. Exposure to certain hydrocarbons tends to reduce the pressure capability of the polyethylene. It is also evidenced by a reduction in tensile strength and an increase in physical dimensions (swelling) due to adsorption of the hydrocarbon by the polyethylene structure. Continued exposure can lead to permeation of the polyethylene pipe wall and possible leaching into the flow stream. The degree of permeation is a function of pressure, temperature, the nature of the hydrocarbons, and the polymer structure of the piping material. Each of these parameters should be considered before using polyethylene pipe to transport hydrocarbons in an above-ground installation. Various references are available concerning the effects of chemical exposure on polyethylene pipe. PPI Technical Report No. 19 contains further information regarding resistance of polyethylene pipe to various chemical environments.[6]

Ultraviolet Exposure

When polyethylene pipe is utilized outdoors in above-ground applications, it will be subjected to extended periods of direct sunlight. The ultraviolet component in sunlight can produce a deleterious effect on the polyethylene unless the material is sufficiently protected. Weathering studies have shown that pipe produced with a minimum 2.0% concentration of finely divided and evenly dispersed carbon black is protected from the harmful effects of UV radiation for indefinite periods of time.[18] Polyethylene pipe that is protected in this manner is the principal material selected for above-ground installations. Black pipe (containing 2.0% minimum carbon black) is normally recommended for above-ground use. Consult the manufacturer's recommendations for any non-black pipe that is either used or stored above ground.

Mechanical Impact or Loading

Any piping material that is installed in an exposed location is subject to the rigors of the surrounding environment. It can be damaged by the movement of vehicles or other equipment, and such damage generally results in gouging, deflecting or flattening of the pipe surfaces. If an above-ground installation must be located in a region of high traffic or excessive mechanical abuse (along a roadway, etc.), the pipe requires extra protection. It may be protected by building a berm or by encasing the pipe where damage is most likely. Other devices may be used, as appropriate to the situation. Design criteria for the installation of buried flexible thermoplastic pipe should be used for those areas where the above-ground polyethylene system must pass under a roadway or other access, and where an underground installation of a portion of the system is necessary.[7,8] In general, in an installation in which any section of polyethylene pipe has been gouged in excess of 10% of the minimum wall thickness, the gouged portion should be removed and replaced. When the polyethylene pipe has been excessively or repeatedly deflected or flattened, it may exhibit stress-whitening, crazing, cracking, or other visible damage, and any such regions should be removed and replaced with new pipe material.

Design Methodology

As previously discussed, above-ground piping systems can be subjected to variations in temperature. These temperature fluctuations can impact the pressure capability of the exposed piping to some degree. The expansion and contraction characteristics of polyethylene pipe must also be addressed in light of the anticipated variations in temperature. Further, the installation characteristics of the proposed above-ground system must be analyzed in some detail. Each of these concerns will be discussed in the sections which follow.

Pressure Capability

The temperature effects on elasticity and pressure rating for polyethylene pipe have been discussed previously in this chapter. A series of temperature design factors were presented in Table 1, along with moduli of elasticity for polyethylene at several specific temperatures. These values may, in turn, be used in the design methodology to estimate the true long-term pressure capability of a polyethylene pipe at a specific service temperature.

EXAMPLE 1

What is the pressure capability of an SDR 11 series of PE 3408 polyethylene pipe used to transport water at 73°F (23°C)?

From Eq. 2

$$P = \frac{2(HDB)(DF)(F_T)}{(SDR-1)}$$

$$P = \frac{2(1600)(0.50)(1.0)}{(11-1)}$$

P = 160 psi at 73°F (23°C)

What is the pressure capability at 100°F (38°C)?

From Table 1, the 100°F (38°C) pressure design factor is 0.78.

$$P = \frac{2(1600)(0.50)(0.78)}{(11-1)}$$

P = 125 psi at 100°F (38°C)

Example 1 assumes that exposure of the pipe to sunlight, combined with the thermal properties of the material flowing within the pipe, has resulted in a normal average operating temperature for the system at 100°F (38°C). Exposure of the pipe to sunlight can result in extremely high outside surface temperatures, particularly if the pipe is black.[9] In the majority of cases, the material flowing within the pipe is substantially cooler than the exterior of the exposed above-ground pipe. The cooler nature of the material flowing through the pipe tends to moderate the surface temperature of the exposed pipe. This results in a pipe wall temperature that is intermediate to the surface of the pipe and that of the flow stream. Obviously, the longer the period of irradiation of the pipe by sunlight, the greater the potential will be to raise the temperature of the flow stream.

Several texts related to temperature design criteria and flow are included in the literature references of this chapter.[10,11]

Expansion and Contraction

As noted in the Design Criteria section of this chapter, temperature changes can produce a substantial change in the physical dimensions of polyethylene pipe. This is evidenced by a coefficient of expansion or contraction that is notably higher than that of many other piping materials. The design methodology for above-ground installation must take this potential for expansion or contraction into consideration.

The expansion or contraction for an unrestrained polyethylene pipe can be calculated by using Equation 3.

Pipe Length vs. Temperature Change

(3) $$L = \alpha (T_2 - T_1) L$$

WHERE
ΔL = Theoretical length change (in.)
$\Delta L > 0$ is expansion
$\Delta L < 0$ is contraction
α = Coefficient of linear expansion = 1.1×10^{-4} in/in/°F
T_1 = Initial temperature (°F)
T_2 = Final temperature (°F)
L = Length of pipe (in.) at initial temperature, T_1

EXAMPLE 2

A 100 foot section of 10-inch (10.75-inch OD) SDR 11 (PE3408) material is left unrestrained overnight. If the initial temperature is 70°F (21°C), determine the length of the pipe section at dawn the next morning if the pipe stabilizes at a nighttime temperature of 30°F (-1°C).

Using Equation 3,

$$L = (1.1 \times 10^{-4})(30° - 70°)(100 \text{ ft})(12 \text{ in/ft}) = -5.28 \text{ Inches}$$

The negative sign indicates a contraction, so the final length is 99 ft., 6.72 in.

As shown in Example 2, the contraction or expansion due to temperature change can be quite significant. However, this calculated change in length assumes both an unrestrained movement of the pipe and an instantaneous drop in temperature. Actually, no temperature drop is instantaneous, and obviously, the ground on which the pipe is resting creates a retarding effect on the theoretical movement due to friction. Practical field experience for polyethylene pipe has shown that the actual contraction or expansion that occurs as a result of temperature change is approximately one-half that of the theoretical amount.

Field experience has also shown that changes in physical length are often further mitigated by the thermal properties or heat-sink nature of the flow stream within the pipe. However, conservative engineering design warrants that consideration be given to the effects of temperature variation when the flow stream is static or even when there is no flow stream.

In cases where polyethylene pipe will be exposed to temperature changes, it is common practice to control the pipe movement by judiciously placing restraining devices. Typical devices include tie-down straps, concrete anchors, thrust blocks, etc. The anchor selection must consider the stresses developed in the pipe wall as a result of the anticipated temperature changes. Equation 4 illustrates how these stresses may be determined.

(4) Longitudinal Stress vs. Temperature Change

$$\sigma_T = \alpha (T_2 - T_1)E$$

WHERE
σ_T = Theoretical longitudinal stress (psi)(Negative for contraction; positive for expansion)
α = Coefficient of expansion or contraction (see Eq. 3)
T_1 = Initial temperature (°F)
T_2 = Final temperature (°F)
E = Apparent short-term modulus of elasticity (see Table 1) at median temperature (T_m)
$T_m = (T_2 + T_1)/2$

(5) Longitudinal Force vs. Temperature Change

$$F_T = \sigma_T (A)$$

WHERE
F_T = Theoretical longitudinal force (lbs)
σ_T = Theoretical longitudinal stress (psi) from Eq. 4
A = Pipe wall cross-sectional area (in²)

EXAMPLE 3

Assuming the same conditions as Example 2, what would be the maximum theoretical force developed on the unrestrained end of the 100 foot section if the other end is restrained effectively? Assume that the cross-sectional area of the pipe wall is approximately 30 in², the temperature change is instantaneous, and the frictional resistance against the soil is zero.

$$\sigma_T = \alpha(T_2 - T_1)E$$

$= (1.1 \times 10^{-4})(30° - 70°)(200,000)$

$= -880$ psi

$$F_T = (\sigma_T)(A)$$

$= -800$ psi $\times$ 30 in²

$= -26,400$ lbs

As previously mentioned, for these conditions where the temperature change is gradual, the actual stress level is approximately half that of the theoretical value. This would account for an actual force at the free end of about -13,200 lbs. To illustrate the differences between the expansion and contraction characteristics of polyethylene pipe versus those of steel, consider the following example:

EXAMPLE 4

Assume the same conditions as Example 2 for 10-inch Schedule 40 steel pipe. The pipe wall has a cross-sectional area of 11.90 in², the value of α_{steel} is 6.5 $\times$ 10⁻⁶, and the value of E for this material is 30,000,000.[14]

$$\varepsilon_T = \alpha_{steel}(T_2 - T_1)E$$

$= (6.5 \times 10^{-6})(30° - 70°)(3 \times 10^7)$

$= -7,800$ psi

$$F_T = (\sigma_T)(A)$$

$= -7,800$ psi $\times$ 11.90 in²

$= -92,820$ lbs

Thus, as shown by Examples 3 and 4, even though the coefficient of thermal expansion is high in comparison to other materials, the comparatively low modulus of elasticity results in correspondingly reduced thermal stresses.

These design considerations provide a general introduction to the studies of temperature effects on polyethylene pipe in above-ground applications. They do not include other factors such as the weight of the installed pipe, frictional resistance of pipe lying on-grade, or grade irregularities. All of these factors affect the overall expansion or contraction characteristics, and individual pipe manufacturers should be consulted for further detail.

Installation Characteristics

There are two basic types of above-ground installations. One of these involves "stringing-out" the pipe over the naturally-occurring grade or terrain. The second involves suspending the pipe from various support structures available along the pipeline right-of-way. Figure 2 illustrates some typical installations for both types. Each type of installation involves different design methodologies, so the installation types are discussed separately.

On-Grade Installations

As indicated previously, pipe subjected to temperature variation will expand and contract in response to temperature variations. The designer has two options available to counteract this phenomenon. Basically the pipe may be installed in an unrestrained manner, thus allowing the pipe to move freely in response to temperature change. Or the pipe may be anchored by some means that will control any change of physical dimensions; anchoring can take advantage of polyethylene's unique stress relaxation properties to control movement and deflection mechanically.[12]

Free Movement

An unrestrained pipe installation requires that the pipe be placed on a bed or right-of-way that is free of material that may abrade or otherwise damage the exterior pipe surface. The object is to let the pipe "wander" freely without restriction or potential for point damage. This installation method usually entails "snaking" the polyethylene pipe along the right-of-way. The excess pipe then allows some slack that will be taken up when the temperature drops and the pipe contracts.

Figure 2 Typical Above-Ground Installations with Plastic Pipe

Figure 2a On-grade Installation of Polyethylene Pipe in an Industrial Application. Note "snaking" along right of way.

Chapter 8
Above-Ground Applications for Polyethylene Pipe

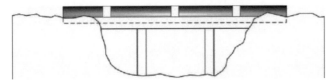

Figure 2b Continuous Support of Polyethylene Pipe at Ravine Crossing

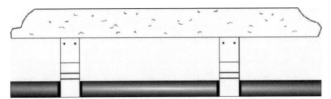

Figure 2c Intermittent Support of Polyethylene Pipe Suspended from Rigid Structure

In all likelihood, a free-moving polyethylene pipe must eventually terminate at or connect to a rigid structure of some sort. It is highly recommended that transitions from free-moving polyethylene pipe to a rigid pipe appurtenance be fully stabilized so as to prevent stress concentration within the transition connection.

Figure 3 illustrates some common methods used to restrain the pipe at a distance of one to three pipe diameters away from the rigid termination. This circumvents the stress-concentrating effect of lateral pipe movement at termination points by relieving the stresses associated with thermal expansion or contraction within the pipe wall itself.

Figure 3 Typical Anchoring Methods at Rigid Terminations of Free-Moving Polyethylene Pipe Sections

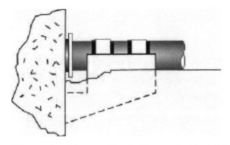

Figure 3a Connection to Concrete Vault Using Grade Beam

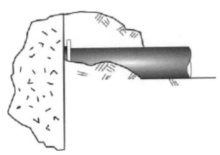

Figure 3b Connection to Rigid Structure Using Consolidated Earthen Berm

Restrained Pipelines

The design for an above-ground installation that includes restraint must consider the means by which the movement will be controlled and the anchoring or restraining force needed to compensate for, or control, the anticipated expansion and contraction stresses. Common restraint methods include earthen berms, pylons, augered anchors, and concrete cradles or thrust blocks.

The earthen berm technique may be either continuous or intermittent. The pipeline may be completely covered with a shallow layer of native earth over its entire length, or it may be stabilized at specific intervals with the earthen berms between the anchor locations. Typical earthen berm configurations are presented in Figure 4.

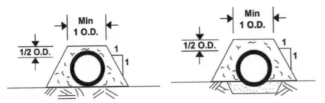

Figure 4 Earthern Berm Configurations

The continuous earthen berm serves not only to stabilize the pipe and restrain its movement but also to moderate temperature fluctuations. With less temperature fluctuation the tendency for pipe movement is reduced.

An intermittent earthen berm installation entails stabilization of the pipe at fixed intervals along the length of the pipeline. At each point of stabilization the above-ground pipe is encased with earthen fill for a distance of one to three pipe diameters. The economy of this method of pipeline restraint is fairly obvious.

Other means of intermittent stabilization are available which provide equally effective restraint of the pipeline with a greater degree of ease of operation and maintenance. These methods include pylons, augered anchors [13], or concrete cradles. These restraint techniques are depicted schematically in Figures 5 through 7.

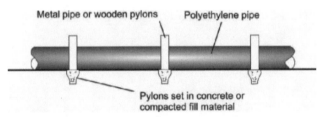

Figure 5 Pylon Type Stabilization

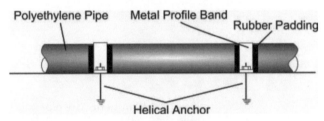

Figure 6 Augered Anchor Stabilization

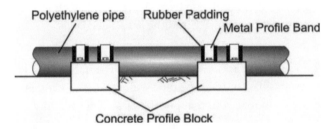

Figure 7 Concrete Cradle or Thrust Block Stabilization

A pipeline that is anchored intermittently will deflect laterally in response to temperature variations, and this lateral displacement creates stress within the pipe wall. The relationships between these variables are determined as follows:

Lateral Deflection (Approximate from Catenary Eq.)

(6) $\Delta y = L \sqrt{0.5 \, \alpha \, (\Delta T)}$

WHERE
Δy = Lateral deflection (in.)
L = Distance between anchor points (in.)
α = Coefficient of expansion/contraction
    = 0.0001 in/in/°F
ΔT = Temperature change ($T_2 - T_1$) in °F

(7) Bending Strain Development

$$\varepsilon = \frac{D\sqrt{96\,\alpha\,(\Delta T)}}{L}$$

WHERE
ε = Strain in pipe wall (%)
D = Outside diameter of pipe (in)
α = Coefficient of expansion/contraction
    = 0.0001 in/in/°F
$\Delta T = (T_2 - T_1)$ in °F
L = Length between anchor points (in)

As a general rule, the frequency of stabilization points is an economic decision. For example, if lateral deflection must be severely limited, the frequency of stabilization points increases significantly. On the other hand, if substantial lateral deflection is permissible, fewer anchor points will be required, and the associated costs are decreased.

Allowable lateral deflection of polyethylene is not without a limit. The upper limit is determined by the maximum permissible strain in the pipe wall itself. This limit is a conservative 5% for the majority of above-ground applications, as determined by Equation 7.

Equations 6 and 7 are used to determine the theoretical lateral deflection or strain in overland pipelines. Actual deflections and strain characteristics may be significantly less due to the friction imposed by the prevailing terrain, the weight of the pipe and flow stream, and given that most temperature variations are not normally instantaneous. These factors allow for stress relaxation during the process of temperature fluctuation.

EXAMPLE 5
Assume that a 10-inch (OD) SDR 11 polyethylene pipe is strung out to grade and anchored at 100-foot intervals. What is the maximum theoretical lateral deflection possible, given a 50°F (27.8°C) temperature increase? What strain is developed in the

pipe wall by this temperature change? What if the pipe is anchored at 50-foot intervals?

Calculations for 100-foot intervals:

$$\Delta y = L \sqrt{0.5\alpha (\Delta T)}$$

$= 100 \times 12 \, [0.5(0.0001)(50)]^{1/2}$
$= 60$ inches lateral displacement

$$\varepsilon = \frac{D \sqrt{96\alpha(\Delta T)}}{L}$$

$$= \frac{10 \sqrt{(96)\,(0.0001)(50)}}{100(12)}$$

$= 0.58\%$ strain

Calculations for 50-foot intervals:

$$\Delta y = L\sqrt{0.5\alpha(\Delta T)}$$

$= 50 \times 12[0.5(0.0001)(50)]^{1/2}$
$= 30$ inches lateral displacement

$$\varepsilon = \frac{D\sqrt{96\alpha(\Delta T)}}{L}$$

$$= \frac{10\sqrt{96\,(0.0001\,)(50)}}{50(12)}$$

$= 1.2\%$ strain

From the calculations in Example 5, it is apparent that lateral deflections which appear significant may account for relatively small strains in the pipe wall. The relationship between lateral deflection and strain rate is highly dependent on the selected spacing interval.

Supported or Suspended Pipelines

When polyethylene pipeline installations are supported or suspended, the temperature and corresponding deflection characteristics are similar to those discussed above for unsupported pipelines with intermittent anchors. There are two additional parameters to be considered as well: beam deflection and support or anchor configuration.

Support or Suspension Spacing

Support spacing for polyethylene pipe is determined much the same as for other types of suspended pipelines.[14] The design methodology involves simple-beam or continuous-beam analysis of the proposed installation and is based on limiting bending stress.

(8) Support Spacing Requirements

$$L = \left[\frac{3(OD^4 - ID^4)\sigma_m \pi}{8qOD} \right]^{1/2}$$

WHERE
L = Center-to-center span (in)
OD = Outside diameter (in)
ID = Inside diameter (in)
σ_m = Maximum bending stress (psi)
   = 100 psi for pressurized pipelines
   = 400 psi for non-pressurized pipelines
q = Load per unit length (lb/in.)

(9) Load per Unit Length

$$q = \frac{W}{12} + \frac{\pi\sigma(ID)^2}{6912}$$

WHERE
q = Load per unit length (lb/in)
W = Weight of pipe (lbs/ft)
σ = Density of Internal fluid (lb/ft³)
π = 3.1416

This calculation gives a conservative estimate of the support span in cases where the pipe is not completely restrained by the supports. (The pipe is free to move within the supports.) A more complex analysis of the bending stresses in the pipe may be performed by treating the pipe as a uniformly loaded beam with fixed ends. The actual deflection that occurs between spans may be determined on the basis of this type of analysis, as shown in Equation 10.

(10) Simple Beam Deflection Analysis[15]
Based on Limiting Deflection

$$d = \frac{5qL^4}{384E_L I}$$

WHERE
d = Deflection or sag (in)
L = Span length (in)
q = Load per unit length (lb/in)
E_L = Apparent long-term modulus of elasticity at average long-term temperature from Table 1
I = Moment of inertia (in⁴)
  = $(\pi/64)(OD^4 - ID^4)$

Simple beam analysis reflects the deflection associated with the proposed support spacing configuration and the modulus of elasticity at a given service temperature. It does not take into consideration the increased or decreased deflection that may be attributed to expansion or contraction due to thermal variations. These phenomena are additive - Equation 11 illustrates the cumulative effect.

(11) Cumulative Deflection Effects

Total deflection = beam deflection + thermal expansion deflection

$$= d + \Delta y$$

$$d = \frac{5qL^4}{384E_L I} + \sqrt{0.5\alpha(\Delta T)}$$

Simple beam analysis assumes one support point at each end of a single span. Most supported pipelines include more than one single span. Normally, they consist of a series of uniformly spaced spans with relatively equal lengths. The designer may analyze each individual segment of a multiple-span suspended pipeline on the basis of simple beam analysis. However, this approach may prove overly conservative in the majority of multiple-span supported pipelines. Equation 12 presents a more realistic approach to deflection determination on the basis of continuous beam analysis.

(12) Continuous Beam Analysis

$$d = \frac{fqL^4}{E_L I}$$

WHERE
d = Deflection or sag (in)
f = Deflection coefficient
q = Load per unit length (lbs/in)
L = Span length (in)
E_L = Apparent long-term modulus of elasticity at average long-term temperature from Table 1
I = Moment of inertia (in⁴)
  = $(\pi/64)(OD^4 - ID2\ pt)$

The deflection coefficient, f, is a function of the number of spans included and whether the pipe is clamped securely, fixed, or simply guided (not fixed) within the supports. Practical values for the deflection coefficient, f, are provided in Table 2.

TABLE 2
Deflection Coefficients, f, for Various span Configurations[17]

1 Span	2 Spans	3 Spans	4 Spans
N-N	N-N-N	N-N-N-N	N-N-N-N-N
		1 2 1	1 2 2 1
f=0.013	f=0.0069	f1=0.0069	f1=0.0065
		f2=0.0026	f2=0.0031
F–N	F-N-N	F-N-N-N	F-N-N-N-N
	1 2	1 2 2	1 2 2 2
f=0.0054	f=0.0026	f1=0.0026	f1=0.0026
	f2=0.0054	f2=0.0054	f2=0.0054
F-F	F-N-F	F-N-N-F	F-N-N-N-F
		1 2 1	1 2 2 1
f=0.0026	f=0.0026	f1=0.0026	f1=0.0026
		f2=0.0031	f2=0.0031
	F-F-F	F-F-F-F	F-F-F-F-F
	f=0.0026	f=0.0026	f=0.0026

F = Fixed Securely   N = Not Fixed

As was the case for simple beam analysis, continuous beam analysis addresses the deflection resulting from a given span geometry at a specified service temperature. The equation does not take into consideration the additional deflection associated with expansion or contraction due to temperature variations. Equation 13 combines the effect of deflection due to span geometry (using continuous beam analysis) with deflection resulting from expansion due to a temperature increase. A total span deflection of ½ to 1 inch is generally considered as a maximum.

(13) Total Span Deflection Based on Continous Beam Analysis and Thermal Response

$$\text{Total Deflection (in)} = \frac{fqL^4}{E_L I} + L\sqrt{0.5\,\alpha(\Delta T)}$$

WHERE
f = Factor from Table 2
q = Load per unit length from Eq. 9 (lbs/in)
L = Span length from Eq. 8 (in)
E_L = Apparent long-term modulus of elasticity at average long-term temperature from Table 1
I = Moment of inertia (in.^{4})
= $(\pi/64)(OD^4 \text{ pt} - ID^4 \text{ pt})$

Anchor and Support Design

Proper design of anchors and supports is as important with polyethylene piping as it is with other piping materials. A variety of factors must be considered.

Some installations of polyethylene pipe have the pipe lying directly on the earth's surface. In this type of installation, the surface under the pipe must be free from boulders, crevices, or other irregularities that could create a point-loading situation on the pipe.

On-grade placement over bed rock or "hard pan" should be avoided unless a uniform bed of material is prepared that will cushion the pipe. If the polyethylene pipe rests directly on a hard surface, this creates a point loading situation and can increase abrasion of the outer pipe surface as it "wanders" in response to temperature variations.

Intermittent pipe supports should be spaced properly, using the design parameters discussed in the preceding pages. Where excessive temperatures or unusual loading is encountered, continuous support should be considered.

Supports that simply cradle the pipe, rather than grip or clamp the pipe, should be from one-half to one-pipe diameter in length and should support at least 120 degrees of the pipe diameter. All supports should be free from sharp edges.

The supports should have adequate strength to restrain the pipe from lateral or longitudinal deflection, given the anticipated service conditions. If the design allows free movement during expansion, the sliding supports should provide a guide without restraint in the direction of movement. If on the other hand, the support is designed to grip the pipe firmly, the support must either be mounted flexibly or have adequate strength to withstand the anticipated stresses.

Heavy fittings or flanges should be fully supported and restrained for a distance of one full pipe diameter, minimum, on both sides. This supported fitting represents a rigid structure within the flexible pipe system and should be fully isolated from bending stresses associated with beam sag or thermal deflection.

Figure 8 includes some typical pipe hanger and support arrangements that are appropriate for use with polyethylene pipe, and Figure 9 shows some anchoring details and cradle arrangements.

Pressure-Testing

It is common practice to pressure-test a pipe system prior to placing it in service. For the above-ground systems described in this chapter, this test should be conducted hydrostatically. Hydrostatic testing procedures are described in a number of publications, including PPI Technical Report 31.[8] The Plastics Pipe Institute does not recommend pneumatic pressure testing of an above-ground installation.[16] An ASTM test method for leakage testing of polyethylene pipe installations is under development and may be applicable.

Figure 8 Typical Pipe Hangers and Supports

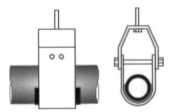

Figure 8.1 Pipe Stirrup Support

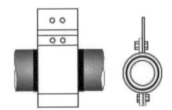

Figure 8.2 Clam Shell Support

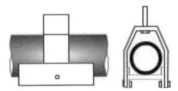

Figure 8.3 Suspended I-Beam or Channel-Continuous Support

Figure 9 Typical Anchoring and Cradling Details

Conclusion

Polyethylene pipe has been used to advantage for many years in above-ground applications. The unique light weight, joint integrity, and overall toughness of polyethylene has resulted in the above-ground installation of polyethylene pipe in various mining, oil, gas production and municipal distribution applications. Many of these systems have provided years of cost-effective service without showing any signs of deterioration.

The key to obtaining a quality above-ground polyethylene piping system lies in careful design and installation. This chapter is intended to serve as a guide by which the designer and/or installer may take advantage of the unique properties of polyethylene pipe for these types of applications. In this way, excellent service is assured, even under the demanding conditions found with above-ground installations.

References

1. ASTM D2837, Standard Method for Obtaining the Hydrostatic Design Basis for Thermoplastic Materials, *Annual Book of Standards*, American Society for Testing and Materials (ASTM), Philadelphia, PA,
2. Plastics Pipe Institute, Report TR-3, Policies and Procedures for Developing Recommended Hydrostatic Design Stresses, Washington, DC.
3. ASTM D3035, Standard Specifications for Polyethylene (PE) Plastic Pipe (DR-PR) Based on Controlled Outside Diameter, *Annual Book of Standards*, American Society for Testing and Materials ASTM), Philadelphia, PA,
4. Arctic Town Gets Royal Flush. (1984, January 5). *Engineering News Record*, New York.
5. Bringing Modern Utilities to Town Beyond the Arctic Circle. (1985, December). *Public Works*.
6. Plastics Pipe Institute. (1991). Report TR19, Thermoplastic Piping for the Transport of Chemicals, Washington, DC.
7. ASTM D2321, Standard Practice for Underground Installation of Flexible Thermoplastic Sewer Pipe, *Annual Book of Standards*, American Society for Testing and Materials (ASTM), Philadelphia, PA,
8. Plastics Pipe Institute. (1988). Report TR-31, Underground Installation of Polyolefin Piping, Washington, DC.
9. Gachter, R., & H. Muller. (1983). *Plastics Addition Handbook*, McMillan Publishing Co., New York, NY.
10. Parker, J. D., James H. Boggs, & Edward F. Click. (1969). *Introduction to Fluid Mechanics and Heat Transfer*, Addison-Wesley Publishing Co., Reading, MA.
11. VanWylen, Gordon J., & Richard E. Sonntag. (1973). *Fundamentals of Classical Thermodynamics*, John Wiley & Sons, New York, NY.
12. Ferry, John D. (1982). Viscoelastic Properties of Polymers, John Wiley & Sons, New York, NY, 1980. *Pipeline Anchoring Encyclopedia*, A. B. Chance Company Bulletin 30-8201.
13. *Pipeline Anchoring Encyclopedia*. (1982). A. B. Chance Company bulletin 30-8201.
14. *Steel Pipe — A Guide for Design and Installation*. (1985). AWWA Manual M11, American Water Works Association, Denver, CO.
15. Moffat, Donald W. (1974). *Plant Engineering Handbook of Formulas, Charts, and Tables*, Prentice-Hall, Inc., Englewood Cliffs, NJ.
16. Plastics Pipe Institute. (1989). *Recommendation B. Thermoplastic Piping for the Transport of Compressed Air or Other Compressed Cases*, Washington, DC.
17. *Manual of Steel Construction*, 6th Edition, American Institute of Steel Construction, Chicago, IL.
18. Gilroy, H. M., *Polyolefin Longevity for Telephone Service*, AT&T Bell Laboratories, Murray Hill, NJ.

References, Equations

Eq 1. The Plastics Pipe Institute. (1976). *Plastics Piping Manual*, Wayne.
Eq 2. *Managing Corrosion with Plastics*. (1983). Volume 5, National Association of Corrosion Engineers.
Eq 3. Roark Raymond J., & Warren C. Young. (1973). *Formulas for Stress & Strain*, McGraw-Hill Co., New York, NY.
Eq 4. Ibid.
Eq 5. Baumeister, T., & L. S. Marks. (1967). *Standard Handbook for Mechanical Engineers*, 7th Edition, McGraw-Hill Book Co., New York, NY.
Eq 6. Roark, Raymond J., & Warren C. Young. (1973). *Formulas for Stress & Strain*, McGraw-Hill Book Co., New York, NY.
Eq 7. Crocker. (1945). *Piping Handbook*, Grunnell Co., Providence, RI.
Eq 8. This is a basic equation utilized to determine the total weight of a pipe filled with fluid.
Eq 9. Shigley, J. E. (1972). *Mechanical Engineering Design*, 2nd Edition, McGraw-Hill Book Co., New York.
Eq 10. Ibid.
Eq 11. Ibid.
Eq 12. Ibid.

Chapter 9

Polyethylene Joining Procedures

Introduction
An integral part of any pipe system is the method used to join the system components. Proper engineering design of a system will take into consideration the type and effectiveness of the techniques used to join the piping components and appurtenances, as well as the durability of the resulting joints. The integrity and versatility of the joining techniques used for polyethylene pipe allow the designer to take advantage of the performance benefits of polyethylene in a wide variety of applications.

General Provisions
Polyethylene pipe or fittings are joined to each other by heat fusion or with mechanical fittings. Polyethylene may be joined to other materials by means of compression fittings, flanges, or other qualified types of manufactured transition fittings. There are many types and styles of fittings available from which the user may choose. Each offers its particular advantages and limitations for each joining situation the user may encounter. Contact with the various manufacturers is advisable for guidance in proper applications and styles available for joining as described in this document. There will be joining methods discussed in this document covering both large and small diameter pipe. Large diameter pipe is considered to be sizes 3" IPS (3.500" OD) and larger. Those persons who are involved in joining gas piping systems must note certain qualification requirements of the U.S. Department of Transportation Pipeline Safety Regulations.[10]

Thermal Heat Fusion Methods
There are three types of heat fusion joints currently used in the industry; Butt, Saddle, and Socket Fusion. Additionally, there are two methods for producing the socket and saddle heat fusion joints.

The principle of heat fusion is to heat two surfaces to a designated temperature, then fuse them together by application of a sufficient force. This force causes the melted materials to flow and mix, thereby resulting in fusion. When fused according to the pipe and/or fitting manufacturers' procedures, the joint area becomes as strong as, or stronger than, the pipe itself in both tensile and pressure properties. As soon as the joint cools to near ambient temperature, it is ready for handling. The following sections of this chapter provide a general procedural guideline for each of these heat fusion methods.

Butt Fusion

The most widely used method for joining individual lengths of polyethylene pipe and pipe to polyethylene fittings is by heat fusion of the pipe butt ends as illustrated in Figure 1. This technique produces a permanent, economical and flow-efficient connection. Quality butt fusion joints are produced by using trained operators and quality butt fusion machines in good condition.

The butt fusion machine should be capable of:

- Aligning the pipe ends
- Clamping the pipes
- Facing the pipe ends parallel with each other
- Heating the pipe ends
- Applying the proper fusion force that results in fusion

Figure 1 A Standard Butt Fusion Joint

The six steps involved in making a butt fusion joint are:

1. Clamp and align the pipes to be joined
2. Face the pipe ends to establish clean, parallel surfaces
3. Align the pipe profile
4. Melt the pipe interfaces
5. Join the two profiles together by applying the proper fusion force
6. Hold under pressure until the joint is cool

Figure 2 Typical Butt Fusion Machine for Smaller Diameter Pipe
(Butt Fusion machines are available to fuse pipe up to 65 inches in diameter)

Most pipe manufacturers have detailed parameters and procedures to follow. The majority of them help develop and have approved the PPI Technical Report TR-33 for the generic butt fusion joining procedure for polyethylene pipe.[15]

Optional Bead Removal

In some pipe systems, engineers may elect to remove the inner or outer bead of the joint. External beads are removed with run-around planing tools, which are forced into the bead, then drawn around the pipe. Power planers may also be used, but care must be taken not to cut into the pipe's outside surface.

It is uncommon to remove internal beads, as they have little or no effect on flow, and removal is time-consuming. Internal beads may be removed from pipes after each fusion with a cutter fitted to a long pole. Since the fusion must be completely cooled before bead removal, assembly time is increased slightly.

Saddle/Conventional Fusion

The conventional technique to join a saddle to the side of a pipe, illustrated in Figure 3, consists of simultaneously heating both the external surface of the pipe and the matching surface of the "saddle" type fitting with concave and convex shaped heating tools until both surfaces reach proper fusion temperature. This may be accomplished by using a saddle fusion machine that has been designed for this purpose.

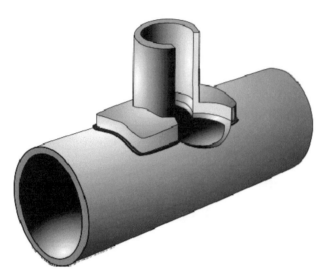

Figure 3 Standard Saddle Fusion Joint

Saddle fusion, using a properly designed machine, provides the operator better alignment and force control, which is very important to fusion joint quality. The Plastics Pipe Institute recommends that saddle fusion joints be made only with a mechanical assist tool unless hand fusion is expressly allowed by the pipe and/or fitting manufacturer.[16]

There are eight basic sequential steps that are normally used to create a saddle fusion joint:

1. Clean the pipe
2. Install heater saddle adapters
3. Install the saddle fusion machine on the pipe
4. Prepare the surfaces of the pipe and fitting
5. Align the parts
6. Heat both the pipe and the saddle fitting
7. Press and hold the parts together
8. Cool the joint and remove the fusion machine

Most pipe manufacturers have detailed parameters and procedures to follow. The majority of them help develop and have approved the PPI Technical Report TR-41 for the generic saddle fusion joining procedure for polyethylene pipe.[16]

Socket Fusion

This technique consists of simultaneously heating both the external surface of the pipe and the internal surface of the socket fitting until the material reaches fusion temperature, inspecting the melt pattern, inserting the pipe end into the socket, and holding it in place until the joint cools. Figure 4 illustrates a typical socket fusion joint. Mechanical equipment is available to hold the fitting and should be used for sizes larger than 2" CTS to attain the increased force required and to assist in alignment.

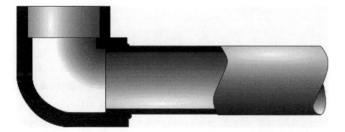

Figure 4 Standard Socket Fusion Joint

Follow these general steps when performing socket fusion:

1. Select the equipment
2. Square and prepare the pipe ends
3. Heat the parts
4. Join the parts
5. Allow to cool

Equipment Selection

Select the proper size tool faces and heat the tools to the fusion temperature recommended for the material to be joined. For many years, socket fusion tools were manufactured without benefit of any industry standardization. As a result, variances of heater and socket depths and diameters, as well as depth gauges, do exist. More recently, ASTM F1056[7] was written, establishing standard dimensions for these tools. Therefore, mixing various manufacturers' heating tools or depth gauges is not recommended unless the tools are marked "F1056," indicating compliance with the ASTM specification and, thereby, consistency of tooling sizes.

Square and Prepare Pipe

Cut the end of the pipe square. Chamfer the pipe end for sizes 1¼"-inch diameter and larger. (Chamfering of smaller pipe sizes is acceptable and sometimes specified in the instructions.) Remove any scraps, burrs, shavings, oil, or dirt from the surfaces to be joined. Clamp the cold ring on the pipe at the proper position, using the integral depth gauge pins or a separate (thimble type) depth gauge. The cold ring will assist in re- rounding the pipe and provide a stopping point for proper insertion of the pipe into the heating tool and coupling during the fusion process.

Heating

Check the heater temperature. Periodically verify the proper surface temperature using a pyrometer or other surface temperature measuring device. If temperature indicating markers are used, do not use them on a surface that will come in contact with the pipe or fitting. Bring the hot clean tool faces Into contact with the outside surface of the end of the pipe and with the inside surface of the socket fitting, in accordance with pipe and fitting manufacturers' instructions. Procedures will vary with different materials. Follow the instructions carefully.

Joining

Simultaneously remove the pipe and fitting from the tool using a quick "snap" action. Inspect the melt pattern for uniformity and immediately insert the pipe squarely and fully into the socket of the fitting until the fitting contacts the cold ring. Do not twist the pipe or fitting during or after the insertion, as is a practice with some joining methods for other pipe materials.

Cooling

Hold or block the pipe in place so that the pipe cannot come out of the joint while the mating surfaces are cooling. These cooling times are listed in the pipe or fitting manufacturer's instructions.

Electrofusion

This technique of heat fusion joining is somewhat different from the conventional fusion joining thus far described. The main difference between conventional heat fusion and electrofusion is the method by which the heat is applied. In conventional heat fusion joining, a heating tool is used to heat the pipe and fitting surfaces. The electrofusion joint is heated internally, either by a conductor at the interface of the joint or, as in one design, by a conductive polymer. Heat is created as an electric current is applied to the conductive material in the fitting. Figure 5 illustrates a typical electrofusion joint, and Figure 6 illustrates an electrofusion control box and fitting.

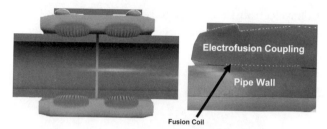

Figure 5 Typical Electrofusion Joint

General steps to be followed when performing electrofusion joining are:

1. Prepare the pipe (scrape, clean)
2. Mark the pipe
3. Align and restrain pipe and fitting per manufacturer's recommendations
4. Apply the electric current
5. Cool and remove the clamps
6. Document the fusion procedures

Prepare the Pipe (Clean and Scrape)

Assure the pipe ends are cut square when joining couplings. The fusion area must be clean from dirts or contaminants. This may require the use of water or 90% isopropyl alcohol (NO ADDITIVES OR NOT DENATURED). Scraping: The pipe surface in the fusion area must be removed to expose clean virgin material. This may be achieved by a various manufactured tools.

Mark the Pipe

Mark the pipe for stab depth of couplings and proper fusion location of saddles. (Caution should be taken to assure that a non-petroleum marker is used.)

Align and Restrain Pipe or Fitting Per the Manufacturer's Recommendations

Align and restrain fitting to pipe per manufacturer's recommendations. Place the pipe(s) and fitting in the clamping fixture to prevent movement of the pipe(s) or fitting. Give special attention to proper positioning of the fitting on the prepared pipe surfaces. Rerounding of pipe may be required with larger diameters.

Apply Electric Current

Connect the electrofusion control box to the fitting and to the power source. Apply electric current to the fitting as specified in the manufacturer's instructions. Read the barcode which is supplied with the electrofusion fitting. If the control does not do so automatically, turn off the current when the proper time has elapsed to heat the joint properly.

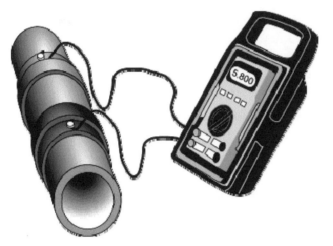

Figure 6 Typical Electrofusion Control Box and Leads with Clamps and Fittings

Cool Joint and Remove Clamps

Allow the joint to cool for the recommended time. If using clamps, premature removal from the clamps and any strain on a joint that has not fully cooled can be detrimental to joint performance.

Consult the fitting manufacturer for detailed parameters and procedures.

Documenting Fusion

The Electrofusion control box that applies current to the fitting also controls and monitors the critical parameters of fusion, (time, temperature, & pressure). The control box is a micro- processor capable of storing the specific fusion data for each joint. This information can be downloaded to a computer for documentation and inspection of the days work.

Heat Fusion Joining of Unlike Polyethylene Pipe and Fittings

Research has indicated that polyethylene pipe and fittings made from unlike resins can be heat-fused together to make satisfactory joints. Some gas companies have been heat-fusion joining unlike polyethylenes for many years with success. Guidelines for heat fusion of unlike materials are outlined in TN 13, issued by the Plastics Pipe Institute. Refer to Plastics Pipe Institute Technical Reports TR-33 and TR-41 and the pipe and fitting manufacturers for specific procedures.

Mechanical Connections

As in the heat fusion methods, many types of mechanical connection styles and methods are available. This section is a general description of these types of fittings.

The Plastics Pipe Institute recommends that the user be well informed about the performance attributes of the particular mechanical connector being utilized. Fitting selection is important to the performance of a piping system. Product performance and application information should be available from the fitting manufacturer to assist in the selection process as well as instructions for use and performance limits, if any. Additional information for these types of products is also contained in a variety of specifications such as ASTM F1924, F1973, AWWA C219, and UNI-B-13.

Polyethylene pipe, conduit and fittings are available in outside diameter controlled Iron Pipe Sizes (IPS), Ductile Iron Pipe Sizes (DIPS), Copper Tubing Sizes (CTS) and Metric Sizes. There are also some inside diameter controlled pipe sizes (SIDR-PR). Before selecting mechanical fittings, establish which of the available piping system sizes and types are being installed to ensure proper fit and function. The pipe manufacturer can provide dimensional information, and the fitting manufacturer can advise on the correct fitting selection for the application.

Mechanical Compression Couplings for Small Diameter Pipes

This style of fitting comes in many forms and materials. The components, as depicted in Figure 7, are generally a body; a threaded compression nut; an elastomer seal ring or O-ring; a stiffener; and, with some, a grip ring. The seal and grip rings, when compressed, grip the outside of the pipe, effecting a pressure-tight seal and, in most designs, providing pullout resistance which exceeds the yield strength of the polyethylene pipe. It is important that the inside of the pipe wall be supported by the stiffener under the seal ring and under the gripping ring (if incorporated in the design), to avoid deflection of the pipe. A lack of this support could result in a loss of the seal or the gripping of the pipe for pullout resistance. This fitting style is normally used in service lines for gas or water pipe 2" IPS and smaller. It is also important to consider that three categories of this type of joining device are available. One type is recommended to provide a seal only, a second provides a seal and some restraint from pullout, and a third provides a seal plus full pipe restraint against pullout.

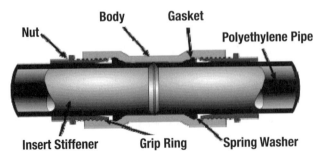

Figure 7 Typical Compression Nut Type Mechanical Coupling for Joining Polyethylene to Polyethylene

Stab Type Mechanical Fittings

Here again many styles are available. The design concept, as illustrated in Figure 8, is similar in most styles. Internally there are specially designed components including an elastomer seal, such as an "O" ring, and a gripping device to effect pressure sealing and pullout resistance capabilities. Self-contained stiffeners are included in this design. With this style fitting the operator prepares the pipe ends, marks the stab depth on the pipe, and "stabs" the pipe in to the depth prescribed for the fitting being used. These fittings are available in sizes from ½"CTS through 2" IPS and are all of ASTM D2513[(2)] Category I design, indicating seal and full restraint against pullout.

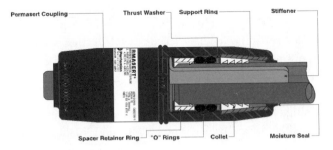

Figure 8 Stab Type Fitting

Mechanical Bolt Type Couplings for Large Diameter Pipes

There are many styles and varieties of "Bolt Type" couplings available to join polyethylene to polyethylene or other types of pipe such as PVC, steel and cast iron in sizes from 1¼" IPS and larger. Components for this style of fitting are shown in Figure 9. As with the mechanical compression fittings, these couplings work on the general principle of compressing an elastomeric gasket around each pipe end to be joined, to form a seal. The gasket, when compressed against the outside of the pipe by tightening the bolts, produces a pressure seal. These couplings may or may not

incorporate a grip ring, as illustrated, that provides pullout resistance sufficient to exceed the yield strength of the polyethylene pipe. When PE pipe is pressurized, it expands slightly and shortens slightly. In a run of PE pipe, the cumulative shortening may be enough to disjoin unrestrained mechanical joints that are in-line with the PE pipe. This can be a particular concern where transitioning from PE pipe to Ductile Iron pipe. Disjoining can be prevented by installing external joint restraints (gripping devices or flex restraints; see Figure 16) at mechanical connections, or by installing
in-line anchors or a combination of both. Additional restraint mechanisms are available to supplement the pull resistance of these types of fittings if needed. The fitting manufacturer can help guide the user with that information. Use of a stiffener is needed in this fitting style to support the pipe under the area of the seal ring and any gripping devices incorporated for pullout resistance.

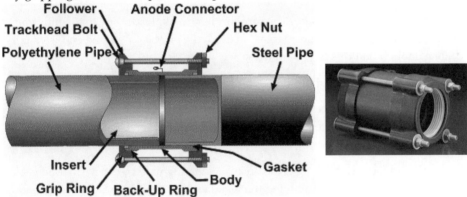

Figure 9 Mechanical Bolt Type Coupling for Joining Steel Pipe to Polyethylene or for Joining Two Polyethylene Pipes

Flanged Connections

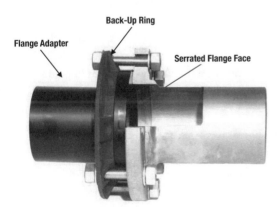

Figure 10 Typical Application of Polyethylene Flange Adapter or Stub End

Polyethylene Flange Adapters and Stub Ends

When joining to metal or to certain other piping materials, or if a pipe section capable of disassembly is required, polyethylene flange adapters, as depicted in Figure 10, are available. The "Flange Adapter" and its shorter version, the "Stub End," are designed so that one end is sized the same as the plastic pipe for butt fusion to the plastic pipe. The other end has been especially made with a flange-type end that, provides structural support, which nullifies the need for a stiffener and, with a metal back-up ring, permits bolting to the non-plastic segment of a pipe line — normally a 150-pound ANSI flange.[1]

The procedures for joining would be:

1. Slip the metal ring onto the plastic pipe section, far enough away from the end to not interfere with operation of the butt fusion equipment.

2. If a stub end is used, first butt-fuse a short length of plastic pipe to the pipe end of the stub end. If a "flange adapter" is used, the plastic pipe-sized end is usually long enough that this step is unnecessary.

3. Butt fuse the flange adapter to the plastic pipe segment.

4. Position the flanged face of the adapter at the position required so that the back-up ring previously placed on the plastic pipe segment can be attached to the metal flange.

5. Install and tighten the flange bolts in an alternating pattern normally used with flange type connections, drawing the metal and plastic flange faces evenly and flat. Do not use the flanges to draw the two sections of pipe together.

At lower pressure, typically 80 psi or less, a gasket is usually not required. At greater pressure, the serrated surface of the flange adapter helps hold the gasket in place. The flange face serration's should be individual closed concentric serration's as opposed to a continuous spiral groove which could act as a leak path. Standard Back-Up Rings are AWWA C207 Class D for 160 psi and lower pressure ratings, or Class 150 for higher pressure. Back-up ring materials are steel, primer coated steel, epoxy coated steel, or stainless steel. Ductile iron and fiberglass are also available. In below ground service, coatings and cathodic protection may be appropriate to protect metal back-up rings from corrosion. One edge of the back-up ring bore must be radiused or chamfered. This edge fits against the back of the sealing surface flange.

An all-polyethylene flange without a back-up ring is not recommended because polyethylene flanges require uniform pressure over the entire sealing surface. Absent a back-up ring, a polyethylene flange will leak between the bolts.

Flange adapters differ from stub-ends by their overall length. A flange adapter is longer allowing it to be clamped in a fusion machine like a pipe end. The back-up ring is fitted to the flange adapter before fusion, so external fusion bead removal is not required.

A stub end is short and requires a special stub-end holder for butt fusion. Once butt fused to the pipe, the external bead must be removed so the back-up ring can be fitted behind the sealing surface flange. In the field, flange adapters are usually preferred over stub-ends.

Flange Gasket

A flange gasket may not be required between polyethylene flanges. At lower pressures (typically 80 psi or less) the serrated flange sealing surface may be adequate. Gaskets may be needed for higher pressures and for connections between polyethylene and non-polyethylene flanges. If used, gasket materials should be chemically and thermally compatible with the internal fluid and the external environment, and should be of appropriate hardness, thickness and style. Elevated temperature applications may require higher temperature capability. Gasket thickness should be about 1/8"-3/16" (3-5mm) and about 55-75 durometer Shore D hardness. Too soft or too thick gaskets may bow out under pressure. Overly hard gaskets may not seal. Common gasket styles are full-face or drop-in. Full-face style gaskets are usually applied to larger sizes, because flange bolts hold a flexible gasket in place while fitting the components together. Drop-in style gaskets are usually applied to smaller pipe sizes.

Flange Bolting

Mating flanges are usually joined together with hex bolts and hex nuts, or threaded studs and hex nuts. Bolting materials should have tensile strength equivalent to at least SAE Grade 3 for pressure pipe service, and to at least SAE Grade 2 for non-pressure service. Corrosion resistant materials should be considered for underground, underwater, or other corrosive environments. Flange bolts are sized 1/8" smaller than the blot hole diameter. Flat washers should be used between the nut and the back-up ring.

Flange bolts must span the entire width of the flange joint, and provide sufficient thread length to fully engage the nut.

Flange Assembly

Mating flanges must be aligned together before tightening. Tightening misaligned flanges can cause flange failure. Surface or above grade flanges must be properly supported to avoid bending stresses. Below grade flange connections to heavy appurtenances such as valves or hydrants, or to metal pipes, require a support foundation of compacted, stable granular soil (crushed stone), or compacted cement stabilized granular backfill, or reinforced concrete. Flange connections adjacent to pipes passing through structural walls must be structurally supported to avoid shear loads.

Prior to fit-up, lubricate flange bolt threads, washers, and nuts with a non-fluid lubricant. Gasket and flange sealing surfaces must be clean and free of significant cuts or gouges. Fit the flange components together loosely. Hand-tighten bolts and re-check alignment. Adjust alignment if necessary. Flange bolts should be tightened to the same torque value by turning the nut. Tighten each bolt according to the patterns and torques recommended by the flange manufacturer. Polyethylene and the gasket (if used) will undergo some compression set. Therefore, retightening is recommended about an hour or so after torquing to the final torque value the first time. In pattern sequence, retighten each bolt to the final torque value. For high pressure or environmentally sensitive or critical pipelines, a third tightening, about 4 hours after the second, is recommended.

Special Cases

When flanging to brittle materials such as cast iron, accurate alignment, and careful tightening are necessary. Tightening torque increments should not exceed 10 ft.-lbs. Polyethylene flange adapters and stub ends are not full-face, so tightening places a bending stress across the flange face. Over-tightening, misalignment, or uneven tightening can break brittle material flanges.

When joining a polyethylene flange adapter or stub end to a flanged butterfly valve, the inside diameter of the pipe flange should be checked for valve disk rotation clearance. The open valve disk may extend into the pipe flange. Valve operation may be restricted if the pipe flange interferes with the disk. If disk rotation clearance is a problem, a tubular spacer may be installed between the mating flanges, or the pipe flange bore may be chamfered slightly. At the sealing surface, chamfering must not increase the flange inside diameter by more than 10%, and not extend into the flange more than 20% of the flange thickness. Flange bolt length must be increased by the length of the spacer.

Mechanical Flange Adapters

Mechanical Flange Adapters are also available and are shown in Figure 11A. This fitting combines the mechanical bolt type coupling shown in Figure 9 on one end with the flange connection shown in Figure 10 on the other. This fitting can provide a connection from flange fittings and valves to plain end pipes. The coupling end of this fitting must use a stiffener when used to join polyethylene pipe. Mechanical flange adapters may or may not include a self-restraint to provide restraint against pipe pullout as part of the design. Alternative means of restraint should be used when joining polyethylene pipe if the mechanical flange adapter does not provide restraint. Contact the manufacturer of these fittings for assistance in selecting the appropriate style for the application.

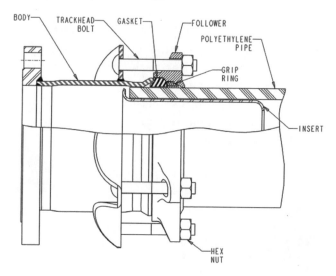

Figure 11A Bolt Type Mechanical Flange Adapter

Mechanical Joint (MJ) Adapters

PE pipe can be connected to traditional hydrants, valves and metal pipes using an MJ Adapter. A gland ring is placed behind the adapter before fusing, which can be connected to a standard ANSI/AWWA mechanical joint. When the gland ring is used, restraining devices are not required on the PE pipe.

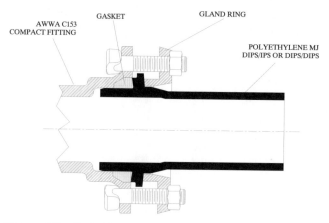

Figure 11B Typical Application of Polyethylene MJ Adapter

Transition Fittings

Other methods are available that allow joining of plastic to metal. Transition fittings are available which are pre-assembled at the manufacturer's facility. These transition fittings are normally pull-out resistant, seal tight with pressure and have tensile values greater than that of the plastic pipe part of a system. However, the user should insist on information from the manufacturer to confirm design capabilities or limitations. Transition fittings are available in all common pipe sizes and polyethylene materials from CTS and larger with a short segment of plastic pipe for joining to the plastic pipe section. The metal end is available with a bevel for butt welding, with male or female pipe threads, or is grooved for a Victaulic[14] style, or flanged for connecting to an ANSI 150-pound flange.[1]

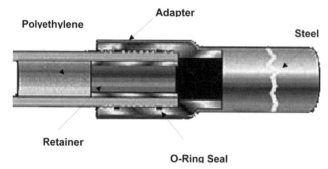

Figure 12 Standard Fitting for Plastic Pipe to Steel Pipe Transition

Mechanical Joint Saddle Fittings

Mechanical joint saddle fittings have at least one mechanical joint which may connect the outlet to the service or branch pipe, or may connect the fitting base to the main, or both connections may be mechanical joints. Mechanical joint saddle fittings are made from plastics, metals, and other materials.

Figure 13 Mechanical Saddle

For mechanical joint outlets, the service or branch pipe is either supported with a tubular stiffener in the pipe ID, or the pipe end is fitted over a spigot (insert) end of the fitting. The outlet joint is completed using mechanical compression around the service or branch pipe OD. Depending upon design, gaskets may or may not be used. Observe the fitting manufacturer's instructions in making the outlet connection.

Plastic outlet pipes must be protected against shear or bending loads by installing protective sleeves or bridging sleeves, or special care must be taken to ensure that embedment materials are properly placed and compacted around the outlet.

The connection between the saddle base and the main may be by hot plate saddle fusion, or by electrofusion, or by mechanical connection. Hot plate saddle fusion and electrofusion have been previously discussed.

Mechanical saddle base connections are clamped or strapped to the side or top of the main pipe. Typically, gaskets or o-rings are used to seal between the saddle base and the main pipe OD surface to prevent leakage when the main wall is pierced. Once secured to the main per the fitting manufacturer's instructions, the main may be pierced to allow flow into the service or branch pipe.

Some mechanical joint saddle fittings can have an internal cutter to pierce the main pipe wall (Fig. 13). "Tapping tees or tapping saddles" (Fig. 14) are generally suitable for installation on a "live" or pressurized main (hot tapping). Branch saddles or service saddles that do not have internal cutters may also be hot tapped using special tapping equipment. Contact equipment manufacturer for information.

Figure 14 Tapping Tee with Cutter

Repair Clamps

Third party damage to polyethylene or any pipe material is always a possibility. Repairs can be made by cutting out the damaged section of pipe and replacing the section by use of heat fusion or mechanical fitting technology discussed earlier. Within limits, repairs can also be made with clamp-on repair saddles as depicted in Figure 15. Such devices do have limitations for use. They are intended only to repair locally damaged pipe such as gouges or even punctures of the pipe wall. A clamp length of not less than 1½" times the nominal pipe diameter is recommended. The procedure is basically to clean the pipe area where the clamp will be placed, and bolt the clamp in place according to the fitting manufacturer's instructions. As with all fittings, limitations on use should be verified with the fitting manufacturer.

Figure 15 Clamp-on Repair Saddle

Other Applications

Restraining Polyethylene Pipe

Restraining of polyethylene is not required in a totally fused system. When concerns of thermal contraction or slippage due to terrain arise, PE may be restrained by use of Electrofusion Flex Restraints fused onto the pipe or a PE water stop fused inline. These fittings serve as an anchor to allow concrete to be formed around the pipe encapsulating the anchor. Other methods of restraining polyethylene pipe include a wall anchor fused in the line with the proper sized reinforced concrete anchor around it or adding restraint harnesses to several existing bell and spigot joints of the existing system to prevent pullout. Contact the pipe manufacturer for details.

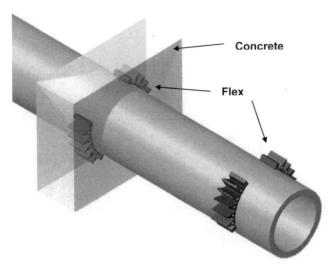

Fig. 16 Electrofusion Flex Pipe Restraint

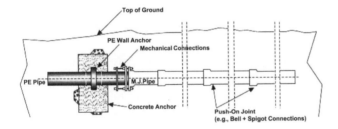

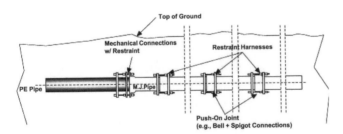

Fig. 17 Illustration of Method I-PE Wall Anchor

Squeeze-off

Regardless of the joining method applied in the installation of polyethylene pipe, it may become necessary to shut off the flow in the system. With PE pipe materials, squeeze-off of the pipe with specially-designed tools is a common practice. Use squeeze-off tools per ASTM F 1563 and follow the squeeze-off procedures in ASTM F 1041.

Fig. 18 Squeeze-Off Tool

Summary

The applications for polyethylene piping products continue to expand at an accelerating rate. Gas distribution lines, potable water systems, submerged marine installations, gravity and force main sewer systems, and various types of above-ground exposed piping systems are but a few of the installations for which polyethylene pipe and fittings have been utilized.

As piping products applications expand, so does the use of new and existing joining methods expand.

A key element to this continued success is the diversity of methods available to join polyethylene pipe and fittings. The integrity of the butt and socket fusion joining technique has been proven by the test of time in a variety of applications. The manufacturers of polyethylene pipe and fittings have made every effort to make the systems as comprehensive as possible by producing a variety of fittings and components to insure compatibility with alternate piping materials and system appurtenances.

The purpose of this chapter has been to provide the reader with an overview of the various methods by which polyethylene piping materials may be joined. As a result the reader has developed a further appreciation for the flexibility, integrity, and overall utility afforded in the design, installation, and performance of polyethylene piping systems and components.

It should be noted that this document does not purport to address the safety considerations associated with the use of these procedures. Information on safe operating procedures can be obtained from the manufacturers of the various types of joining equipment or polyethylene products

References
1. ASME/ANSI B16.5. (1996). *American National Standard on Pipe Flanges and Flanged Fittings*, American National Standards Institute, New York, NY.
2. ASTM D2513, Standard Specifications for Thermoplastic Gas Pressure Pipe, Tubing, and Fittings, *Annual Book of Standards*, ASTM, West Conshohoken, PA.
3. ASTM D2657, Standard Practice for Heat Fusion Joining of Polyolefin Pipe and Fittings, *Annual Book of Standards*, ASTM, West Conshohoken, PA.
4. ASTM D3140, Standard Practice for Flaring Polyolefin Pipe and Tubing, *Annual Book of Standards*, ASTM, West Conshohoken, PA.
5. ASTM F894, Standard Specification for Polyethylene (PE) Large Diameter Profile Wall Sewer and Drain Pipe, *Annual Book of Standards*, ASTM, West Conshohoken, PA.
6. ASTM F1041, Standard Guide for Squeeze-Off of Polyolefin Gas Pressure Pipe and Tubing, *Annual Book of Standards*, ASTM, West Conshohoken, PA.
7. ASTM F1056, Standard Specification for Socket Fusion Tools for Use in Socket Fusion Joining Polyethylene Pipe or Tubing and Fittings, ASTM, West Conshohoken, PA.
8. AWWA C901, Polyethylene (PE) Pressure Pipe and Tubing, 1/2 in. Through 3 in., for Water Service, American Water Works Association, Denver, CO.
9. Caution Statement on Sidewall Heat Fusion Without Use of Mechanical Assist Tooling, *Statement T*. Plastics Pipe Institute, Washington, DC.
10. *Code of Federal Regulations*, Title 49, Part 192, Subpart F., Pipeline Safety Regulations, Washington, DC.
11. General Guidelines for the Heat Fusion of Unlike Polyethylene Pipes and Fittings, Report TN-13, Plastics Pipe Institute, Washington, DC.
12. *IAPMO*, International Association of Plumbing and Mechanical Officials, Walnut, CA.
13. PPFA. Plastics Pipe and Fittings Association, Glen Ellyn, IL.
14. *Victaulic General Catalog on Mechanical Piping Systems.* (1988). Victaulic Company of America, Easton, PA.
15. Generic Butt Fusion Procedure for Polyethylene Gas Pipe, PPI TR-33, Plastics Pipe Institute, Inc., Washington, DC.
16. Generic Saddle Fusion Procedure for Polyethylene Gas Pipe, PPI TR-41, Plastics Pipe Institute, Inc., Washington, DC.

Chapter 10

Marine Installations

Introduction

Since the early 1960's, just a few years after its first introduction, polyethylene (PE) piping has been increasingly used for various marine applications such as effluent outfalls, river and lake crossings, and fresh and salt-water intakes. Immunity to galvanic corrosion is a major reason for selecting PE. The combination of air and water, but particularly seawater, can be very corrosive to ordinary metallic piping materials. But other beneficial features, as follows, combine to make PE piping particularly well-suited for marine applications:

Light weight – For a given pipe diameter and equivalent performance requirements, the weight of PE pipe is around one tenth of the weight of concrete pipe and less than one half that of cast iron. Handling of PE requires a minimum of heavy equipment.

It floats – Because PE's density is about 96% of that for fresh water, and about 94% of that for sea water, PE pipe floats even when full of water. Long lengths can be assembled on shore where the empty pipe may be weighted to an extent that allows air-filled pipe to be floated to its intended location, and in most cases, is also sufficiently weighted to keep it anchored at its final submerged location after the air has been replaced with water.

Integral, "bottle-tight" joints – By means of the butt fusion method, continuous lengths of PE pipe can be readily assembled without the need of mechanical fittings. The resultant heat fusion joints are as strong as the pipe, and they eliminate the risk of joint leakage.

Flexibility –The flexibility of PE pipe allows it to be gradually sunk and to adapt to the natural topography of underwater surfaces. This results in a more simplified sinking procedure, and it also means that the flexible pipeline can normally be placed directly on the natural bottom without any trenching or other form of preparation of continuous level support.

Ductility (strainability) – Because of its relatively high strain capacity, PE piping can safely adjust to variable external forces generated by wave and current action. High strain capacity also allows the PE piping to safely shift or bend to accommodate itself to altered bedding that can result by the underscouring that may sometimes occur with strong wave and current actions.

Conventional, non-flexible materials such as concrete or iron can only afford relatively small deformations before risking leakage at, or structural failure of, the joints. As the exact magnitude of the maximum forces that can act on rigid pipes is difficult to predict, installations using piping that only allows relatively small deformation at the joints, or limited bending strain in the pipe, requires a large "safety factor," such as a relatively heavy loading to stabilize the pipe against movement, or the trenching of the pipe into sea bed sediments so as to stabilize it against movement that can result from heavy sea action. Such construction techniques tend to be more difficult, time-consuming and relatively expensive. In contrast, the flexibility and ductility of PE allows it to adapt to unconsolidated river and sea bottoms, and also to safely shift or bend under the forces resulting from occasionally strong currents or other actions. For most marine installations, PE piping needs only to be sufficiently weighted to keep it at the intended location and to prevent it from floating. This results in easier and less costly installations and in a submerged piping system that is capable of delivering very reliable and durable service. By choosing PE pipes, many projects have been accomplished which would not have been economically realistic with traditional piping materials. The lower overall cost of PE piping installations allows for the option of installing several small outfalls rather than one large one. Multiple outfalls can achieve greater environmental protection by the discharging of smaller quantities of effluent at separated points
of discharge, and their use often results in lower onshore pretreatment cost.

A marine pipeline installation may involve considerable risk to
the pipeline integrity both during installation and while in service. Guidance provided herein on the design and installation of PE

piping is limited to those issues that are specific or are related to this material. It is not the intent of this chapter to cover the many other design, construction and safety issues that need to be considered in a marine installation.

The primary focus of this chapter is the design and installation of underwater lines by the "float-and-sink" method that is made possible through the use of the light-in-weight and flexible PE pipe. Under certain conditions – such as when it is not possible to delay navigation long enough to launch and sink a pipeline – it may be necessary, or it may be more practical, to use a variation of the "float-and-sink" method that is herein described. In one variation, one or more separate long-segments of the pipeline with a flange at each end are assembled and floated. These segments are then sunk, properly positioned and bolted together by divers. Another alternative method is the "bottom-pull" method, which is briefly described at the end of Section 2.8. However, regardless of which method is used, the general design and installation principles that apply to the "float-and-sink" method also apply to alternate methods.

Other marine applications for which PE piping has proven to be very suitable include temporary water surface pipelines, lines installed over marshy soils and lines used in dredging operations. These are described briefly. Design and installation for these marine applications are conducted in accordance with essentially the same criteria and principles as described for the "float-and-sink" method.

The Float-and-Sink Method – Basic Design and Installation Steps

In nearly all underwater applications, the design and installation of PE piping is comprised of the following basic steps:

1. Selection of an appropriate pipe diameter
2. Selection of an appropriate pipe SDR (i.e., an appropriate wall thickness) in consideration of the anticipated installation and operating conditions
3. Selection of the design, weight and frequency of spacing of the ballast weights that will be used to sink and then hold the pipe in its intended location
4. Selection of an appropriate site for staging, joining and launching the pipe

5. Preparing the land-to-water transition zone and, when required, the underwater bedding
6. Assembly of the individual lengths of pipe into a continuous string of pipe
7. Mounting of the ballast weights (This step may be done in conjunction with the next step.)
8. Launching the joined pipe into the water
9. Submersion of the pipeline into the specified location
10. Completion of the land-to-water transition

General guidance for the conduct of each of these steps follows. Since the specific conduct of each step can be affected by the choice of design and installation options discussed in other steps, the reader should review the entire chapter before deciding on the most applicable design and installation program.

Step 1   Selection of an Appropriate Pipe Diameter

Selection of an appropriate pipe diameter involves the estimation of the minimum flow diameter that is needed to achieve the design discharge rate. Guidance for doing this is provided in the design chapter of this Handbook.

A confirmation is then performed after the required pipe dimension ratio (DR) is determined in accordance with Step 2 which follows. Since the actual internal diameter of a pipe that is made to a standard outside diameter is dependent on the choice of pipe DR (see Table in the Annex to the design chapter), the nominal pipe diameter/DR combination that is finally selected needs to have an actual inside diameter that is at least as large as the above determined minimum required flow diameter.

Step 2   Determination of the Required SDR

The DR of the PE pipe, in combination with the pipe material's assigned maximum hydrostatic design stress, should allow the pipe to safely resist the maximum anticipated sustained net internal pressure at the maximum anticipated operating temperature. Information, including temperature and environmental de-rating factors, for determining the appropriate pipe DR is presented in the design chapter of this Handbook. As an added "safety factor" it is common practice to pressure rate the pipe for the maximum anticipated operating temperature of either the internal or external environment, whichever is higher.

A check should be made to ensure that the selected pipe pressure rating is also sufficient to safely withstand any momentary pressure surges above normal operating pressure. Pressure surges tend to occur during pump start-ups or shut-downs, and also during sudden pump stops caused by emergencies, such as loss of power. Guidance for selecting a PE pipe with sufficient surge pressure strength is also presented in the design chapter of this Handbook.

A sudden pump stop can sometimes also result in flow separation, giving rise to a momentary reduction in pressure along some portion of the pipeline. Since underwater pipelines can be subject to relatively large external hydrostatic pressure, flow separation can sometimes lead to a significant net negative internal pressure. A check needs to be made to ensure that the pipe DR that has been selected based on maximum internal pressure considerations is also adequate to safely resist buckling, or pipe collapse, under the largest net negative internal pressure that could ever develop from whatever cause. Guidance for this design check is also provided in the design chapter of this Handbook. The ballast weights that are attached to PE pipe for purposes of its submersion also fulfill an important role as ring stiffeners that tend to enhance a pipe's inherent resistance to buckling. Common design practice is to accept this benefit as an added "safety factor," but not to directly consider it in the design procedure for selection of a pipe of appropriate ring stiffness.

Step 3 Determination of the Required Weighting, and of the Design and the Spacing of Ballast Weights

The determination of these parameters is made in accordance with the following sub-steps.

Step 3a Maximum Weighting that Allows Weighted Pipe to be Floated into Place

The buoyant or vertical lift force exerted by a submerged PE pipe is equal to the sum of the weight of the pipe and its contents minus the weight of the water that the pipe displaces. This relationship can be expressed mathematically as follows:

(1) $F_B = [W_P + W_C] - W_{DW}$

WHERE
F_B = buoyant force, lbs/foot of pipe
W_P = weight of pipe, lbs/foot of pipe
W_C = weight of pipe contents, lbs/foot of pipe
W_{DW} = weight of water displaced by pipe, lbs/foot of pipe

Since the density of PE (~59.6 lbs/cubic foot) is only slightly lower than that of fresh water (~62.3 lbs/cubic foot) the pipe contributes somewhat towards net buoyancy. However, the major lift force comes from the air-filled inner volume of the pipe. Since, for a pipe of given outside diameter, the size of the inner volume is determined by the pipe's wall thickness – the greater the thickness, the smaller the inner volume – and since a pipe's actual wall thickness can be expressed in terms of the pipe's diameter ratio (DR), Equation 1 can be rearranged as shown in Equation 2. The resultant net buoyancy force can be determined from the pipe's actual outside diameter, its DR (or SDR), the extent to which the pipe is filled with air, the density of the water into which the pipe is submerged, and the densities of the pipe and of the liquid inside the pipe:

(2)
$$F_B = \left[0.00545 D_o^2 \rho_w\right] \left[4.24 \frac{(DR-1.06)}{(DR)^2} \frac{\rho_p}{\rho_w} + \left(1 - \frac{2.12}{DR}\right)^2 (1-R)\frac{\rho_c}{\rho_w} - 1\right]$$

WHERE
F_B = buoyant force, lbs/foot of pipe
D_o = external diameter of pipe, in
DR = pipe dimension ratio, dimensionless
R = fraction of inner pipe volume occupied by air
P_w = density of the water outside the PE pipe, lbs/cu. ft
P_p = density of the pipe material, lbs/ cu. ft.
P_c = density of pipe contents, lbs/ cu. ft.
The derivation of Equation 2 is presented in Appendix A-1.

A more succinct way of expressing the principle embodied in Equation 2 is as follows:

(3) $F_B = W_{DW} \, [\text{"K"}]$

WHERE
$W_{DW} = 0.000545 \, D_0^5 \rho w$

Stated in words, the resultant buoyant force (F_B) is equal to the potential theoretical buoyant force (W_{DW}) times a buoyancy reduction factor ("K") that takes into account inner pipe volume, degree of air filling and the densities of the pipe and the liquid inside the pipe.

The manner by which the buoyancy reduction factor "K" is affected by a pipe's DR and the extent to which its inner pipe volume is filled with air, R, is indicated by the calculation results reported in Table 1. The values in this table have been computed based on the following densities: 62.3 lbs/ cu. ft for water both inside and outside the pipe, and 59.6 lbs/cu. ft for the PE pipe material. Using these K-values for approximation of the net buoyant force of a submerged pipeline in which a portion of the line is occupied by air greatly simplifies the calculations involved.

TABLE 1
Typical values of "K" in equation 3.0

"K" is the fraction of maximum potential buoyancy. The exact value of "K" is determined by the particular combination of pipe diameter ratio (SDR), pipe material and liquid densities and the extent (R) to which a PE pipe is filled with air*

Pipe SDR	Value of "K" as a function of R, the fraction of inner pipe volume that is occupied by air					
	R = 0.10	R = 0.15	R = 0.20	R = 0.25	R = 0.30	R = 1.0 (100% Air)
9	-0.078	-0.107	-0.136	-0.166	-0.195	-0.604
11	-0.081	-0.113	-0.146	-0.178	-0.211	-0.667
13.5	-0.084	-0.119	-0.155	-0.190	-0.226	-0.723
17	-0.087	-0.125	-0.163	-0.202	-0.240	-0.776
21	-0.089	-0.130	-0.170	-0.210	-0.251	-0.817
26	-0.091	-0.133	-0.176	-0.218	-0.260	-0.850
32.5	-0.093	-0.137	-0.180	-0.224	-0.268	-0.879

* The "K" values in this table have been computed using Equation 2 and based on the following assumptions: a density of 62.3 lbs/cu ft for water outside and inside the pipe and 59.6 lbs/cu ft for the PE pipe material. The minus sign before each resultant value of "K" indicates a net upward, or buoyant force.

Step 3b Determining the Maximum Weighting That Still Allows PE Pipe To Float

When a PE pipe that is completely filled with air is weighted so that the submerged weighting is equal to W_{DW} (the weight of the water that is displaced by the outer volume of the pipe) times the appropriate value of "K" (e.g., the value given in the last column of Table 1), that pipe achieves neutral buoyancy – it neither sinks nor floats. Therefore, "K" represents the fraction of pipe displacement that, when counteracted by the placement of external weighting on the pipe, results in neutral buoyancy. With the objective in mind of facilitating a marine installation by the floating of a PE pipe so that it may readily be stored above water and then towed and maneuvered to its intended location, the weighting that is attached to the pipe needs to be limited to an amount that still allows an air-filled pipe to freely float on top of the water. To this end, the practice is to limit the weighting of an air-filled PE pipe to about 85% of the pipe displacement times the "K" value that corresponds to that pipe's DR and the densities of the pipe material and the water, for example, the "K" values reported in the last column of Table 1. This practice results in the limiting of the weighting of an air-filled pipe that is to be installed by the "float-and-sink" method to a maximum that can vary, depending on the pipe's DR, from about 57 to 75% of the pipe's displacement.

Step 3c Determining the Required Minimum Weighting for the Anchoring of a Submerged Pipe in its Intended Location

Fortunately, as indicated by analysis and confirmed by experience[1,2], in most cases a weighting of 25 to 50% of the pipe displacement is quite sufficient to maintain a properly anchored submerged PE pipe after it has been filled with water. The lower weighting has been found satisfactory in cases, like in lake crossings, where current and wave action are relatively mild, while the larger weighting is used in sea installations where sea actions are stronger. However, even for pipes that are exposed to normal sea conditions close to the shore, it has been found that a weighting of about 70% of the pipe displacement is quite satisfactory[1]. As indicated by the values shown in Table 1, this extent of weighting still allows most PE pipes to float when air-filled.

In an article summarizing the state of the art in utilizing plastics pipe for submarine outfalls, Janson[3] reports that, based on past practical experience and theoretical studies, a 40-inch ocean outfall was installed in Sweden where, for depths greater than 40 feet, the pipe was weighted to 25% of its displacement; and in the surf zone, where the waves break and the water depth is about 10 feet, the loading was increased to 60% of the displacement. Closer to the shore, where wave action is at its strongest, it is common to protect the pipe by trenching it. In respect to trenched pipe, Janson also reports that, when a trench is refilled with fine-grained soil, the buried pipe can sometimes float from the trench, apparently a reaction resulting from the fluidization of the fill by strong wave action. This reference further reports that the possibility of floating from fine-grained backfill can be avoided by weighting the pipe to at least 40% of its displacement.

Calculation techniques have been developed for the determination of the required weighting of plastic pipes depending on anticipated current and wave action. A brief overview of the technical considerations upon which these calculations are based is included in Appendix A-2. References for further information are also provided.

In cases where it is indicated that the pipeline, or certain sections of the line, should be weighted to a greater extent than that which allows the pipe to float while filled with air, the attachment of the required ballast weights can be conducted in two stages: preliminary weighting is conducted so as to still allow the pipe to be floated into position, and then the additional required weights are added where required after the completion of the submerging of the pipe. Another option is to temporarily increase the pipe's buoyancy by the use of empty tanks or drums, or large blocks of rigid plastic foamed material that are then released as the pipe is being submerged. A further option, which is illustrated in Figure 1, is to attach the required ballast weights onto the pipe from a barge from which the pipe is slid to the bottom by means of a sled that has been designed to ensure that the bending of the pipe is less than that which might risk buckling (See the discussion on pipe submersion).

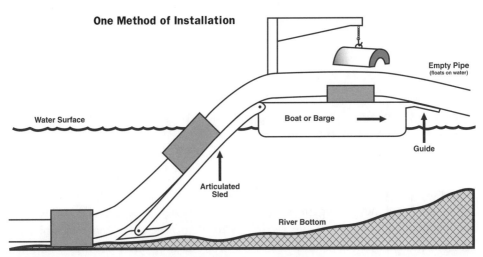

Figure 1 Submerging a heavily weighted pipe from a barge

Step 3d Ensuring that the Required Weighting Shall Not Be Compromised by Air Entrapment

As suggested by the "K" values in Table 1 that apply to pipes that are partially filled with air, even a modest amount of air entrapment can result in a lift force that can significantly reduce the quality of pipe anchorage. For example, if a pipeline is weighted to 25% of the water it displaces and in a section of that pipeline enough air accumulates to occupy just 10% of the pipe's inner volume, the lift produced by that amount of air will reduce the effective weighting in that portion of the pipeline to about only 15% of the pipe displacement. Such reduction is sure to compromise the stability of that pipe section against wave and current actions. Accordingly, one important objective in the design of the piping system to prevent the entrance and accumulation of air in all portions of the submerged section. In outfall systems, one effective means for achieving this objective is to utilize a surge or "drop" chamber into the system design, as illustrated in Figure 2. Another precautionary measure is to ensure that there are no localized high points along the submerged pipeline that could accumulate air or gases, particularly during periods of low or no flow rate.

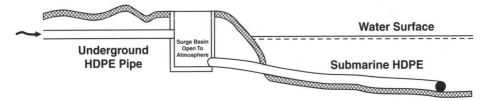

Figure 2 A surge chamber may be used to prevent air from entering a pipeline

In cases where the possibility of some accumulation of air or gas – which may be given off by chemical reactions – cannot be avoided, or where the line may at some time be emptied, it is necessary to add enough ballast weighting to offset the additional negative buoyancy so as to always hold the pipe in its intended location.

Step 3e Determining the Spacing and the Submerged Weight of the Ballasts To Be Attached to the Pipe

The objectives for limiting the spacing between ballast weights are essentially the same as those for establishing the support spacing requirements for above-ground suspended pipelines. In both cases the pipes are subject to a distributed loading – in the case of submerged pipelines, by the combined effect of current, lift and wave actions. The objective of the design is to limit resultant pipe deflection so that the resultant maximum fiber bending stresses and strains are within safe limits. An additional reason for limiting deflection in submerged pipelines is to reduce the chances of forming pockets in which air or gas can accumulate. The lift created by air-filled pockets can, if large enough, compromise the quality of the anchoring of the submerged pipe. Information on conducting the required calculations and on the appropriate limiting values for bending stress and strain is included in the chapter on design. Because of the concern of trapping air, support spacing for submerged pipes is normally delimited by allowable pipe deflection – considerably greater deflection would generally be permitted under the criteria of maximum bending stress or strain.

Listed in Table 2 are commonly used ballast spacings. To satisfy the objective for minimizing air entrapment, the spans in this table are somewhat shorter than for pipes that are suspended above ground. An added benefit of shorter spans is that they better distribute anchoring loads on the sea bottom, which often offers only moderate load bearing capacity. Additionally, these shorter spans minimize the chance of pipe shifting, help smooth out the submersion process and they lead to ballasts that are more manageable both in size and in weight.

TABLE 2
Commonly Used Values for the Spacing of Ballasts

Nominal Pipe Diameter, in	Approximate Spacing (L), ft
Up to 12	5 to 10
Over 12, up to 24	7.5 to 15
Over 24, up to 64	10 to 20

Source: Committee Report: Design and Installation of PE Pipe, Journal AWWA, Volume 91, Issue 2, February, 1999, page 92.

The required submerged weight of the ballasts can be determined from the following:

(4) $$B_W = \frac{W_s}{L}$$

WHERE
B_w = weight of ballast in water, lbs
W_s = required submerged weighting by ballasts, lbs per foot
L = center to center spacing between ballasts, feet

The resultant dry weight of the ballast depends on the density of the ballast material as compared to that of the water into which the ballast is to be submerged:

(5) $$B_A = B_W \frac{\rho_B}{(\rho_B - \rho_W)}$$

WHERE
B_A = weight of ballast in air, lbs
P_B = density of ballast, lbs/cu. ft (~144 lbs/cu ft for plain concrete, ~ 150 for reinforced)
P_W = density of water, lbs/ cu ft (~62.3 lbs/cu ft for fresh water, ~64.0 lbs/cu ft for sea water)

Since the weight of a ballast cannot be closely predicted or readily adjusted, it is more practical to tune in the final weighting to the required value by adjusting the distance between ballasts of known weight. To this end the following formula, derived by combining Equations 4 and 5, may be used:

(6) $$L = \frac{W_s}{B_a} \frac{\rho_B}{(\rho_B - \rho_W)}$$

Step 3f Design and Construction of Ballast Weights

To prevent cracking of ballasts when handling, tightening and moving PE pipe, they are typically made of suitably reinforced concrete. Ballasts can be made to different shapes, although a symmetrical design such as round, square, or hexagonal is preferred to avoid twisting during submersion. Flat-bottomed ballasts are preferred if the submerged piping is likely to be subjected to significant currents, tides or wave forces because they help prevent torsional movement of the pipe.

Also, when such conditions are likely to occur, the ballasts should place the pipeline at a distance of at least one-quarter of the pipe diameter above the sea or river bed. The lifting force caused by rapid water movement that is at a right angle to a pipe that rests on, or is close to a sea or river-bed is significantly greater than that which acts on a pipe that is placed at a greater distance from the bed. This means that ballasts designed to give an open space between the pipe and the bed will give rise to smaller lifting forces.

For example, in accordance with the calculation procedure developed by Janson (See Appendix A-2), the lifting force that develops on a 12-in PE pipe that is resting directly on a sea bed and that is at an angle of 60° to the direction of a strong current that is flowing at a rate of about 10 feet per second is approximately 100 lbs per foot. When this pipe is raised above the sea bed so that the space between the bottom of the pipe and the sea bed is one-quarter of the pipe's outside diameter, the lifting force is reduced to about 25 lbs per foot.

The ballasts should comprise a top and bottom section that, when mated together over a minimum gap between the two halves, the resultant inside diameter is slightly larger than the outside diameter of the pipe. This slightly larger inside diameter is to allow the placement of a cushioning interlining to protect the softer plastic pipe from being damaged by the hard ballast material. Another function of the interlining is to provide frictional resistance that will help prevent the ballasts from sliding along the pipe during the submersion process. Accordingly, slippery interlining material such as polyethylene film or sheeting should not be used. Some suggested interlining materials include several wraps of approximately 1/8-in thick rubber sheet or approximately 1/4-in thick neoprene sponge sheet.

The purpose of the minimum gap between the two halves of the ballasts is to allow the two halves to be tightened over the pipe so as to effect a slight decrease in pipe diameter and thereby enhance the hold of the ballast on the pipe.

Additionally, experience has shown that in certain marine applications where tidal or current activity may be significant, it is feasible for the pipe to "roll" or "twist". This influence combined with the mass of the individual ballasts may lead to a substantial torsional influence on the pipe. For these types of installations, an asymmetric ballast design in which the bottom portion of the ballast is heavier than the upper portion of the ballast is recommended. Typical design considerations for this type of ballast are shown in Appendix A-3-2.

Suitable lifting lugs should be included in the top and bottom sections of the ballasts. The lugs and the tightening hardware should be corrosion resistant. Stainless steel strapping or corrosion-resistant bolting is most commonly used. Bolting is preferable for pipes larger than 8-in in diameter because it allows for post-tightening prior to submersion to offset any loosening of the gripping force that may result from stress-relaxation of the pipe material.

Examples of various successfully used ballast designs are shown in Appendix A-3.

Figure 3 Two-piece Concrete Anchors in Storage at Marine Job-Site

Step 4 Selection of an Appropriate Site for Staging, Joining and Launching the Pipe

The site for staging, joining and launching the pipe should preferably be on land adjacent to the body of water in which the pipeline is to be installed and near the point at which the pipe is to enter the water. Also, the site should be accessible to land delivery vehicles. If these requirements are not easily met, the pipe may be staged, joined and weighted at another more accessible location and then floated to the installation site. Long lengths of larger diameter PE pipe have been towed over substantial distances. However, considerable precautions should be exercised for insuring the stability of the towed materials in light of marine traffic, prevailing currents or impending weather considerations.

To facilitate proper alignment of the pipe-ends in the fusion machine and to leave enough room for the attachment of the ballast weights, the site near the water should be relatively flat. It is best to allow a minimum of two pipe lengths between the machine and the water's edge. The site should also allow the pipe to be stockpiled conveniently close to the joining machine.

The ground or other surface over which the pipe is to be moved to the water should be relatively smooth and free of rocks, debris or other material that may damage the pipe or interfere with its proper launching. When launching a pipe with ballast weights already attached, provision should be made for a ramp or a rail skidway arrangement to allow the ballasts to move easily into the water without hanging up on the ground. As elaborated under the launching step, the end of a pipe that is moved into the water needs to sealed to prevent water from entering and, thereby, compromising its capacity to float freely.

Step 5 Preparing the Land-to-Water Transition Zone and, When Required, the Underwater Bedding

At some point in time before the start of the submersion procedure, usually before the pipe is launched, a trench needs to be prepared in which to place the pipe between the point where it leaves the shore and the first underwater location beyond which the pipe is completely submerged without the need for external protection. The trench needs to be deep and long enough to protect the pipe from wave action, tidal scour, drifting ice and boat traffic. Special care should be employed in the design and construction of the land-to-water transition in ocean outfalls where occasional rough seas can result in very strong waves and in the scouring of the material below and around the pipe.

Unless weighted to a relatively high extent, say to at least 40% of the pipe displacement, a pipe lying in a land-to-water transition trench that has been filled with fine silt or sand could float up when that zone is subjected to strong wave action. One method of controlling this tendency would be to utilize increased weighting via enhanced ballast design. Alternatively, the submerged pipe could be placed on a bed of prepared backfill and subsequently surrounded by graded material in accordance with ASTM D2774. This ASTM standard provides that plastic pipe installed underground will be bedded and backfilled using material with a particle size in the range of ½" to 1 ½" depending on the outside pipe diameter. However, it may be necessary to place a layer of even larger particle sized fill (1 ½" to 4") over the graded material to avoid movement of the stone backfill in some tidal zones or areas of strong current activity. Protection and stabilization of the pipe installation may be further enhanced by the placement of a 1 to 2 foot cover of blast rock over the completed installation.

With regard to the preparation of the underwater support generally, no dredging of filling needs to be carried out because the ballasts act to keep the pipe above the bottom material. The principal requirement is that the pipe should not rest or come in contact with large stones. To this end, larger stones that project above the bottom and that could come in contact with the pipe should be removed, as well as those that lie within about 3 pipe diameters on either side of the pipe.

Step 6 Assembly of Individual Lengths of Pipe Into Long Continuous Lengths

The butt fusion of individual lengths into a long string of pipe should be conducted by trained personnel and by means of appropriate equipment. The heat fusion parameters – e.g., temperature, interfacial pressure, heating and cooling times – should be as recommended by the pipe manufacturer for the particular pipe material and the joining conditions, including outdoor temperature and wind. (See Chapter on Polyethylene Joining Procedures.)

Upon the completion of the heat fusing of an added individual length to the pipeline, the resultant longer pipe string is further moved into the water. As discussed elsewhere, the pipe should always be moved to the water using suitable mechanical equipment that will cause no damage to the pipe or to the pipe ends.

Ballast weights can be mounted before the pipe string reaches the water. If circumstances make it more practical, the ballasts can also be attached on the floating pipe from a floating barge by a scheme such as illustrated in Figure 4.

Step 7 Mounting the Ballasts on the Pipe

Since the process of heat fusing a new pipe section on a string of pipe usually takes less time than the attaching of ballasts, the later procedure can be quickened by increasing the number of work stations. It is also helpful to stockpile the ballasts adjacent to each work station. Adequate lift equipment needs to be on hand to move the ballasts from the stockpile to the pipe location and to lift the pipe to allow the ballasts to be positioned under it. This equipment can also be used to lift and pull the pipe into the water. A suitable ramp or skidway should be provided to move weighted pipe into the water with a minimum of drag. (See discussion on launching the pipeline.)

For mounting ballasts on the floating pipe it is necessary to have low-profile equipment such as a barge or raft that is of sufficient size to accommodate the required lifting equipment and to carry sufficient ballasts to allow for efficient operation. In this method the barge is brought alongside the floating pipe, the pipe is lifted to install one or more ballasts, and after their installation the pipe is returned to the water and a new section is moved onto the barge or the barge is advanced along the floating string of pipe. In either case, the working surface or platform of the barge should be as close as possible to the water to reduce the need for a high lifting of the weighted pipe.

The steps involved in the mounting of ballasts include the following:

1. The placing of the protective/friction inducing material around the pipe. This can be done by first placing a pad over the lower half of the ballast and then placing a similar pad over the top of the pipe before the upper half of the ballast is lowered into position.

2. Lifting the pipe and positioning the lower half of the ballast under the pipe

3. Lowering the pipe so that it sits in the lower half of the ballast

4. Positioning and then lowering the upper half of the ballast so it sits on top of the pipe

5. Applying the strapping or tightening the bolts so that the ballasts are held fast to the pipe. (Note: before submersion, retightening of the bolts may be necessary to overcome any loss of gripping that may result from the stress-relaxation effect).

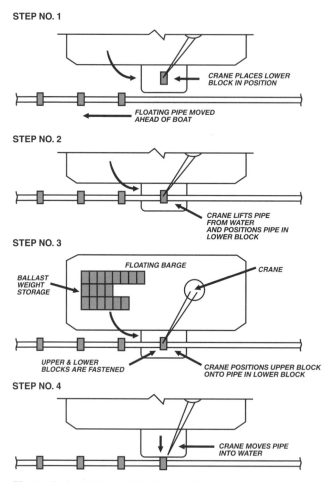

Figure 4   Installation of ballast weights from a raft or barge

Step 8 Launching the Pipeline into the Water

As previously cautioned, pipe that is launched into the water needs to have its ends closed, or its outlets located sufficiently high above the water, to prevent any water from entering the pipe. When the pipe is launched in the form of shorter strings of pipe that will later be joined to each other to produce the required overall length of submerged pipe, each separate section needs to have both ends sealed to prevent water from entering.

Effluent outfalls usually terminate in one or more diffuser sections. Diffusers can be of different designs such as a "Y" or "T" outlet, a pipe length in which holes have been drilled on top of the pipe within 10 and 2 o'clock, or a pipe length onto which vertical risers consisting of short sections of smaller diameter PE pipe have been fused. Diffusers are often designed for connection to the pipe by means of flange assemblies. The connection can be made prior to launching, or by divers after the pipeline has been submerged. When a diffuser is attached prior to launching, it is necessary to float the diffuser higher up over the water by means of some additional buoyancy. This is necessary to prevent water from entering the pipe through the diffuser openings. This additional buoyancy is released as the pipe is sunk into position.

Extreme care should be taken in the submersion of a marine line with an engineered diffuser attached to the pipeline which is being sunk in place. The sinking process can create considerable stresses on the fittings that may be inherent to the design of the diffuser itself such as flanges, tees and/or other mechanical connections. A preferred method when placing a highly engineered diffuser into an HDPE marine pipeline is to first sink the flanged effluent pipe and then submerge the diffuser separately in easily controlled segments which may be connected to the main effluent pipe underwater using qualified diving contractors.

A pipe end that does not terminate in a diffuser section is best closed against entering water by attaching a blind flange assembly. The flange assembly consists of a PE stub end that is butt fused to the pipe end on which has been bolted a slip-on metal flange. A number of required tapped holes are drilled on the blind flange so as to allow for the installation of valves and other fittings required to control the sinking operation. (See the section on submersion of the pipeline.)

Figure 5 Unballasted PE Pipeline Being Floated Out to Marine Construction Barge Where Ballast Weights are Installed

Pipe with attached ballast weights should be moved into the water by means of a ramp or skidway arrangement that allows the ballasts to move easily into the water without hanging up on the ground. The ramp or skidway must extend sufficiently into the water so that when the pipe leaves this device the ballast weight is fully supported by the floating pipe. Pipe without ballast weights may be moved over the ground provided it is free of rocks, debris or any other material that may damage the pipe. When this is not practical, wooden dunnage or wooden rollers may be placed between the pipe and the ground surface.

The pipe should be moved using suitable equipment. The pipe may be moved by lifting and then pulling it using one piece of equipment while using another piece of equipment to simultaneously push the pipe from its inboard end. PE pipe should only be lifted using wide-band nylon slings, spreader slings with rope or band slings, or any other means that avoids the development of concentrated point loading. Under no conditions should the flange assemblies be used to pull the pipe.

Prior to the launching of the pipe into the water, a strategy should be worked out to control the floating pipeline as it moves into the water and to store it away from navigational traffic until such time as the entire length is ready for submerging. For this purpose, suitable marine equipment – such as boats that have adequate tugging power and maneuverability – may need to be on hand. Other means for controlling the pipe can be a system of heavy block anchors that are positioned on either side of the proposed site into which the pipe will be submerged. In the case of river crossings, a system of guide cables that are anchored on the opposite shore can serve to control the position of the pipeline, particularly when the pipeline is subject to strong river flow.

In the case of river crossings when navigational traffic prohibits the float-and sink procedure, a "bottom-pull" procedure, illustrated in Figure 6, has been successfully used. When using this procedure, only sufficient ballast is added to the pipe to ensure that the pipe follows the river bottom as it is winched from one shore to the other. After the completion of the "bottom-pull," additional ballast can added or the pipeline can be adequately backfilled to produce the required anchoring and to offset any lift that may be created by currents or river flow.

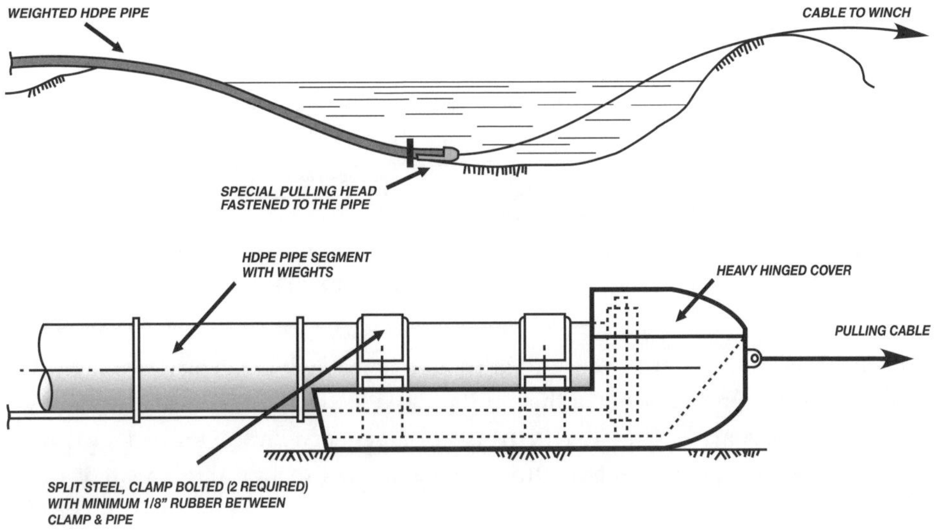

Figure 6 "Bottom-Pull" Installation of PE Pipe

Step 9 Submersion of the Pipeline Using the Float-and-Sink Method

To prepare the pipe for submersion, it is first accurately positioned over its intended location. The sinking operation basically consists of the controlled addition of water from the on-shore end of the pipe and the release of the entrapped air from the opposite end. The sinking is conducted so that it starts at the shore where the pipe enters the body of water and then gradually progresses into deeper waters. To achieve this, an air pocket is induced by lifting the floating pipe close to the shore. As the water is allowed to enter the pipe from the shore side, the added weight causes this initial air pocket to move outward and the intermediate section of pipe between the air pocket and the shore end to sink. As additional water is added, this pocket moves to deeper waters causing the sinking to progress to its terminal point in the body of water. This controlled rate of submersion minimizes pipe bending and it allows the pipeline to adjust and conform to the bottom profile so that it is evenly supported along its entire length (See Figure 7).

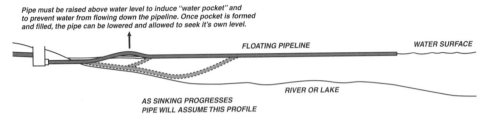

Figure 7 An induced water pocket initiates the submersion of the pipe and, as the pocket enlarges, it allows the submerging to gradually progress forward

A potential risk during the submersion operation is that, when the pipe sinking occurs too quickly, the bending of the pipe between the water-filled and air-filled portions may be sharp enough to risk the development of a kink, a form of localized pipe buckling. As a pipe is bent, its circumferential cross-section at the point of bending becomes increasingly ovalized. This ovalization reduces the pipe's bending moment of inertia, thus decreasing the bending force. Upon sufficient ovalization, a hinge or kink can form at the point of maximum bending an event that also leads to a sudden reduction of the bending force. Since the formation of a kink impedes the submersion process and can also compromise the pipe's flow capacity and structural integrity – in particular, the pipe's resistance to collapse under external pressure – it is essential that during submersion the bending of the pipeline be limited to an extent that will not risk the formation of a localized kink. The pipe bending radius at which buckling is in risk is given by the following expression:

(7)
$$R_b = D_o \frac{(DR-1)}{1.12}$$

WHERE
R_b = bending radius at which buckling can be initiated, in
D_o = outside pipe diameter, in
DR = pipe diameter ratio = average outside diameter divided by minimum wall thickness, dimensionless

Janson's relationship for determination of minimum buckling radius (Eq. 7) was derived on the basis of a maximum pipe deflection (ovalization) due to bending of the pipe of 7% and a maximum strain limit in the pipe wall of 5%. In actuality, the short term strain limit for modern polyethylene pipe materials is somewhat higher, on the order of 7-10%. Further, we know that these pipe materials are capable of long-term service at higher degrees of ovalization in buried pipe installations. (Please see the chapter on Pipeline Design.) As a result, the values presented in Table 3 are considered conservative guidelines for the short-term bending radius of polyethylene pipe during submersion of most marine pipelines. The designer may want to utilize a higher minimum bending radius to compensate for additional factors such as extremely strong currents, tidal activity, prevailing marine traffic, frequency of ballast placement, or other installation variables associated with a specific installation.

TABLE 3
Pipe Diameter Multipliers for the Determining of Minimum Bending Radii

Pipe DR	Multiplier*
11	8.9
13.5	11.2
17	14.3
21	17.8
26	22.3
32.5	28.1

* The minimum buckling radius of a pipe, in inches, is equal to the pipe's outside diameter, in inches, times the listed multiplier

It is essential that the water be introduced into the pipe at a controlled rate. This is done to ensure that the submersion process occurs at a rate that does not result in excessive localized pipe bending that could buckle the pipe. It also allows the pipe to settle properly on the bottom – thus avoiding any bridging between high spots which may make the pipe more vulnerable to movement when subjected to strong currents. Experience has shown that submerging the pipe at a rate in the range of about 800 to 1500 feet per hour has been found to be adequate for most cases. While the pipe is in the bent condition, long stoppage of the submersion procedure must be avoided. Consult with the pipe manufacturer and design engineer for specific submersion techniques for individual installations.

The risk of buckling can be minimized by applying a suitable pulling force during the submerging, such as illustrated by Figure 8.

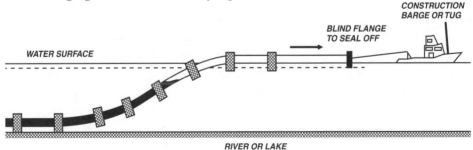

Figure 8   Pulling the pipe during submersion is a means for avoiding excessive bending that could risk buckling of the pipe

As water is being added at the shore-end of the pipe, air must be allowed to escape from the opposite end. In the case of outfall pipelines that terminate in one or more diffuser sections, the air is released through the diffuser outlets. When a pre-attached diffuser is used, it is necessary to support it with some additional buoyancy as a precaution against the water entering the pipe and causing that section of the pipeline to sink prematurely. Extreme care should be taken in the ballasting and submersion of elaborate diffuser systems that are sunk in concert with the main effluent pipe as the submersion process can create significant stresses on the tees, elbows or other fittings used in the design of the diffuser system. The preferred method is to submerge the flange or valved main effluent pipe and the diffuser separately and join the two sections underwater using qualified diving contractors.

When the end of a pipe that is being submerged terminates with a flange connection, air release can best be accomplished by installing a valved outlet in the blind flange outlet. To ensure that water will not enter through this outlet, a length of hose may be connected to the outlet, and the free end is held above water on a boat or by means of a float. After the completion of the submersion, a diver can remove the hose.

Should a problem be encountered during the submersion, the availability of a valved outlet on the outboard end of the pipeline allows the sinking procedure to be reversed. Compressed air can be pumped into the submerged line to push the water out and thus allow the line to be raised. Because compressed air packs a lot of potential energy – which, when suddenly released through a failure of a piping component, could present a serious safety hazard – the rule of thumb is to limit air pressure to not more than one-half the pipe's pressure rating for water.

Under certain methods, such as the bottom-pull method that is described above, the necessary ballast to offset floatation during the installation of a water filled PE pipe can be of a temporary nature – for example, steel reinforcing bars that are strapped on the outside of the pipe. This temporary ballast can be removed after the installation of permanent anchoring. Permanent anchoring can consist of an appropriate quantity of stable backfill that is placed on pipe that has been installed in a trench, or it can consist of tie-down straps that are installed by augering or other procedures that result in the permanent anchoring of the pipeline. However, when considering an alternate means for anchoring a pipeline, it should be kept in mind that, as discussed earlier, a pipeline lying on the sea or river floor is subject to greater lift action by currents or waves than a pipeline that lies even a short distance above the bottom.

Step 10 Completing the Construction of the Land-to-Water Transition

After the pipeline has been submerged, the portion of the pipeline that has been lowered into a land-to-water transition trench should be backfilled with specified material and to the required depth of cover.

Post-Installation Survey

Upon completion of the installation of a submerged pipeline, it is advisable to have the complete line surveyed by a competent diver to ensure that:

- The pipeline is located within the prescribed right-of-way
- The ballasts holding the pipeline are all properly sitting on the bottom contour and that the line is not forced to bridge any changes in elevation
- The pipe is not resting on any rocks, debris or material that could cause damage
- Any auxiliary lines, such as hoses, ropes, buoyancy blocks or any other equipment used during the installation has been removed
- Where required, the pipe has been backfilled and the backfilling was done properly
- All other installation requirements established by the designer for the subject application have been complied with, as established by the designer

Other Kinds of Marine Installations

Because of its flexibility, light-weight and toughness PE piping has also emerged as a choice material for other types of marine applications. The basic design and installation principles described above for the "float-and-sink" method are, with some modifications, also valid for other types of marine applications. A brief description of some other kinds of marine applications is presented in the paragraphs that follow.

Winter Installations

Where ice conditions permit, PE pipe may be submerged from the surface of a frozen lake or river. After a long pipe length is assembled by means of heat fusion it can be easily pulled alongside the right-of-way. The heat fusion process needs to be performed in an adequately heated tent, or other shelter, to ensure fusion joint quality. Once the heat fusion has been completed, the ballast weights can be mounted. An ice trench is then cut with a saw, the ice blocks are moved out of the way and the pipeline is pushed into the trench. The submersion is carried out in accordance with the procedure previously described.

Installations in Marshy Soils

Installation of pipe in marshy or swampy soils represents one of the most demanding applications for any design engineer. Generally, marshy soils do not provide the firm and stable foundation that is required by rigid, more strain sensitive traditional piping materials.

Due to its flexibility and butt fusion joining technology, PE piping can readily adapt itself to shifting and uneven support without sacrifice of joint integrity. As soil conditions vary, the PE pipe can accommodate these irregularities by movement within the fluid-like soil envelope. Of course, care must be taken to consider any line, grade or external hydrostatic design requirements of the pipeline based on the operating conditions of the system. However, with these design aspects in mind, it is possible to utilize the engineering features of PE pipe to design a cost-effective and stable piping system that can provide years of satisfactory service in this highly variable environment.

In certain situations, the high water table that is characteristic of these soils can result in significant buoyant forces that may raise the pipe from the trench in which it has been installed. When this possibility presents itself, a ballast system may be designed using the same guidelines presented in this chapter which can prevent or minimize pipe flotation.

Water Aeration Systems

Smaller diameter submerged PE pipe, with small holes drilled into the top of the pipe has been used for the de-icing of marinas. Compressed air that bubbles out of these pipes raises warmer water that melts ice that forms on the water surface. When the system is operating, the submerged pipe is full of air, and the ballast weight design should be adequate to prevent the line from floating. Ballast also needs to be spaced frequently enough to minimize the upward deflection that results from the buoyancy force.

Dredging

PE piping is a natural choice for use in marine dredging operations. Its flexibility, combined with its light weight, buoyant nature and overall durability, provides for a piping material which has been successfully used for years in the demanding rigors of dredging operations. Generally, these types of applications require that the HDPE pipe be fused into manageable lengths that can be easily maneuvered within the dredge site. These individual lengths are then mechanically joined together using flanges or quick-connect type fittings to create a pipeline structure of suitable length for the project. As the dredge operation proceeds, pipe segments may be added or removed to allow for optimum transport of the dredge material.

Dredging operations can vary significantly in type of slurry, scale or operation and overall design. As such, a detailed analysis of dredge design using HDPE pipe is beyond the scope of this writing. However, the reader should note that as the particulate size and nature varies from project to project, it is possible to ballast the pipe so that it still floats and can be managed from the surface using tow boats or booms. This is accomplished by analysis of the composition of the dredge material and the design and attachment of suitable floats to the HDPE discharge or transport pipe.

Temporary Floating Lines

PE piping has also been used for temporary crossings of rivers and lakes. Its natural buoyancy allows a PE pipeline to float on or near the water surface. The principal design and installation requirement for floating line applications is to work out a system to maintain the pipe in its intended location when it is subject to currents, winds and wave action. To this end, cable restraints are generally used. The cables need to hold the pipe by means of stable collars that do not slip along the axis of the pipe and that cause no damage to the pipe material.

Conclusion

Modern HDPE piping materials are a natural choice for marine installations. The overall durability and toughness of these products, combined with the innovative and cost-effective installation methods that they facilitate, are compelling reasons for their use in effluent discharge systems, water intake structures and potable water or sanitary sewer force main marine crossings, as well as more temporary marine systems such as dredging operations.

The dependable butt fusion system of joining PE pipe, supplemented by the availability of a wide array of mechanical fittings, means that the design engineer has an abundance of tools available by which to design a leak-free piping system that lends itself to the most demanding marine installations. This same system of joining allows for the cost-effective installation of long lengths of pipe via the float and sink method, directional drilling or pull-in-place techniques. Utilizing the unique features of the PE piping system allows the designer to investigate installation methods that minimize the necessity of costly pipe construction barges or other specialized equipment. These same installation techniques may minimize the economic impact associated with marine traffic disruption.

This chapter provides an overall design perspective for some of the more typical applications of HDPE pipe in marine environments. Its intent is to provide the designer with a basic understanding of the utility that PE pipe brings to the designer of these challenging installations. More elaborate design investigation and methodology may be required depending on the specifics of the project under consideration. However, through a basic understanding of the benefits of PE pipe in marine installations and a fundamental understanding of the installation flexibility that they provide, it can be seen that PE pipe systems are a proven choice for modern, durable marine piping structures.

References

1. Janson, Lars-Eric. (1990, Sept 10-13). Review of Design and Experience With Thermoplastic Outfall Piping, ASTM STP 1093, Compendium of papers presented at the Buried Plastic Pipe Technology Symposium, Dallas, TX.
2. Berndtson, B. (1992, Sept 21-24). Polyethylene Submarine Outfall Lines, Paper presented at Plastics Pipes VIII, a symposium held in Koningshof, the Netherlands, and included in the Book of Proceedings for this symposium as published by The Plastics and Rubber Institute, London.
3. Janson, Lars-Eric. (1986). The Utilization of Plastic Pipe for Submarine Outfalls—State of The Art, *Water Science and Technology*, Great Britain, Vol. 18, No. 11, pp 171-176.
4. Janson, Lars-Eric. (1996). *Plastics Pipe for Water Supply and Sewage Disposal*, published by Borealis, Sven axelsson AB/Affisch & Reklamtryck AB, Boras, Sewede.

Appendix A-1

Derivation of the Equation for the Determining of the Buoyant Force Acting on a Submerged PE Pipe (Equation 2 in the Text)

The first bracketed term in Equation 2, namely $[0.00054 D_0^2 P_w]$, is one commonly used form of the formula for obtaining a numerical value for the term W_{DW} in Equation 1, the weight of water that is displaced by the submerged PE pipe. This displaced weight is equivalent to the lift force acting on a submerged pipe that has an infinitely thin wall and that is completely filled with air. The sum of the three terms within the second set of brackets expresses the reduction of this potential lift force in consequence of the weight of the pipe (the first term) and that of its contents (the second term). As is evident from inspection of Equation 2, the extent to which the inner volume of a pipe is occupied by air (represented by the fraction R) exerts the more significant effect on resultant pipe buoyancy. Since a decrease in pipe DR (i.e., an increase in pipe wall thickness) results in a decrease in potential air volume space, a lower DR tends to reduce the potential buoyancy that can result from air filling.

1. The net buoyant (upward acting force) acting on a submerged PE pipe is:

(1) $$F_B = [W_p + W_c] - W_{DW}$$

WHERE

F_B = buoyant force, lbs/foot of pipe
W_p = weight of pipe, lbs/foot of pipe
W_c = weight of pipe contents, lbs/foot of pipe
W_{DW} = weight of the water displaced by the pipe, lbs/foot of pipe

2. W_p, the weight of pipe is:

$$W_p = V_p P_p$$

WHERE
V_p = volume occupied by pipe material per foot of pipe
P_p = density of pipe material, lbs/ cu. ft

Since
$$V_p = \frac{\pi}{144} D_m t_a$$

WHERE
D_m = mean pipe diameter of the pipe, in
t_a = average wall thickness, in

And since
$$DR = \frac{D_o}{t_m}$$

WHERE
D_o = outside pipe diameter, in
t_m = minimum wall thickness, in

Then, by assuming that the average wall thickness (t_a) is 6% larger than the minimum (t_m), it can be shown that:

(2) $$W_p = \frac{1.06\pi}{144}\left(\frac{D_o}{DR}\right)^2 (DR - 1.06)\rho_p$$

3. W_c, the weight of the pipe contents is equal to the volume occupied by the liquid inside the pipe times the density of the liquid:
$$W_c = V_L \rho_L$$

WHERE
V_L = the volume occupied by the liquid, cu ft/linear ft
P_L = the density of the liquid inside the pipe, lbs/cu ft

If the fraction of the inside volume of the pipe (V_I) is expressed as R and as the formula for the inside volume is as follows:

$$V_I = \frac{\pi D_I^2}{4} \frac{1}{144}$$

WHERE
D_I = inside diameter of the pipe, in

And also, since $D_I = D_o - 2t_a$ (where t_a is 1.06 t_m as previously assumed) it can then be shown that:

(3) $$Wc = \frac{\pi}{144}\frac{\rho_L}{4}\left[D_o\left(1 - \frac{2.12}{DR}\right)\right]^2 (1 - R)$$

4. W_{DW}, the weight of the water displaced by the pipe is determined by means of the following formula:

(4)
$$W_{DW} = \frac{\pi D_o^2}{4} \cdot \frac{1}{144} \rho_W$$

WHERE
ρ_W = the density of the displaced water, lbs/cu ft

5. By substituting Equations 2, 3 and 4 into Equation 1, and by simplifying the resultant relationship, the following formula (Equation 2 in the text) is obtained:

$$F_B = \left[0.0054 D_o^2 \rho_W\right]\left[4.24\frac{(DR-1.06)}{(DR)^2}\frac{\rho_p}{\rho_w} + \left(1-\frac{2.12}{DR}\right)^2 (1-R)\frac{\rho_c}{\rho_w} - 1\right]$$

Appendix A-2

Water Forces Acting on Submerged PE Piping

The following is a brief introduction to the technology for the estimating of the magnitude of the lateral forces that can act on a submerged pipe in consequence of water currents and wave action. As this technology is relatively complex and it is still emerging, the objective of this introduction is to provide basic information and references that can provide initial guidance for the proper design of PE piping for submerged applications. It is the responsibility of the designer to determine the design requirements and appropriate design protocol for the specific anticipated conditions and design requirements of a particular project. In addition to the information and references herein provided, the reader should consult the technical staff of PPI member companies for further information, including references to engineering companies that have experience in the use of PE piping for submerged applications.

Submerged pipes can be subject to lateral forces generated by currents or by wave action. A principal design objective is to ensure that the resultant lateral forces do not subject the pipe to excessive deflection, nor to fiber stresses or strains that could challenge the pipe material's capabilities. Thus, the capacity to estimate with some reasonable accuracy the potential maximum lateral stresses to which a submerged pipe may be subjected is an important element for achieving a successful design.

Currents impinging on a submerged pipe can cause two principal forces: a drag force in the direction of the current; and a vertical force at right angles to the drag force. The magnitude of these forces depends on the angle between the direction of the current flow and the pipe. They are at their maximum when the current flow is at a right angle to the pipe. As this angle (Θ) is reduced, the resultant force is reduced by $\sin^2 \Theta$.

For the purpose of estimating the drag and lift forces that a current can exert on a submerged pipe, Janson developed the graphical solution that is herein reproduced as Figure A-2-1. This graph is applicable to the condition where the current velocity, expressed in feet per second, times the pipe diameter, expressed in feet, is equals to or is greater than 0.5 m²/sec (2.7 ft²/sec).

Janson's nomagraph is based on the assumption that certain design variables are known. These design variables are as follows:

D = external diameter of pipe, in meters (feet)
l = distance from the bottom, in meters (feet)
u_m = mean velocity of water, in m/sec (ft/sec)
h = depth of water, in meters (feet)
k = hydraulic roughness of the water bed, meters (feet)
Θ = angle between the direction of the current and that of the pipe, degrees
λ = ratio of l/h, dimensionless
= 0 for pipe placed on seafloor or bed of body of water

Janson determined that for values of D x U_m > 0.50 m²/sec, a nomograph could be constructed which allowed for a relatively quick approximation of the drag and/or lift forces for which an underwater HDPE piping installation must be designed.

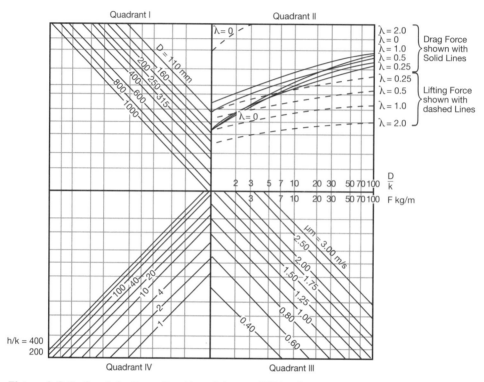

Figure A-2-1 Graph for the estimation of drag and lifting forces on underwater pipes when the flow rate of the current times the pipe diameter is 0.5 m²/sec, or greater [4]

Consider the following example:
A 315 mm HDPE pipe is to be placed directly on the floor of a body of water that is flowing at approximately 3 m/sec and at 90 degrees to longitudinal axis of the pipe. The depth of the water is 10 meters and the pipe will be placed directly on a bed of gravel for which we will assume a hydraulic roughness of 10 cm.

Step 1 First, check to see if the nomograph is applicable

$D \times u_m = 0.315 \text{ m} \times 3 \text{ m/sec} = 0.96 \text{ m}^2/\text{sec}$

So, the nomograph can be utilized.

Step 2 Determine the two key dimensionless design ratios, D/k and h/k

GIVEN THAT
D = 315 mm = 0.315 meter
and
k = 10 mm = 0.10 meter

Then
D/k = 0.315 m / 0.10 m = 3.2
h/k = 10 m / 0.10 m = 100

Step 3 Determine the Drag Force
Utilizing the nomograph in Figure A-2-1, start at the horizontal axis between quadrant II and III. On the D/k axis locate the point 3.2 from the calculation in step 2. Draw a line vertically up to the solid curve (drag force) for $\lambda = 0$ (the pipe will rest on the bed of the body of water). Now draw a horizontal line from quadrant II into quadrant I to the line for diameter, in this case 315 mm. At the point of intersection with this line, draw another line downward to the line for h/k = 100 shown in quadrant IV. At that point of intersection, then draw another line horizontally back across to quadrant III to the line for flow velocity, in this example 3m/sec. From this point draw a line upward to the original axis and read drag velocity directly from nomograph. The result is 20 kg/m.

Step 4 Determine the Lift Force
Generally speaking, the lift force for a pipe laying on the floor of a body of water is eight times that of the drag force. In this case, the lift force generated is approximately 160 kg/m.

Alternatively, the lift force could have also been approximated from the nomograph by starting on the same axis between quadrant II and III and proceeding up to the dashed line for $\lambda = 0$ in quadrant II. The dashed line represent the curves for lift force relationships. From the intercept with the dashed curve for $\lambda = 0$, the procedure of is the same as that described for determination of the drag force from the nomograph.

Consider another example:
Now, using the scenario outlined in the preceding example, assume that the pipe is oriented in the water such that the angel of impact, θ, is 60 degrees.

Solution:

The revised angle of impact suggest that the drag force may be reduced by a factor, $\sin^2\theta$.

$\sin^2\theta = \sin^2 \text{pt } 60° = 0.75$.

Using this, we get a net drag force as follows:

Drag Force$_{(90)}$ x $\sin^2\theta$ = 20 kg/m x 0.75 = 15 kg/m

English Units

Janson's nomograph was originally published in metric units. However, the curves presented in quadrants II and IV are dimensionless. By converting quadrants I and III and the horizontal axis to English units then the nomograph may be used for pipe sized and installed accordingly. For ease of reference, Janson's nomograph is recreated using English units in figure A-2-2 below.

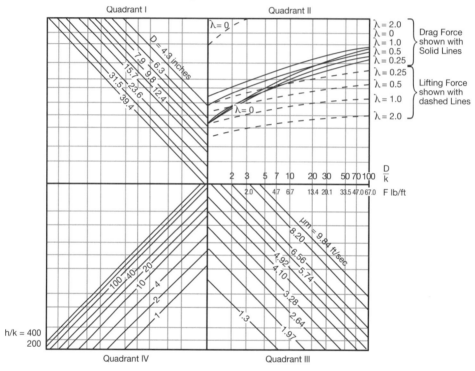

Figure A-2-2 Graph for the estimation of drag and lifting forces on underwater pipes when the flow rate of the current times the pipe diameter is 2.7 ft²/sec, or greater

Consider the previous example restated in English units

A 12" IPS HDPE (325 mm) pipe is to be placed directly on the floor of a body of water that is flowing at approximately 9.8 ft/sec (3 m/sec) and at 90 degrees to longitudinal axis of the pipe. The depth of the water is 33 feet (10 meters) and the pipe will be placed directly on a bed of gravel for which we will assume a hydraulic roughness of 4 inches (10 cm).

Step 1 First, check to see if the nomograph is applicable

$$D \times u_m = 1 \text{ ft} \times 9.8 \text{ ft/sec} = 9.8 \text{ ft}^2/\text{sec} = 0.91 \text{ m}^2/\text{sec} > 0.50 \text{ m}^2/\text{sec}$$

So, the nomograph can be utilized.

Step 2 Determine the two key dimensionless design ratios, D/k and h/k

GIVEN THAT
D = 12.75 inches = 1.06 foot
and
k = 4 inches = 0.33 foot
Then
D/k = 1.06/0.33 = 3.2
h/k = 33/ 0.33 m = 100

Step 3 Determine the Drag Force
Utilizing the English version of the nomograph in Figure A-2-2, start at the horizontal axis between quadrant II and III. On the D/k axis locate the point 3.1 from the calculation in step 2. Draw a line vertically up to the solid curve (drag force) for $\lambda = 0$ (the pipe will rest on the bed of the body of water). Now draw a horizontal line from quadrant II into quadrant I to the line for diameter, in this case 12 inch. At the point of intersection with this line, draw another line downward to the line for
h/k = 100 shown in quadrant IV. At that point of intersection, then draw another line horizontally back across to quadrant III to the line for flow velocity, in this example 9.8 ft/sec. From this point draw a line upward to the original axis and read drag velocity directly from nomograph. The result is 13.5 lbf/ft. The reader should keep in mind that this is only an approximation and is not intended to displace a more detailed engineering analysis of a specific marine installation design.

Step 4 Determine the Lift Force
As with the previous example, the lift force for a pipe laying on the floor of a body of water is eight times that of the drag force. In this case, the lift force generated is approximately 108 lbf/ft.

The lift force may be approximated from Figure A-2-2 by starting on the same axis between quadrant II and III and proceeding up to the dashed line for $\lambda = 0$ in quadrant II. The dashed line represent the curves for lift force relationships. From the intercept with the dashed curve for $\lambda = 0$, the procedure of is the same as that described for determination of the drag force from the nomograph.

APPENDIX A-3

Some Designs of Concrete Ballasts

Concrete ballast designs may take on a variety of different sizes, shapes and configurations depending on job-site needs, installation approach and/or availability of production materials. Table A-3-1 below provides some typical designs for concrete ballasts and details some suggested dimensional considerations based on pipe size, density of unreinforced concrete at 144 lb/ft³ and per cent air entrapment in a typical underwater installation. The reader is advised to consider these dimensions and weights for reference purposes only after a careful analysis of the proposed underwater installation in accordance with the guidelines presented in this chapter.

TABLE A-3-1
Suggested Concrete Weight Dimensions (All dimensions in inches)

Nominal Pipe Size	Mean Outside Diameter (inches)	Spacing of Weights To Offset % Air (feet)			Approx. Weight of Concrete Block (pounds)		Approximate Block Dimensions (inches)						Bolt Dimensions (inches)	
		10%	15%	20%	In Air	In Water	"D"	"X"	"Y"	"T"	"S" (min)	"W"	Dia.	Length
3 IPS	3.50	10	6 ¾	5	12	7	4	9	3 ¾	2 ½	1 ½	2 ½	¾	12
4 IPS	4.50	10	6 ¾	5	20	10	5	11	4 ¾	2 ½	1 ½	3	¾	12
5 IPS	5.56	10	6 ¾	5	30	18	6	12	5 ¼	3 ½	1 ½	3	¾	12
6 IPS	6.63	10	6 ¾	5	35	20	7 1/8	13	5 ¾	3 ½	1 ½	3	¾	12
7 IPS	7.13	10	6 ¾	5	45	26	7 5/8	13 ½	6	4 ¼	1 ½	3	¾	12
8 IPS	8.63	10	6 ¾	5	55	30	9 ¼	15 ¼	6 7/8	4 ¼	1 ½	3	¾	12
10 IPS	10.75	10	6 ¾	5	95	55	11 ¾	19 ¼	8 5/8	4 ½	2	4	¾	12
12 IPS	12.75	10	6 ¾	5	125	75	13 ¼	21 ¼	9 5/8	5	2	4	¾	13
13 IPS	13.38	10	6 ¾	5	175	100	13 7/8	24	11	5 ¼	2	5	¾	13
14 IPS	14.00	15	10	7 ½	225	130	14 ½	24 ½	11 ¼	6 ½	2	5	1	13
16 IPS	16.00	15	10	7 ½	250	145	16 ½	26 ½	12 ¼	6 ½	2	5	1	13
18 IPS	18.00	15	10	7 ½	360	210	18 ½	28 ½	13 ¼	8 ¼	2	5	1	13
20 IPS	20.00	15	10	7 ½	400	235	20 ½	30 ½	14 ¼	8 ¼	2	6	1	13
22 IPS	22.00	15	10	7 ½	535	310	22 ½	34 ½	16 ¼	8 ½	2	6	1	13
24 IPS	24.00	15	13 ½	7 ½	610	360	24 ½	36 ½	17 ¼	8 ¾	2	6	1	13
28 IPS	28.00	20	13 ½	10	900	520	28 ½	40 ¼	19 ¼	11 ¼	2	6	1	13
32 M	31.59	20	13 ½	10	1140	660	32	44	21	12 ¼	2	6	1	13
36 IPS	36.00	20	13 ½	10	1430	830	36 ½	48 ½	23 ¼	13 ½	2	6	1	13
40 M	39.47	20	13 ½	10	1770	1020	40 1/8	52	25	15 ¼	2	6	1	13
42 IPS	42.00	20	13 ½	10	1925	1125	42 ½	54 ½	26 ¼	15	2	6	1	13
48 IPS	47.38	20	13 ½	10	2500	1460	48 ¼	60 ¼	29 1/8	17	2	6	1 1/8	13
55 M	55.30	20	13 ½	10	3390	1980	55 ¾	68	33	18 ¾	2	6 1/8	1 1/8	15
63 M	63.21	20	13 ½	10	4450	2600	63 ¾	78	38	18 ½	2	7 1/8	1 1/8	15

Notes to Table A-3-1

1. Suggested underpad material: 1/8" black or red rubber sheet, 1/4" neoprene sponge padding width to be "T"+ 2" minimum to prevent concrete from contacting pipe surface.
2. Concrete interior surface should be smooth (3000 psi – 28 days).
3. Steele pipe sleeves may be used around the anchor bolts (1" for 3/4" bolt, etc.). Hot dip galvanize bolts, nuts, washers and sleeves.

4. A minimum gap, "S", between mating blocks **must** be maintained to allow for tightening on the pipe.
5. To maintain their structural strength some weights are more than the required minimum.
6. Additional weight may be required for tide or current conditions.
7. Weights calculated for fresh water.
8. All concrete blocks should be suitably reinforced with reinforcing rod to prevent cracking during handling, tightening, and movement of weighted pipe.
9. See Table II for alternative weight design and suggested reinforcement for use with 28" to 48" HDPE pipe.

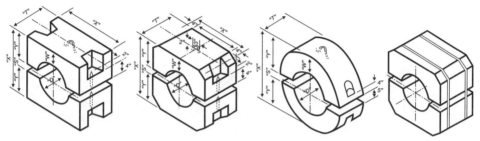

Figure A-3-1 Schematics of Concrete Ballast Designs

TABLE A-3-2
Suggested Dimensions and Reinforcing for Bottom-heavy Concrete Weights (For Extra Stability)
All dimensions in inches

Nominal Pipe Size	Mean Outside Diameter (inches)	Spacing of Weights To Offset % Air (feet)			Approx. Weight of Concrete Block (pounds)		Approximate Block Dimensions (inches)						Bolt Dimensions (inches)	
		10%	15%	20%	In Air	In Water	"D"	"X"	"Y"	"Z"	"R"	"T"	Dia.	Length
28 IPS	28.00	20	13 ½	10	900	520	28 ½	44	19 ½	26 ½	48	7 ½	1	54
32 M	31.59	20	13 ½	10	1140	660	32 1/8	48	21	28	51	8 ½	1	57
36 IPS	36.00	20	13 ½	10	1430	830	36	52	23	30 ½	55 ½	9 3/8	1	61 ½
40 M	39.47	20	13 ½	10	1770	1020	40 1/8	56	25	33	60	10 ¼	1	66
42 IPS	42.00	20	13 ½	10	1925	1125	42 ½	59	26 ½	34 ½	63	10	1 1/8	69
48 M	47.38	20	13 ½	10	2500	1460	48 ¼	64	29	39	70	11 ½	1 1/8	76
55 M	55.30	20	13 ½	10	3390	1980	55 ¾	72	33	43	78	12 ¾	1 1/8	84
63 M	63.21	20	13 ½	10	4450	2600	63 ¾	80	37	47	86	14 ½	1 1/8	92

Notes to Table A-3-2
1. Minimum cover of rebar to be 2 ½".
2. Rebar to be rail steel or equivalent.
3. Anchor bolt material to be ASTM A307.
4. It may be desirable to increase the amount of reinforcing used in the 55" and 63" pipe weights.
5. See recommended bore detail below.

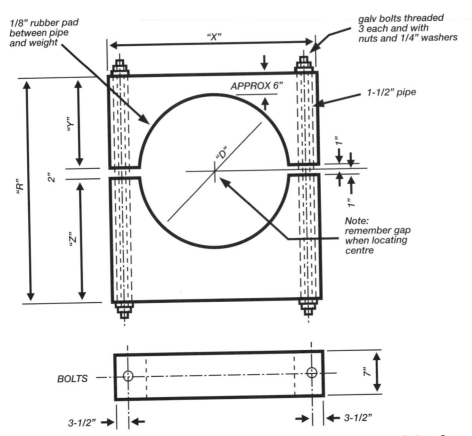

Figure A-3-2 Typical Detail of Concrete Ballast Showing 1-inch Gap Between Ballast Sections

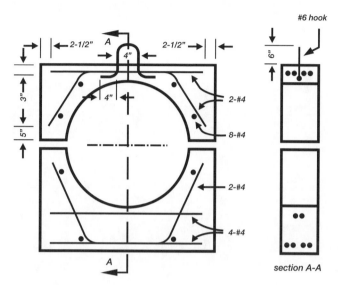

Figure A-3-2 Typical Rebar Detail in Concrete Ballast Design

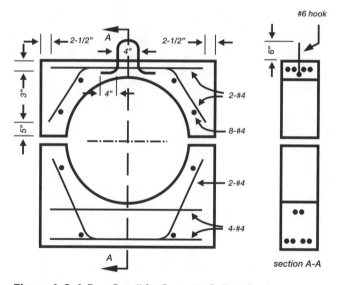

Figure A-3-4 Bore Detail for Concrete Ballast Design

Chapter 11

Pipeline Rehabilitation by Sliplining with Polyethylene Pipe

Introduction

An integral part of the infrastructure is the vast network of pipelines, conduits, and culverts in North America. These are among the assets we take for granted, since most are buried and we never see them. We do not see them deteriorate either, but we know that they do. Television inspection of the interiors of this structure often reveals misaligned pipe segments, leaking pipe joints, or other failing pipe integrity.

The effects of continued deterioration of a pipeline could be quite drastic and costly. A dilapidated gravity sewer system permits substantial infiltration of groundwater, which increases the volume of flow and reduces the available hydraulic capacity of the existing line. So the old pipeline often increases treatment and transportation costs for the intended flow stream[24]. Continued infiltration may also erode the soil envelope surrounding the pipe structure and cause eventual subsidence of the soil.

The case for positive-pressure pipelines is somewhat different, but the results are equally unacceptable. In this situation, continued leakage through the existing pipeline allows exfiltration of the contents of the flow stream that eventually leads to extensive property damage or water resource pollution. Also, in many cases, the contents of the flow stream are valuable enough that their loss through exfiltration becomes another economic factor. Butt fusion results in a monolithic pipe structure that is absent bell/spigot joints and therefore will not draw in contaminated ground water during vacuum conditions.

When the harmful results of pipeline deterioration become apparent, we must either find the most economical method that will restore the original function or abandon the function. Excavation and replacement of the deteriorating structure can prove prohibitively expensive and will also disrupt the service for which the original line is intended[18]. An alternate method for restoration is "sliplining" or "insertion renewal" with polyethylene pipe. More than 30 years of field experience shows that this is a proven cost-effective means that provides a new pipe structure with minimum disruption of service, surface traffic, or property damage that would be caused by extensive excavation.

The sliplining method involves accessing the deteriorated line at strategic points within the system and subsequently inserting polyethylene pipe lengths, joined into a continuous tube, throughout the existing pipe structure. This technique has been used to rehabilitate gravity sewers[11, 24], sanitary force mains, water mains, outfall lines, gas mains[2, 13], highway and drainage culverts[18], and other piping structures with extremely satisfactory results. It is equally appropriate for rehabilitating a drain culvert 40-feet long under a road or straight sewer line with manhole access as far as
1/2 mile apart. The technique has been used to restore pipe as small as 1-inch, and there are no apparent maximum pipe diameters.

Mechanical connections are used to connect PE pipe systems to each other and to connect PE pipe systems to other pipe materials and systems. The reader can refer to the Handbook chapter that is titled 'Polyethylene Joining Procedures' for additional information on Mechanical Connections and Mechanical Joint (MJ) Adapters.

Design Considerations

The engineering design procedure required for a sliplining project consists of five straightforward steps:

1. Select a pipe liner diameter.
2. Determine a liner wall thickness.
3. Determine the flow capacity.
4. Design necessary accesses such as terminal manholes, headwall service and transition connections.
5. Develop the contract documents.

Select a Pipe Liner Diameter

To attain a maximum flow capacity, select the largest feasible diameter for the pipe liner. This is limited by the size and condition of the original pipe through which it will be inserted. Sufficient clearance will be required during the sliplining process to insure trouble-free insertion, considering the grade and direction, the severity of any offset joints, and the structural integrity of the existing pipe system.

The selection of a polyethylene liner that has an outside diameter 10% less than the inside diameter of the pipe to be rehabilitated will generally serve two purposes. First, this size differential usually provides adequate clearance to accommodate the insertion process. Second, 75% to 100% or more of the original flow capacity may be maintained. A differential of less than 10% may provide adequate clearance in larger diameter piping structures. It is quite common to select a 5% to 10% differential for piping systems with greater than 24-inch diameters, assuming that the conditions of the existing pipe structure will permit insertion of the liner.

Determine a Liner Wall Thickness

Non-Pressure Pipe

In the majority of gravity pipeline liner projects, the principal load that will act on the polyethylene pipe is the hydrostatic load that is created when the water table rises above the crown (top) of the liner.

The generic Love's equation (Eq. 1) shows that the ability of a free-standing pipe to withstand external hydrostatic loading is essentially a function of the pipe wall moment of inertia and the apparent modulus of elasticity of the pipe material. The critical buckling pressure, P_c, for a specific pipe construction can be determined by using equation Eq. 1.

(1) Love's Equation

$$P_c = \frac{24EI}{(1-v^2) \times D_m^3} \times f$$

WHERE
P_c = Critical buckling pressure, psi
E = Apparent modulus of elasticity
    = 30,000 psi for HDPE at 73.4°F (23°C), 50 year loading
I = Pipe wall moment of inertia, in4/in
    = t3/12 for solid wall polyethylene
v = Poisson's ratio, 0.45 for polyethylene
D_m = Mean diameter, inches (inside diameter plus one wall thickness)
f = Ovality compensation factor, dimensionless (see Figure 10-1)
D = Pipe average diameter, in

where % Deflection = $\dfrac{D - D_{min}}{D} \times 100$

D_{min} = Pipe minimum diameter, in

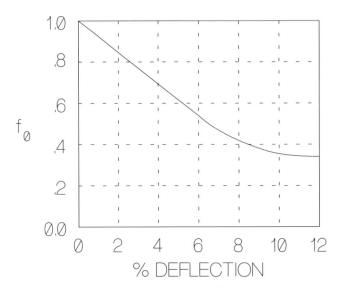

Figure 1 % Deflection vs. Ovality Correction Factor, f

To compute the buckling pressure of a dimension ratio (DR) series polyethylene pipe (i.e., a grouping of solid wall pipes of different diameters but with the same ratio of specified outside diameter to minimum wall thickness), the following variation of Love's equation[22], Eq. 2, is used.

(2) Love's Equation for DR Solid Wall Pipe

$$P_c = E \times \left(\dfrac{2}{1-v^2}\right) \times \left(\dfrac{1}{DR-1}\right)^3 \times f$$

WHERE
DR = Dimension ratio, dimensionless (OD/t)
OD = Actual outside diameter, inches
t = Minimum wall thickness, inches

The process of calculating the buckling resistance of a free-standing pipe is iterative in that, once the critical buckling resistance of a trial choice has been determined, it can be compared to the anticipated hydrostatic load. If the pipe's calculated buckling resistance is significantly larger than the anticipated hydrostatic loading, the procedure can be used to evaluate a lesser wall thickness (with the advantages of lighter weight materials and lower costs). The prudent practice is to select a design buckling resistance that provides an adequate safety factor (SF) over the maximum anticipated hydrostatic load.

(3) Safety Factor, SF

$$SF = \frac{P_c}{\text{Anticipated Hydrostatic Load}}$$

For an example of the calculations that can be made with Equations 2 and 3, consider a 22-inch DR 26 solid-wall polyethylene liner placed within a 24-inch clay tile pipe and subjected to a maximum excess hydrostatic load of 3 feet of water table.

1. Calculate the equivalent hydrostatic load in psi.
   Water load = 3 ft x 62.4 lb/ft^3 x 1 ft^2/144 in^2 = 1.3 psi

2. Calculate the critical buckling pressure, P$_c$, using Eq. 2 assuming a 50-year hydrostatic loading and 3% deflection. P$_c$ = 3.6 psi

3. Calculate the Safety Factor, SF, from Eq. 3 for this load assumption.
   SF = 3.6/1.3 = 2.8

A safety factor of 2.0 or greater is often used for frequent or long-term exposure to such loads. If a larger safety factor is preferred, repeat the procedure for a heavier wall configuration or consider the enhancement of the pipe's buckling strength by the effects of external restraint.

Love's equation assumes that the liner being subjected to the indicated hydrostatic load is free-standing and is not restrained by any external forces. Actually, the existing pipe structure serves to cradle the flexible liner, enhancing its collapse resistance. Maximum external reinforcement can be provided, where required, by placing a stable load-bearing material such as cement, fly ash, polyurethane foam, or low-density grout in the annular space between the liner and the existing pipe. Studies show that filling the annular cavity will enhance the collapse resistance of a polyethylene pipe by at least a four-fold factor and often considerably more, depending on the load-bearing capabilities of the particular fill material. Contact the pipe suppliers for additional information.

For solid wall pipe, the significant variable that determines adequate wall stiffness is the pipe DR. It is a simple matter to specify the DR once the amount of the loading on the pipe is determined. A typical manufacturer's recommendation for safe long-term (50-year) external pressure loading might follow the guidelines in the following table, which was derived according to the procedure shown in ASTM F585.

TABLE 1
Critical Height of Water above Pipe, No Grout vs Grout

DR	Height of Water (ft) Above Pipe (50 yrs), No Grout	Height of Water (ft) Above Pipe (50 yrs), Grout
32.5	2.0	10.0
26	4.0	20.0
21	8.0	40.0
17	16.0	80.0
13.5	32.0	160.0
11	64.0	320.0

The figures in this table represent a Safety Factor, SF, of 1.0 and a diametrical ovality of 3%. Grouted strength of the pipe was derived by applying a multiplier of 5 to the non-grouted value[32]. If the existing sewer will not provide structural integrity to earth and live loads, a more conservative Safety Factor should be used.

For profile wall pipe the variable that determines adequate wall stiffness is a function of the pipe wall moment of inertia and pipe inside mean diameter. The following equation can be used to estimate maximum allowable long-term (50-year) height of water above the pipe with no grout:

(4)
$$H = \frac{0.9 \times RSC}{D_m}$$

WHERE
H = Height of water, feet
RSC = Measured Ring Stiffness Constant
D_m = Mean diameter, inches

This equation contains a Safety Factor (SF) of 2.0 based on pipe with a maximum 3% deflection.

For grout with a minimum compressive strength of 500 psi at 24 hours (1,800 psi at 28 days), the allowable long-term (50-year) height of water above the pipe may be determined from the following equation:

(5)
$$H = 5 \times \frac{(0.9 \times RSC)}{D_m}$$

This equation contains a Safety Factor (SF) of 2.0.

Pressure Pipe

A liner, which will be exposed to a constant internal pressure or to a combination of internal and external stresses must be analyzed in a more detailed manner. The guidelines for a detailed loading analysis such as this are available from a variety of resources that discuss in detail the design principles concerned with underground installation of flexible piping materials.[3,15,16,19,26,29]

In those installations where the liner will be subjected to direct earth loading, the pipe/soil system must be capable of withstanding all anticipated loads. These include earth loading, hydrostatic loading, and superimposed loads. The structural stability of a polyethylene liner under these conditions is determined largely by the quality of the external support. For these situations, refer to any of the above referenced information sources that concern direct burial of thermoplastic pipe. A polyethylene liner that has been selected to resist hydrostatic loading will generally accommodate typical external loading conditions if it is installed properly.

Other Loading Considerations

Filling of the entire annular space is rarely required. If it is properly positioned and sealed off at the termination points, a polyethylene liner will eliminate the sluice path that could contribute to the continued deterioration of most existing pipe structures. With a liner, a gradual accumulation of silt or sediment occurs within the annular space, and this acts to eliminate the potential sluice path.

On occasion, deterioration of the original pipe may continue to occur even after the liner has been installed.[18] This situation may be the result of excessive ground-water movement combined with a soil quality that precludes sedimentation within the annular space. Soil pH and resistivity can also help deteriorate the host culvert or pipe. As a result, uneven or concentrated point loading upon the pipe liner or even subsidence of the soil above the pipe system may occur. This can be avoided by filling the annular space with a cement-sand mixture, a low-density grout material[10], or fly ash.

Determine the Flow Capacity

The third step in the sliplining process is to assess the impact of sliplining on the hydraulic capacity of the existing pipe system. This is accomplished by using commonly-accepted flow equations to compare the flow capacity of the original line against that of the smaller, newly-installed polyethylene liner. Two equations widely used for this calculation are the Manning Equation (Eq. 6) and the Hazen-Williams Approximation for other than gravity flow systems (Eq. 7).[2,5]

(6) Manning Equation for Gravity Flow

$$Q = \frac{1.486 \times A \times R^{0.667} \times S^{0.5}}{n}$$

WHERE
Q = Flow, ft³/sec
A = Flow area, ft² (3.14 x ID²/4)
R = Hydraulic radius, feet (ID/4 for full flow)
S = Slope, ft/ft
n = Manning flow factor for piping material, 0.009 for smooth wall PE
ID = Inside diameter, feet

For circular pipe flowing full, the formula may be simplified to

$$Q = \frac{0.463 \times ID^{2.667} \times S^{0.5}}{n}$$

(7) Hazen-Williams Approximation for Other Than Gravity Flow

$$H = \frac{1044 \times G^{1.85}}{C_H^{1.85} \times ID^{4.865}}$$

WHERE
H = Friction loss in ft of H₂O/100 ft
G = Volumetric flow rate, gpm
 = 2.449 x V x ID²
V = Flow velocity, ft/sec
ID = Inside diameter, inches
C_H = Hazen Williams flow coefficient, dimensionless
 = 150 for smooth wall polyethylene

The insertion of a smaller pipe within the existing system may appear to reduce the original flow capacity. However, in the majority of sliplining applications, this is not the case. The polyethylene liner is extremely smooth in comparison to most piping materials. The improved flow characteristic for clear water is evidenced by a comparatively low Manning Flow Coefficient, n of 0.009, and a Hazen-Williams coefficient, C_H, of 150.

While a reduction in pipe diameter does occur as a consequence of sliplining, it is largely compensated by the significant reduction in the Manning Flow Coefficient. As a result, flow capacity is maintained at or near the original flow condition.[18] Manning Flow Coefficients and Hazen-Williams Flow Coefficients for a variety of piping materials are listed in Table 2a and 2b. These factors may be used to approximate the relative flow capacities of various piping materials.

TABLE 2A
Typical Manning Flow Coefficients for Water Flowing through Common Piping Materials

Polyethylene (solid wall)	0.009
PVC	0.009
Cement-lined Ductile Iron	0.012
New Cast Iron, Welded Steel	0.014
Wood, Concrete	0.016
Clay, New Riveted Steel	0.017
Old Cast Iron, Brick	0.020
CSP	0.023
Severely Corroded Cast Iron	0.035

TABLE 2B
Typical Hazen-Williams Flow Coefficients for Water Flowing through Common Piping Materials[31]

Polyethylene (solid wall)	150
PVC	150
Cement-lined Ductile Iron	140
New Cast Iron, Welded Steel	130
Wood, Concrete	120
Clay, New Riveted Steel	110
Old Cast Iron, Brick	100
Severely Corroded Cast Iron	80

Quite often the hydraulic capacity of a gravity flow pipe can actually be improved by an insertion renewal. For example, consider the following illustrations of calculations using the Manning Equation (Eq. 6).

Calculation for Flow Rate, Q, through a 24-inch ID Concrete Pipe at 1% slope (1 ft/100 ft)

$$Q = \frac{1.486 \times 3.14 \times 1^2 \times 0.5^{0.667} \times 0.01^{0.5}}{0.016} = 18.3 \text{ ft}^3/\text{sec } (8{,}248 \text{ gpm})$$

Calculation of Flow Rate, Q, through a 22-inch OD Polyethylene Pipe with a 20.65-Inch ID at 1% slope (1 ft/100 ft)

$$Q = \frac{1.486 \times 3.14 \times 0.8604^2 \times 0.429^{0.667} \times 0.01^{0.5}}{0.009} = 21.8 \text{ ft}^3/\text{sec } (9{,}800 \text{ gpm})$$

Comparison of the two calculated flow rates shows that sliplining this 24-inch concrete pipe with the smaller polyethylene pipe actually improves the capacity by 1,000 gallons per minute. This will often be the situation. Occasionally, the theoretical flow capacity of the liner may appear to be equivalent to or slightly less than that of the original system. In many such cases, the presence of the liner eliminates the infiltration associated with the deterioration of the original pipe and the corresponding burden this places on the existing flow capacity. So an apparently small reduction in theoretical flow capacity may, in reality, prove to be quite acceptable since it eliminates the infiltration and the effect this produces on available hydraulic capacity.

Design the Accesses

The polyethylene liner will need to be connected to existing system components or appurtenances. Proper planning for a rehabilitation project must include the specific engineering designs by which these connections will be made.

Gravity flow pipeline rehabilitation often requires that the individual liner lengths be terminated at manholes or concrete headwalls that already exist within the system that is being sliplined. The annular space at these locations must provide a water-tight seal against continued infiltration in the void area that exists between the liner and the original pipe where they connect to these structures.

Typically, the required seal can be made by compacting a ring or collar of Okum saturated with non-shrink grout into the void area to a distance equal to one-half to one full liner diameter. The annular space is then "dressed" with a non-shrink elastomeric grout. The face of the elastomeric grout may then be covered with a quick-set chemical-resistant concrete. The same concrete material may then be used to reconstruct an invert in the manhole. This type of seal is shown in Figure 2.

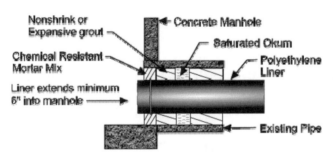

Figure 2 Typical Manhole Seal for Gravity Flow Applications

For those installations where a new manhole or headwall will be set, the amount of elastomeric grout may be minimized by fusing a water-stop or stub end onto the liner length before it is finally positioned. This fitting may then be embedded within the poured headwall or grouted into the new manhole. Some typical connecting arrangements for newly constructed appurtenances are shown in Figure 3. The connection described (water stop/wall anchor grouted in place) can also work on existing structures.

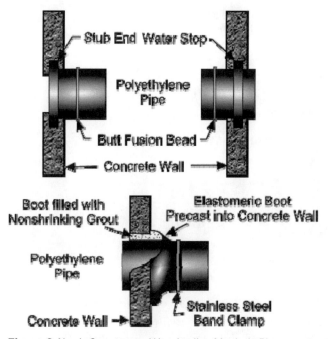

Figure 3 Newly Constructed Headwall or Manhole Placements

Deteriorated lateral service connections are a leading cause of infiltration in gravity flow pipelines.[19] An integral part of the insertion process is rebuilding these connections. This aspect of sliplining assures maximum reduction of infiltration, provides for long-term structural stability of the service, and minimizes the potential for continued deterioration of the existing pipe system.

Individual home services or other laterals may be connected to the liner by using any of several different connection methods. For example, upon relaxation of the liner, sanitary sewer connections may be made to the polyethylene liner by using a strap-on service saddle or a side-wall fusion fitting. Either of these options provides a secure water-tight connection to the liner and allows for effective renewal of the riser with no reduction in the inside diameter of the service. Both of these types of connection are shown in Figure 4.

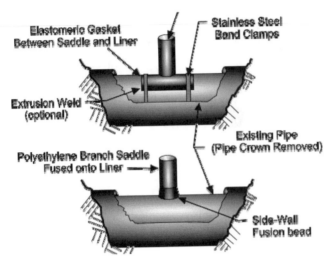

Figure 4 Lateral Service Connections for Sliplining Gravity Pipelines

Rehabilitation of pressure pipelines often requires that connections be made to lateral pressure-rated piping runs. Connections to these lines should be designed to insure full pressure capability of the rehabilitated system. Several alternatives are available to meet this requirement. These include in-trench fusion of molded or fabricated tees, sidewall fusion of branch saddles, insertion of spool pieces via electrofusion and insertion of low-profile mechanical connectors. One of these options is illustrated schematically in Figure 5. Performance requirements and installation parameters of the rehabilitation project most often dictate the selection of one specific connection design.

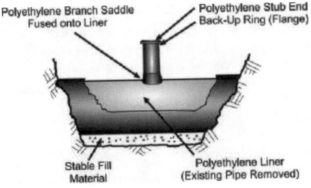

Figure 5 Typical Lateral Service Connection for Sliplining Pressure Pipelines

Develop the Contract Documents

When the rehabilitation design has been completed, attention will be focused on writing the specifications and contract documents that will ensure a successful installation. Reference documents for this purpose include: ASTM D3350[4], ASTM F585[5], ASTM F714[6], and ASTM F894.[7] To assist further in the development of these documents, a model sliplining specification is available from the Plastics Pipe Institute, "Guidance and Recommendations on the Use of Polyethylene (PE) Pipe for the Sliplining of Sewers."

The Sliplining Procedure

The standard sliplining procedure is normally a seven-step process. While the actual number of steps may vary to some degree in the field, the procedure remains the same for all practical purposes.[23,24] The procedures for rehabilitation of gravity and positive pressure pipelines are essentially the same. Some subtle differences become apparent in the manner by which some of the basic steps are implemented. The seven basic steps are as follows:

1. Inspect the existing pipe.
2. Clean and clear the line.
3. Join lengths of polyethylene pipe.
4. Access the original line.
5. Installation of the liner.
6. Make service and lateral connections.
7. Make terminal connections and stabilize the annular space.

1. Inspect the Existing Pipe

The first step for a sliplining project is the inspection of the existing pipe. This will determine the condition of the line and the feasibility of insertion renewal. During this step, identify the number and the locations of offset pipe segments and other potential obstructions.

Use a remote controlled closed circuit television camera to inspect the pipe interior. As the unit is pulled or floated through the original pipe, the pictures can be viewed and recorded with on-site television recording equipment.

2. Clean and Clear the Line

The existing pipeline needs to be relatively clean to facilitate placement of the polyethylene liner. This second step will ensure ease of installation. It may be accomplished by using cleaning buckets, kites or plugs, or by pulling a test section of polyethylene liner through the existing pipe structure.

Obviously, to attempt a liner insertion through a pipeline obstructed with excess sand, slime, tree roots or deteriorated piping components would be uneconomical or even impossible. Step 2 is often undertaken in conjunction with the inspection process of Step 1.

3. Weld Lengths of Polyethylene Pipe

Polyethylene pipe may be joined by butt fusion technology, gasketed bell and spigot joining methods, or by extrusion welding. The specific method to be used will be determined by the type of polyethylene pipe being inserted into the existing pipe structure. Solid wall polyethylene pipe is usually joined using butt fusion techniques. Polyethylene profile walled pipe, on the other hand, can be joined by integral gasketed bell and spigot joining methods or by the extrusion welding technique. Consult the manufacturer for the recommended procedure.

Butt Fusion — Solid Wall Pipe

Individual lengths of solid wall polyethylene pipe are joined by using the butt fusion process technique. The integrity of this joining procedure is such that, when it is performed properly, the strength of the resulting joint equals or exceeds the structural stability of the pipe itself. This facilitates the placement of a leak-free liner throughout the section of the existing system under rehabilitation.

The external fusion bead, formed during the butt fusion process, can be removed following the completion of joint quality assurance procedures by using a special tool prior to the insertion into the existing system. The removal of the bead may be necessary in cases of minimal clearance between the liner and the existing pipeline, but otherwise not required.

Pulling Lengths

Individual pulling lengths are usually determined by naturally occurring changes in grade or direction of the existing pipe system. Severe changes in direction that exceed the minimum recommended bending radius of the polyethylene liner may be used as access points. Likewise, severe offset joints, as revealed during the television survey, are commonly used as access points. By judicious planning, potential obstructions to the lining procedure may be used to an advantage.

There is a frequent question regarding the maximum pulling length for a given system. Ideally, each pull should be as long as economically possible without exceeding the tensile strength of the polyethylene material. It is rare that a pull of this magnitude is ever attempted. As a matter of practicality, pulling lengths are more often restricted by physical considerations at the job site or by equipment limitations.[23]

To ensure a satisfactory installation, the designer may want to analyze what is considered the maximum pulling length for a given situation. Maximum pulling length is a function of the tensile strength and weight of the polyethylene liner, the temperature at which the liner will be manipulated, the physical dimensions of the liner, and the frictional drag along the length of the polyethylene pipe liner.

Equations 8 and 9 are generally accepted for determination of the maximum feasible pulling length. One of the important factors in these calculations is the tensile strength of the particular polyethylene pipe product, which must be obtained from the manufacturer's literature.

(8) Maximum Pulling Force, MPF[1]

$$MPF = f_y \times f_t \times T \times \pi \times OD \left(\frac{1}{DR} - \frac{1}{DR^2} \right)$$

WHERE
MPF = Maximum pulling force, lb-force
f_y = Tensile yield design (safety) factor, 0.40
f_t = Time under tension design (safety) factor, 0.95*
T = Tensile yield strength, psi
　= 3,500 psi for PE3408 at 73.4°F
OD = Outside diameter, inches
DR = Dimension Ration, dimensionless
* The value of 0.95 is adequate for pulls up to 12 hours.

(9) Maximum Pulling Length, MPL

$$MPL = \frac{MPF}{W \times CF}$$

WHERE
MPL = Maximum straight pulling length on relatively flat surface, ft
MPF = Maximum pulling force, lb-force (Eq. 10-8)
W = Weight of pipe, lbs/ft
CF = Coefficient of friction, dimensionless
　　= 0.1, flow present through the host pipe
　　= 0.3, typical for wet host pipe
　　= 0.7, smooth sandy soil

Profile Wall Pipe

Polyethylene profile pipe is typically joined using gasketed bell and spigot techniques or extrusion welding.

When extrusion welding is used, the calculations for maximum pulling force and maximum pulling length are similar to those used for solid wall polyethylene pipe except that the "effective" cross sectional area of the pipe may be conservatively presumed to be ¼ of the cross sectional area between the pipe OD and ID. Consult the manufacturer's information for specific dimensional data.

The integral bell and spigot method does not require fusion joining. Instead, the pipe segments are joined by pushing or "jacking" the spigot on the leading end of each pipe into the trailing bell on the preceding pipe as it is installed into the deteriorating pipe structure. In this case the major design consideration is the maximum push length, which can be calculated by using Eq. 10 and Eq. 11.

(10) Maximum Pushing Force, MPHF

$$MPHF = S \times (ID + T) \times \pi \times PS$$

WHERE
MPHF = Maximum pushing force, lb-force
S = Inner and outer wall thickness of pipe, inches
ID = Inside diameter, inches
T = Outer wall thickness, inches
  = Use for closed profile pipe only. For open profile, T=0.
PS = Allowable pushing stress, psi
   = 400 psi for profile wall, bell and spigot
   = 800 psi for welded polyethylene

(11) Maximum Pushing Length, MPHL

$$MPHL = \frac{MPHF}{W \times CF}$$

WHERE
MPHL = Maximum pushing length, ft
MPHF = Maximum pushing force, lb-force
W = Weight of pipe, lbs/ft
CF = Coefficient of friction, dimensionless
   = 0.1, flow present through the host pipe
   = 0.3, typical for wet host pipe
   = 0.7, smooth sandy soil

Profile polyethylene pipes that are joined by extrusion welding can be pulled or pushed, "jacked," into place (see Figures 6 and 7). The maximum pull length can be calculated by using equations 10 and 11. Consult the manufacturers for additional information.

4. Access the Original Line

Excavation of the access pits is the next step in the insertion renewal procedure. Access pits will vary considerably in size and configuration, depending on a number of project-related factors such as:

- Depth of the existing pipe
- Diameters of the liner and the existing pipe
- Stiffness of liner pipe
- Prevailing soil conditions
- Equipment availability
- Traffic and service requirements
- Job site geography

For example, a fairly large access pit may be required when attempting to slipline a large diameter system that is buried deep in relatively unstable soil. In contrast, the access pit for a smaller diameter pipeline that is buried reasonably shallow (5 to 8 feet) may be only slightly wider than the liner itself. In actual practice, the simpler situation is more prevalent. An experienced contractor will recognize the limiting factors at a particular job site and utilize them to the best economic advantage, thus assuring a cost-effective installation.

A typical access pit for sliplining with pre-fused or welded lengths of solid wall polyethylene pipe is illustrated in Figure 6. Figure 7 is a schematic of an access method that may be used with profile pipe.

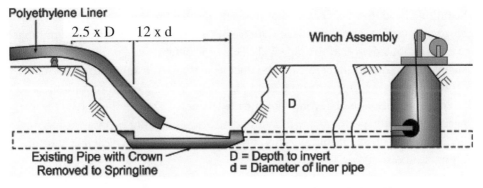

Figure 6 Typical Sliplining Access Pit for Prefused Lengths of Polyethylene Liner

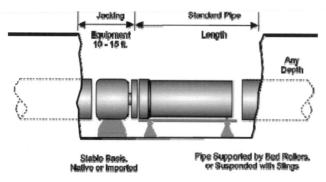

Figure 7 Typical Sliplining Access Pit for Bell and Spigot Polyethylene Liner

5. Installation of the Liner

Insertion of the polyethylene liner may be accomplished by one of several techniques. Prefused or welded lengths of solid wall polyethylene pipe may be "pulled" or "pushed" into place. Gasket-Jointed profile pipe, on the other hand, must be installed by the push method to maintain a water-tight seal.

The "Pulling" Technique

Prefused or welded lengths of polyethylene liner may be pulled into place by using a cable and winch arrangement. The cable from the winch is fed through the section of pipe that is to be sliplined. Then the cable is fastened securely to the liner segment, thus permitting the liner to be pulled through the existing pipe and into place.

Figure 6 is a schematic of an installation in which the liner is being pulled through the existing pipe from the left side toward a manhole at the right. This procedure requires some means, such as a pulling head, to attach the cable to the leading edge of the liner. The pulling head may be as simple or as sophisticated as the particular project demands or as economics may allow.

The pulling head may be fabricated of steel and fastened to the liner with bolts. They are spaced evenly around the circumference of the profile so that a uniform pulling force is distributed around the pipe wall. This type of fabricated pulling head will usually have a conical shape, aiding the liner as it glides over minor irregularities or through slightly offset joints in the old pipe system. The mechanical pulling head does not normally extend beyond the Outside Diameter (O.D.) of the polyethylene liner and is usually perforated to accommodate flow as quickly as possible once the liner is inserted inside the old system. Three practical styles of typical mechanical pulling heads are shown in Figure 8.

Figure 8 Fabricated Mechanical Pulling Heads

A less sophisticated but cost-effective approach is to fabricate a pulling head out of a few extra feet of liner that has been fused onto a single pipe pull. Cut evenly spaced wedges into the leading edge of the extra liner footage, making it look like the end of a banana being peeled. Collapse the ends toward the center and fasten them together with bolts or all-thread rods. Then attach the cable to secondary bolts that extend across the collapsed cross section. This simple technique is illustrated in Figure 9.

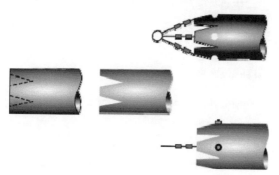

Figure 9 Field-Fabricated Pulling Heads

As the polyethylene liner is pulled into the pipeline, a slight elongation of the liner may occur. A 24-hour relaxation period will allow the liner to return to its original dimensions. After the relaxation period, the field fabricated pulling head may be cut off. It is recommended the liner be pulled past the termination point by 3-5%. This allows the liner to be accessible at the connection point after the relaxation period.

The pull technique permits a smooth and relatively quick placement of the liner within an old pipe system. However, this method may not be entirely satisfactory when attempting to install a large-diameter heavy-walled polyethylene pipe. This is especially true when the load requires an unusually large downstream winch. A similar problem may exist as longer and larger pulls are attempted so that a heavier pulling cable is required. When the pull technique is not practical, consider the advantages that may be offered by the push technique.

The "Push" Technique

The push technique for solid wall or welded polyethylene pipe is illustrated schematically in Figure 10. This procedure uses a choker strap, placed around the liner at a workable distance from the access point. A track-hoe, backhoe, or other piece of mechanical equipment pulls the choker to push the liner through the existing pipe. With each stroke of the backhoe, the choker grips the pipe and pushes the leading edge of the liner further into the deteriorated pipe. At the end of each stroke, the choker must be moved back on the liner, usually by hand. The whole process may be assisted by having a front-end loader or bulldozer simultaneously push on the trailing end of the liner segment.

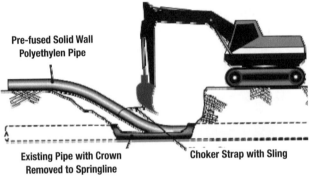

Figure 10 Pushing Technique for Solid Wall Polyethylene Pipe

Gasketed polyethylene pipe typically uses the push technique to affect the pipe seal, as well as to position the liner. The push technique for gasketed pipe is shown schematically in Figure 10. This process inserts the liner without the necessity for having a high capacity winch and cable system.

The Combination Technique
The pushing and pulling techniques can sometimes be combined to provide the most efficient installation method. Typically, this arrangement can be used when attempting the placement of unusually heavy walled or long lengths of polyethylene liner.

Flow Control
For most insertion renewal projects it is not necessary to eliminate the entire flow stream within the existing pipe structure. Actually, some amount of flow can assist positioning of the liner by providing a lubricant along the liner length as it moves through the deteriorated pipe structure. However, an excessive flow can inhibit the insertion process. Likewise, the disruption of a flow stream in excess of 50% of pipe capacity should be avoided.

The insertion procedure should be timed to take advantage of cyclic periods of low flow that occur during the operation of most gravity piping systems. During the insertion of the liner, often a period of 30 minutes or less, the annular space will probably carry sufficient flow to maintain a safe level in the operating sections of the system being rehabilitated. Flow can then be diverted into the liner upon final positioning of the liner. During periods of extensive flow blockage, the upstream piping system can be monitored to avoid unexpected flooding of drainage areas.

Consider establishing a flow control procedure for those gravity applications in which the depth of flow exceeds 50%. The flow may be controlled by judicious operation of pump stations, plugging or blocking the flow, or bypass pumping of the flow stream.

Pressurized piping systems will require judicious operation of pump stations during the liner installation.

6. Make Service and Lateral Connections

After the recommended 24-hour relaxation period following the insertion of the polyethylene liner, each individual service connection and lateral can be added to the new system. One common method of making these connections involves the use of a wrap-around service saddle. The saddle is placed over a hole that has been cut through the liner and the entire saddle and gasket assembly is then fastened into place with stainless steel bands. Additional joint integrity can be obtained by extrusion welding of the lap joint created between the saddle base and the liner. The service lateral can then be connected into the saddle, using a readily available flexible coupling[11]. Once the lateral has been connected, following standard direct burial procedures can stabilize the entire area.

For pressure applications, lateral connections can be made using sidewall fusion of branch saddles onto the liner. As an alternate, a molded or fabricated tee may be fused or flanged into the liner at the point where the lateral connection is required (see Figures 3 and 4). Mechanical fittings are also a viable option; refer to the chapter on polyethylene joining procedures in this Handbook.

7. Make Terminal Connections and Stabilize the Annular Space Where Required

Making the terminal connections of the liner is the final step in the insertion renewal procedure. Pressurized pipe systems will require connection of the liner to the various system appurtenances. These terminal connections can be made readily through the use of pressure-rated polyethylene fittings and flanges with fusion technology. Several common types of pressurized terminal connections are illustrated in Figure 11. All of these require stabilization of the transition region to prevent point loading of the liner. Mechanical Joint (MJ) Adapters can be used. Refer to the chapter on polyethylene joining procedures in this Handbook.

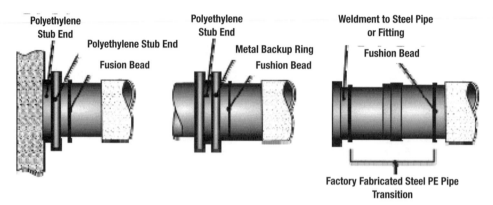

Figure 11 Terminal and Transition Connections for Pressurized Insertion Renewal Projects

Gravity lines do not typically require pressure-capable connections to the other system appurtenances. In these situations, the annular space will be sealed to prevent migration of ground water along the annulus and, ultimately, infiltration through the manhole or headwall connection. The typical method for making this type of connection is shown in Figure 11. Sealing materials should be placed by gravity flow methods so that the liner's buckling resistance is not exceeded during installation. Consideration should be given to the specific load bearing characteristics of the fill material in light of the anticipated loading of the liner.

Other Rehabilitation Methods

Rehabilitation by sliplining is only one (but probably the most popular) of a number of methods using polyethylene pipe currently available for pipeline rehabilitation. As mentioned in the introduction to this chapter, sliplining has been in use for more than thirty years.

Several other methods of rehabilitation that use polyethylene piping will be described briefly here. Please note that, due to rapidly advancing technology, this listing may become incomplete very quickly. Also note that any reference to proprietary products or processes is made only as required to explain a particular methodology.

Swagelining

A continuous length of polyethylene pipe passes through a machine where it is heated. It then passes through a heated die, which reduces the outside diameter (OD). Insertion into the original pipeline then follows through an insertion pit. The liner pipe relaxes (pressurization may be used to speed the process) until the OD of the liner matches the inside diameter (ID) of the original pipeline. Grouting is not required.

Rolldown

This system is very similar to swagelining except OD reduction is by mechanical means and expansion is through pressurization.

Titeliner

A system that is very similar to the swagelining and rolldown systems.

Fold and Form

Continuous lengths of polyethylene pipe are heated, mechanically folded into a "U" shape, and then coiled for shipment. Insertion is made through existing manholes. Expansion is by means of a patented heat/pressure procedure, which utilizes steam. The pipe is made, according to the manufacturer, to conform to the ID of the original pipeline; therefore, grouting is not required.

Pipe Bursting

A technique used for replacing pipes made from brittle materials, e.g. clay, concrete, cast iron, etc. A bursting head (or bursting device) is moved through the pipe, simultaneously shattering it, pushing the shards aside, and drawing in a polyethylene replacement pipe. This trenchless technique makes it possible to install pipe as much as 100% larger than the existing pipe.

Pipe Splitting

A technique, similar to pipe bursting, used for pipes made from ductile materials, e.g. steel, ductile iron, plastic, etc. A "splitter" is moved through the existing pipe, simultaneously splitting it with cutter wheels, expanding it, and drawing in a polyethylene replacement pipe. This trenchless technique is generally limited to replacement with same size or one pipe size (ie., 6" to 8") larger replacement pipe.

Summary

This chapter has provided an introductory discussion on the rehabilitation of a deteriorated pipe structure by insertion renewal with continuous lengths of polyethylene pipe. It also includes a brief description of other rehabilitation methods that utilize polyethylene piping. The sliplining or insertion renewal procedure is a cost-effective means by which a new pipeline is obtained with a minimum interference with surface traffic. An inherent benefit of the technology is the installation of a new, structurally sound, leak-free piping system with improved flow characteristics. The resulting pipe structure allows for a flow capacity at or near that of the deteriorating pipe system while eliminating the potential for infiltration or exfiltration. And the best feature of all is the vastly improved longevity of the plastic pipe, especially compared to the decay normally associated with piping materials of the past.

The continuing deterioration of this country's infrastructure necessitates innovative solutions to persistent and costly problems. Insertion renewal, or sliplining, is a cost-effective means by which one aspect of the infrastructure dilemma may be corrected without the expense and long-term service disruption associated with pipeline replacement.

References

1. ASTM 1804, Standard Practice for Determining Allowable Tensile Load for Polyethylene (PE) Gas Pipe During Pull-In Installation, *Annual Book of Standards*, American Society for Testing and Materials (ASTM).
2. ASTM D2321, Standard Practices for Underground Installation of Flexible Thermoplastic Sewer Pipe, *Annual Book of Standards*, American Society for Testing and Materials (ASTM).
3. ASTM D2774, Standard Practices for Underground Installation of Thermoplastic Pressure Piping, *Annual Book of Standards*, American Society for Testing and Materials (ASTM).
4. ASTM D3350, Standard for Polyethylene Pipe and Fittings Materials, *Annual Book of Standards*, American Society for Testing and Materials (ASTM).
5. ASTM F585, Standard Practices for Insertion of Flexible Polyethylene Pipe into Existing Sewers, *Annual Book of Standards*, American Society for Testing and Materials (ASTM).
6. ASTM F714, Standard Specification for Polyethylene (SDR-PR) Based Outside Diameter, *Annual Book of Standards*, American Society for Testing and Materials (ASTM).
7. ASTM F894, Standard Specification for Polyethylene (PE) Large Diameter Profile Wall, Sewer and Drain Pipe, *Annual Book of Standards*, American Society for Testing and Materials (ASTM).
8. Diskin, Joe. (1987, May). Plastic Pipe Insertion, (NF), *Pipeline and Gas Journal*.
9. *Driscopipe System Design*. (1985, January). A Publication of Phillips Driscopipe, Inc.
10. Elastizell Product Catalog. (1984). *The Elastizell Advantage*. A Publication of the Plastizell Corporation of America.
11. *Existing Sewer Evaluation and Rehabilitation*. (1983). American Society of Civil Engineers and the Water Environment Federation, New York, NY.
12. *Fernco Product Catalog 185*. (1985, January). A Publication of Ferncom, Inc.
13. Gross, Sid. (1985, May). Plastic Pipe in Gas Distribution, Twenty-Five Years of Achievement, *Gas Industries*.
14. Gross, Sid. (1987, September). Choosing Between Liners for Sewer Rehabilitation, Public Works.
15. *Handbook of PVC Pipe*. (1978, February). A Publication of the Uni-Bell Plastic Pipe Association, Dallas, TX.
16. Howard, A.E., & Selander, C.E., (1977, September). Laboratory Load Tests on Buried Thermosetting, Thermoplastic and Steel Pipe, *Journal AWWA*.
17. Jenkins, C.F. & Kroll, A.E. (1981, March 30-April 1). External Hydrostatic Loading of Polyethylene Pipe, Proceedings of the International Conference on Underground Plastic Pipe, sponsored by the Pipeline Division of the American Society of Civil Engineers.
18. Lake, Donald W., Jr. (1979, December). Innovative Rehabilitation of Embankment Conduits. Paper presented to the 1984 winter meeting of the American Society of Agricultural Engineers.
19. Plastics Pipe Institute. (1979, September). Technical Report TR-31, Underground Installation of Polyolefin Piping.
20. *Polyolefin Piping*. (1985, May). A Publication of the Plastics Pipe Institute, Washington, DC.
21. *Recommended Specifications for Sewer Connection Rehabilitation*. (1984, January). A Publication of the National Society of Service Companies.
22. *Renewing Sewers with Polyolefin Pipe*. (1985, June). A Publication of the Plastics Pipe Institute, Washington, DC.
23. Sandstrum, S.D. (1986, August). Sewer Rehabilitation with Polyethylene Pipe. Paper presented at the Second Annual Sewer Maintenance and Rehabilitation Conference, Center for Local Government Technology, Oklahoma State University.
24. Schock, David B. (1982, August 6). New Pipes Do Job Better, *Mount Pleasant Morning Sun*.
25. Spangler, M.G. (1941). The Structural Design of Flexible Pipe Culverts, Bulletin 153, Iowa State Engineering Experiment Station.
26. *Spiral Engineered Systems*, A Publication of Spirolite, a Subsidiary of the Chevron Corporation.
27. Watkins, R.K. (1977). *Buried Structures*, Foundation Engineering Handbook, Utah State University.
28. Watkins, R.K. (1977). *Principles of Structural Performance of Buried Pipes*, Utah State University.
29. DOT FHWA-1P-85-15.
30. *Mark's Handbook for Mechanical Engineers*, 10th Edition.
31. Larock, B.E., Jeppson, R.W., & Watters, G.Z. (2000). *Hydraulics of Pipeline Systems*, Boca Raton: CRC Press.
32. Lo, King H., & Zhang, Jane Q. (1994, February 28-March 2). Collapse Resistance Modeling of Encased Pipes, Buried Plastic Pipe Technology: 2nd Volume, Papers presented at the ASTM symposium, New Orleans.

Chapter 12

Horizontal Directional Drilling

Introduction

The Horizontal Directional Drilling (HDD) Industry has experienced so much growth in the past two decades that HDD has become commonplace as a method of installation. One source reported that the number of units in use increased by more than a hundred-fold in the decade following 1984. This growth has been driven by the benefits offered to utility owners (such as the elimination of traffic disruption and minimal surface damage) and by the ingenuity of contractors in developing this technology. To date, HDD pipe engineering has focused on installation techniques, and rightfully so. In many cases, the pipe experiences its maximum lifetime loads during the pullback operation.

The purpose of this chapter is to acquaint the reader with some of the important considerations in selecting the proper polyethylene pipe. Proper selection of pipe involves consideration not only of installation design factors such as pullback force limits and collapse resistance, but also of the long-term performance of the pipe once installed in the bore-hole. The information herein is not all-inclusive; there may be parameters not discussed that will have significant bearing on the proper engineering of an application and the pipe selection. For specific projects, the reader is advised to consult with a qualified engineer to evaluate the project and prepare a specification including design recommendations and pipe selection. The reader many find additional design and installation information in ASTM F1962, "Standard Guide for Use of Maxi- Horizontal Directional Drilling for Placement of Polyethylene Pipe or Conduit Under Obstacles, Including River Crossings," and in the ASCE Manual of Practice 108, "Pipeline Design for Installation by Directional Drilling."

Background

Some of the earliest uses of large diameter polyethylene pipe in directional drilling were for river crossings. These are major engineering projects requiring thoughtful design, installation, and construction, while offering the owner the security of deep river bed cover with minimum environmental damage or exposure, and no disruption of river traffic. Polyethylene pipe is suited for these installations because of its scratch tolerance and the fused joining system which gives a zero-leak-rate joint with design tensile capacity equal to that of the pipe.

To date, directional drillers have installed polyethylene pipe for gas, water, and sewer mains; electrical conduits; and a variety of chemical lines. These projects involved not only river crossings but also highway crossings and right-of-ways through developed areas so as not to disturb streets, driveways, and business entrances.

Polyethylene Pipe for Horizontal Directional Drilling

This chapter gives information on the pipe selection and design process. It is not intended to be a primer on directional drilling. The reader seeking such information can refer to the references of this chapter. Suggested documents are the "Mini-Horizontal Directional Drilling Manual" and the "Horizontal Directional Drilling Good Practices Guidelines" published by the North American Society for Trenchless Technology (NASTT).

Horizontal Directional Drilling Process

Knowledge of the directional drilling process by the reader is assumed, but some review may be of value in establishing common terminology. Briefly, the HDD process begins with boring a small, horizontal hole (pilot hole) under the crossing obstacle (i.e. a highway) with a continuous string of steel drill rod. When the bore head and rod emerge on the opposite side of the crossing, a special cutter, called a back reamer, is attached and pulled back through the pilot hole. The reamer bores out the pilot hole so that the pipe can be pulled through. The pipe is usually pulled through from the side of the crossing opposite the drill rig.

Pilot Hole

Pilot hole reaming is the key to a successful directional drilling project. It is as important to an HDD pipeline as backfill placement is to an open-cut pipeline. Properly trained crews can make the difference between a successful and an unsuccessful drilling program for a utility. Several institutions provide operator-training programs, one of which is Michigan State University's Center for Underground Infrastructure Research and Education (CUIRE). Drilling the pilot hole establishes the path of the drill rod ("drill-path") and subsequently the location of the PE pipe. Typically, the bore-head is tracked electronically so as to guide the hole to a pre-designed configuration. One of the key considerations in the design of the drill-path is creating as large a radius of curvature as possible within the limits of the right-of-way, thus minimizing curvature. Curvature induces bending stresses and increases the pullback load due to the capstan effect. The capstan effect is the increase in frictional drag when pulling the pipe around a curve due to a component of the pulling force acting normal to the curvature. Higher tensile stresses reduce the pipe's collapse resistance. The drill-path normally has curvature along its vertical profile. Curvature requirements are dependent on site geometry (crossing length, required depth to provide safe cover, staging site location, etc.) But, the degree of curvature is limited by the bending radius of the drill rod and the pipe. More often, the permitted bending radius of the drill rod controls the curvature and thus significant bending stresses do not occur in the pipe. The designer should minimize the number of curves and maximize their radii of curvature in the right-of-way by carefully choosing the entry and exit points. The driller should also attempt to minimize extraneous curvature due to undulations (dog-legs) from frequent over-correcting alignment or from differences in the soil strata or cobbles.

Pilot Hole Reaming

The REAMING operation consists of using an appropriate tool to open the pilot hole to a slightly larger diameter than the carrier pipeline. The percentage oversize depends on many variables including soil types, soil stability, depth, drilling mud, borehole hydrostatic pressure, etc. Normal over-sizing may be from 120% to 150% of the carrier pipe diameter. While the over-sizing is necessary for insertion, it means that the inserted pipe will have to sustain vertical earth pressures without significant side support from the surrounding soil.

Prior to pullback, a final reaming pass is normally made using the same sized reamer as will be used when the pipe is pulled back (swab pass). The swab pass cleans the borehole, removes remaining fine gravels or clay clumps and can compact the borehole walls.

Drilling Mud

Usually a "drilling mud" such as fluid bentonite clay is injected into the bore during cutting and reaming to stabilize the hole and remove soil cuttings. Drilling mud can be made from clay or polymers. The primary clay for drilling mud is sodium montmorillonite (bentonite). Properly ground and refined bentonite is added to fresh water to produce a "mud." The mud reduces drilling torque, and gives stability and support to the bored hole. The fluid must have sufficient gel strength to keep cuttings suspended for transport, to form a filter cake on the borehole wall that contains the water within the drilling fluid, and to provide lubrication between the pipe and the borehole on pullback. Drilling fluids are designed to match the soil and cutter. They are monitored throughout the process to make sure the bore stays open, pumps are not overworked, and fluid circulation throughout the borehole is maintained. Loss of circulation could cause a locking up and possibly overstressing of the pipe during pullback.

Drilling muds are thixotropic and thus thicken when left undisturbed after pullback. However, unless cementitious agents are added, the thickened mud is no stiffer than very soft clay. Drilling mud provides little to no soil side-support for the pipe.

Pullback

The pullback operation involves pulling the entire pipeline length in one segment (usually) back through the drilling mud along the reamed-hole pathway. Proper pipe handling, cradling, bending minimization, surface inspection, and fusion welding procedures need to be followed. Axial tension force readings, constant insertion velocity, mud flow circulation/exit rates, and footage length installed should be recorded. The pullback speed ranges usually between 1 to 2 feet per minute.

Mini-Horizontal Directional Drilling

The Industry distinguishes between mini-HDD and conventional HDD, which is sometimes referred to as maxi-HDD. Mini-HDD rigs can typically handle pipes up to 10″ or 12″ and are used primarily for utility construction in urban areas, whereas HDD rigs are typically capable of handling pipes as large as 48″. These machines have significantly larger pullback forces ranging up to several hundred thousand pounds.

General Guidelines

The designer will achieve the most efficient design for an application by consulting with an experienced contractor and a qualified engineer. Here are some general considerations that may help particularly in regard to site location for PE pipes:

1. Select the crossing route to keep it to the shortest reasonable distance.
2. Find routes and sites where the pipeline can be constructed in one continuous length; or at least in long multiple segments fused together during insertion.
3. Although compound curves have been done, try to use as straight a drill path as possible.
4. Avoid entry and exit elevation differences in excess of 50 feet; both points should be as close as possible to the same elevation.
5. Locate all buried structures and utilities within 10 feet of the drill-path for mini-HDD applications and within 25 feet of the drill-path for maxi-HDD applications. Crossing lines are typically exposed for exact location.

6. Observe and avoid above-ground structures, such as power lines, which might limit the height available for construction equipment.

7. The HDD process takes very little working space versus other methods. However, actual site space varies somewhat depending upon the crossing distance, pipe diameter, and soil type.

8. Long crossings with large diameter pipe need bigger, more powerful equipment and drill rig.

9. As pipe diameter increases, large volumes of drilling fluids must be pumped, requiring more/larger pumps and mud-cleaning and storage equipment.

10. Space requirements for maxi-HDD rigs can range from a 100 feet wide by 150 feet long entry plot for a 1000 ft crossing up to 200 feet wide by 300 feet long area for a crossing of 3000 or more feet.

11. On the pipe side of the crossing, sufficient temporary space should be rented to allow fusing and joining the polyethylene carrier pipe in a continuous string beginning about 75 feet beyond the exit point with a width of 35 to 50 feet, depending on the pipe diameter. Space requirements for coiled pipe are considerably less. Larger pipe sizes require larger and heavier construction equipment which needs more maneuvering room (though use of polyethylene minimizes this). The initial pipe side "exit" location should be about 50' W x 100' L for most crossings, up to 100' W x 150' L for equipment needed in large diameter crossings.

12. Obtain "as-built" drawings based on the final course followed by the reamer and the installed pipeline. The gravity forces may have caused the reamer to go slightly deeper than the pilot hole, and the buoyant pipe may be resting on the crown of the reamed hole. The as-built drawings are essential to know the exact pipeline location and to avoid future third party damage.

Safety

Safety is a primary consideration for every directionally drilled project. While this chapter does not cover safety, there are several manuals that discuss safety including the manufacturer's Operator's Manual for the drilling rig and the Equipment Manufacturer's Institute (EMI) Safety Manual: *Directional Drilling Tracking Equipment*.

Geotechnical Investigation

Before any serious thought is given to the pipe design or installation, the designer will normally conduct a comprehensive geotechnical study to identify soil formations at the potential bore sites. The purpose of the investigation is not only to determine if directional drilling is feasible, but to establish the most efficient way to accomplish it. With this information the best crossing route can be determined, drilling tools and procedures selected, and the pipe designed. The extent of the geotechnical investigation often depends on the pipe diameter, bore length and the nature of the crossing. Refer to ASTM F1962 and ASCE MOP 108 for additional information.

During the survey, the geotechnical consultant will identify a number of relevant items including the following:

a. Soil identification to locate rock, rock inclusions, gravelly soils, loose deposits, discontinuities and hardpan.
b. Soil strength and stability characteristics
c. Groundwater

(Supplemental geotechnical data may be obtained from existing records, e.g. recent nearby bridge constructions, other pipeline/cable crossings in the area.)

For long crossings, borings are typically taken at 700 ft intervals. For short crossings (1000 ft or less), as few as three borings may suffice. The borings should be near the drill-path to give accurate soil data, but sufficiently far from the borehole to avoid pressurized mud from following natural ground fissures and rupturing to the ground surface through the soil-test bore hole. A rule-of-thumb is to take borings at least 30 ft to either side of bore path. Although these are good general rules, the number, depth and location of boreholes is best determined by the geotechnical engineer.

Geotechnical Data For River Crossings

River crossings require additional information such as a study to identify river bed, river bed depth, stability (lateral as well as scour), and river width. Typically, pipes are installed to a depth of at least 20 ft below the expected future river bottom, considering scour. Soil borings for geotechnical investigation are generally conducted to 40 ft below river bottom.

Summary

The best conducted projects are handled by a team approach with the design engineer, bidding contractors and geotechnical engineer participating prior to the preparation of contract documents. The geotechnical investigation is usually the first step in the boring project. Once the geotechnical investigation is completed, a determination can be made whether HDD can be used. At that time, design of both the HDPE pipe and the installation can begin. The preceding paragraphs represent general guidance and considerations for planning and designing an HDD polyethylene pipeline project. These overall topics can be very detailed in nature.

Individual HDD contractors and consultant engineering firms should be contacted and utilized in the planning and design stage. Common sense along with a rational in-depth analysis of all pertinent considerations should prevail. Care should be given in evaluating and selecting an HDD contractor based upon successful projects, qualifications, experience and diligence. A team effort, strategic partnership and risk-sharing may be indicated.

Product Design: DR Selection

After completion of the geotechnical investigation and determination that HDD is feasible, the designer turns attention to selecting the proper pipe. The proper pipe must satisfy all hydraulic requirements of the line including flow capacity, working pressure rating, and surge or vacuum capacity. These considerations have to be met regardless of the method of installation. Design of the pipe for hydraulic considerations can be found elsewhere such as in AWWA C906 or the pipe manufacturer's literature and will not be addressed in this chapter. For HDD applications, in addition to the hydraulic requirements, the pipe must be able to withstand (1) pullback loads which include tensile pull forces, external hydrostatic pressure, and tensile bending stresses, and (2) external service loads (post-installation soil, groundwater, and surcharge loads occurring over the life of the pipeline). Often the load the pipe sees during installation such as the combined pulling force and external pressure will be the largest load experienced by the pipe during its life. The remainder of this document will discuss the DR selection based on pullback and external service loads. (Polyethylene pipe is classified by DR. The DR is the "dimension ratio" and equals the pipe's outer diameter divided by the minimum wall thickness.)

While this chapter gives guidelines to assist the designer, the designer assumes all responsibility for determining the appropriateness and applicability of the equations and parameters given in this chapter for any specific application. Directional drilling is an evolving technology, and industry-wide design protocols are still developing. Proper design requires considerable professional judgment beyond the scope of this chapter. The designer is advised to consult ASTM F 1962 when preparing an HDD design.

Normally, the designer starts the DR selection process by determining the DR requirement for the internal pressure (or other hydraulic requirements). The designer will then determine if this DR is sufficient to withstand earth, live, and groundwater service loads. If so, then the installation (pullback) forces are considered. Ultimately, the designer chooses a DR that will satisfy all three requirements: the pressure, the service loads, and the pullback load.

Although there can be some pipe wall stresses generated by the combination of internal pressurization and wall bending or localized bearing, generally internal pressure and external service load stresses are treated as independent. This is permissible primarily since PE is a ductile material and failure is usually driven by the average stress rather than local maximums. There is a high safety factor applied to the internal pressure, and internal pressurization significantly reduces stresses due to external loads by re-rounding. (One exception to this is internal vacuum, which must be combined with the external pressure.)

Figure 1 Borehole Deformation

Design Considerations for Net External Loads

This and the following sections will discuss external buried loads that occur on directionally drilled pipes. One important factor in determining what load reaches the pipe is the condition of the borehole, i.e. whether it stays round and open or collapses. This will depend in great part on the type of ground, the boring techniques, and the presence of slurry (drilling mud and cutting mixture). If the borehole does not deform (stays round) after drilling, earth loads are arched around the borehole and little soil pressure is transmitted to the pipe. The pressure acting on the pipe is the hydrostatic pressure due to the slurry or any groundwater present. The slurry itself may act to keep the borehole open. If the borehole collapses or deforms substantially, earth pressure will be applied to the pipe. The resulting pressure could exceed the slurry pressure unless considerable tunnel arching occurs above the borehole. Where no tunnel arching occurs, the applied external pressure is equal to the combined earth, groundwater, and live-load pressure. For river crossings, in unconsolidated river bed soils, little arching is anticipated . The applied pressure likely equals the geostatic stress (sometimes called the prism load). In consolidated soils, arching above the borehole may occur, and the applied pressure will likely be less than the geostatic stress, even after total collapse of the borehole crown onto the pipe. If the soil deposit is a stiff clay, cemented, or partially lithified, the borehole may stay open with little or no deformation. In this case, the applied pressure is likely to be just the slurry head or groundwater head.

In addition to the overt external pressures such as slurry head and groundwater, internal vacuum in the pipe results in an increase in external pressure due to the removal of atmospheric pressure from inside the pipe. On the other hand, a positive internal pressure in the pipe may mediate the external pressure. The following equations can be used to establish the net external pressure or, as it is sometimes called, the differential pressure between the inside and outside of the pipe.

Depending on the borehole condition, the net external pressure is defined by either Eq. 1 (deformed/collapsed borehole) or Eq. 2 (open borehole):

(1) $P_N = P_E + P_{GW} + P_{SUR} - P_I$

(2) $P_N = P_{MUD} - P_I$

WHERE
P_N = Net external pressure, psi
P_E = External pressure due to earth pressure, psi
P_{GW} = Groundwater pressure (including the height of river water), psi
P_{SUR} = Surcharge and live loads, psi
P_I = Internal pressure, psi (negative in the event of vacuum)
P_{MUD} = Hydrostatic pressure of drilling slurry or groundwater pressure, if slurry can carry shear stress, psi
(Earth, ground water, and surcharge pressures used in Eq. 1 are discussed in a following section of this chapter.)

(3) $$P_{MUD} = \frac{g_{MUD} H_B}{144 \frac{in^2}{ft^2}}$$

WHERE
g_{MUD} = Unit weight of slurry (drilling mud and cuttings), pcf
H_B = Elevation difference between lowest point in borehole and entry or exit pit, ft
(144 is included for units conversion.)

When calculating the net external pressure, the designer will give careful consideration to enumerating all applied loads and their duration. In fact, most pipelines go through operational cycles that include (1) unpressurized or being drained, (2) operating at working pressure, (3) flooding, (4) shutdowns, and (5) vacuum and peak pressure events. As each of these cases could result in a different net external pressure, the designer will consider all phases of the line's life to establish the design cases.

In addition to determining the load, careful consideration must be given to the duration of each load. PE pipe is viscoelastic, that is, it reacts to load with time-dependent properties. For instance, an HDD conduit resists constant groundwater and soil pressure with its long-term stiffness. On the other hand, an HDD force-main may be subjected to a sudden vacuum resulting from water hammer. When a vacuum occurs, the net external pressure equals the sum of the external pressure plus the vacuum. Since surge is instantaneous, it is resisted by the pipe's short-term stiffness, which can be four times higher than the long-term stiffness.

For pressure lines, consideration should be given to the time the line sits unpressurized after construction. This may be several months. Most directionally drilled lines that contain fluid will have a static head, which will remain in the line once filled. This head may be subtracted from the external pressure due to earth/groundwater load. The designer should keep in mind that the external load also may vary with time, for example, flooding.

Earth and Groundwater Pressure

Earth loads can reach the pipe when the borehole deforms and contacts the pipe. The amount of soil load transmitted to the pipe will depend on the extent of deformation and the relative stiffness between the pipe and the soil. Earth loading may not be uniform. Due to this complexity, there is not a simple equation for relating earth load to height of cover. Groundwater loading will occur whether the hole deforms or not; the only question is whether or not the slurry head is higher and thus may in fact control design. Thus, what loads reach the pipe will depend on the stability of the borehole.

The designer may wish to consult a geotechnical engineer for assistance in determining earth and groundwater loads, as the loads reaching the pipe depend on detailed knowledge of the soil.

Stable Borehole - Groundwater Pressure Only

A borehole is called stable if it remains round and deforms little after drilling. For instance, drilling in competent rock will typically result in a stable borehole. Stable boreholes may occur in some soils where the slurry exerts sufficient pressure to maintain a round and open hole. Since the deformations around the hole are small, soil pressures transmitted to the pipe are negligible. The external load applied to the pipe consists only of the hydrostatic pressure due to the slurry or the groundwater, if present. Equation 4 gives the hydrostatic pressure due to groundwater or drilling slurry. Standing surface water should be added to the groundwater.

(4)
$$P_{GW} = \frac{g_W H_W}{144 \frac{in^2}{ft^2}}$$

WHERE
P_{GW} = Hydrostatic fluid pressure due to ground and surface water, psi
g_w = Unit weight of water, pcf
H_W = Height to free water surface above pipe, ft (144 is included for correct units conversion.)

Borehole Deforms/Collapse With Arching Mobilized

When the crown of the hole deforms sufficiently to place soil above the hole in the plastic state, arching is mobilized. In this state, hole deformation is limited. If no soil touches the pipe, there is no earth load on the pipe. However, when deformation is sufficient to transmit load to the pipe, it becomes the designer's chore to determine how much earth load is applied to the pipe. At the time of this writing, there have been no published reports giving calculation methods for finding earth load on directionally drilled pipes. Based on the successful performance of directionally drilled PE pipes, it is reasonable to assume that some amount of arching occurs in many applications. The designer of HDD pipes may gain some knowledge from the approaches developed for determining earth pressure on auger bored pipes and on jacked pipes. It is suggested that the designer become familiar with all of the assumptions used with these methods.

O'Rourke et. al. published an equation for determining the earth pressure on auger bored pipes assuming a borehole approximately 10% larger than the pipe. In this model, arching occurs above the pipe similar to that in a tunnel where zones of loosened soil fall onto the pipe. The volume of the cavity is eventually filled with soil that is slightly less dense than the insitu soil, but still capable of transmitting soil load. This method of load calculation gives a minimal loading. The method published here is more conservative. It is based on trench type arching as opposed to tunnel arching and is used by Stein to calculate loads on jacked pipe. In Stein's model, the maximum earth load (effective stress) is found using the modified form of Terzhaghi's equation given by Eq. 6. External groundwater pressure must be added to the effective earth pressure. Stein and O'Rourke's methods should only be considered where the depth of cover is sufficient to develop arching (typically exceeding five (5) pipe diameters), dynamic loads such as traffic loads are insignificant, the soil has sufficient internal friction to transmit arching, and conditions are confirmed by a geotechnical engineer.

Using the equations given in Stein, the external pressure is given below:

(5)
$$P_{EV} = \frac{Kg_{SE}H_C}{144\frac{in^2}{ft^2}}$$

$$\text{(6)} \quad k = \frac{1 - \exp\left(-2\dfrac{KH_C}{B}\tan\left(\dfrac{d}{2}\right)\right)}{2\dfrac{KH_C}{B}\tan\left(\dfrac{d}{2}\right)}$$

WHERE
P_{EV} = external earth pressure, psi
g_{SE} = effective soil weight, pcf
H_C = depth of cover, ft
k = arching factor
B = "silo" width, ft
d = angle of wall friction, degrees (For HDD, d = f)
f = angle of internal friction, degrees
K = earth pressure coefficient given by:

$$K = \tan^2\left(45 - \frac{f}{2}\right)$$

The "silo" width should be estimated based on the application. It varies between the pipe diameter and the borehole diameter. A conservative approach is to assume the silo width equals the borehole diameter. (The effective soil weight is the dry unit weight of the soil for soil above the groundwater level, it is the saturated unit weight less the weight of water for soil below the groundwater level.)

Borehole Collapse with Prism Load

In the event that arching in the soil above the pipe breaks down, considerable earth loading may occur on the pipe. In the event that arching does not occur, the upper limit on the load is the weight of the soil prism ($P_E = g_{SE}H_C$) above the pipe. The prism load is most likely to develop in shallow applications subjected to live loads, boreholes in unconsolidated sediments such as in some river crossings, and holes subjected to dynamic loads. The "prism" load is given by Eq. 7.

$$\text{(7)} \quad P_E = \frac{g_{SE}H_C}{144\dfrac{in^2}{ft^2}}$$

WHERE
P_E = earth pressure on pipe, psi
g_{SE} = effective weight of soil, pcf
H_C = soil height above pipe crown, ft
(Note: 144 is included for units conversion.)

Combination of Earth and Groundwater Pressure

Where groundwater is present in the soil formation, its pressure must be accounted for in the external load term. For instance, in a river crossing one can assume with reasonable confidence that the directionally drilled pipe is subjected to the earth pressure from the sediments above it combined with the water pressure.

Case 1 Water level at or below ground surface

(8)
$$P_E + P_{GW} = \frac{g_B H_W + g_D(H_C - H_W) + g_W H_W}{144 \frac{in^2}{ft^2}}$$

Case 2 Water level at or above ground surface (i.e. pipe in river bottom)

(9)
$$P_E + P_{GW} = \frac{g_B H_C + g_W H_W}{144 \frac{in^2}{ft^2}}$$

WHERE
H_W = Height of Ground water above pipe springline, ft
H_C = height of cover, ft
g_B = buoyant weight of soil, pcf
g_W = weight of water, pcf
g_D = dry unit weight of soil, pcf

Live Loads

Wheel loads from trucks or other vehicles are significant for pipe at shallow depths whether they are installed by open cut trenching or directional drilling. The wheel load applied to the pipe depends on the vehicle weight, the tire pressure and size, vehicle speed, surface smoothness, pavement and distance from the pipe to the point of loading. In order to develop proper soil structure interaction, pipe subject to vehicular loading should be installed at least 18" or one pipe diameter (whichever is larger) under the road surface. Generally, HDD pipes are always installed at a deeper depth so as to prevent frac-outs from occurring during the boring.

For pipes installed under rigid pavement and subjected to H20 loadings, Table 1 gives the vertical earth pressure at the pipe crown as determined by AISI [3]. Live loads under flexible pavement and unpaved roads can be calculated. (See Spangler and Handy in references.)

TABLE 1
H20 Loading Under Rigid Pavement (AISI)

Height of Cover (ft)	(ft) Load (psf)
1	1800
2	800
3	600
4	400
5	250
6	200
7	175
8	100

The live-load pressure can be obtained from Table 1 by selecting the load based on the height of cover and converting the load to units of "psi" by dividing the load in "psf" by 144.

Performance Limits

Hydrostatic Buckling or Collapse   **Ring Deformation**

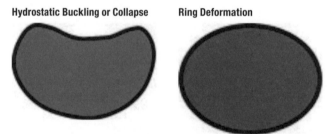

Figure 2 Performance Limits of HDD Pipe Subjected to Service Loads

Performance Limits of HDD Installed Pipe

The design process normally consists of calculating the loads applied to the pipe, selecting a trial pipe DR, then calculating the safety factor for the trial DR. If the safety factor is adequate, the design is sufficient. If not, the designer selects a lower DR and repeats the process. The safety factor is established for each performance limit of the pipe by taking the ratio of the pipe's ultimate strength or resistance to the applied load.

External pressure from earth load, groundwater, vacuum and live load applied to the HDD pipe produces (1) a compressive ring thrust in the pipe wall and (2) ring bending deflection. The performance limit of unsupported PE pipe subjected to compressive thrust is ring buckling (collapse). The performance limit of a PE pipe subjected to ring bending (a result of non-uniform external load, i.e. earth load) is ring deflection. See Figure 2.

Time-Dependent Behavior

Both performance limits are proportional to the apparent modulus of elasticity of the PE material. For viscoelastic materials like PE, the modulus of elasticity is a time-dependent property, that is, its value changes with time under load. A newly applied load increment will cause a decrease in apparent stiffness over time. Unloading will result in rebounding or an apparent gain in stiffness. These changes occur because the molecular structure rearranges itself under load. The result is a higher resistance to short term loading than to long-term loading. Careful consideration must be given to the duration and frequency of each load, so that the performance limit associated with that load can be calculated using PE material properties representative of that time period. The same effects occur with the pipe's tensile strength. For instance, during pullback, the pipe's tensile yield strength decreases with pulling time, so the safe (allowable) pulling stress is a function of time.

For viscoelastic materials, the ratio of the applied stress to strain is called the apparent modulus of elasticity, because the ratio varies with load rate. Typical values for the apparent modulus of elasticity at 73°F (23°C) are presented in Table 2. Consult the manufacturer for specific applications.

TABLE 2
Apparent Modulus of Elasticity and Safe Pull Tensile Stress @ 73° F

Typical Apparent Modulus of Elasticity			Typical Safe Pull Stress		
Duration	HDPE	MDPE	Duration	HDPE	MDPE
Short-term	110,000 psi (800 MPa)	87,000 psi (600 MPa)	30 min	1,300 psi (9.0 MPa)	1,000 psi (6.9 MPa)
10 hours	57,500 psi (400 MPa)	43,500 psi (300 MPa)	60 min	1,200 psi (8.3 MPa)	900 psi (6.02 MPa)
100 hours	51,200 psi (300 MPa)	36,200 psi (250 MPa)	12 hours	1,150 psi (7.9 MPa)	850 psi (5.9 MPa)
50 years	28,200 psi (200 MPa)	21,700 psi (150 MPa)	24 hours	1,100 psi (7.6 MPa)	800 psi (5.5 MPa)

Ring Deflection (Ovalization)

Non-uniform pressure acting on the pipe's circumference such as earth load causes bending deflection of the pipe ring. Normally, the deflected shape is an oval. Ovalization may exist in non-rerounded coiled pipe and to a lesser degree in straight lengths that have been stacked, but the primary sources of bending deflection of directionally drilled pipes is earth load. Slight ovalization may also occur during pullback if the pipe is pulled around a curved path in the borehole. Ovalization reduces the pipe's hydrostatic collapse resistance and creates tensile bending stresses in the pipe wall. It is normal and expected for buried PE pipes to undergo ovalization. Proper design and installation will limit ovalization (or as it is often called "ring deflection") to prescribed values so that it has no adverse effect on the pipe.

Ring Deflection Due to Earth Load

As discussed previously, insitu soil characteristics and borehole stability determine to great extent the earth load applied to directionally drilled pipes. Methods for calculating estimated earth loads, when they occur, are given in the previous section on "Earth and Groundwater Pressure."

Since earth load is non-uniform, the pipe will undergo ring deflection, i.e. a decrease in vertical diameter and an increase in horizontal diameter. The designer can check to see if the selected pipe is stiff enough to limit deflection and provide an adequate safety factor against buckling. (Buckling is discussed in a later section of this chapter.)

The soil surrounding the pipe may contribute to resisting the pipe's deflection. Formulas used for entrenched pipe, such as Spangler's Iowa Formula, are likely not applicable as the HDD installation is different from installing pipe in a trench where the embedment can be controlled. In an HDD installation, the annular space surrounding the pipe contains a mixture of drilling mud and cuttings. The mixture's consistency or stiffness determines how much resistance it contributes. Consistency (or stiffness) depends on several factors including soil density, grain size and the presence of groundwater. Researchers have excavated pipe installed by HDD and observed some tendency of the annular space soil to return to the condition of the undisturbed native soil. See Knight (2001) and Ariaratnam (2001). It is important to note that the researched installations were located above groundwater, where excess water in the mud-cuttings slurry can drain. While there may be consolidation and strengthening of the annular space soil particularly above the groundwater level, it may be weeks or even months before significant resistance to pipe deflection develops. Until further research establishes the soil's contribution to resisting deflection, one option is to ignore any soil resistance and to use Equation 10 which is derived from ring deflection equations published by Watkins and Anderson (1995). (Coincidentally, Equation 10 gives the same deflection as the Iowa Formula with an E' of zero.)

$$\text{(10)} \quad \frac{y}{D} = \frac{0.0125 P_E}{\dfrac{E}{12(DR-1)^3}}$$

WHERE
y = ring deflection, in
D = pipe diameter, in
P_E = Earth pressure, psi
DR = Pipe Dimension Ratio
E = modulus of elasticity, psi
* To obtain ring deflection in percent, multiply y/D by 100.

Ring Deflection Limits (Ovality Limits)

Ovalization or ring deflection (in percent) is limited by the pipe wall strain, the pipe's hydraulic capacity, and the pipe's geometric stability. Jansen observed that for PE, pressure-rated pipe, subjected to soil pressure only, "no upper limit from a practical design point of view seems to exist for the bending strain." On the other hand, pressurized pipes are subject to strains from both soil induced deflection and internal pressure. The combined strain may produce a high, localized outer-fiber tensile stress. However, as the internal pressure is increased, the pipe tends to re-round and the bending strain is reduced. Due to this potential for combined strain (bending and hoop tensile), it is conservative to limit deflection of pressure pipes to less than non-pressure pipes. In lieu of an exact calculation for allowable deflection limits, the limits in Table 3 can be used.

TABLE 3
Design Deflection Limits of Buried Polyehtylene Pipe, Long Term, %*

DR or SDR	21	17	15.5	13.5	11	9	7.3
Deflection Limit (% y/D) Non-Pressure Applications	7.5	7.5	7.5	7.5	7.5	7.5	7.5
Deflection Limit (%y/D) Pressure Applications	7.5	6.0	6.0	6.0	5.0	4.0	3.0

* Deflection limits for pressure applications are equal to 1.5 times the short-term deflection limits given in Table X2.1 of ASTM F-714.

Design deflections are for use in selecting DR and for field quality control. (Field measured deflections exceeding the design deflection do not necessarily indicate unstable or over-strained pipe. In this case, an engineering analysis of such pipe should be performed before acceptance.)

Unconstrained Buckling

Uniform external pressure applied to the pipe either from earth and live load, groundwater, or the drilling slurry creates a ring compressive hoop stress in the pipe's wall. If the external pressure is increased to a point where the hoop stress reaches a critical value, there is a sudden and large inward deformation of the pipe wall, called buckling. Constraining the pipe by embedding it in soil or cementitious grout will increase the pipe's buckling strength and allow it to withstand higher external pressure than if unconstrained. However, as noted in a previous section it is not likely that pipes installed below the groundwater level will acquire significant support from the surrounding mud-cuttings mixture and for pipe above groundwater support may take considerable time to develop. Therefore, until further research is available it is conservative to assume no constraint from the soil. The following equation, known as Levy's equation, may be used to determine the allowable external pressure (or negative internal pressure) for unconstrained pipe.

(11)
$$P_{UA} = \frac{2E}{(1-m^2)} \left(\frac{1}{DR-1}\right)^3 \frac{f_o}{N}$$

WHERE
P_{ua} = Allowable unconstrained pressure, psi
E = Modulus of elasticity (apparent), psi
m = Poisson's Ratio
    Long-term loading - 0.45
    Short-term loading - 0.35
DR = Dimension ratio (Do/t)
f_o = ovality compensation factor (see figure 3)
N = Safety factor, generally 2.0 or higher

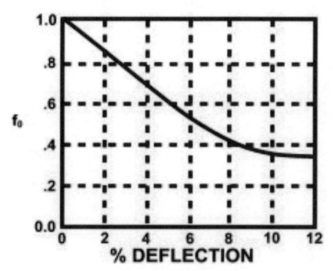

Figure 3 Ovality Compensation Factor

Note that the modulus of elasticity and Poisson's ratio are a function of the duration of the anticipated load. If the safety factor in Levy's equation is set equal to one, the equation gives the critical buckling pressure for the pipe. Table 4 gives values of the critical buckling (collapse) pressure for different DR's of PE pipe. For design purposes, the designer must reduce the values by a safety factor and by ovality compensation. When using this table for determining pipe's resistance to buckling during pullback, an additional reduction for tensile stresses is required, which is discussed in a later section of this chapter. When selecting a modulus to use in Equation 11 consideration should be given to internal pressurization of the line. When the pressure in the pipe exceeds the external pressure due to earth and live load, groundwater and/or slurry, the stress in the pipe wall reverses from compressive to tensile stress and collapse will not occur.

TABLE 4
Critical Buckling (Collapse) Pressure for unconstrained HDPE Pipe*@ 73° F

Service Pipe Life	Units	7.3	9	11	13.5	15.5	17	21
Short term	psi	1003	490	251	128	82	61	31
	ft H₂O	2316	1131	579	297	190	141	72
	in Hg	2045	999	512	262	168	125	64
100 hrs	psi	488	238	122	62	40	30	15
	ft H2O	1126	550	282	144	92	69	35
	in Hg	995	486	249	127	82	61	31
50 yrs	psi	283	138	71	36	23	17	9
	ft H2O	653	319	163	84	54	40	20
	in Hg	577	282	144	74	47	35	18

(Table does not include ovality compensation or safety factor.)
* Full Vacuum is 14.7 psi, 34 ft water, 30 in Hg.
* Axial Tension during pullback reduces collapse strength.

Multipliers for Temperature Rerating

$$\frac{60°F\ (16°C)}{1.08} \qquad \frac{73.4°F\ (23°C)}{1.00} \qquad \frac{100°F\ (38°C)}{0.78} \qquad \frac{120°F\ (49°C)}{0.63}$$

Wall Compressive Stress

The compressive stress in the wall of a directionally drilled PE pipe rarely controls design and it is normally not checked. However, it is included here because in some special cases such as directional drilling at very deep depths such as in landfills it may control design.

The earth pressure applied to a buried pipe creates a compressive thrust stress in the pipe wall. When the pipe is pressurized, the stress is reduced due to the internal pressure creating tensile thrust stresses. The net stress can be positive or negative depending on the depth of cover. Buried pressure lines may be subject to net compressive stress when shut down or when experiencing vacuum. These are usually short-term conditions and are not typically considered significant for design, since the short-term design stress of polyolefins is considerably higher than the long-term design stress. Pipes with large depths of cover and operating at low pressures may have net compressive stresses in the pipe wall. The following equation can be used to determine the net compressive stress:

(12)
$$S_c = \frac{P_s D_o}{288 t} - \frac{PD}{2t}$$

WHERE
S_C = Compressive wall stress, psi
P_S = Earth load pressures, psf
D_O = Pipe outside diameter, in

t = Wall thickness, in
P = (Positive) internal pressure, psi
D = Mean diameter, D_o-t, in

The compressive wall stress should be kept less than the allowable compressive stress of the material. For PE3408 HDPE pipe grade resins, 1000 psi is a safe allowable stress.

EXAMPLE CALCULATIONS An example calculation for selecting the DR for an HDD pipe is given in Appendix A.

Installation Design Considerations

After determining the DR required for long-term service, the designer must determine if this DR is sufficient for installation. Since installation forces are so significant, a lower DR (stronger pipe) may be required.

During pullback the pipe is subjected to axial tensile forces caused by the frictional drag between the pipe and the borehole or slurry, the frictional drag on the ground surface, the capstan effect around drill-path bends, and hydrokinetic drag. In addition, the pipe may be subjected to external hoop pressures due to net external fluid head and bending stresses. The pipe's collapse resistance to external pressure given in Equation 2 is reduced by the axial pulling force. Furthermore, the drill path curvature may be limited by the pipe's bending radius. (Torsional forces occur but are usually negligible when back-reamer swivels are properly designed.) Considerable judgment is required to predict the pullback force because of the complex interaction between pipe and soil. Sources for information include experienced drillers and engineers, programs such as DRILLPATH (1) and publications such as ASTM F1962, and the Pipeline Research Council (PRCI) Manual, Installation of Pipelines by Horizontal Directional Drilling, and Engineering Design Guide. Typically, pullback force calculations are approximations that depend on considerable experience and judgment.

The pullback formulas given herein and in DRILLPATH and ASTM F1962 are based on essentially an "ideal" borehole. The ideal borehole behaves like a rigid tunnel with gradual curvature, smooth alignment (no dog-legs), no borehole collapses, nearly complete cuttings removal, and good slurry circulation. The ideal borehole may be approached with proper drilling techniques that achieve a clean bore fully reamed to its final size before pullback. The closer the bore is to ideal; the more likely the calculated pullback force will match the actual.

Because of the large number of variables involved and the sensitivity of pullback forces to installation techniques, the formulas presented in this document are for guidelines only and are given only to familiarize the designer with the interaction that occurs during pullback. Pullback values obtained should be considered only as qualitative values and used only for preliminary estimates. The designer is advised to consult with an experienced driller or with an engineer familiar with calculating these forces. The following discussion assumes that the entry and exit pits of the bore are on the same, or close to the same, elevation.

Pullback Force

Large HDD rigs can exert between 100,000 lbs. to 500,000 lbs. pull force. The majority of this power is applied to the cutting face of the reamer device/tool, which precedes the pipeline segment into the borehole. It is difficult to predict what portion of the total pullback force is actually transmitted to the pipeline being inserted.

The pulling force which overcomes the combined frictional drag, capstan effect, and hydrokinetic drag, is applied to the pull-head and first joint of HDPE pipe. The axial tensile stress grows in intensity over the length of the pull. The duration of the pullload is longest at the pull-nose. The tail end of the pipe segment has zero applied tensile stress for zero time. The incremental time duration of stress intensity along the length of the pipeline from nose to tail causes a varying degree of recoverable elastic strain and viscoelastic stretch per foot of length along the pipe.

The DR must be selected so that the tensile stress in the pipe wall due to the pullback force, does not exceed the permitted tensile stress for the pipe material. Increasing the pipe wall thickness will allow for a greater total pull-force. Even though the thicker wall increases the weight per foot of the pipe, the pullback force within the bore itself is not significantly affected by the increased weight. Hence, thicker wall pipe generally reduces stress. The designer should carefully check all proposed DR's.

Frictional Drag Resistance

Pipe resistance to pullback in the borehole depends primarily on the frictional force created between the pipe and the borehole or the pipe and the ground surface in the entry area, the frictional drag between pipe and drilling slurry, the capstan effect at bends, and the weight of the pipe. Equation 13 gives the frictional resistance or required pulling force for pipe pulled in straight, level bores or across level ground.

(13) $$F_P = mW_B L$$

WHERE
F_P = pulling force, lbs
m = coefficient of friction between pipe and slurry (typically 0.25) or between pipe and ground (typically 0.40)
W_B = net downward (or upward) force on pipe, lb/ft
L = length, ft

When a slurry is present, W_B is the upward buoyant force of the pipe and its contents. Filling the pipe with fluid significantly reduces the buoyancy force and thus the pulling force. Polyethylene pipe has a density near that of water. If the pipe is installed "dry" (empty) using a closed nose-pull head, the pipe will want to "float" on the crown of the borehole leading to the sidewall loading and frictional drag through the buoyancy-per-foot force and the wetted soil to pipe coefficient of friction. Most major pullbacks are done "wet". That is, the pipeline is filled with water as it starts to descend into the bore (past the breakover point). Water is added through a hose or small pipe inserted into the pullback pipe. (See the calculation examples.)

Note: The buoyant force pushing the empty pipe to the borehole crown will cause the PE pipe to "rub" the borehole crown. During pullback, the moving drill mud lubricates the contact zone. If the drilling stops, the pipe stops, or the mud flow stops, the pipe - slightly ring deflected by the buoyant force - can push up and squeeze out the lubricating mud. The resultant "start-up" friction is measurably increased. The pulling load to loosen the PE pipe from being "stuck" in the now decanted (moist) mud can be very high. This situation is best avoided by using thicker (lower DR) pipes, doing "wet" pulls, and stopping the pull only when removing drill rods.

Capstan Force

For curves in the borehole, the force can be factored into horizontal and vertical components. Huey et al.[3] shows an additional frictional force that occurs in steel pipe due to the pressure required by the borehole to keep the steel pipe curved. For bores with a radius of curvature similar to that used for steel pipe, these forces are insignificant for PE pipe. For very tight bends, it may be prudent to consider them. In addition to this force, the capstan effect increases frictional resistance when pulling along a curved path. As the pipe is pulled around a curve or bend creating an angle q, there is a compounding of the forces due to the direction of the pulling vectors. The pulling force, F_C, due to the capstan effect is given in Eq. 14. Equations 13 and 14 are applied recursively to the pipe for each section along the pullback distance as shown in Figure 4. This method is credited to Larry Slavin, Outside Plant Consulting Services, Inc. Rockaway, N.J.

(14) $$F_c = e^{mq}(mW_B L)$$

WHERE
e = Natural logarithm base (e=2.71828)
m = coefficient of friction
q = angle of bend in pipe, radians
W_B = weight of pipe or buoyant force on pipe, lbs/ft
L = Length of pull, ft

$$F_1 = \exp(m_g a)(m_g W_p (L_1 + L_2 + L_3 + L_4))$$

$$F_2 = \exp(m_b a)(F_1 + m_b W_b L_2 + W_b H - m_g W_p L_2 \exp(m_g a))$$

$$F_3 = F_2 + m_b W_b L_3 - \exp(m_b a)(m_g W_p L_3 \exp(m_g a))$$

$$F_4 = \exp(m_b b)(F_3 + m_b W_b L_4 - W_b H - \exp(m_b a)(m_g W_p L_4 \exp(m_g a)))$$

WHERE
H = Depth of bore (ft)
Fi = Pull Force on pipe at Point i (lb)
Li = Horizontal distance of Pull from point to point (ft)
m = Coeff. of friction (ground (g) and borehole (b))
w = Pipe weight (p) and Buoyant pipe weight (lb/ft)
a, b = Entry and Exit angles (radians)

Figure 4 Estimated Pullback Force Calculation

Hydrokinetic Force

During pulling, pipe movement is resisted by the drag force of the drilling fluid. This hydrokinetic force is difficult to estimate and depends on the drilling slurry, slurry flow rate pipe pullback rate, and borehole and pipe sizes. Typically, the hydrokinetic pressure is estimated to be in the 30 to 60 kPa (4 to 8 psi) range.

(15) $$F_{HK} = p \frac{\pi}{8} (D_H^2 - OD^2)$$

WHERE
F_{HK} = hydrokinetic force, lbs
p = hydrokinetic pressure, psi
D_H = borehole diameter, in
OD = pipe outside diameter, in

ASCE MOP 108 suggests a different method for calculating the hydrokinetic drag force. It suggests multiplying the external surface area of the pipe by a fluid drag coefficient of 0.025 lb/in² after Puckett (2003). The total pull back force, F_T, then is the combined pullback force, F_P, plus the hydrokinetic force, F_{HK}. For the example shown in Figure 4, F_P equals F_4.

Tensile Stress During Pullback

The maximum outer fiber tensile stress should not exceed the safe pull stress. The maximum outer fiber tensile stress is obtained by taking the sum of the tensile stress in the pipe due to the pullback force, the hydrokinetic pulling force, and the tensile bending stress due to pipe curvature. During pullback it is advisable to monitor the pulling force and to use a "weak link" (such as a pipe of higher DR) mechanical break- away connector or other failsafe method to prevent over-stressing the pipe.

The tensile stress occurring in the pipe wall during pullback is given by Eq. 16.

(16)
$$S_t = \frac{F_T}{\pi t (D_{OD} - t)} + \frac{E_T D_{OD}}{2R}$$

WHERE
S_T = Axial tensile stress, psi
F_T = Total pulling force, lbs
t = Minimum wall thickness, in
D_{OD} = Outer diameter of pipe, in
E_T = Time-dependent tensile modulus, psi
R = Minimum radius of curvature in bore path, in

The axial tensile stress due to the pulling force should not exceed the pipe's safe pull load. As discussed in a previous section, the tensile strength of PE pipe is load-rate sensitive. Time under load is an important consideration in selecting the appropriate tensile strength to use in calculating the safe pull load. During pullback, the pulling force is not continually applied to the pipe, as the driller must stop pulling after extracting each drill rod in order to remove the rod from the drill string. The net result is that the pipe moves the length of the drill rod and then stops until the extracted rod is removed. Pullback is an incremental (discrete) process rather than a continuous process. The pipe is not subjected to a constant tensile force and thus may relax some between pulls. A one-hour modulus value might be safe for design, however, a 12-hour value will normally minimize "stretching" of the pipeline. Table 5 and Table 6 give safe pull loads for HDPE pipes based on a 12-hour value. Allowable safe pullback values for gas pipe are given in ASTM F-1807, "Practice for Determining Allowable Tensile Load for Polyethylene (PE) Gas Pipe during Pull-In Installation".

After pullback, pipe may take several hours (typically equal to the duration of the pull) to recover from the axial strain. When pulled from the reamed borehole, the pull-nose should be pulled out about 3% longer than the total length of the pull. The elastic strain will recover immediately and the viscoelastic stretch will "remember" its original length and recover overnight. One does not want to come back in the morning to discover the pull-nose sucked back below the borehole exit level due to stretch recovery and thermal-contraction to an equilibrium temperature. In the worst case, the driller may want to pull out about 4% extra length (40 feet per 1000 feet) to insure the pull-nose remains extended beyond the borehole exit.

TABLE 5
Safe Pull Load @ 12 hours for HDPE Pipes (Iron Pipe Size)

IPS Size	DR = Nom. OD	9 Lbs.	11 Lbs.	13.5 Lbs.	17 Lbs.
1.25	1.660	983	823	683	551
1.5	1.900	1288	1078	895	722
2	2.375	2013	1684	1398	1128
3	3.500	4371	3658	3035	2450
4	4.500	7226	6046	5018	4050
6	6.625	15661	13105	10876	8779
8	8.625	26544	22212	18434	14880
10	10.750	41235	34505	28636	23115
12	12.750	58006	48538	40282	32516
14	14.000	69937	58522	48568	39204
16	16.000	91347	76437	63435	51205
18	18.000	115611	96741	80285	64806
20	20.000	142729	119433	99118	80008
22	22.000	172703	144514	119932	96809
24	24.000	205530	171983	142729	115211
26	26.000	241213	201841	167509	135213
28	28.000	279750	234088	194271	156815
30	30.000	321141	268724	223015	180017
32	32.000	N.A.	305748	253741	204819
34	34.000	N.A.	345161	286450	231222
36	36.000	N.A.	386962	321141	259224
42	42.000	N.A.	N.A.	437109	352833
48	48.000	N.A.	N.A.	N.A.	460843
54	54.000	N.A.	N.A.	N.A.	N.A.

TABLE 6
Safe Pull Load @ 12 hours for HDPE Pipes (Ductile Iron Pipe Size)

DIPS Size	DR = Nom. OD	9 bs.	11 Lbs.	13.5 Lbs.	17 Lbs.
4	4.800	8221	6879	5709	4608
6	6.900	16988	14215	11797	9523
8	9.050	29225	24455	20295	16382
10	11.100	43964	36788	30531	24644
12	13.200	62173	52025	43176	34851
14	15.300	83529	69895	58006	46822
16	17.400	108032	90399	75022	60558
18	19.500	135682	113536	94224	76057
20	21.600	166480	139306	115611	93321
24	25.800	237516	198748	164942	133141
30	32.000	N.A.	305748	253741	204819
36	38.300	N.A.	N.A.	363487	293406
42	44.500	N.A.	N.A.	N.A.	396087
48	50.800	N.A.	N.A.	N.A.	516177

External Pressure During Installation

During pullback it is reasonable to assume that the borehole remains stable and open and that the borehole is full of drilling slurry. The net external pressure due to fluid in the borehole, then, is the slurry head, P_{MUD}. This head can be offset by pulling the pipe with an open nose or filling the pipe with water for the pullback. However, this may not always be possible, for instance when installing electrical conduit. In addition to the fluid head in the borehole, there are also dynamic sources of external pressure:

1. If the pulling end of the pipe is capped, a plunger action occurs during pulling which creates a mild surge pressure. The pressure is difficult to calculate. The pipe will resist such an instantaneous pressure with its relatively high short-term modulus. If care is taken to pull the pipe smoothly at a constant speed, this calculation can be ignored. If the pipe nose is left open, this surge is eliminated.

2. External pressure will also be produced by the frictional resistance of the drilling mud flow. Some pressure is needed to pump drilling mud from the reamer tool into the borehole, then into the pipe annulus, and along the pipe length while conveying reamed soil debris to the mud recovery pit. An estimate of this short term hydrokinetic pressure may be calculated using annular flow pressure loss formulas borrowed from the oil well drilling industry. This external pressure is dependent upon specific drilling mud properties, flow rates, annular opening, and hole configuration. This is a short-term installation condition. Thus, HDPE pipe's short-term external differential pressure capabilities are compared to the

actual short-term total external pressure during this installation condition. Under normal conditions, the annular-flow back pressure component is less than 4-8 psi.

In consideration of the dynamic or hydrokinetic pressure, P_{HK}, the designer will add additional external pressure to the slurry head:

(17) $P_N = P_{MUD} + P_{HK} - P_I$

Where the terms have been defined previously.

Resistance to External Collapse Pressure During Pullback Installation

The allowable external buckling pressure equation, Eq.11, with the appropriate time-dependent modulus value can be used to calculate the pipe's resistance to the external pressure, P_N, given by Eq.17 during pullback. The following reductions in strength should be taken:

1. The tensile pulling force reduces the buckling resistance. This can be accounted for by an additional reduction factor, f_R. The pulling load in the pipe creates a hoop strain as described by Poisson's ratio. The hoop strain reduces the buckling resistance. Multiply Eq.11 by the reduction factor, f_R to obtain the allowable external buckling pressure during pullback.

(18) $F_R = \sqrt{(5.57 - (r+1.09)^2)} - 1.09$

(19) $r = \dfrac{S_T}{2S}$

WHERE

S_T = calculated tensile stress during pullback (psi)

S = safe pull stress (psi)

Since the pullback time is typically several hours, a modulus value consistent with the pullback time can be selected from Table 2.

Bending Stress

HDD river crossings incorporate radii-of-curvature, which allow the HDPE pipe to cold bend within its elastic limit. These bends are so long in radius as to be well within the flexural bending capability of polyethylene pipe. PE3408 of SDR 11 can be cold bent to 25 times its nominal OD (example: for a 12" SDR 11 HDPE pipe, the radius of curvature could be from infinity down to the minimum of 25 feet, i.e., a 50-foot diameter circle). Because the drill stem and reaming rod are less flexible, normally polyethylene can bend easily to whatever radius the borehole steel drilling and reaming shafts can bend because these radii are many times the pipe

OD. However, in order to minimize the effect of ovaling some manufacturers limit the radius of curvature to a minimum of 40 to 50 times the pipe diameter. As in a previous section, the tensile stress due to bending is included in the calculations.

Thermal Stresses and Strains

HDD pipeline crossings are considered to be fully restrained in the axial direction by the friction of the surrounding soil. This is generally accepted to be the case, though, based on uncased borings through many soil types with the progressive sedimentation and borehole reformation over a few hours to several months. This assumption is valid for the vast majority of soil conditions, although it may not be completely true for each and every project. During pipe installation, the moving pipeline is not axially restrained by the oversize borehole. However, the native soil tends to sediment and embed the pipeline when installation velocity and mud flow are stopped, thus allowing the soil to grip the pipeline and prevent forward progress or removal. Under such unfortunate stoppage conditions, many pipelines may become stuck within minutes to only a few hours.

The degree to which the pipeline will be restrained after completed installation is in large part a function of the sub-surface soil conditions and behavior, and the soil pressure at the depth of installation. Although the longitudinal displacement due to thermal expansion or contraction is minimal, the possibility of its displacement should be recognized. The polyethylene pipe should be cut to length only after it is in thermal equilibrium with the surrounding soil (usually overnight). In this way the "installed" versus "operating" temperature difference is dropped to nearly zero, and the pipe will have assumed its natural length at the existing soil/water temperature. Additionally, the thermal inertia of the pipe and soil will oppose any brief temperature changes from the flow stream. Seasonal temperature changes happen so slowly that actual thermally induced stresses are usually insignificant within polyethylene for design purposes.

Torsion Stress

A typical value for torsional shear stress is 50% of the tensile strength. Divide the transmitted torque by the wall area to get the torsional shear stress intensity. During the pullback and reaming procedure, a swivel is typically used to separate the rotating cutting head assembly from the pipeline pull segment. Swivels are not 100% efficient and some minor percent of torsion will be transmitted to the pipeline. For thick wall HDPE pipes of SDR 17, 15.5, 11, 9 and 7, this torsion is not significant and usually does not merit a detailed engineering analysis.

EXAMPLE CALCULATIONS Example Calculations are given in Appendix A.

References

Ariaratnam, S.T. (2001). *Evaluation of the Annular Space Region in Horizontal Directional Drilling Installations*, Arizona State University.

Huey, D.P., Hair, J.D., & McLeod, K.B. (1996). Installation Loading and Stress Analysis Involved with Pipelines Installed by Horizontal Directional Drilling, No-Dig '96 Conf., New Orleans, LA.

Janson, L.E. (1991). Long-Term Studies of PVC and PE Pipes Subjected to Forced Constant Deflection, Report No. 3, KP-Council, Stockholm, Sweden.

Kirby, M.J., Kramer, S.R., Pittard, G.T., & Mamoun, M. (1996). Design Guidelines and Procedures for Guided Horizontal Drilling, Proceedings of the International No-Dig '96 Conf., New Orleans, La.

Knight, M.A., Duyvestyn, G., & Gelinas, M. (2001, Sept). Excavation of surface installed pipeline, J. *Infrastructural Systems*, Vol. 7, no 3, ASCE.

O'Rourke, T.D., El-Gharbawy, S.L., & Stewart, H.E. (1991). Soil Loads at Pipeline Crossings, ASCE Specialty Conference on Pipeline Crossings, Denver, CO.

Petroff, L.J. (1997). Design Guidelines for Directional Drilled PE Pipe, ASCE Congress on Trenchless Technology, Boston, MA.

Petroff, L.J. & Dreher, P.L. (1997). Design Considerations for HDPE for Trenchless Applications, No-Dig '97, Seattle, WA.

Puckett, J.S. (2003). *Analysis of Theoretical versus Actual HDD Pulling Loads*, ASCE International Conference on Pipeline Engineering and Construction, Baltimore, MD.

Rybel V. Directional Drilling for Trenchless Construction, CH2M Hill, Corvalis, Oregon.

Sener, E.M. & Stein, R. (1995). *Mini-Horizontal Directional Drilling Manual*, North American Society for Trenchless Technology, Chicago, Il.

Spangler, M. G. & Handy, R. L. (1973). *Soil Engineering*, Intext, Harper and Row, New York, NY.

Stein, D., Mollers, &K., Bielecki, R. (1989). *Microtunnelling*, Ernst & Sohn, Berlin.

Svetlik, H. (1995, March). Design Considerations for HDPE Pipe Used in Directional Drilling, *No-Dig Engineering*, Vol.2, No.3.

Watkins, R.K. (1994). Structural Mechanics of Buried Pipes, Buried Pipes Seminar, University of Colorado at Denver.

Watkins, R.K. & Anderson, L.R. (1995). Structural Mechanics of Buried Pipes, Dept. of Civil and Environmental Engineering, Utah State University, Logan, UT.

APPENDIX A

Design Calculation Example for Service Loads (Post-Installation)

Example 1
A 6" IPS DR 11 HDPE pipe is being pulled under a railroad track. The minimum depth under the track is 10 ft. Determine the safety factor against buckling.

GIVEN PARAMETERS

OD = 6.625 in
Nominal Pipe OD

DR = 11 Pipe
Dimension Ratio

H = 10 ft.
Max. Borehole Depth

g_s = 120 lbf/ft³
Unit Weight of Soil

P_{Live} = 1,100 lbf/ft²
E-80 Live Load

PE Material Parameters

Wheel loading from train will be applied for several minutes without relaxation. Repetitive trains crossing may accumulate. A conservative choice for the apparent modulus is the 1000-hour modulus.

$E_{mid} = 43{,}700$ psi

$\mu = 0.45$ Long-Term Poisson's Ration

Soil and Live Load Pressure on Pipe (Assuming that the earth load equals the prism load is perhaps too conservative except for a calculation involving dynamic surface loading.)

$P = (g_s H + P_{Live})\, 1\text{ ft}^2/144\text{ in}^2$

$P = 15.97$ psi

Ring Deflection resulting from soil and live load pressures assuming no side support is given by equation 10.

$$\%\frac{y}{D} = \frac{0.0125 P}{\dfrac{E_{mid}}{12(DR-1)^3}}$$

% y/D = 5.482 Percent deflection from soil loads

Determine critical unconstrained buckling pressure based on deflection from loading and safety factor using Eq. 11

$f_o = 0.56$ Ovality compensation factor for 5.5% ovality from Figure 3

$$P_{UC} = \frac{2 E_{mid}}{(1-m^2)} \left(\frac{1}{DR-1}\right)^3 f_o$$

$P_{UC} = 61.37$ psi
Critical unconstrained buckling pressure (no safety factor)

$$SF_{cr} = \frac{P_{UC}}{P}$$

$SF_{cr} = 3.84$ Safety factor against buckling

Example 2
A 6" IPS DR 13.5 HDPE pipe is being pulled under a small river for use as an electrical duct. At its lowest point, the pipe will be 18 feet below the river surface. Assume the slurry weight is equal to 75 Ib/cu.ft. The duct is empty during the pull. Calculate a) the maximum pulling force and b) the safety factor against buckling for the pipe. Assume that the pipe's ovality is 3% and that the pulling time will not exceed 10 hours.

Solution
Calculate the safe pull strength or allowable tensile load.

OD = 6.625in. - Pipe outside diameter
DR = 13.5 - Pipe dimension ratio
T_{allow} = 1150 psi - Typical safe pull stress for HDPE for 12-hour pull duration

$$F_s = \pi T_{allow} \, OD^2 \left(\frac{1}{DR} - \frac{1}{DR^2}\right)$$

F_s = 1.088 x 10⁴ lbf
Safe pull strength for 6" IPS DR 13.5 HDPE pipe assuming 10-hour maximum pull duration

Step 1
Determine the critical buckling pressure during Installation for the pipe (include tensile reduction factor assuming the frictional drag during pull results in 1000 psi longitudinal pipe stress)

E = 57,500 ps - Apparent modulus of elasticity (for 10 hours at 73 degrees F)
μ = 0.45 - Poisson's ratio (long term value)
fo = 0.76 - Ovality compensation factor (for 3% ovality)
R = 0.435 - Tensile ratio (based on assumed 1000 psi pull stress calculation)

$$f_R = \sqrt{5.57 - (r+1.09)^2} - 1.09$$

f_R = 0.71
Tensile Reduction Factor

$$Pcr = \frac{2E}{(1-\mu^2)} \left(\frac{1}{DR-1}\right)^3 \cdot f_o \cdot f_R$$

P_{CR} = 39.90
Critical unconstrained buckling pressure for DR 13.5 pipe without safety factor

Step 2
Determine expected loads on pipe (assume only static drilling fluid head acting on pipe, and borehole intact with no soil loading)

gslurry = 75 lbf/ft³, drilling fluid weight
H = 18 ft, Maximum bore depth

$$P_{slurry} = Hg_{slurry}(\frac{1ft^2}{144in^2})$$

P_{slurry} = 9.36 psi
Total static drilling fluid head pressure if drilled from surface

Step 3
Determine the resulting safety factor against critical buckling during installation

$$SF_{CR} = \frac{P_{CR}}{P_{slurry}}$$

SF_{CR} = 4.25
Safety factor against critical buckling during pull

Example 3
Determine the safety factor for long-term performance for the communication duct in Example 2. Assume there are 10 feet of riverbed deposits above the borehole having a saturated unit weight of 110 Ib/ft³. (18 feet deep, 3% initial ovality)

Solution

Step 1
Determine the pipe soil load (Warning: Requires input of ovality compensation in step 4.

E long = 28,200 psi - Long-term apparent modulus
g_w = 62.4 lbf/ft.³ - Unit weight of water
H = 18 ft Max. - Borehole depth
g_s = 110 lbf/ft.³ - Saturated unit weight of sediments
GW = 18 ft - Groundwater height
C = 10ft. - Height of soil cover
OD = 6.625 in - Nominal pipe OD
DR = 13. - Pipe dimension ratio
μ = 0.4 - Long-term Poisson's ratio

$$P_{soil} = (g_S - g_W) C (\frac{1ft^2}{144in^2})$$

P_{soil} = 3.30 psi
Prism load on pipe from 10' of saturated cover (including buoyant force on submerged soil)

Step 2
Calculate the ring deflection resulting from soil loads assuming no side support.

$$\%(y/D) = \frac{0.0125 \times P_{soil} \times 100}{[\frac{E_{long}}{12(DR-1)^3}]}$$

% (y/D) = 3.43 Percent deflection from soil loads
t = OD/DR t = 0.491 in

Step 3
Determine the long-term hydrostatic loads on the pipe

$$P_W = (\frac{GW}{2.31 \text{ ft/psi}}) + P_{soil}$$

Pw = 11.09
External pressure due to groundwater head

$$g_{slurry} = 75 \text{ lb/cu.ft.}^3$$

Unit weight of drilling fluid

$$P_{slurry} = g_{slurry} H (\frac{1\text{ft}^2}{144\text{in}^2})$$

P_slurry = 9.37 psi
External pressure due to slurry head

$$P_W > P_{slurry}$$

Therefor use PW for buckling load

Step 4
Determine critical unconstrained buckling pressure based on deflection from loading

f₀ = 0.64 5% Ovality Compensation based on 3% initial ovality and 2% deflection

$$P_{UC} = \frac{2E_{long}}{(1-m^2)} (\frac{1}{DR-1})^3 f_o$$

P_UC = 23.17 psi
Critical unconstrained buckling pressure (no safety factor)
SF_CR = 2.08

$$SF_{CR} = \frac{P_{UC}}{P_W} \qquad SF_{CR} = 2.08$$

Safety Factor against buckling pressure of highest load (slurry)

APPENDIX B

Design Calculations Example for Pullback Force

Example 1

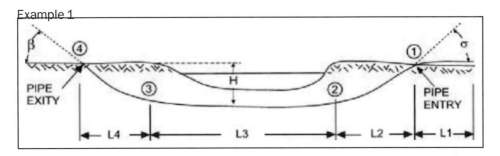

Find the estimated force required to pull back pipe for the above theoretical river crossing using Slavin's Method. Determine the safety factor against collapse. Assume the HDPE pipe is 35 ft deep and approximately 870 ft long with a 10 deg. entry angle and a 15 deg. exit angle. Actual pullback force will vary depending on backreamer size, selection, and use; bore hole staying open; soil conditions; lubrication with bentonite; driller expertise; and other application circumstances.

PIPE PROPERTIES
Outside Diameter
OD = 24 in - Long-term Modulus - E_{long} = 28,250 psi

Standard Dimension Ratio
DR = 11 - 24 hr Modulus - E_{24hr} = 56,500 psi

Minimum wall thickness
t = 2.182 in - Poisson's ratio (long term) - μ = 0.45 - Safe Pull Stress (24 hr) - spb = 1,100 psi

PATH PROFILE
H = 35 ft Depth of bore

g_{in} = 10 deg Pipe entry angle

g_{ex} = 15 deg Pipe exit angle

L_1 = 100 ft Pipe drag on surface (This value starts at total length of pull, approximately 870 ft. then decreases with time. Assume 100 ft remaining at end of pull)

L_{cross} = 870 ft

PATH LENGTH (DETERMINE L2 AND L4)
Average Radius of Curvature for Path at Pipe Entry g_{in} is given in radians

$$R_{avgin} = 2H/g_{in}^2$$

R_{avgin} = 2.298 x 10³ ft
Average Radius of Curvature for Path at Pipe Exit

$$R_{agex} = 2H/g_{ex}^2$$

R_{agex} = 1.021 x 10³ ft
Horizontal Distance Required to Achieve Depth or Rise to the Surface at Pipe Entry

$$L_2 = 2H/g_{in}$$

L_2 = 401.07 ft
Horizontal Distance Required to Achieve Depth or Rise to the Surface at Pipe Exit

WHERE
L_2 & L_4 = horizontal transition distance at bore exit & entry respectively.

DETERMINE AXIAL BENDING STRESS
R = R_{avgex} - Min. Radius for Drill path
R = 1.021 x 10³ ft
OD = 24 in
Radius of curvature should exceed 40 times the pipe outside diameter to prevent ring collapse.
r = 40 OD
r = 80 ft Okay
R > r

Bending strain
e_a = OD/2R
e_a = 9.79 x 10⁻⁴ in/in

WHERE
e_a = bending strain, in/in
OD = outside diameter of pipe, in
R = minimum radius of curvature, ft

Bending stress
$S_a = E_{24hr} e_a$
sa = 55.32 psi

WHERE
S_a = bending stress, psi

FIND PULLING FORCE
Weight of Empty Pipe
$P_w = 3.61 \times 10^{-2}$ lbf/in³
$g_a = 0.95$
$g_b = 1.5$
$w_a = \pi OD^2 (DR-1/DR^2) r_w\, g_a\, 12$ in/ft
$w_a = 61.54$ lbf/ft

Net Upward Buoyant Force on Empty Pipe Surrounded by Mud Slurry
$w_b = \pi(OD^2/4)\, r_w\, g_b - w_a$
$w_b = 232.41$ lbf/ft

WHERE
r_w = density of water, lb/in3
g_a = specific gravity of the pipe material
g_b = specific gravity of the mud slurry
w_a = weight of empty pipe, lbf/ft
w_b = net upward buoyant force on empty pipe surrounded by mud slurry

DETERMINE PULLBACK FORCE ACTING ON PIPE
See figure:
$L_1 = 100$ ft - $v_a = 0.4$
$L_2 = 401.07$ ft - $v_b = 0.25$
$L_3 = 200$ ft - $\sigma = g_{in}$ - $\sigma = 10$ deg
$L_4 = 267.38$ - $\beta = g_{ex}$ - $\beta = 15$ deg
$L_3 = L_{cross} - L_2 - L_4$ - $L_3 = 201.549$ ft
$T_A = \exp(v_a\, \sigma)\,[v_a\, w_a\, (L_1 + L_2 + L_3 + L_4)]$
$T_A = 2.561 \times 10^4$ lbf
$T_B = \exp(v_b\, \sigma)\,(T_A + v_b\,[w_b]\,L_2 + w_b\,H - v_a\, w_a\, L_2\, \exp(v_b\, \sigma))$
$T_B = 4.853 \times 10^4$ lbf
$T_C = T_B + v_b\,[w_b]\,L_3 - \exp(v_b\, \sigma)\,(v_a\, w_a\, L_3.\exp(v_a\, \sigma))$
$T_C = 5.468 \times 10^4$ lbf
$T_D = \exp(v_b\, \sigma)\,[T_C + v_b\,[w_b]\,L_4 - w_b\,H - \exp(v_b\, \sigma)\,(v_a\, w_a\, L_4\, \exp(v_b\, \sigma))]$
$T_D = 5.841 \times 10^4$ lbf

WHERE
T_A = pull force on pipe at point A, lbf
T_B = pull force on pipe at point B, lbf
T_C = pull force on pipe at point C, lbf

TD = pull force on pipe at point D, lbf
L1 = pipe on surface, ft
L2 = horizontal distance to achieve desired depth, ft
L3 = additional distance traversed at desired depth, ft
L4 = horizontal distance to rise to surface, ft
v_a = coefficient of friction applicable at the surface before the pipe enters bore hole
v_b = coefficient of friction applicable within the lubricated bore hole or after the (wet) pipe exits
σ = bore hole angle at pipe entry, radians
β = bore hole angle at pipe exit, radians
(refer to figure at start of this appendix)

HYDROKINETIC PRESSURE
ΔP = 10 psi

Dh = 1.5 OD

Dh = 36in

$\Delta T = \Delta P\, (\pi/8)\, (Dh^2 - OD^2)$

$\Delta T = 2.82 \times 10^3\, lbf$

WHERE:
ΔT = pulling force increment, lbf

ΔP = hydrokinetic pressure, psi

Dh = back reamed hole diameter, in

Compare Axial Tensile Stress with Allowable Tensile Stress During Pullback of 1,100 psi:
Average Axial Stress Acting on Pipe Cross-section at Points A, B, C, D

$$s_1 = (T_i + \Delta T)\ \left(\frac{1}{\pi OD^2}\right)\ \left(\frac{DR^2}{DR-1}\right)$$

s1 = 190.13 psi <1,100 psi OK

s2 = 343.40 psi <1,100 psi OK

s3 = 384.55 psi <1,100 psi OK

s4 = 409.48 psi <1,100 psi OK

WHERE
T_i = TA, TB, TC, TD (lbf)

s_i = corresponding stress, psi

Breakaway links should be set so that pullback force applied to pipe does not exceed 1,100 psi stress.
ID = OD - 2t

$Fb = s_{pb}\, (\pi/4)(OD^2 - ID^2)$

$Fb = 1.64 \times 10^5\, lbf$

DETERMINE SAFETY FACTOR AGAINST RING COLLAPSE DURING PULLBACK
External Hydraulic Load
External static head pressure

P_{ha} = (1.5) (62.4 lbf/ft3) (H)

P_{ha} = 22.75 psi

Combine static head with hydrokinetic pressure

$P_{effa} = P_{ha} + \Delta P$

P_{effa} = 32.75 psi

CRITICAL COLLAPSE PRESSURE
Resistance to external hydraulic load during pullback

f_o = 0.76 Ovality compensation factor (for 3% ovality)

$r = S_4/2S_{Pb}$

$r = 0.186$

Tensile ratio (based on 1,100 psi pull stress calculation)

Tensile reduction factor

P_{CR} = 96.41 psi

SAFETY FACTOR AGAINST COLLAPSE
SF = Pcr/Pha

F = 4.238

WHERE
P_{ha} = applied efflective pressure due to head of water of drilling

P_{cr} = calculated critical buckling pressure due to head of water of drilling fluid, psi

SF = Safety Factor

chapter 13

HVAC Applications

Introduction
The performance and use characteristics of polyethylene pipe make it an ideal choice for use in certain HVAC – Heating, Ventilation, and Air Conditioning – applications. Typically, HVAC is thought of as flexible vent pipes, steam pipes, etc. However, since the 1980's polyethylene pipe's flexibility, strength, and ease of use has had a major impact on HVAC applications such as geothermal heat pumps and radiant heating.

This chapter presents information and design criteria for the use of polyethylene pipe in applications such as:

Ground Source Heat Pumps - basic use and standards, configuration, joining methods and installation considerations.

Solar Applications – use of PE pipe for solar water heating applications.

Vacuum Systems – use and design limitations.

Ground Source Heat Pump Systems

Due to polyethylene pipe's versatility, flexibility, durability, leakproof fusion joints, and ease of use, it has become a key component in the success of Ground Source Heat Pumps.

There are two basic types of heat pumps – air source and ground source. An air source system utilizes temperature variations with the air to gain operating efficiency. A ground source, or Geothermal Heat Pump (GHP) system uses an electric pump to circulate fluid from the heat pump cycle through a series of polyethylene pipes buried in the ground to take advantage of the relatively constant ground temperatures. These pipes are known as Ground Heat Exchangers. In simple terms, in the summer the heat pump's refrigerant cycle transfers heat from the building into the circulating fluid. The fluid is then circulated through the ground heat exchanger where the ground acts as a heat sink, cooling the fluid before it returns to the building. In the winter, the system works in reverse. The heat pump uses the earth to warm the circulating fluid, which is then transferred back to the inside heat exchanger. In addition to heating and cooling the air, a desuperheater can be added to this cycle that can provide most, if not all, hot water for use in the building as well.

The properties that control this process are based on the ability of the PE pipe to transfer heat either out of, or into, the system. The heat transfer by conduction mechanism that governs this system is the same as any heat exchanger. It is assumed that the ground is at a steady state condition. This type of heat transfer mechanism is governed by the basic equation:

$$Q = k \times A \times \Delta T$$

WHERE
Q = Heat transfer, BTU/hr
k = Thermal transfer coefficient, BTU/hr*ft2*°F
A = Surface area, ft2
ΔT = Temperature differential, °F

Polyethylene itself is typically considered an insulator and holds heat rather well. However, in this application, the benefits of the polyethylene pipe far outweigh this performance characteristic. There are many other variables that need consideration when designing a GHP system. Most manufacturers have software available to aid in the determination of the size of the unit and the footage of pipe needed for the geothermal heat exchanger.

Geothermal heat pumps are very economical to operate and can save a substantial amount of money in operating costs over the life of the system. It has been reported[1] that a traditional furnace uses one unit of energy and returns less than one unit back as heat. A ground source heat pump uses one unit of energy and returns as much as three units back as heat. The polyethylene pipe acting as the heat transfer medium with the ground helps make this possible.

Types of Ground Heat Exchangers

The polyethylene pipe used in the ground heat exchanger can be configured several different ways depending on the size of the system, surrounding land, or availability of a large open water source. The two basic types are open systems and closed systems.

Open systems require a suitable supply of water where open discharge is possible. This type of system uses the HDPE pipe to bring fresh water to the heat pump, and then discharges the water back into the water supply. Only fresh water is used, and there is no need for a special heat transfer or antifreeze solution. A key PE pipe design consideration for an open system is the fact that the system will have a suction and discharge loop. This means the pipe may need to handle negative vacuum pressures and positive pumping pressures.

The more common type of GHP installation is a closed loop system. A closed system is just that — a "closed loop" recirculating system where the HDPE pipe circulates an "antifreeze" solution continuously. This type of system can be installed several different ways such as: a pond loop system, a vertical loop system, or a horizontal (slinky) loop system. Each of these types of installation utilizes the basic performance benefits and versatility of HDPE pipe to get the most beneficial type installation for the surrounding conditions.

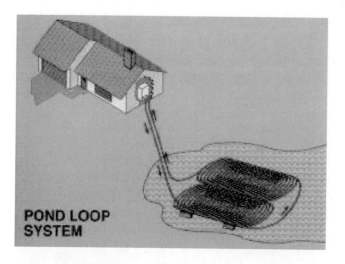

POND LOOP SYSTEM

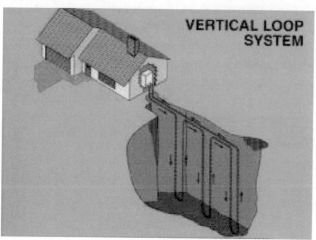

VERTICAL LOOP SYSTEM

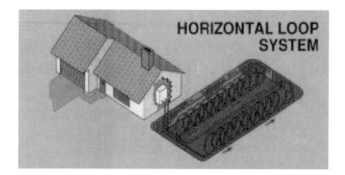

Pipe Specifications and Requirements

Polyethylene pipe is the material of choice for the ground heat exchanger in a ground source heat pump system. The International Ground Source Heat Pump Association (IGSHPA) has developed some design and installation standards for the HDPE pipe that is required for a geothermal heat exchanger. Specifications for polyethylene pipe and fittings used in the geothermal heat exchanger are as follows:

1. **General** – All pipe and heat-fused materials shall be manufactured from virgin polyethylene extrusion compound material in accordance with ASTM D 3350. Pipe shall be manufactured to outside diameters, wall thickness, and respective tolerances as specified in ASTM D 3035, or D 2447. Fittings shall be manufactured to diameters, wall thickness, and respective tolerances as specified in ASTM D 2683 for socket fittings, ASTM D 3261 for butt fittings, and ASTM F 1055 for electrofusion fittings.

2. **Material** – The material shall maintain a 1600 psi hydrostatic design basis at 73.4°F (23°C) per ASTM D 2837, and shall be listed in PPI's TR-4 as a PE 3408 compound. The material shall be a high density extrusion compound having a minimum cell classification of 345434 with a UV stability of C, D, or E as specified in ASTM D 3350 with the following exception: this material shall exhibit zero failures (F_0) when tested for a minimum of 192 hours under ASTM D 1693, condition C, as required in ASTM D 3350.

3. **Dimensions** – Pipe with a nominal diameter of less than 1 ¼" shall be manufactured in accordance with ASTM D 3035 with a maximum SDR of 11.

   Pipe manufactured with a nominal diameter of 1 ¼" and larger shall be made in accordance with ASTM D 3035 with a maximum DR of 15.5, or ASTM D 2447 schedule 40. If the pipe is used in a vertical bore application, it shall be manufactured in accordance with ASTM D 3035 with a maximum DR of 11.

   Pipe 3" nominal diameter and larger shall be manufactured in accordance with ASTM D 3035, with a maximum DR of 17, or D 2447 schedule 40.

4. **Markings** – Sufficient information shall be permanently marked on the length of pipe as defined by the appropriate ASTM pipe standard.

5. **Certification materials** – Manufacturer shall supply a notarized document confirming compliance with the above standards.

This specification is from the IGSHPA *"Closed Loop/Geothermal Heat Pump Design and Installation Standards 2000."* For the most current revision contact IGSHPA at Oklahoma State University.

TABLE 1
Maximum Allowable Operating Pressures (MAOP) at 73.4°F and 140°F for Specified DR's in Ground Heat Exchangers

Pipe	MAOP (psig)		
DR	73.4°F (HDB=1600 psi)	140°F (HDB=800 psi)	140°F (HDB=1000 psi)
9	200	100	125
11	160	80	100
15.5	110	55	69
17	100	50	63

Notes:
1) PE 3408 at 73.4°F HDB of 1600 psi
2) DF = 0.5
3) HDPE pipe is not rated for service above 140°F

The recommended specification takes into account the optimum performance based on the need to make sure the pipe and fittings can handle the pressures and stresses involved in the application, as well as the heat transfer requirements for the heat exchanger itself. Heavier wall pipe may be able to handle higher pressures and stresses, but the thicker wall lowers the heat transfer efficiency with the ground. All of these parameters must be balanced. When designing the PE pipe heat exchanger, maximum operating pressures and temperatures, as well as head and surge pressures must be taken into account.

For closed-loop geothermal heat exchangers, even though a high stress crack-resistant polyethylene is required, it is appropriate to make sure the antifreeze solution used in the heat exchanger does not adversely affect the stress crack performance of the pipe and fittings. The antifreeze solution manufacturer should be able to supply this information.

More information on the design of PE pipe systems for pressure, surges, flow capacities, etc. can be found in the design of polyethylene piping systems chapter of this Handbook.

Pipe Joining Methods

Polyethylene pipe can be joined by several different methods. One of the best attributes of PE pipe is its ability to be heat fused, producing a 100% leakproof joint that is as strong, or stronger, than the pipe itself. Extensive information on joining PE pipe can be found in the Polyethylene Joining Procedures chapter of this handbook.

IGSHPA recommends acceptable methods for joining as 1) a heat fusion process, or 2) stab-type mechanical fittings to provide a leak-free union between the pipe ends that is stronger than the pipe itself. This type of mechanical joint is also known as a Category 1 mechanical joint according to ASTM D 2513.

In addition, it is recommended that fusion transition fittings with threads must be used to adapt to copper pipe or fittings. Fusion transition fittings with threads or barbs must be used to adapt to high strength hose. Barbed fittings are not permitted to be connected directly to the polyethylene pipe, with the exception of stab-type fittings as described above. All mechanical connections must be accessible.

Since mechanical connections must be accessible, fusion joints are typically used wherever possible. Butt, socket or electro-fusion is used to join individual sections of pipe. "U-bend" fusion fittings are used for creating the return line in vertical bores. In fact, it is common for polyethylene pipe made for geothermal heat exchangers to be double wrapped on a coil and the "u-bend" fitting fused on at the factory. This makes insertion into a vertical bore very quick and easy. Sidewall fusion can be used to join parallel pipe loops to a header. All fittings must be pressure rated for the expected operating and surge pressures, and joined according to the manufacturer's recommended procedures. This is a critical feature since this joint will be at the bottom of a well and grouted into place. Repairing a leaking joint would be very difficult. However, due to polyethylene pipe's ability to create very strong fusion joints, this concern is easily overcome.

Pipe Installation

As discussed previously, there are several types of installation choices for ground source heat pumps. It is important to follow the GHP manufacturer's requirements for the type of unit being used. This will define the amount of pipe needed for the particular installation and environment. However, there are some general guidelines for polyethylene pipe that will help assure a successful installation.

Generally, it is desired to keep the diameter of the HDPE pipe as small as possible, but not so small that pumping power to circulate the antifreeze solution becomes too great, thus losing the operating efficiency of the GHP. The smaller the diameter, the higher the surface to volume ratio will be, and the better chance for turbulent flow inside the pipe. Both of these conditions promote more efficient heat transfer. Most ground heat exchangers are constructed from ¾" to 2" pipe. The headers will be 1 ¼" to 2", and the individual loops will be ¾", 1" or 1 ¼". The amount of pipe utilized varies depending on environmental conditions and how much heating or cooling capacity is needed. As an example, a typical 3-ton ground heat exchanger may use 200 feet of headers and 400 feet for each parallel loop.

If trenching for a horizontal installation or header system, avoid sharp bends around corners. Pipe manufacturers have a minimum bend radius that will assure that the pipe is not overstressed. If a sharp corner is needed, utilize an elbow fitting. Remove any sharp rocks from backfill material. Long-term contact between the polyethylene pipe and a sharp object could lead to premature failure of the pipe. Even though the polyethylene pipe is very stress crack-resistant, it is a good idea to minimize this type of contact. The addition of sand in the bottom of the trench will help minimize incidental contact with sharp objects. It is also possible to plow the pipe directly into the ground using a vibratory plow. This works well up to 3-4 feet depth in areas with loose or unstable soils, and where there is not an excessive amount of rocks that could impinge on the pipe over time.

Vertical bores for ground heat exchangers are typically much simpler than drilling a water well. Generally casing is not needed if the borehole is sufficiently stable long enough to get the pipe loop installed. It is sometimes more economical to have several shallow bores rather than one deep bore. However, the bores need to be more than 50 ft. to be assured of reaching depths where ground temperatures are cooler and constant. Vertical bores must be backfilled appropriately to be sure the pipe loops have intimate contact with the soil or grout. If there are air gaps around the pipe, the heat transfer by conduction will be negatively affected.

For both types of installations leave a significant portion (3-5% of total length) of pipe extending from the bores or trenches to compensate for any relaxation from stretching, or contraction from temperature changes. Final connections to the header can be made after the system comes to steady state, usually within 24 hours.

Pressure Testing Ground Heat Exchanger

After installation of pipe is completed, but prior to backfilling and/or grouting, it is necessary to flush, purge and pressure test the system. Flushing any dirt or foreign matter that entered the piping during construction is necessary in order to minimize excessive wear on pumps and seals. Purging of any air pockets will make sure that all loops are flowing as intended and heat transfer will be optimized. Flushing and purging can be done at the same time.

Before charging the system with antifreeze, it is necessary to pressure test the system with water (not air) to make sure all of the joints and connections were done correctly. IGSHPA recommends that the heat exchanger be isolated and tested to 150% of the pipe design pressure, or 300% of the system operating pressure, whichever is less, when measured from the lowest point in the loop being tested. No leaks shall occur within a 30-minute test period. At this time flow rates and pressure drops can be compared to calculated design values. A minimum flow velocity of 2 ft/min. must be maintained for a minimum of 15 minutes to remove all air from the heat exchanger.

Since the PE pipe can expand slightly during this high level of pressurization, a certain amount of make-up water may be required. This is normal and does not indicate a leak in the system. If the pressure does not stabilize, then this may be an indication of a leak. Follow pipe manufacturer's guidelines for pressure testing of system.

For additional information of Ground Source Heat Pump design and installation contact:
International Ground Source Heat Pump Association (IGSHPA) www.igshpa.okstate.edu.
American Society of Heating, Refrigeration, and Air-Conditioning Engineers (ASHRAE) www.ashrae.org.

Solar Applications

The use of solar energy was virtually nonexistent 25 years ago, but has grown to become a significant industry in the United States. Most solar applications are geographically concentrated in the states with a high percentage of sunshine - California, Arizona, New Mexico, Colorado, and Florida.

Solar heating systems vary in size. The very simplest consist of nothing more than a black pipe lying in the sun connected to a swimming pool circulating pump. The more complex systems utilize collectors with 1, 2, or 3 layers of glazing plus piping and pumps. In addition, the later systems may include heat transfer fluids, heat storage tanks, heat exchangers, and temperature and pressure controls. PE piping can play a major role in this application. Its combination of flexibility, high temperature properties, and resistance to freeze damage and corrosion are major advantages to this end-use. There are, however, precautions that should be taken to prevent misuse.

Not all polyethylene pipe is recommended for solar heating applications. Check with the manufacturer before use.

Features and Benefits

The performance benefits of polyethylene pipe in solar heating are:

Ease of Installation - Minimizing the overall cost of solar heating is important to make them viable alternatives and to expand customer acceptance. Polyethylene pipe and tubing is available in many sizes and lengths. Its versatility and flexibility allows installations to be made with the most cost-effective design.

Freeze Tolerant - Frozen lines can be a major problem. Although collectors are protected, supply lines need to be protected from freezing or they should be made of materials that are resistant to damage if water freezes. Polyethylene pipe can normally handle a full-freeze situation without cracking or splitting.

High Temperature Resistance - For continuous use, polyethylene pipe must be suitable for high temperature environments. Polyethylene materials for use at elevated temperatures are listed in PPI's TR-4. Currently, the maximum rated temperature for polyethylene is 140°F (60°C). For use at higher temperatures contact the manufacturer for recommendations.

Collector Technologies

The most significant use of solar heating has been for swimming pool, domestic hot water, and space heating. Solar collectors are classified according to their water discharge temperatures: low temperature, medium temperature, and high temperature. Low temperature systems generally operate at a temperature of 110°F and have a maximum stagnation temperature of 180°F. Medium temperature collectors typically have discharge temperatures of 180-200°F, but can generate stagnation temperatures of 280°F, or more, for several hours. High temperature collectors routinely operate at temperatures of at least 210°F and can generate stagnation temperatures of more than 400°F. Pipe or tube made of polyethylene can be used directly with low temperature collectors with no special precautions. In addition, PE piping is being used extensively inside unglazed collectors where temperatures rarely exceed 110°F on a frequent basis.

To protect against ultraviolet exposure damage and to increase efficiency, plastic piping for use in collector panels should contain a minimum of 2% carbon black of proper particle size and with good dispersion. The carbon black has a two-fold benefit. One, the right kind of carbon black in the proper levels and adequately dispersed protects the PE from UV degradation for up to 40 years. Two, the carbon black aids in the absorption and retention of solar radiation, making the pipe more efficient in the collection of solar energy. Check with the pipe manufacturer for recommendations on long-term UV exposure resistance.

Plastic piping should not be used in conjunction with high temperature collectors such as the evacuated tube or concentrating types because of their extreme temperatures. In between these two extremes are the systems with medium temperature collectors that constitute the bulk of the market. These glazed collectors are used for domestic hot water and space heating systems. Depending on the type of collector and system design, some special precautions should be taken. The major types of medium temperature systems are described in the following paragraphs along with appropriate precautions. Medium temperature systems are either passive or active types.

Passive systems use no pumps or mechanical equipment to transport the heated water. The breadbox (passive) design uses a tank placed under a glazing material. The tank is painted flat black or coated with a selective absorber to increase the solar energy absorption. The collector may be the primary storage tank or the storage tank may be in the house. In the later case, when a preset temperature is reached, water flows by gravity to the storage tank in the home and fresh water from the main is added to bring the system up to volume. In the thermosyphon passive design, a storage tank is mounted above a collector and cold water flows down into the collector. As the water is heated in the collector, it rises through thermosyphon action back up to the storage tank. Because of the large volume of water in the collector, passive solar systems are not subject to high stagnation temperatures. Thus, polyethylene piping can be used throughout, including a hook-up directly to the collector system.

Active solar systems utilize a pump to move heat transfer fluids through the collector. Some utilize potable water as the heat transfer fluid (open systems) while others use solutions such as ethylene glycol, propylene glycol, silicone oils, or hydrocarbon oils (closed systems). Hydrocarbon oil or silicone oils are generally not recommended with polyethylene pipe. In closed systems, heat is transferred from the heat transfer fluid to potable water by means of a heat exchanger in the hot water storage tank. There are many types of heat transfer fluids, and it is necessary to verify with the manufacturer of the pipe that the fluid being used is compatible and will not negatively affect the long-term performance of the pipe or other system components.

Precautions

The extreme conditions encountered during stagnation can be a problem in active medium temperature collectors. As mentioned earlier, stagnation temperatures can exceed 280°F in most active medium temperature collectors. Under no circumstances should any PE piping be used inside the collector, or in the system where it will be exposed to such temperatures unless that material has been qualified for service at those elevated temperatures.

Installation

In general, solar collector manufacturers do not provide piping for the system.

The installer most likely will purchase the piping from the local plumbing supply wholesaler or solar supply house. Installers are usually plumbers, but in some areas like California, solar specialists also do installations. A qualified plumbing supply house may also perform installations. The installation requires knowledge of carpentry to provide roof support or mounting, electricity to install the control system, and plumbing to install the piping system and to tie it in to the storage tank and the existing domestic water supply. Always be sure the installation meets the requirements of the local building, plumbing and mechanical codes.

Vacuum Systems

Even though polyethylene pipe is normally used for internal pressure applications, it can also be used for vacuum systems as well. Some of the advantages of using PE pipe for vacuum lines are flexibility and heat fusion joining.

The increased flexibility means the pipe or tubing can more easily bend around curves removing the need for additional fittings. However, minimum bend radius restriction from the manufacturer should be followed. As a general rule-of-thumb, the minimum bend radius for PE pipe is 30 times the outside diameter (i.e. 25 OD's), and 100 OD's where there are fittings in line. The minimum bend radius may be more for thinner walled pipe.

Heat fusion joining gives a completely leak-proof joint, whereas a mechanical fitting may leak under a vacuum situation. Fusion joining information can be found in the Polyethlyne Joining Procedures chapter of this handbook, or from the pipe manufacturer.

Critical Buckling Under Vacuum

A vacuum situation can be designed the same way as an unconstrained external hydrostatic loading application. A full vacuum is 14.7 psi (1 atm). A polyethylene piping system can be designed to handle whatever amount of vacuum is required, but physically a vacuum can never be more than 14.7-psi. However, if other conditions such as groundwater are present, the total effective loading may be higher and should be taken into consideration.

To see if a PE pipe can handle this amount of loading, the unconstrained buckling equation is used:

$$P_{uc} = \frac{2E}{(1-\mu^2)} \left(\frac{1}{DR-1}\right)^3 \frac{fo}{N}$$

WHERE
P_{uc} = Unconstrained critical bucking pressure, psi
E = Elastic modulus (time dependent) at the average expected service temperature
μ = Poisson's Ratio = 0.35 for short-term loading
DR = Dimension ratio
f_0 = ovality compensation factor, 0.62 for 5% ovality
N = Safety factor, normally 2.0

The value of the elastic modulus for polyethylene is time dependent as shown in Table 2. Most vacuum situations will be short-term, and the initial short-term elastic modulus can be used. If it is determined that the vacuum will be continuous for a long period of time, then a reduced effective modulus should be used. However, with polyethylene these reduced time-dependent physical properties are for continuous times only – the times are not additive. If the vacuum is to be applied for one hour then released for a time, then reapplied, the times are not added together. The initial
short-term modulus can be used in the calculations. For more information on the effects on the elastic modulus with time and stress see the engineering properties chapter of this Handbook.

TABLE 2
PE Elastic Modulus as a Function of Time

Duration of Load	Apparent Modulus of Elasticity, E
Short-term	110,000 psi (800 MPa)
1 hour	90,000 psi (620 MPa)
10 hours	57,500 psi (400 MPa)
100 hours	51,200 psi (350 MPa)
50 years	28,200 psi (200 MPa)

Note: Values shown are for a typical PE 3408 material.

Let's assume that the vacuum is to be applied for periods of less than one hour. Therefore the short-term modulus can be used. Also, f_0 is approximately 0.62 for 5% ovality, and a typical engineering safety factor is 2. By plugging in these values and solving for DR, to handle a full, short-term vacuum without collapse, the SDR needed is 17. If severe bends or tensile stresses are being applied to the pipe at the time of vacuum, a heavier wall pipe may be needed. Table 3 shows level of vacuum other SDR's can handle under the same conditions.

TABLE 3
Short-Term Vacuum Levels for Various SDR's

SDR	Short-Term Vacuum (psi)
17 or less	14.7
21	9.7
26	5.0
32	2.6
41	1.2

Due to polyethylene's flexibility and elastic memory, if the piping is completely flattened by a vacuum, once the vacuum is released the piping can return to its original shape. This will depend on how long the pipe remained flattened.

References

1. Ground Source Heat Pump. (1996, February). *Popular Science*, p.73, Times Mirror Magazines, Inc.
2. *Do It Naturally!* Brochure, International Ground Source Heat Pump Association, Oklahoma State University.
3. Closed-Loop/Geothermal Heat Pump Systems, Design and Installation Standards 2000. (2000). ed. P. Albertson, International Ground Source Heat Pump Association, Oklahoma State University.
4. *Geothermal Heat Pumps*, Introductory Guide, Rural Electric Research (RER) Project 86-1A, Oklahoma State University Division of Engineering Technology, Stillwater, OK.
5. *ASTM Annual Book*, Volume 8.04, Plastic Pipe and Building Products, American Society for Testing and Materials, Philadelphia, PA.
6. Plastics Pipe Institute: Various Technical Reports, Technical Notes, Model Specifications, Washington, DC.

Chapter 14

Duct and Conduit

Introduction

The general purpose of conduit, or duct, is to provide a clear, protected pathway for a cable, or for smaller conduits, sometimes called innerducts. Advances in cable technologies, as well as the expense of repairing sensitive cable materials like fiber optic cable, have driven preferences for protective conduit over that of direct burial. Polyethylene (PE) conduit provides mechanical protection to fragile cable materials like fiber optic and coaxial cables, as well as protection from moisture or chemicals and even, in some cases, animals. Furthermore, the permanent pathway provided by conduit also facilitates replacement projects or future installations of additional cable or duct.

Buried conduit evolved from terracotta tile, cast concrete and Transite to plastics in the 1960s. Originally, PVC was utilized, but ultimately, PE has emerged as the material of choice due to its distinct advantages in installation options, versatility and toughness.

PE conduit can be installed below ground by a variety of methods, including open trench, plowing, continuous trenching and directional drilling. Also, its flexibility and availability in continuous coiled lengths facilitates installation into existing conduits or ducts as innerduct. In addition PE conduit provides many above ground or aerial options.

Conduit Specifications

The following specifications are utilized by the industry for the production of Conduit and Raceways:

- Telecommunication Conduits – ASTM F2160 Standard Specification for Solid-Wall High Density Polyethylene (HDPE) Conduit Based on Controlled Outside Diameter (OD).

- Power Conduits – ASTM F2160 Standard Specification for Solid-Wall High Density Polyethylene (HDPE) Conduit Based on Controlled Outside Diameter (OD)

- NEMA TC7 Smooth-wall Coilable Polyethylene Electric Plastic Conduit
- Electrical Conduits – UL 651A Type EB and A Rigid PVC Conduit and HDPE Conduit – UL 651B Continuous Length HDPE Conduit
- Premise Raceways – UL 2024 Optical Fiber Cable Raceway

Applications

PE conduit serves two primary industries: communications (telephone, CATV, data transmission) and electrical (power transmission).

In the communications industry, the advent of fiber optic cable has had a tremendous impact due to its significantly higher data-carrying capacity, particularly due to the explosion of the Internet. In telecommunications service (phone, data transmission), fiber optic cable is used, along with traditional copper cable. In cable television service (CATV), fiber optic is also growing rapidly in addition to (or replacing) coaxial cable. This progression toward fiber optic cable has made the need for protection more critical, since these materials are highly sensitive to moisture and mechanical stress. Damage can be very expensive in terms of interrupted service and replacement costs. Also, these cables are installed in very long, continuous runs which require a clear, protected pathway, as well as a leak-free system for air-assisted ("blow-in") installations. In addition to fiber optic, coaxial cables have seen improvements to increase bandwidth, making these materials more mechanically sensitive.

In the electrical industry, a critical requirement is on maintaining uninterrupted service, as consumers and businesses are even less tolerant of power outages than they are of phone or CATV service interruptions. Although many direct-buried power cable systems are designed for 30- or 40-year lifetimes, they are susceptible to external influences like rock impingement and often require frequent repairs. Conduit is finding favor over direct burial in these applications due to improved protection, but it must be continuous and facilitate quick repair operations. PE conduit is used to carry both primary (substation to transformer) and secondary (transformer to end-user) cables. Some of these installations also contain fiber optic cables placed alongside the power cables to connect with load-monitoring sensors located throughout the network.

Advantages of PE Conduit

High Density Polyethylene (HDPE) is the most commonly used PE material for conduit. HDPE conduit delivers significant physical property advantages over other conduit materials:

- **Ductility** - tough, HDPE conduit will better resist brittleness with age or cold weather.

- **Low temperature impact resistance** - PE withstands low temperature impact better than any other material. This is illustrated by impact testing on HDPE conduit conditioned at 4°F as compared to other materials conditioned at 73°F.
- **Permanent flexibility** - HDPE conduit bends and flexes without breakage, even with ground heaves or shifts, over a wide range of temperatures.
- **Temperature versatility** - HDPE conduit can be installed over an ambient temperature range of -30°F to 180°F. Power conductors rated at 90°C and medium voltage cable rated at 105°C are permitted for use with PE Conduit.

Installation

Flexible HDPE conduit can be wound onto reels several thousand feet long, does not require manufactured bends, and can be easily navigated around unexpected obstructions (in the ground or within existing ducts), simplifying installation. The few joints that are required can be made reliably through a number of options.

HDPE conduit is suitable for all methods of duct and cable installation, including trenching, direct plow and installation into existing main pathways (conduit pulling, sliplining and pipe bursting). Also, the flexible nature of PE conduit facilitates directional bore installations to breach obstacles like rivers or highways. Cable can consistently be pulled or blown into HDPE duct in great distances and at fast rates due to its low coefficient of friction. Special PE products and accessories are also available for above ground or aerial applications.

Features

A variety of PE conduit products are available for special applications.

- **Multiple ducts** of different color/stripe combinations and sizes can be delivered on one common reel, for a more efficient installation.
- Pre-installed **Cable-in-Conduit (CIC)** saves time and labor by allowing one-step placement of both cable and duct. The integrity of the cable is protected during the installation process by the HDPE duct. Testing prior to and after the duct has been extruded around cable is performed to ensure no performance loss. Cable-in-Conduit can be provided with fiber, coaxial, twisted pair and electrical cables.
- **Corrugated** innerduct is flexible, lightweight with a low coefficient of friction.
- **Ribbed conduit** (longitudinally or spiral) provides friction reduction in cable installation.
- **Self-supporting duct** includes a suspension strand already built into the duct for greater dimensional stability and ease of installation in the aerial plant. Deployment of ducts aerially allows for enhanced protection for the cable and allows for less costly cable repairs and capacity upgrade options.

Material Selection

The primary physical property advantages of HDPE conduit are flexibility, ductility and chemical resistance. Other physical attributes critical to the performance of HDPE conduit are tensile strength and stress crack resistance. However, the designing or specifying engineer should be aware that not all PE materials deliver the same level of performance in these areas, and it is critical to ensure that the material meets all the demands of the installation and service conditions. This section will briefly discuss these material considerations, but a more thorough discussion of PE technology is provided in the chapter on engineering properties of polyethylene in this Handbook.

Physical Properties

Cell Classification

The Cell Classification (ASTM D 3350) is a 6-digit numeric "code" which describes an HDPE conduit material's performance level in six key physical characteristics. This 6-digit classification often includes a single letter suffix representing a color or UV stabilizer category. This cell classification is used in specifications such as ASTM F2160 Standard Specification for Solid Wall High Density Polyethylene (HDPE) Conduit Based on Controlled Outside Diameter (OD). Each property is broken into 4-6 specific performance ranges.

Density - HDPE density generally has the greatest effect on many physical properties. For example, higher densities favor increased tensile strength and stiffness, while lower densities generally favor impact resistance, and flexibility and stress crack resistance. Density also affects coefficient of friction (COF – see Section 7), with higher density typically related to lower COF. Therefore, some degree of compromise may be necessary to balance properties required for a particular application.

Melt Index - Melt Index (MI), a measurement of a polymer's molten flow properties (ASTM D 1238), is related to molecular weight, or the length of the individual polymer chains. Generally, lower melt indices represent higher molecular weights while higher values indicate lower molecular weights. For any given PE resin, a lower melt index (higher molecular weight) will normally have superior physical properties.

Flexural Modulus - Flexural modulus is a measure of a plastic's stiffness, or its resistance to bending or deflection under applied load. In HDPE conduit, these stiffness characteristics generally affect load-bearing capability, bending radius, and tendency to ovalize (when coiled or bent). Flexural modulus should be taken into account when determining the appropriate wall thickness for an installation.

Tensile Strength/Yield Strength - Tensile yield strength, or the point at which a stress causes a material to deform beyond its elastic region (irreversible deformation), is a critical property for many conduit installation methods involving pulling (e.g., directional drilling). Yield strength can limit the rates or lengths of such installations (see page 7 Design Considerations), and it is an important consideration in determining allowable pull loads. It is important to note that both flexural modulus and tensile strength are affected by temperature (both decrease with increasing temperature).

Slow Crack Growth - ASTM D1693 Environmental Stress Cracking Resistance (ESCR) or ASTM F1473 Polyethylene Notched Tensile (PENT) can measure properties of slow crack growth. For PE conduit applications, ESCR is utilized. As one of the most important properties affecting the service life of PE conduit, stresses due to bends and rock impingement can cause inferior conduit materials to crack and fail, particularly at higher temperatures. ESCR is a laboratory test which measures a material's ability to resist cracking under these conditions. As mentioned above, higher densities generally have a negative effect on ESCR, and, as a general practice, base resins with densities below 0.950 have ESCR properties suitable for conduit applications.

Hydrostatic Strength Classification - The hydrostatic strength classification describes the material's resistance to failure under internal pressure; this property is primarily used for pressure piping applications and is not required for conduit. HDPE conduit materials are represented by a "0" (not pressure rated) in this category.

Other Important Physical Properties
Chemical Resistance - PE is highly resistant to a wide range of chemical agents even at elevated temperatures. However, when installing in potentially aggressive environments, the user should refer to PPI Technical Report TR-19, *Thermoplastic Piping for the Transport of Chemicals*, which provides chemical resistance data for PE with a wide range of chemicals.

Impact Resistance - Impact resistance is related to the pipe's ability to absorb impact and resist cracking during handling and installation, particularly in cold weather. An advantage of HDPE over many other materials is its ductility at low temperatures. For example, PE's glass transition temperature (the temperature below which it is more brittle and glassy) is well below 0°F, at approximately -166°F (~-110°C), while for PVC it is well above room temperature, at about 176°F (~80°C). Like ESCR, impact resistance is strongly influenced by density, with lower densities generally favoring greater impact resistance.

There are a number of impact tests for materials, like Izod or Charpy (see the chapter on engineering properties in this Handbook), but generally finished pipe and fittings are tested by a falling weight (tup) impact test (for example, ASTM D2444) at low temperature — typically -4°F (-20°C). This test, commonly used in Quality Assurance, is a pass/fail test, in which any cracking or breaking is considered a failure.

Stabilization

Unprotected PE, like virtually all other polymers, is vulnerable to degradation due to prolonged exposure to heat, oxygen or ultraviolet (UV) radiation, resulting in embrittlement and reduced service life. To prevent these damaging effects, PE conduit materials are typically formulated with a variety of stabilizing additives, ranging from antioxidants to UV stabilizers, to maintain required long-term performance. For a more in-depth discussion on both antioxidants and UV protection, see the chapter on engineering properties in this Handbook. Regardless of the type of UV protection used, the conduit must be adequately protected from UV attack to withstand normal storage conditions and special use intervals. Adequate protection for conduit destined for underground installation is to provide for at least one year's protection from outdoor storage. If longer storage times are possible or anticipated, the user may specify additional stabilization, or, preferably, should provide for a covered storage environment. Otherwise, if the conduit exceeds one year of UV exposure, it should be tested to ensure it meets all physical property requirements (cell classification, impact resistance) prior to installation.

The most common means for UV protection is to employ carbon black at a minimum loading of 2%. For long-term aerial exposure in self-supporting Figure-8 duct designs, due to the heightened mechanical stress level, the carbon black should be more finely divided and dispersed, having an average particle size of less than or equal to 20 nanometers, in accordance with ASTM F 2160.

PE non-black materials, however, require special stabilizers in addition to their normal pigments, generally UV blockers or Hindered Amine Light Stabilizers (HALS).

Colorants for Conduit

PE conduit is produced in a variety of solid and striped colors, which serve to help identify the duct for either its end use application (e.g., fiber optic cable, power, etc.) or owner. In determining the color of the conduit, its striping or the marking of the conduit or a combination thereof, it is recommended for safety reasons that the color yellow not be utilized since this is the uniform color code for natural gas applications.

Design Considerations

Conduit vs. Pipe

In general, plastic conduits and plastic pipes are very similar in structure and composition, but deployment is where they differ.

- Conduits do not have long-term internal pressure.

  External forces are unchecked; if ovalized during installation, it may not recover during service.

  Long-term stress rupture is not a factor. (Hydrostatic Design Basis is not required in material selection.)

- Conduit ID is chosen by cable occupancy, where internal clearances are critical; whereas, for piping applications, ID is based on volumetric flow requirements.

- Path of installation for conduit is very important – radius of curvature, vertical and horizontal path deviations (undulations) and elevation changes all significantly affect cable placement.

Cable Dimension Considerations

Determination of a conduit's dimensions begins with the largest cable, or group of cables or innerducts, intended for occupancy. From a functional viewpoint, selection of diameter can be broken down into the following general considerations:

1. The inside diameter of the conduit is determined by the cable diameter and placement method (pulling or air-assisted pushing).

2. Pulling cables into underground conduits requires sufficient free clearance and is typically further distinguished by classifying the cables into two groups: power and coax (short lengths) and fiber (long lengths). Additionally, electrical cable fill is controlled by the National Electric Code (Chapter 9), whereas, dielectric, or fiber optic cables, are not.

3. Long pulling lengths require low volume fill, i.e. 36% max.

4. Short pulling lengths may be filled up to 53%, or up to the latest NEC limitations for groups of cables.

5. Push-blow installation methods for long length fiber cables utilize higher volume fills, i.e. up to 70% max.

6. Innerducts are smaller diameter conduits, intended for placement into larger conduits or casings. Their purpose is to subdivide the larger conduit space into discrete continuous pathways for incorporation of fiber optic cables. Diameters of conduits and innerducts are often specially designed to maximize the conduit fill.

Using these guidelines, one can determine the minimum ID of the conduit or innerduct. When over-sizing a conduit for power, coaxial or multi-pair telecom cables, the more room the better. This rule does not necessarily apply for push-blow methods of installation. Here, it is found to be more difficult to push a cable with additional clearance since a cable tends to form a helix, which transfers some of the axial load laterally into the wall causing friction. The air velocity moving over the cable can also be maximized with a minimum volume of air when the free volume is low. Higher air velocities result in improved drag forces on the cable, thus aiding with its placement.

Conduit Wall Determination

Conduit and duct products come in a wide range of sizes, spanning 1/4-inch (5mm) to 24-inch (610mm) bore casings. The standard dimension ratio, SDR, of a conduit is defined as the ratio of the average conduit diameter divided by the minimum wall thickness. Wall thickness typically ranges between SDR 9 to SDR 17. (Larger SDR numbers indicate a thinner wall thickness.)

Conventions exist that work off of either the average outside diameter (SDR) or the average inside diameter (SIDR). Internally sized (SIDR) are usually chosen when the inside diameter clearance must be very carefully controlled. This usually does not apply to most duct installations because, as noted above, the free clearance between the cable and the inner wall of the conduit is not usually that close. Bore casings, on the other hand, offer situations that can benefit from close ID control because many times several innerducts are tightly fit into a casing. In this latter case, the conduit wall can be increased or decreased relative to service conditions without jeopardizing the inside clearance fit. Internally sized dimension tables tend to preserve the minimum ID above the nominal conduit size, whereas, externally sized conduits often fall below the nominal ID as the wall thickness increases.

For most conduit installations, SDR sizing is utilized because the OD control lends itself to better joint formation using external couplers. This becomes very important when air-assisted placement methods are used for placing the cable. On the other hand, large diameter conduits (4 and above) typically undergo butt fusion as a means of joining.

Determination of the wall thickness becomes a function of either the method by which the conduit is placed, or the nature of environmental stresses that it will be exposed to over the service life. ASTM F 2160, *"Standards Specification for Solid Wall High Density Polyethylene (HDPE) Conduit Based on Controlled Outside Diameter (OD)"*, explains the conduit sizing systems fully.

Installation Method vs. Short-Term and Long-Term Stress

The viscoelastic nature of HDPE results in differences in the observed mechanical properties as a function of time (and/or temperature). The apparent stress/strain behavior of the material is time dependent under the influence of a sustained load. This is referred to as "creep" properties. In this regard, we can distinguish between "short-term" properties, such as those exhibited during a laboratory tensile test at a strain (stretching) rate of two inches per minute, as compared with "long-term" properties typical of conduit placement and sustained service loads.

Knowledge of the load-bearing capability of HDPE as a function of loading rate allows one to select appropriate strength values to substitute into design equations. Loads are applied to conduits both by the environment that they are placed into and by the placement means under which they are installed; the chief difference being the duration over which the load is applied. For example, a common means to install multiple conduits is to directly plow them into the ground using either a railroad plow or tractor-drawn plow. During this installation process, a certain amount of bending and tensile stress is encountered over a rather short period of time (only seconds to minutes). Whereas, after the plow cavity collapses about the conduit, the ground continues to settle upon stones that may be pressing directly against the conduit, thus setting up a long-term compressive load. For this application, we see that we would require both long-term and short-term moduli to assess the deflection resistance. Initially the conduit may offer resistance to ovalization, but in time, the resin may yield under the sustained load, resulting in a reduced pathway for the cable.

Numerous approaches to placing conduits have evolved over the years. Each method presents its own unique set of challenges with respect to the potential for conduit damage, or installation related predicaments. Perhaps one way to compare the potential sensitivity to damage of the various methods is the following table. Here the potential for damage is depicted by a numerical scale ranging from 0 to 5, where 5 is the most severe condition, resulting in yielding and permanent deformation of the conduit; 4 is the potential for loads greater than 75% of yield stress; 3 represents loads greater than 50%; 2 representing greater than 25%; 1 less than 25%, and 0 representing no significant load at all. The shaded areas depict the most severe condition.

TABLE 1
Relative Damage Sensitivity vs. Installation Method

Installation Method	Short-Term Loading				Long-Term Loading		Recommended SDR Range
	Tensile	Bending	Crushing	Impact	Crushing	Tensile	
Conduit*	3 - 5	3	2	1	1	1 - 2	9.0 – 13.5
Horizontal Bore	4 - 5	2	3 - 4	0	3 - 5	1	9.0 – 11.0
Direct Plow	2	3	4 - 5	1 - 2	4 - 5	1	9.0 –11.0
Continuous Trench	2	2	3 - 4	1 - 2	3 - 4	1	9.0 – 11.0
Open Trench	0	0	1 - 3	1	1 - 3	1	11.0 – 17.0
Aerial	1 - 2	3 - 5	2 - 3	1	1	2	11.0 – 13.5

* The term "conduit" in this chart refers to the placement of HDPE innerducts into a buried 4" to 6" PVC conduit typical of the underground telecom plant. The SDR recommendation range attempts to select safe SDR's based upon the potential for stressful conditions.

It should be noted that the above table is not intended to be representative of all conduits installed by these methods, but is indicative of what can happen when the wrong diameter, wall or material is used. Check with supplier for specific design recommendations.

Perhaps the most serious and least controlled problem for cable placement is that of ovalization or kinking of the conduit. This condition can be brought about through tensile yielding, severe bending, excessive sidewall loading, or probably more frequently, the crushing action of rocks in the underground environment. In direct plow or bore applications, one gets little feedback from the process to indicate that a potential problem is developing. For these applications, the most robust conduit design should be considered.

Below Ground Installations

Open Trench / Continuous Trenching

Conduits intended for buried applications are commonly differentiated into two classes, rigid and flexible, depending on their capacity to deform in service without cracking, or otherwise failing. PE conduit can safely withstand considerable deformation and is, therefore, classified as a flexible conduit.

Flexible conduits deform vertically under load and expand laterally into the surrounding soil. The lateral movement mobilizes the soil's passive resistance forces, which limit deformation of the conduit. The accompanying vertical deflection permits soil-arching action to create a more uniform and reduced soil pressure acting on the conduit. PE stress relaxes over time to decrease the bending moment in the conduit wall and accommodates local deformation (strain) due to imperfections in the embedment material, both in the ring and longitudinal directions.

The relationship between pipe stiffness, soil modulus (stiffness), compaction and vertical loading is documented by the work of Spangler and others. The pipe stiffness, as measured in ASTM D2412 and Spangler's Iowa formula provide a basis for prediction of conduit deflection as related to dimension ratio and resin modulus. It should be noted, however, that creep affects the pipe stiffness, so the long-term modulus should be used. Additional information pertaining to soil embedment materials, trench construction and installation procedures can be found in the chapter on "Underground Installation of Polyethylene Piping" in this Handbook.

Flexible conduit can occasionally fail due to stress cracking when localized forces (for example, from a large sharp rock) exceed the material's ability to relax and relieve stress. However, PE resins suitable for conduit applications should have adequate stress relieving properties to avoid these failures. Therefore, the design process should include consideration of the conduit resin's stress crack resistance, as well as the selection of appropriate embedment material and compaction.

Direct Plow

Flexible conduit materials need adequate compressive strength to safely resist the compressive stresses generated by external loading. However, the usual design constraint is not material failure due to overstraining, but, rather, excessive deflection or buckling under anticipated earth and ground water pressures. Deflection or buckling is more probable when the embedment material does not provide adequate side support. For example, pipe installed by directional drilling and plowing typically does not receive side support equivalent to that provided by the embedment material used in trench installations where bed and backfill can be "engineered" to provide a specific level of lateral support.

Plowing installations often encounter rocky soils, which would induce significant crush loads for conduits 2-inch diameter and smaller. In these cases, SDR 11 is the minimum wall thickness that should be used, and if rocky conditions were likely, SDR 9 would be more appropriate.

Pipe stiffness, as calculated per ASTM D2412, gives a measure of flexural stiffness of the pipe. Pipe stiffness equals the ratio of the applied load in units of lbs/lineal inch to the corresponding deflection in units of inches at 5% deflection. It should be understood, however, that although two conduits, 6-inch and 1.25-inch diameter, may possess the same pipe stiffness, the amount of soil load required to induce a 5% deflection in each is considerably different. As a result, the sensitivity of smaller diameter conduits to underground obstructions is that much greater. Another physical parameter for smaller conduits, crush strength, is often employed to establish limits of crush resistance. Unfortunately, there is no universally agreed upon criterion or test method for crush testing. Typically, the conduits are subjected to an increasing load, similarly applied as in ASTM D2412, but to a far greater deflection—on the order of 25 to 50% of the inside diameter. This deflection-limiting load is then reported on a per-foot basis.

Table 2 illustrates the difference in the load required to induce a 5% deflection in conduits having different diameters but common pipe stiffness values. These values were generated assuming a flexural modulus of 150,000 psi for the resin. Units for pipe stiffness are in pounds/inch of length/inch of deflection, whereas those for the crush are presented as pounds per foot. It is apparent that a fixed external load more easily deflects smaller diameter conduits. It is also important to remember that, in long-term loading, the resin will maintain only about 22 to 25% of its original modulus; thus, smaller thin-wall conduits can be quite susceptible to localized loads brought about by buried obstructions.

Conduit Network Pulling

In the telephone and electrical utility industries, the underground plant is often comprised of a network of 3", 4", and 6" conduit banks. These "rigid" conduits are composed of clay tile, cement conduit, or more recently, PVC constructions. They are usually separated by manhole vaults or buried pull-boxes. Distances between, and placement of manholes and pull-boxes is largely a function of the following constraints:

1. Location of branch circuit intersections
2. Lengths of cables (or innerducts) available on reels
3. Access to, or limited by physical obstructions
4. Path difficulty for placement of cable or innerducts
5. Surface environment
6. Method of cable placement (mid-assist access)

In addition, Department of Transportation (DOT) regulations often require additional protection and support structure for buried conduits in road bores and traffic areas. Although steel casings have been used in the past, it is becoming more prevalent to horizontally bore under roadways (or waterways) and pull back an HDPE casing into which HDPE innerducts are installed.

Pull placement of innerducts has obvious similarity to traditional cable placement methods. Several good references on this subject exist, including *Guide For Installation* of *Extruded Dielectric Insulated Power Cable Systems Rated 69KV Through 138KV, Underground Extruded Power Cable Pulling Guide, AEIC Task Group 28* and *IEEE Guide Distribution Cable Installation Methods In Duct Systems.*

There are a number of variables that influence loading and selection of innerducts when pulling into conduit structures:

- Diameter of conduit and innerduct, and number of innerducts to be installed – clearance fit
- Length and direction changes of conduit run, sweeps
- Composition of conduit and coefficient of friction
- Jam combinations
- Pull speed and temperature
- Elevation and innerduct weight

Horizontal Directional Bore

For directional drilling the design process should include consideration of tensile forces and bend radii created during these processes. Flexible conduits installed in continuous lengths are susceptible to potential tensile failures when pulled into place, so allowable tensile forces should be determined to avoid neck-down from tensile yield. The engineer should also account for the conduit's allowable bend radius, especially on bends with no additional support given to the conduit, to prevent ovalization and kinking from installation. For additional information, please refer to the chapter on horizontal directional drilling in this Handbook.

TABLE 2
Pipe Stiffness (PS) vs. Crush Strength

Conduit Size	OD In.	SDR 9 Wall In.	SDR 9 PS Lb/in. in.	SDR 9 Crush Lb./6 in.	SDR 11 Wall In.	SDR 11 PS Lb/in. in.	SDR 11 Crush Lb./6 in.	SDR 13.5 Wall In.	SDR 13.5 PS Lb/in. in.	SDR 13.5 Crush Lb./6 in.	SDR 15.5 Wall In.	SDR 15.5 PS Lb/in. in.	SDR 15.5 Crush Lb./6 in.	SDR 17 Wall In.	SDR 17 PS Lb/in. in.	SDR 17 Crush Lb./6 in.
1	1.315	.146	1310	804	.120	671	433	.097	344	231	.085	220	151	.077	164	114
1.25	1.660	.184	1310	1020	.151	671	547	.123	344	292	.107	220	190	.098	164	144
1.5	1.900	.211	1310	1160	.173	671	626	.141	344	33	.123	220	218	.112	164	165
2	2.375	.264	1310	1450	.216	671	782	.176	344	417	.153	220	272	.140	164	206
2.5	2.875	.319	1310	1760	.261	671	947	.213	344	50	.185	220	330	.169	164	249
3	3.5	.389	1310	2140	.318	671	1150	.259	344	615	.226	220	402	.206	164	304
4	4.5	.500	1310	2750	.409	671	1480	.333	344	790	.290	220	516	.265	164	390
6	6.625	.736	1310	4050	.602	671	2180	.491	344	1160	.427	220	760	.390	164	575

Table 2 is for comparative purposes only. Pipe stiffness values are based on 150,000-psi flexural modulus. Crush values are estimated from empirical data for 6" long conduit samples compression tested in accordance with ASTM D2412 to 50% deflection.

Installation Methods

This section discusses various conduit installation options in general terms and should not be interpreted as a step-by-step guide or "operations manual." The user should contact the equipment manufacturer for more detailed instruction, as operating procedures will vary with equipment.

NOTE: The consequences of striking gas or power lines (above and below ground) during installation can be dangerous, possibly deadly. Before digging, it is critical to ensure that all existing underground service lines (gas, water, power, etc.) in the vicinity are located and marked. It is recommended to contact the local "Call Before You Dig" agency to ensure these provisions are made. Furthermore, prior to installation, consult NEC, NFPA and NESC codes, as well as any applicable local codes.

General Considerations

Mechanical Stress

Regardless of the installation method, mechanical stress is of great concern during conduit placement. Exceeding the maximum allowable pulling tension or the minimum allowable bending radii can damage conduit. Consult the conduit supplier for allowable pulling tensions.

Pulling Tension

During conduit pulling placement, attention should be given to the number of sweeps, bends or offsets and their distribution over the pull.

Tail loading is the tension in the cable caused by the mass of the conduit on the reel and reel brakes. Tail loading is controlled by two methods. Using minimal braking during the pay-off of the conduit from the reel at times can minimize tension; no braking is preferred. Rotating the reel in the direction of pay-off can also minimize tail loading.

Breakaway swivels should be placed on the conduit to ensure that the maximum allowable tension for that specific conduit type is not exceeded. The swivel is placed between the winch line and pulling grip. A breakaway swivel is required for each conduit.

Bending Radii

Conduit is often routed around corners during placement, and pulling tension must be increased to complete the pull. It is important to determine the minimum radius to which the conduit can be bent without mechanically degrading the performance of the conduit. See Table 3.

TABLE 3
Minimum Bend Radius as a function of Diameter and Standard Dimension Ratio

Size	OD In.	SDR 9		SDR 11		SDR 13.5		SDR 15.5		SDR 17	
		Wall In.	Min. Radius In.	Wall In.	Min. Radius In.	Wall In.	Min. Radius In.	Wall In.	Min. Radius In.	Wall In.	Min. Radius In.
1	1.315	.146	15.4	.120	20.1	.097	25.9	.085	30.6	.077	34.1
1.25	1.660	.184	17.1	.151	22.3	.123	28.9	107	34.2	.098	38.1
1.5	1.900	.211	18.2	.173	23.8	.141	30.8	.123	36.4	.112	40.6
2	2.375	.264	20.0	.216	26.3	.176	34.2	.153	40.5	.140	45.2
2.5	2.875	.319	21.8	.261	28.0	.213	37.3	.180	44.3	169	49.5
3	3.500	.389	23.8	.318	31.4	.259	40.9	.226	48.5	.206	54.2
4	4.500	.500	26.4	.409	35.0	.333	45.8	.290	54.5	.265	61.0
6	6.625	.736	30.9	.602	41.3	.491	54.4	.427	64.9	.390	72.8

Ovalization is independent of tensile strength or modulus, but is controlled by diameter, wall thickness and bending radius. The radii listed above are estimated, as the minimum unsupported bending radius required producing a 5% ovalization. The values in the above table are calculated based on minimum wall thickness and are a first approximation to ovality in the bending conduit (actual bending radius may be slightly smaller). Ovality is calculated as: Ovality = [(Max. OD – Min. OD)/ Avg. OD] x 100.

Underground Installation

Generally, the three primary underground installation (or "underground plant") methods are trenching, plowing and boring, described in general terms below.

Trenching Methods

As with all methods, there are many variations on trenching installations, but generally the two main variations are the traditional "open trench" method and "continuous" trenching.

Open Trench/Continuous Trench

As the name implies, open trench installations involve digging an open trench and laying the conduit directly into the trench, often along with embedment material to protect the conduit from damage due to the surrounding soil. This installation is accomplished with specialized trenching machines that cut the trench and remove the soil in a single action and can be used to place multiple conduits over long or short distances. This technique, more common in pressure pipe or PVC installations, is described in more detail in the chapter on underground installation in this Handbook.

In Continuous trenching, conduit payoff moves along with the trenching process.

Digging the Trench

The trench should be dug as straight, level and rock free as possible. Avoid curves smaller than the conduit's allowable bend radius. Undercut inside corners to increase the radius of the bend. Should there be a rapid grade change, use back-fill to support the conduit.

Excavate the trench to the desired depth, and remove all rocks and large stones from the bottom of the trench to prevent damage to the conduit. Push some clean fill (fine material, without stones) into the trench to cushion the conduit as it is installed in the trench.

Supplemental trenches should be made to all offset enclosure locations. Trench intersections should be excavated to provide adequate space to make sweeping bends in the conduit.

Fill the trench and compact as required. Tamp the trench to provide compaction that will prevent the trench backfill from settling.

Placing the conduit

An important consideration for open-trench installations of PE conduit is that conduit should be straightened to remove any residual "coil memory," which can create a tortuous path for the cable and create significant challenges to cable installation. Conduit pay off can be accomplished by pulling the conduit into the trench from a stationary reel or by laying the conduit into the trench from a moving reel, usually attached to a trailer.

Spacers should be used when placing multiple ducts in a trench. Spacers prevent the ducts from twisting over and around each other. By keeping the ducts in straight alignment, cable-pulling tensions are reduced. When water is present in the trench, or when using extremely wet concrete slurry, floating of the conduit can be restricted through the use of the spacers.

Backfilling

It is best to place the best quality soil directly on and around the conduit. DO NOT place large rocks directly on the conduit. Allow at least 2 – 4 inches (5 – 10 cm) of clean, uniform soil to cushion the conduit.

A good practice to insure long-term protection of underground facilities is to utilize sand for padding the conduit. It provides a more stable environment for the conduit, prohibiting damage from rocks and allowing water to drain away from conduit easily. More importantly is the protection it can provide during future excavation near your facilities. The apparent change in soil condition provides warning that there is a utility buried there. This should not replace the practice of placing warning tape, but rather should serve as a supplement.

During backfill, warning tape should be placed typically 1 to 3 feet above the conduit.

Plowing

Plowing is the preferred installation for long continuous runs where space permits, for example, in rural areas. Plowing installations use a plow blade (pulled by a tractor or mounted to a railroad car) to split the earth and place the cable at the required depth through a feed tube located directly on the plow blade. The key distinction between plowing and continuous trenching is that trenching involves the actual removal of soil from the trench, whereas plowing only displaces soil while laying in the conduit.

Consult the equipment manufacturer for specific recommendations on plow blade and feed tube designs. It is strongly recommend to have a professionally engineered single or double feed tube plow blade with a tube at least 0.5 inch (1.25 cm) larger than the largest conduit size and a radius no smaller than the minimum bend radius of the largest conduit size. It is recommended that DR 11 or DR 9 be used, depending on conditions and conduit diameter.

Local regulation may require that warning tape be plowed in with the cable. Most plow manufacturers make plow blades that bury cable and tape at the same time.

Plowing Variations

There are several variations of plowing installations. A few are described briefly below:

- **Vibratory Plowing** This method uses a vibrating blade and may allow use of a smaller tractor than that used for static plowing.
- **Rip and Plow** This method may be required when significant obstructions (for example, roots) are anticipated and uses an additional lead plow (without conduit) to rip the ground and clear obstructions several hundred yards ahead of the primary plow with conduit.

- **Pull Plows Method** Instead of installing from a reel traveling with the plow, conduit is pulled from a stationery reel behind the plow through the plowed trench.

Directional Bores

Directional boring allows the installation of conduit under obstacles that do not allow convenient plowing or trenching installations, for example rivers or highways. This unique installation method, which capitalizes on a primary strength of PE conduit—its flexibility, can be accomplished over very long distances.

Directional boring is accomplished using a steerable drill stem to create a pathway for the conduit. The equipment operator can control the depth and direction of the boring. A detailed discussion of this installation method is presented in the chapter on "Polyethylene Pipe for Horizontal Directional Drilling" in this Handbook. Also, consult the equipment supplier for detailed operating procedures and safety precautions.

It is recommended that DR 11 or DR 9 be used, depending on conditions and conduit diameter.

Installation into Existing Conduit

Conduit (or multiple conduits) is often pulled into existing conduit systems as innerduct.

NOTE: ALWAYS test and ventilate manholes prior to entering into them and follow OSHA confined space requirements.

Proofing

An important step that should be taken prior to this type of installation is "proofing" the existing conduit to ensure that all obstructions are cleared and that conduit continuity and alignment is good. It is recommended that a rigid mandrel roughly 90% of the inner diameter of the conduit be used to perform the proof. Proofing conduit is typically performed by pushing a fiberglass fish with a rigid mandrel attached to the end of it through the conduit. Any problem areas should be felt by the person pushing the fiberglass fish and should then be marked on the fish so that the distance to the problem is recorded and if necessary can be located for repair with greater ease. If the fiberglass fish makes its way through the conduit without any difficulties experienced, then the conduit has "proofed out," and no repairs should be necessary.

Before placement of the innerduct inside the conduit can be started, it is important to have all of the necessary equipment to protect the innerduct. The use of sheaves, bending shoes, rolling blocks (45 and 90 degrees) and straight pulleys are required for protection of the innerduct during installation. It is important that they all meet the proper radius for the innerduct size. The use of a pulling lubricant will greatly reduce the tension and stress on the innerduct when pulling innerduct into an existing conduit. Ball bearing swivels are needed for attaching the winch line to the innerduct harness system.

Mid-Assists

On long routes and routes with many turns in them it is important to consider the selection of mid-assist locations. There are different ways of providing mid-assist for innerduct pulls. Typically the use of a winch is required such as a capstan or vehicle drum winch. The introduction of mid-assist capstan winches has made innerduct pulling an easier task, requiring less manpower and communication than traditional drum winching involves. More importantly it provides greater production capabilities.

After Pulling

The stress of pulling innerduct through existing conduit will vary with the length of the route and the number of turns it has to make, as well as the condition of the conduit it is being pulled into and the amount of lubrication used. The effects of the stress will cause the innerduct to elongate (or stretch) in proportion to the amount of stress, but should be less than 2% of the total length placed. Due to this effect, it is important to pull past the conduit system slightly to compensate for recovery to the original length. An allowance of at least one hour needs to be given for the innerduct to "relax" before cutting and trimming it.

Above Ground/Aerial

There are many applications for aerial conduit, which include but are not limited to road crossings, rail crossings, trolley line crossings, and water crossings. They provide for efficient means of supporting cable that can easily be replaced and/or allow for the addition of cables without requiring encroachment in often hazardous or difficult to access spaces.

A critical consideration for aerial applications is UV protection. For this reason, only conduit materials with special carbon black pigments can be used, since constant direct exposure to UV radiation significantly shortens the lifetime of unprotected PE conduit (see Material Selection in this chapter).

Installation

The two preferred methods for aerial installation of conduit are the back-pull/stationary reel method and the drive-off/moving reel method. Circumstances at the construction site and equipment/manpower availability will dictate which placement method will be used.

Design consideration must be given to the expansion/contraction potential of PE conduit. This consideration is more important when lashing conduit than with the use of self-supporting conduit.

Installation – Back-Pull/Stationary Reel Method
The back-pull/stationary reel method is the usual method of aerial conduit placement. This method is also best suited for locations where the strand changes from the field side of the pole to the street side of the pole and where there are excessive obstacles to work around. The conduit is run from the reel up to the strand, pulled back by an over lash cable puller that only travels forward and is held aloft by the cable blocks and rollers. Once the section of conduit is pulled into place it is lashed and then cut.

Installation – Drive-Off/Moving Reel Method
The drive-off/moving reel method may realize some manpower and timesaving in aerial conduit placement and lash-up. This method is used where there is existing strand and is on one side of the poles, typically roadside. In it, the conduit is attached to the strand and payed off a reel moving away from it. The conduit is being lashed as it is pulled.

Self-Supporting Conduit
Installation of self-supporting conduit can be accomplished by both of the above methods, the difference being that the support strand is an integral part of the conduit. This product approach not only simplifies installation by eliminating the step of independently installing a support strand, but it improves the controllability of the expansion-contraction properties of the conduit.

Installation – Over-lashing Existing Cable
Over-lashing conduit onto existing cable plant is similar to installing conduit onto new strand. However, there are some unique aspects.

A sag and tension analysis should be performed to see if the new cable load will overwhelm the strand. Also, over-lashing conduit on top of sensitive coaxial cables may influence the cables signal carrying capability due to rising lashing wire tensions that may result from contraction-induced movement of the conduit. It is best to seek the help of engineering services in planning an aerial plant.

Joining Methods

Introduction
Conduit can be joined by a variety of thermal and mechanical methods. Since conduit does not experience any long-term internal pressure and acts only as a

pathway for power or other cables, the owner of the system may be tempted to neglect the importance of specifying effective couplings. However, an integral part of any conduit system is the type and quality of joining method used. Proper engineering design of a system will consider the type and effectiveness of these joining techniques.

The owner of the conduit system should be aware that there are joint performance considerations that affect the system's reliability well beyond initial installation. Some of those might include:

- **Pull out resistance**, both at installation and over time due to thermal contraction/expansion, must be considered. This is critical for "blow-in" cable installations, which will exert an outward force at joints, less so for pulling installations, which will tend to exert the opposite force.
- **Pressure leak rates**, for "blow-in" installations at pressures of 125 to 150 psig. Consideration must be given to how much leakage can be tolerated without reducing the distance the cable can consistently be moved through the conduit.
- **Infiltration leakage**, allowing water and/or silt to enter the conduit over time, can create obstacles for cable installation and repair or cause water freeze compression of fiber optic cables.
- **Corrosion resistance** is important as conduit systems are often buried in soils exposed to and containing alkali, fertilizers, and ice-thawing chemicals, insecticides, herbicides and acids.
- **Cold temperature brittleness resistance** is required to avoid problems with installation and long-term performance in colder climates.

General Provisions

PE-to-PE joints may be made using heat fusion, electrofusion or mechanical fittings. However, mechanical couplings are often preferred over fusion joints, due to the internal bead of a butt fusion joint, which can interfere with cable installation. PE conduit may be joined to other materials in junction boxes or other hardware utilized by communication and electrical industries, by using mechanical fittings, flanges, or other types of qualified transition fittings. The user may choose from many available types and styles of joining methods, each with its own particular advantages and limitations for any joining situation encountered. Contact with the various manufacturers is advisable for guidance in proper applications and styles available for joining as described in this section.

Mechanical Fittings

PE conduit can be joined by a variety of available styles of mechanical fittings, each with its own particular advantages and limitations in any given application. This section will not address these advantages or limitations but will only offer general descriptions of many of these fitting types and how they might be utilized. ASTM F 2176, "Standard Specification for Mechanical Couplings Used on Polyethylene Conduit, Duct and Innerduct," establishes performance requirements for material, workmanship, and testing of 2-inch and smaller mechanical fittings for PE conduit. PPI recommends that the user be well informed about the manufacturer's recommended joining procedure, as well as any performance limitations, for the particular mechanical connector being used.

Barbed Mechanical Fittings
Barbed fittings are available in various materials and configurations for joining conduit sizes 2-inch and smaller. None of these fittings are offered with sealing capabilities. Installation involves pressing the fitting over ends of the conduit to be joined using a special tool. The inside of these fittings contain sharp, inward-facing barbs which allow the conduit to be pressed in, yet dig into the conduit and resist removal when pulled.

Threaded Mechanical Fittings
Threaded mechanical fittings are available in various materials and configurations for conduit sizes 2-inches and smaller. Some are designed with sealing capabilities while others are not. Internal thread designs of these fittings are typically tapered similar to pipe threads, with a left-hand thread on one end and a right-hand thread on the other to cut thread paths on the conduit's outer surface. This thread design allows the operator to thread the fitting onto the ends of both conduit sections simultaneously. Some variations of threaded fittings may also be pressed on the conduit ends and used as barbed fittings. The user should consult the fitting manufacturer to determine if this alternate installation method is recommended.

Compression Fittings
As with the other mechanical fittings, compression fittings are also available in numerous designs – some designs for conduit as large as 8-inch and others for only 2-inch and below. While compression fittings used in PE pressure piping industries, such as water or gas, require internal stiffeners, conduit systems typically do not, because stiffeners may create obstacles for cable being blown through the conduit. For any fitting style being considered, consult the fitting manufacturer for available sizes and written instructions for use.

Expansion Joints
Expansion joints are designed primarily for aerial conduit installations. The primary purpose of this fitting design is to absorb thermal expansion and contraction in the conduit system created by ambient temperature changes, which can be extreme in

these above ground installations. System designers should determine the number of expansion joints required based on the expansion length provided by the fitting and a calculation of the pipe's overall thermal expansion factor for the overall length of above ground installation.

Heat Fusion

The principle of heat fusion is to heat two surfaces to a designated temperature and fuse them together by application of a force sufficient to cause the materials to flow together and mix. When fused in accordance with the manufacturer's recommended procedure and allowed to cool to nearly ambient temperatures, the joint becomes as strong or stronger than the conduit itself in both tensile and pressure properties.

Three primary heat fusion methods used in joining PE conduit are butt, socket and electrofusion. Butt and socket fusion joints are made using "hot irons" designed specifically for PE joining, and electrofusion supplies heat internally by electric current applied to a special fitting containing a wire coil. More specific information on heat fusion joining practices can be found in the chapter on "Joining" in this Handbook, as well as in ASTM D 2657 for the hot iron methods (butt and socket fusion) and in ASTM F 1290 for electrofusion.

PPI recommends that the user precisely follow the qualified fusion procedures established by the manufacturer of the particular heat fusion and joining equipment being used.

Butt Fusion Joining

Butt fusion joints are produced without need of special fittings, using specially developed butt fusion machines, that secure, face and precisely align the conduit for the flat face hot iron (not shown) fusion process. It should be noted that the butt fusion process produces an internal bead of equal or larger size than the visible outer bead. If internal restrictions are a concern for the cable installation, alternative-joining methods may be more appropriate.

Socket Fusion Joining

This technique requires the use of specially designed hot irons to simultaneously heat both the external surface of the pipe and the internal surface of the socket coupling. Specially designed hand tools are available to maintain alignment and stab depth of the hot irons until the materials reach fusion temperature. These tools also help secure the heated conduit end and coupling as the joint is made. Design requirements for socket fusion can be found in ASTM D 2683 for fittings and in ASTM F 1056 for socket fusion tools. As with butt fusion, socket-fused joints may have an internal bead that can interfere with cable placement.

Electrofusion Joining

Electrofusion is somewhat different from the hot iron fusion method described previously, the main difference being the method by which heat is applied. Electrofusion involves the use of a special electrofusion fitting with an embedded wire coil. Electrical current supplied to the wire coil by an electrofusion control box generates the heat for fusion. Special training in equipment use and maintenance may be needed. For additional information consult the chapter on "Joining" in this Handbook.

Repair Operations

Repair joints, as the name implies, are often designed specifically for use in repair situations. The nature of the damage will often dictate what types of joints are needed for repairs. For example, one type of design, a clamp-on style may be preferred when damage is limited and removal of the cable for repair is not necessary. However, in more severe damage situations, where new cable and conduit sections must be installed, many of the joining methods described earlier in this section may be suitable. Ultimately, the type of repair fitting or joint installed should maintain the integrity of the conduit system, prevent infiltration and provide sufficient resistance to thermal expansion/contraction.

Cable Installation

Installing cable-in-conduit or innerduct can be accomplished in a number of ways. These include:

1. Pulling cable into the conduit using a pull line or rope
2. Blowing cable into the conduit using specialized equipment that installs the cable in conjunction with a high volume jet of air
3. Pre-installed in the conduit by the conduit manufacturer (cable-in-conduit)

Pulling Cable into Conduit

The traditional method of installing cable-in-conduit has been to attach a pull line (or rope) to the cable and pull the cable into the conduit. This placement method requires equipment to do the actual pulling, to apply lubricants to reduce friction, and devices that measure the amount of tension being applied to the cable.

Conduit may be supplied with a preinstalled pull line. This line is either a twisted rope or a woven tape. These pull lines come in a wide variety of tensile strengths that range from 500 - 6000 pounds-force. Pull lines are also available pre-lubricated to reduce friction.

Pull tapes are available with sequential footage marks. This type of tape is useful in determining the progress of the cable pull.

Empty conduit would require a pull line to be installed. Blowing a pull line directly or blowing a lightweight line through the conduit using compressed air accomplishes this. This line is then used to pull a pull line or a winch line into the conduit to pull the cable.

A winch mechanism with a take-up reel is used to pull the pull line with the cable attached. The winch should have a tension meter to monitor the amount of tension being placed on the cable during the pull. This monitor will reduce the risk of damaging a sensitive fiber optic cable during the pull. Check with the cable manufacturer to determine the amount of tension a cable can safely withstand.

The use of cable lubricants is strongly recommended. Cable lubricants reduce the amount of friction during a pull and therefore allow longer cable pulls and reduce the risk of damage to a cable during the pull.

When the cable is attached to the pull line, it is recommended that a swivel be used between the two. This swivel will allow the cable and pull line to move independently in the conduit during the pull and prevent unnecessary twisting of the cable or pull line.

On very long pulls the use of mid-assists is common. Mid-assist equipment can be as simple as a person pulling on the cable midway or it can be a capstan type device that provides a controlled amount of pulling tension to the cable to reduce the tension on the cable and increase the possible length of the pull.

If the conduit is in a manhole, protective devices are needed to guide the cable into the manhole and then into the conduit. These guides protect the cable from scraping on metal or concrete surfaces that could damage the cable sheath.

Cable Blowing or Jetting

In recent years the practice of pulling cable has frequently been replaced with a newer method that uses compressed air to blow the cable into the conduit. Cable blowing requires specialized equipment produced by a number of manufacturers that utilize high volume air compressors. There are two categories of air-assisted cable placement: Low Volume/High Pressure, and High Volume/Low Pressure. In the first case a dart seal is attached to the end of the cable and compressed air is introduced into the duct building pressure behind the seal, thus forcing the dart forward and creating a tensile pull on the end of the cable. At the same time, the cable is pushed into the conduit through a manifold seal using a tractor pusher. The cable then experiences simultaneous push and pull forces. In the second case, the cable is tractor fed into the conduit, again through a manifold seal, but this time has no dart seal. Instead, cable progress is based on the viscous drag of high volume air alone. In these methods of cable installation, much longer lengths of cable can be placed than traditional cable pulling methods, and the tension applied to the cable is significantly reduced.

When blowing cables into conduit, the use of corrugated conduit is not recommended. Corrugated conduit causes turbulence of the air that disrupts the flow of air in the conduit and thus reduces the distance a cable can be blown.

The conduit should also be capable of withstanding the pressure of the air being introduced. Generally the maximum pressure used is in the range of 125 psi.

Caution should be exercised when using compressed air to pressurize the conduit as a loose joint can lead to injury due to the conduit/joint exploding.

Cable Installed by the Conduit Manufacturer (Cable-in-Conduit)

Some producers of conduit have the capability of installing cable while the conduit is being extruded. Each conduit producer has specific size and length limits, and it is necessary to discuss with the producer the type of cable you desire to be installed: its size, type of material and lengths.

Most producers can lubricate the conduit during this process to allow easy movement of the cable in the conduit for future removal and replacement.

Cable can be tested prior to and following installation to guarantee the integrity of the cable. Check with the conduit producer for specific information on testing the cable.

Friction in Conduit Systems

Friction is a critical limiting factor in determining the type and length of cable installation. Although very little information on cable installation is provided in this guide, this section has been made available as a background reference on frictional properties.

Definitions

Friction: the nature of interaction occurring between two surfaces. The basis of friction has its roots in the mechanical and physical-chemical makeup of the interface created by bringing together two surfaces.

Coefficient of friction, COF: the ratio of the force required to move a body relative to the normal, or clamping force, acting to keep the bodies together.

Static COF: the ratio of forces required to bring about the onset of motion between two bodies at rest with each other.

Kinetic COF: the ratio of forces acting on a body already in motion. It is essentially a measure of the effort required to keep the body in motion.

Friction Reduction

Friction reduction can be promoted by reducing mechanical interactions, grounding electrostatic charges, reducing polar interactions, selecting dissimilar polymers, and employing methods and mechanisms which act to dissipate heat. Although many times little can be done to control the composition of cable jacket materials, choices can be made to select friction-reducing conduit designs and lubricating mechanisms.

The use of lubricants is strongly recommended during the placement of the conduit or cable, or may be included in the manufacturing process of the conduit. Typical lubrication methods would include:

Water-soluble lubricants are available in many different forms including low viscosity free-flowing petuitous liquids, creamy consistencies, and stiff gels. Low viscosity liquids are best suited for placement of long lengths of lightweight cables, such as fiber cables. Heavier, cream-like consistencies are useful on lightweight power conductors. Stiff gels are used in vertical applications in buildings, or where high sidewall loads are expected in placement of heavy power cables or innerducts.

Polymeric water-soluble lubricants are commonly used in the field to lubricate the placement of cable, or of the conduits themselves. In this case the lubricant is applied either ahead of, or in conjunction with, the advancing cable. Water-soluble polymer chemistries include a number of different enhancements including surface wetting and cling, modification via fatty acids or their derivatives, or by inclusion of various friction-reducing oils, including silicones.

Conduits may be **pre-lubricated** during the manufacturing process by incorporation of lubricants directly onto the conduit inner wall, or via a lubricant-modified coextruded layer. The most common type of lubricant used for this type of application is silicone polymer, although other agents such as mineral oils, fatty acid derivatives and glycols have also found use.

Prelubrication finds particular value with fiber cable push-blow systems. Because the sidewall loads with these techniques are quite low compared with pulling, and the distances so great, the viscous drag contributed by water-soluble lubes can be detrimental. The ultra-light amount of lubricants employed by factory pre-lubrication methods can be a real advantage.

Geometry of the inner surface of the conduit can also play a role in friction reduction. As the normal load increases, the COF is found to decrease, unless the surface is damaged in such a way so as to increase the contact area, or heat is allowed to build up at a rate faster than it can be conducted away. Ribs formed on the inner conduit wall are a common design feature to reduce friction.

Longitudinal ribbing results in a reduction of the contact surface between the cable and the conduit wall from an area to a line of contact. Decreasing the area of contact under the same sidewall load results in a higher localized normal force. Within a limited range of sidewall loads, the COF is found to go down – at least until the loading causes localized damage to the jacket sheath.

Spiral ribbing further reduces the contact area from a line to a series of points. In addition, because the advancing cable is alternately on and off the ribbing, there is an opportunity for cooling and re-lubrication. Constantly changing the direction of the spiral eliminates the tendency to accumulate spiral-induced torque in the cable.

Transverse ribbing, or corrugated profiles, results in similar friction reducing geometries. However, there is a tendency for field-added lubrication to be scraped off the cable by the corrugations. In addition, the high degree of flexibility requires careful placement of the duct to reduce the buildup of friction due to path curvature.

Field Effects of Friction

Burn-through results when the winch line or cable develops so much frictional heat that it melts its way through the conduit wall. There are a number of factors that exacerbate this condition including: sidewall load, pull speed, conduit and pull-line materials of construction.

Aside from lubrication, sidewall loading may not be easily reduced; however, speed of pulling is controllable by the operator. Because PE and other thermoplastics are such good insulators, frictional heat build-up can go unchecked. Slower pull speeds combined with water-based lubricants can help reduce the rate of heat accumulation.

PVC elbows are commonly used for transitions out of the underground plant. Unfortunately, PVC not only has a higher COF than PE conduit (due largely to hydrogen bonding to the fillers), but also tends to soften with the onset of heating at a much faster rate (due to plasticizers). PE conduit on the other hand, has lower inherent COF (about 0.35 vs. >0.40 for PVC), as well as higher heat capacity due to its semi-crystalline nature.

Pull-line construction also plays a significant role in burn-through. Polypropylene ropes or even HDPE pull-lines exhibit low COF at low sidewall loads, but rapidly cut through both PVC conduit and PE conduit when the load increases. The tendency for these materials to soften, combined with high structural similarity (to PE), limit the pull load range over which they may be used. Polyester and polyaramid pull lines, particularly in tape form, offer greater protection from burn through.

Sidewall loading results any time a cable or pull-line is pulled about a sweep or bend. Dividing the tension in the pull-line by the radius of the bend may approximate the magnitude of the load. Obviously, the smaller the radius is, the greater the magnitude of load.

Speed, as noted above, is a critical variable in the operator's hands that can often spell the difference between success and failure. Speeds, which are too low, can result in a lot of mechanical interaction, whereas an excessively high speed results in heat
build-up.

Compatibility, in conjunction with high sidewall loading, can be a problem – not only for higher relative friction, but also is a key determinant in burn-through.

Contamination with inorganic soils roughens the surfaces of both conduit and cable jacket, thus increasing the mechanical interaction between them. In addition, the embedment of small particles increases hydrogen bonding with water that may be in the conduit, further enhancing the interaction of jacket with conduit.

Placement Planning

Curvature in the conduit run is the greatest deterrent to long pulls. Some curvature is unavoidable due to path layout, e.g. elevation changes, direction changes, etc. On the other hand, sloppy installation techniques can introduce more curvature than would otherwise be planned. For example, open trench work without proper tensioning and bedding can lead to installations that severely limit cable placement.

Equations for calculation of accumulated frictional drag have been derived and can be found in Appendix A. These are combinations of straight section and exponential sweeps. If the cable has appreciable weight, the transition to sweep up or sweep down results in significant differences. In addition, for multiple conductor power cables, certain combinations of cable multiples and free volume result in locking configurations.

Push-blow techniques are also greatly affected by friction. As noted above, pre-lubricated ducts, or very light applications of silicone emulsions, produce the best results. Techniques that rely on air predominantly to accelerate the cable work best with lightweight cables. As cable weight increases, systems with greater pushing power and piston seals provide improved performance.

Insert sizing is different for pulling vs. push-blow installations. In pulling cables, the greater the free volume in the conduit, the better, and maximum fill ratios based on cable and duct diameters are around 60 percent. On the other hand, maximum fill ratios in push-blow installations are closer to 85 percent fill. The reason for this is that if the cable is not allowed to deflect laterally, it can assume a greater axial load. The more free volume existing in the conduit during pushing, the easier it is to deflect, and having done so, the greater the curvature, and the greater the accompanying sidewall loads.

Placement planning for fiber cable installation is critical because the cable lengths are so long. Typically, one would locate a point along the route possessing similar accumulated frictional drag in either direction. Part of the cable is then installed to one end of the run, then the cable is figure-eighted to recover the opposite free end. The free end is then installed into the other end of the run. It is not uncommon to place 3,000 to 6,000 feet over any given span, and to gang placement equipment at mid-assist intervals along the path to deliver over 20,000 feet continuously in one direction. Using proper combinations of conduit design, installation method, lubrication and placing equipment, it is possible for crews to install over 40,000 feet of cable per day.

Special Applications

Corrugated Duct

Corrugated conduit has properties that generally make it easier to work with in difficult and confined environments. Primarily, this is a result of the lack of memory with corrugated and greater flexibility vs. smooth wall conduit. The lack of memory also provides a corrugated conduit that, when installed as an innerduct (inside of another larger conduit), does not spiral and therefore has lower friction when cables are pulled through it.

The greater degree of flexibility makes corrugated conduit easier to handle when used in confined spaces and other restricted environments.

Corrugated conduit is not appropriate for use in direct buried applications because of its limited crush resistance and the difficulty of laying it in a straight path. Corrugated conduit is also not appropriate for use when cables are to be installed using air-assisted placement. Corrugated conduit is relatively thin-walled and may not be able to handle the air pressure of air-assisted placement. The corrugations create air turbulence that is counterproductive to the air-assisted placement systems and significantly reduce the distance cables can be blown through it.

Corrugated conduit should not be installed using directional drilling equipment due to limited tensile strength and the fact that the corrugations will create significant friction during the pullback that will likely cause the conduit to separate.

The ASTM standards that cover SIDR and SDR designs do not apply to corrugated duct. Corrugation equipment varies from producer to producer, and inside and outside diameter may vary from each source of supply. All corrugated conduit specifications are per the producer only. Generally a minimum ID is specified and a maximum OD.

Corrugation design (or profile) greatly affects the properties of the conduit such as crush resistance and tensile strength. Tooling used to produce corrugated conduit does not allow the producer to change the profile or dimensions without costly retooling.

Check with the source of supply for detailed dimensional and performance specifications.

Bridge Structures

Bridge structures can range from a simple conduit placed in the bridge structure when the bridge is built to a major retrofit of an existing bridge that does not contain a conduit or structure in place to secure a single conduit or conduits. Bridge structures, new or old, require specially designed support systems to ensure structural integrity and meet all federal, state and local requirements.

When installing conduit on bridges, it is important to incorporate into the design the expansion and contraction of the bridge. Expansion joints must be installed in the conduit to prevent the conduit from either separating or bending and kinking due to bridge movement. As an alternative to expansion joints, use of a serpentine path have been proven effective in reducing expansion/contraction issues.

Underwater

The term underwater is also referred to as marine, or submarine, applications. The three basic methods of placing a structure conduit are laying the conduit on the bottom, plowing and jetting the conduit into the sub-aqueous terrain, or drilling under the waterway. Each method has it's own unique requirements based on the type of waterway, length, environmental issues, and federal, state and local requirements. There may be instances when all three types of application will be required on the same installation.

Conduit placed on the bottom of waterways should be black to prevent UV damage. For a complete discussion of underwater installations, see the chapter on marine installations in this Handbook.

Premise (Flame Retardant)

In addition to using conduit for installing fiber optic/communication cables in the underground, there are a few other very specialized applications for conduit type products.

With the growing market for data communications systems within buildings, there has been a concurrent growth in the use of fiber optic/communication cables in buildings as well. These installations typically place fiber optic/communication cables in the same cable trays and vertical risers as other communications cables and electrical cables.

Designers and installers have been concerned about identifying and protecting these fiber optic/communication cables. Manufacturers have responded with the development of several types of conduit for building use, or as it is known in the industry, premise wiring.

Premise wiring generally uses plenum air spaces, vertical riser shafts and general-purpose areas to run cables throughout buildings. The types of conduit developed were specifically for these environments. Because fiber optic/communication premise wiring falls into areas generally thought of as electrical, the National Electric Code and Underwriters Laboratories have addressed the characteristics needed by conduits to be safely used in building wiring.

Initial development produced the Plenum Raceway, a specialized conduit that meets stringent Underwriters Laboratories (UL 2024) requirements for minimum flame spread and smoke generation. Plenum Raceway is a corrugated conduit made from plastic materials that do not support flame and produces almost no smoke. At this time PVDF is the material of choice for Plenum Raceway. Products for plenum air spaces are required to carry a Listing Mark to verify that the product has been tested and meets the requirements for installation in the plenum environment.

A riser raceway was developed for premise wiring applications in riser shafts. Riser Raceway meets the Underwriters Laboratories (UL 2024) requirements for vertical flame spread. Riser Raceway is also a corrugated conduit, which is currently produced from either PVC or Nylon materials. Products for riser locations are required to carry a Listing Mark to verify that the product has been tested and meets the requirements for installation in the riser environment. Plenum Raceways are permitted to be placed in a riser application.

A general-purpose raceway was developed for premise wiring applications in general purpose applications. General Purpose Raceway meets the Underwriters Laboratories (UL 2024) requirements for flame spread. General Purpose Raceway is typically a corrugated conduit, which is currently produced from either PVC or Nylon materials. Products for general-purpose locations are required to carry a Listing Mark to verify that the product has been tested and meets the requirements for installation in the riser environment. Plenum and Riser Raceways are permitted to be placed in a general-purpose application.

The uses of Plenum or Riser Raceways do not eliminate the use of a Plenum or Riser rated cable.

As the use of fiber optic/communication cables in premise wiring increases there will likely be other specific needs that may generate other types of conduit for use in building wiring systems.

Electrical/Building Code (Conduit Entry Issues)

Electrical/Building Code regulations vary greatly regarding the placement of conduit into a building. Codes require the use of conduit constructed of a material that is listed for use in specific building areas, and these codes prohibit the use of PE conduit beyond a specific distance after entry through an exterior wall. The greatest variation in local code is the location of the transition from PE conduit to a conduit that meets the code requirement (distance from the exterior of the wall). Check your local codes for local amendments.

Armored (Rodent and Mechanical Protection)

When placing cables in the underground there is occasionally concern about the ability of the conduit to protect the cable(s) inside. Concerns usually are for crush resistance and resistance to cutting and gnawing by animals.

This need led to the development of armored conduit. Armored Conduit is standard PE conduit that has been wrapped with a second layer of metal and jacketed to provide a barrier to the problem of gnawing by animals. Armored Conduit also protects against cuts and abrasions from accidental strikes by persons digging nearby.

Multi-Cell Conduit

Multi-cell conduits are designed to meet special needs and unique job situations. There are a number of designs available to meet most of these special needs. Multi-cell conduit can be a product that is installed as an innerduct inside of existing conduits designed to maximize the available space in a vacant or occupied conduit, or it can be a fully assembled conduit with internal conduits that when installed provides a multi-channel conduit without the need to install any other innerducts. Some multi-cell designs can be direct buried like PE conduit using standard installation methods (plowing or open trenching).

Summary

The information contained in this chapter should help the reader to understand the fundamental properties of polyethylene (PE) conduit. A basic understanding of these properties will aid the engineer or designer in the use of PE conduit and serve to maximize the utility of the service into which it is ultimately installed.

While every effort has been made to present the fundamental properties as thoroughly as possible, it is obvious that this discussion is not all-inclusive. For further information concerning PE conduit, the reader is referred to a variety of sources including the pipe manufacturers' literature, additional publications of the Plastics Pipe Institute and the references at the end of this chapter.

References

ASTM International, D 1238, Standard Test Method for Flow Rates of Thermoplastics by Extrusion Plastometer.

ASTM International, D 1693, Standard Test Method for Environmental Stress-Cracking of Ethylene Plastics.

ASTM International, D 2444, Standard Test Method for Determination of the Impact Resistance of Thermoplastic Pipe and Fittings by Means of a Tup (Falling Weight).

ASTM International, D 2657, Standard Practice for Heat Fusion Joining of Polyolefin Pipe and Fittings.

ASTM International, D 2683, Standard Specification for Socket-Type Polyethylene Fittings for Outside Diameter-Controlled Polyethylene Pipe and Tubing.

ASTM International, D 3350, Standard Specification for Polyethylene Plastics Pipe and Fittings Materials.

ASTM International, F 1056, Standard Specification for Socket Fusion Tools for Use in Socket Fusion Joining Polyethylene Pipe or Tubing and Fittings.

ASTM International, F 1290, Standard Practice for Electrofusion Joining Polyolefin Pipe and Fittings.

ASTM International, F 1473, Standard Test Method for Notch Tensile Test to Measure the Resistance to Slow Crack Growth of Polyethylene Pipes and Resins.

ASTM International, F 2160, Standard Specification for Solid Wall High Density Polyethylene (HDPE) Conduit Based on Controlled Outside Diameter (OD).

ASTM International, F 2176, Standard Specification for Mechanical Couplings Used on Polyethylene Conduit, Duct and Innerduct.

AEIC Task Group 38.

Guide for Installation of Extruded Dielectric Insulated Power Cable Systems Rated 69KV -138KV.

IEEE Guide Distribution Cable Installation Methods in Duct Systems.

National Electrical Code (NEC), Chapter 9.

National Electrical Manufacturers Association, NEMA TC 7, Smooth-Wall Coilable Polyethylene Electrical Plastic Conduit.

Plastics Pipe Institute, Inc., *Handbook of Polyethylene Pipe.*

Plastics Pipe Institute, Inc., TR19, Thermoplastic Piping for the Transport of Chemicals.

Underground Extruded Power Cable Pulling Guide.

Underwriters Laboratories, Inc. UL 651A.

Underwriters Laboratories, Inc., UL 651B, Continuous Length HDPE Conduit.

Underwriters Laboratories, Inc., UL 2024, Optical Fiber Cable Raceway.

Appendix A

Calculation of Frictional Forces

Reference – *Maximum Safe Pulling Lengths for Solid Dielectric Insulated Cables – vol. 2: Cable User's Guide*, EPRI EL=3333-CCM, Volume 2, Research Project 1519-1, Electric Power Research Institute.

Calculations of Pulling Tensions

The following formulae can be employed to determine pulling tensions for a cable installation. Each equation applies to a specific conduit configuration. In order to use the formulae, the cable pull should be subdivided into specific sections. The configuration of each section should be identifiable with one of the graphical depictions accompanying the equations.

The mathematical expression associated with each of the accompanying sketches will yield the cumulative tension (T2) on the leading end of the cable(s) as it exits from a specified section when T1 is the tension in the cable entering that section. The maximum tension obtained when pulling in one direction often differs from that obtained when pulling in the opposite direction due to the location of the bends and the slope of the pull. Therefore, the required tension should be calculated for both directions.

A listing of the symbols employed and their definitions are as follows:

DEFINITIONS OF SYMBOLS

Symbols	Definition	Units
T_1	Section incoming cable tension	Pounds
T_2	Section outgoing cable tension	Pounds
R	Inside radius of conduit bend	Feet
W	Total weight of cables in conduit	Pounds/foot
Θ	Angle subtended by bend for curved sections or angle of slope measured from horizontal for inclined planes	Radians
Θ_a	Offset angle from vertical axis	Radians
Θ_b	Total angle from vertical axis	Radians
K	Effective coefficient of friction	—
L	Actual length of cable in section	Feet
D'	Depth of dip from horizontal axis	Feet
2s	Horizontal length of dip section	Feet

FIGURE 1 PULLING TENSION FORMULAE FOR CABLE IN CONDUIT

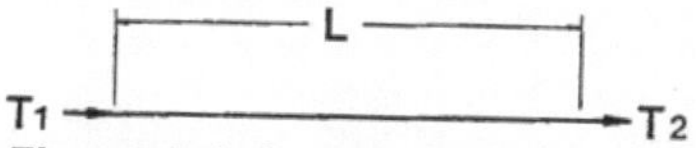

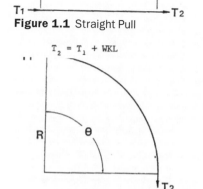

Figure 1.1 Straight Pull

$$T_2 = T_1 + WKL$$

$$T_2 = T_1 \cosh K\Theta + \sqrt{T_1^2 + (WR)^2} \sinh K\Theta$$

Figure 1.2 Horizontal Bend Pull

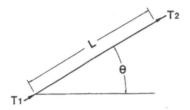

Note: Angle θ measured from horizontal axis

$$T_2 = T_1 + LW(\sin\theta + K\cos\theta)$$

Figure 1.3 Slope - Upward Pull

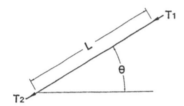

Note: Angle θ measured from horizontal axis

$$T_2 = T_1 - LW(\sin\theta - K\cos\theta)$$

Figure 1.4 Slope - Downward Pull

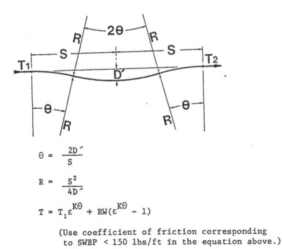

$$\theta = \frac{2D'}{S}$$

$$R = \frac{S^2}{4D'}$$

$$T = T_1 \epsilon^{K\theta} + RW(\epsilon^{K\theta} - 1)$$

(Use coefficient of friction corresponding to SWBP < 150 lbs/ft in the equation above.)

Where D´ is small compared to S (i.e. $\tan\theta/2 = \sin\theta/2 = D'/S$)

For T > RW

$$T_2 = T_1 \epsilon^{4K\theta} + RW[\epsilon^{4K\theta} - 2\epsilon^{3K\theta} + 2\epsilon^{K\theta} - 1]$$

For T ≤ RW

$$T_2 = T_1 + WK2S$$

Figure 1.5 Vertical Dip Pull (Small Angle)

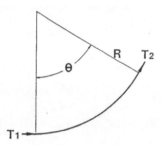

$$T_2 = T_1 \varepsilon^{K\Theta} - \frac{WR}{1+K^2}[2K\sin\Theta - (1-K^2)(\varepsilon^{k\Theta} - \cos\Theta)]$$

Figure 1.6a Concave Bend - Upward Pull, for Angle Θ Measured from Vertical Axis

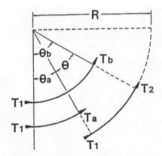

$$\Theta = \Theta_b - \Theta_a$$

$$T_a = T_1 \varepsilon^{K\Theta_a} - \frac{WR}{1+K^2}[2K\sin\Theta_a - (1-K^2)(\varepsilon^{K\Theta_a} - \cos\Theta_a)]$$

$$T_b = T_1 \varepsilon^{K\Theta_b} - \frac{WR}{1+K^2}[2K\sin\Theta_b - (1-K^2)(\varepsilon^{K\Theta_b} - \cos\Theta_b)]$$

$$T_2 = T_b - T_a + T_1$$

Figure 1.6b Concave Bend - Upward Pull, for Angle Θ Offset from Vertical Axis by Angle Θ_a (Derived from Figure 1.6a, above)

Chapter 14
Duct and Conduit

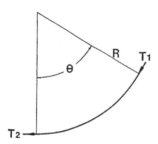

$$T_2 = T_1 \varepsilon^{K\Theta} - \frac{WR}{1+K^2}[2K\varepsilon^{K\Theta}\sin\Theta + (1-K^2)(1-\varepsilon^{K\Theta}\cos\Theta)]$$

Figure 1.7a Concave Bend - Downward Pull, for Angle Θ Measured from Vertical Axis

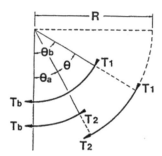

$$\Theta = \Theta_b - \Theta_a$$

$$T_b = T_1\varepsilon^{K\Theta_b} - \frac{WR}{1+K^2}[2K\varepsilon^{K\Theta_b}\sin\Theta_b + (1-K^2)(1-\varepsilon^{K\Theta_b}\cos\Theta_b)]$$

$$T_2 = \left(\frac{T_b + \frac{WR}{1+K^2}[2K\varepsilon^{K\Theta_a}\sin\Theta_a + (1-K^2)(1-\varepsilon^{K\Theta_a}\cos\Theta_a)]}{\varepsilon^{K\Theta_a}}\right)$$

Figure 1.7b Concave Bend - Downward Pull, for Angle Θ Offset from Vertical Axis by Angle Θ_a (Derived from Figure 1.7a, above)

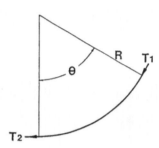

$$T_2 = T_1 \varepsilon^{K\Theta} - \frac{WR}{1+K^2} [2K\varepsilon^{K\Theta} \sin\Theta + (1-K^2)(1-\varepsilon^{K\Theta} \cos\Theta)]$$

Figure 1.8a Concave Bend - Downward Pull, for angle Θ Measured from Vertical Axis

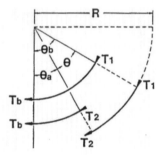

$$\Theta = \Theta_b - \Theta_a$$

$$T_b = T_1 \varepsilon^{K\Theta_b} - \frac{WR}{1+K^2} [2K\varepsilon^{K\Theta_b} \sin\Theta_b + (1-K^2)(1-\varepsilon^{K\Theta_b} \cos\Theta_b)]$$

$$T_2 = \left(\frac{T_b + \frac{WR}{1+K^2} [2K\varepsilon^{K\Theta_a} \sin\Theta_a + (1-K^2)(1-\varepsilon^{K\Theta_a} \cos\Theta_a)]}{\varepsilon^{K\Theta_a}} \right)$$

Figure 1.8b Concave Bend - Downward Pull, for angle Θ Offset from Vertical Axis by Angle Θ_a
(Derived from Figure 1.8a, above)

Chapter 14
Duct and Conduit

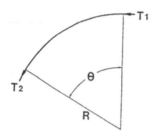

$$T_2 = T_1 \epsilon^{K\Theta} + \frac{WR}{1+K^2} [2K \sin\Theta - (1-K^2)(\epsilon^{K\Theta} - \cos\Theta)]$$

Figure 1.9a Convex Bend - Downward Pull, for Angle Θ Measured from Vertical Axis

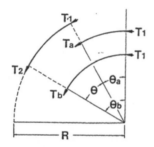

$$\Theta = \Theta_b - \Theta_a$$

$$T_a = T_1 \epsilon^{K\Theta_a} + \frac{WR}{1+K^2} [2K \sin\Theta_a - (1-K^2)(\epsilon^{K\Theta_a} - \cos\Theta_a)]$$

$$T_b = T_1 \epsilon^{K\Theta_b} + \frac{WR}{1+K^2} [2K \sin\Theta_b - (1-K^2)(\epsilon^{K\Theta_b} - \cos\Theta_b)]$$

$$T_2 = T_b - T_a + T_1$$

Figure 1.9b Convex Bend - Downward Pull, for Angle Θ Offset from Vertical Axis by Angle Θa
(Derived from Figure 1.8a, above)

Glossary

Abrasion - Wear or scour by hydraulic traffic.

Abrasion and Scratch Resistance - Ability of a material to resist the infliction of damage in the form of scratches, grooves and other minor imperfections.

Abutment - A wall supporting the end of a bridge or span, and sustaining the pressure of the abutting earth.

Acceptance - By the owner of the work as being fully complete in accordance with the contract documents.

Action - A positively charged ion which migrates through the electrolyte toward the cathode under the influence of a potential gradient.

Addenda - Written or graphic instruments issued prior to the execution of the agreement, which modify or interpret the contract documents, drawings and specifications, by addition, deletions, clarifications or corrections.

Additive - A substance added in a small amount for a special purpose such as to reduce friction, corrosion, etc.

Aerial sewer - An unburied sewer (generally sanitary type) supported on pedestals or bents to provide a suitable grade line.

Aerobic - Presence of unreacted or free oxygen (O_2).

Aggressive - A property of water which favors the corrosion of its conveying structure.

Aggressive Index (AI) - Corrosion index established by the American Water Work Association (AWWA) Standard C-400; established as a criterion for determining the corrosive tendency of the water relative to asbestos-cement pipe; calculated from the pH, calcium hardness (H), and total alkalinity (A) by the formula $AI = pH + \log^{(A \times H)}$

Agreement - The written agreement between the owner and the contractor covering the work to be performed; the contract documents are attached to and made a part of the agreements. Also designated as the contract.

Alkalinity - The capacity of a water to neutralize acids; a measure of the buffer capacity of a water. The major portion of alkalinity in natural waters is caused by (1) hydroxide, (2) carbonates, and (3) bicarbonates.

Anaerobic - An absence of unreacted or free oxygen [oxygen as in H_2O or Na_2SO_4 (reacted) is not "free"].

Angle of Repose - The angle which the sloping face of a bank of loose earth, gravel, or other material, makes with the horizontal.

Anode - (opposite of cathode) The electrode at which oxidation or corrosion occurs.

Apparent Tensile Strength - A value of tensile strength used for comparative purpose that is determined by tensile testing pipe rings in accordance with ASTM D 2290. This differs from true tensile strength of the material due to a bending moment induced by the change in contour of the ring as it is tested. Apparent tensile strength may be at yield, rupture or both.

Apparent Tensile Yield - The apparent tensile strength calculated for the yield condition.

Application for Payments - The form furnished by the engineer which is to be used by the contractor in requesting progress payments, and an affidavit of the contractor that progress payments theretofore received from the owner on account of the work been applied by the contractor to discharge in full all of the contractor's obligations stated in prior applications for payment.

Approval - Accept as satisfactory.

Aqueous - Pertaining to water; an aqueous solution is a water solution.

Areaway - A paved surface, serving as an entry area basement or subsurface portion of a building, which is provided with some form of drainage that may be connected to a sewer line.

ASTM - American Society of Testing and Materials and technical organization formed for the development of standards on characteristics and performance of materials, products, systems and services, and the promotion of related knowledge.

Available Water - Water necessary for the performance of work, which may be taken from the fire hydrant nearest the worksite, given conditions of traffic and terrain which are compatible with the use of the hydrant for performance of work.

Backfill Density - Percent compaction for pipe backfill (required or expected).

Base (course) - A layer of specified or selected material of planned thickness, constructed on the subgrade (natural foundation) or subbase for the purpose of distributing load, providing drainage or upon which a wearing surface or a drainage structure is placed.

Base Resin - Plastic materials prior to compounding with other additives or pigments.

Batter - The slope or inclination from a vertical plane, as the face or back of a wall.

Bedding - The earth or other material on which a pipe or conduit is supported.

Berm - The space between the toe of a slope and excavation made for Bedding – The earth or other material on which a pipe or conduit is supported.

Bid - The offer or process of the bidder submitted on the prescribed form setting forth the prices for the work to be performed.

Bidder - Any person, firm, or corporation submitting a bid for the work.

Biological Corrosion - Corrosion that results from a reaction between the Pipe material and organisms such as bacterial, algae, and fungi.

Bituminous (coating) - Of or containing bitumen; as asphalt or tar.

Bonds - Bid, performance and payment bonds and other instruments of security furnished by the contractor and his surety in accordance with the contract documents and in accordance with the law of the place of the project.

Boring - An earth-drilling process used for installing conduits or pipelines, or obtaining soil samples for evaluation and testing.

Bridge - A structure for carrying traffic over a stream or gully, or other traffic ways including the pavement directly on the floor of the structures. A structure measuring 10 ft. or more in clear span.

Bridge Plank (deck or flooring) - A corrugated steel sub-floor on a bridge to support a wearing surface.

Brittle Failure - A pipe failure mode that exhibits no visible (to the naked eye) material deformation (stretching, elongation, or necking down) in the area of the break.

Brittleness Temperature - Temperature at which 50% of the tested specimens will fail when subjected to an impact blow.

Building Sewer - The conduit which connects building wastewater sources, to the public or street sewer, including lines serving homes, public buildings, commercial establishments, and industrial structures. The building sewer is referred to in two sections: (1) the section between the building line and the property line, frequently specified and supervised by plumbing or housing officials; and (2) the section between the property line and the street sewer, including the connection thereto frequently specified and supervised by sewer, public works, or engineering officials (Referred to also as "house sewer," "building connection,)' "service connection," or "lateral connection").

Buoyancy - The power of supporting a floating body, including the tendency to float an empty pipe (by exterior hydraulic pressure).

Burst Strength - The internal pressure required to cause a pipe or fitting to fail within a specified time period.

Butt Fusion - A method of joining polyethylene pipe where two pipe ends are heated and rapidly brought together under pressure to form a homogeneous bond.

Bypass - An arrangement of pipes and valves whereby the flow may be passed around a hydraulic structure or appurtenance. Also, a temporary setup to route flow around a part of a sewer system.

Bypass Pumping - The transportation of sewage which flows around a specific sewer pipe/line section or sections via any conduit for the purpose of controlling sewage flows in the specified section or sections without flowing or discharging onto public or private property.

Caisson - A watertight box or cylinder used in excavating for foundations or tunnel pits to hold out water so concreting or other construction can be carried on.

Camber - Rise or crown of the center of a bridge, or Bowline through a culvert, above a straight line through its ends.

Cantilever - The part of a structure that extends beyond its support.

Carbon Black - A black pigment produced by the incomplete burning of natural gas or oil, that possesses excellent ultraviolet protective properties.

Catastrophic Rainfall Event - Rainfall event of return frequency far in excess of any sewerage design performance criteria typically, say, a 20 to 200 year storm.

Cathode - The electrode of an electrolytic cell at which reduction is the principal reaction (Electrons flow toward the cathode in the external circuit). Typical cathodic processes are cations taking up electron and being discharged, oxygen being reduced, and the reduction of an element or group of elements from a higher to a lower valence state.

Cathodic Corrosion - An unusual condition (especially with Al, Zn, Pb) in which corrosion is accelerated at the cathode because the cathodic reaction creates an alkaline condition which is corrosive to certain metals.

Cathodic Protection - Preventing corrosion of a pipeline by using special cathodes (and anodes) to circumvent corrosive damage by electric current. Also a function of zinc coatings on iron and steel drainage products - galvanic action.

Cavitation - Formulation and sudden collapse of vapor bubbles in a liquid; usually resulting from local low pressures - as on the trailing edge of a propeller; this develops momentary high local pressure which can mechanically destroy a portion of a surface on which the bubbles collapses.

CCTV - Closed circuit television.

Cell - Electrochemical system consisting of an anode and a cathode immersed in an electrolyte. The anode and cathode may be separate metals or dissimilar areas on the same metal. The call includes the external circuit which permits the flow of electrons from the anode toward the cathode (See Electrochemical Cell).

Cell Classification - Method of identifying thermoplastic materials, such as polyethylene, as specified by ASTM D 3350, where the Cell Classification is based on these six properties for PE are: Density, Melt Index, Flexural Modulus, Tensile Strength at Yield, Environmental Stress Crack Resistance, and Hydrostatic Design Basis.

Cellar Drain - A pipe or series of pipe which collect wastewater which leaks, seeps, or flow into subgrade parts of structures and discharge them into a building sewers or by other means dispose of such wastewater's into sanitary, combined or storm sewers (Referred to also as "basement drain").

Change Order - A written order to the contractor authorizing an addition, deletion or revision in the work, within the general scope of work of the agreement, authorizing an adjustment in the agreement price or agreement time.

Chemical Resistance - Ability to render service in the transport of a specific chemical for a useful period of time at a specific concentration and temperature.

Chimney - The cylindrical, variable height portion of the manhole structure having a diameter as required for the manhole frame. The chimney extends from the top of the corbel or cone to the base of the manhole frame and is used for adjusting the finished grade of the manhole frame.

Circumferential Coefficient of Expansion and Contraction - The fractional change in circumference of a material for a unit change in temperature. Expressed as inches of expansion or contraction per inch of original circumference per °F.

Coefficient of Thermal Expansion and Contraction - The fractional change in length of a material for a unit change in temperature.

Cofferdam - A barrier built in the water so as to form an enclosure from which the water is pumped to permit free access to the area within.

Cohesive Soil - A soil that when unconfined has considerable strength when air-dried, and that has significant cohesion when submerged.

Cold Bend - To force the pipe into a curvature without damage, using no special tools, equipment or elevated temperatures.

Collector Sewer - A sewer located in the public way collects the wastewater's discharged through building sewers and conducts such flows into larger interceptor sewers and treatment works. (Referred to also as "street sewer.")

Combined Sewer - A sewer intended to serve as both a sanitary sewer and a storm sewer, or as both an industrial sewer and a storm sewer.

Compaction - The densification of a soil by means of mechanical manipulation.

Composite Pipe - Pipe consisting of two or more different materials arranged with specific functional purpose to serve as pipe.

Compound - A mixture of a polymer with other ingredients such as fillers, stabilizers, catalysts, processing aids, lubricants, modifiers, pigments, or curing agents.

Compounding - The process where additives and carbon black are homogeneously mixed with the base polyethylene resin in a separate and additional process to produce a uniform compound material for polyethylene pipe extrusion.

Compression Gasket - A device which can be made of several materials in a variety of cross sections and which serves to secure a tight seal between two pipe sections (e.g., "0" rings).

Conductivity - A measure of the ability of a solution to carry an electrical current. Conductivity varies both with the number and type of ions the solution carries.

Conduit - A pipe or other opening, buried or aboveground, for conveying hydraulic traffic, pipelines, cables or other utilities.

Consolidation - The gradual reduction in the volume of a soil mass resulting from an increase in compaction.

Contamination - The presence of a substance not intentionally incorporated in a product.

Contract Documents - The Agreement, Addenda, Instructions to Bidders, Contractor's Bid, the Bonds, the Notice of Award, the General Conditions, the Supplementary Conditions, Special Conditions, Technical Conditions, the Specifications, Drawings, Drawing Modifications, and Notice to Proceed, all make up the Contract Documents.

Contract Price - The total moneys payable to the Contractor under the Contract Documents.

Contract Time - The number of calendar days stated in the Agreement for the completion of the work.

Contracting Officer - The owner (guarantee) - The Individual who is authorized to sign the contract documents on behalf of the owner's governing body.

Contractor - The person, firm or corporation with whom the owner has executed the agreement.

Corbel or Cone - That portion of a manhole structure, which slopes upward, and inward from the barrel of the manhole to the required chimney or frame diameter. "Corbel" refers to section built of brick or block, while "cone" refers to a precast section.

Core Area - That part of a sewer network containing the critical sewers, and other sewers where hydraulic problems are likely to be most severe and require detailed definition within a flow simulation model.

Corrosion - The destruction of a material or its properties because of a reaction with its (environment) surroundings.

Corrosion Fatigue - Fatigue type cracking of metal caused by repeated or fluctuating stresses in a corrosive environment characterized by shorter life than would be encountered as a result of either the repeated or fluctuating stress alone or the corrosive environment alone.

Corrosion Index - Measurement of the corrosivity of a water (e.g. Langelier Index, Ryznar Index, Aggressive Index, etc.)

Corrosion Rate - The speed (usually an average) with which corrosion progresses (it may be linear for a while); often expressed as though it was linear, in units of mdd (milligrams per square decimeter per day) for weight change, or mpy (milligrams per year) for thickness changes.

Corrosion Resistance - Ability of a material to withstand corrosion in a given corrosion system.

Corrosion-erosion - Corrosion which is increased because of the abrasive action of a moving stream; the presence of suspended particles greatly accelerates abrasive action.

Cracks - Crack lines visible along the length and/or circumference.

Crazing - Apparent fine cracks at or under the surface of a plastic.

Creep - The dimensional change, with time, of a material under continuously applied stress after the initial elastic deformation. The time dependent part of strain due to a constant stress.

Crew - The number of persons required for the performance of work at a site as determined by the contractor in response to task difficulty and safety considerations at the time or location of the work.

Critical Pressure - The minimum internal compressed gas pressure at which rapid crack propagation (RCP) can be sustained along a section of plastic pipe.

Critical Sewers - Sewers with the most significant consequences in the event of structural failure.

Crosslink - The formation of a three dimensional polymer by means of interchain reactions resulting in changes in physical properties.

CTS – Copper tube sizing convention for PE tubing.

Density, Base Resin – The mass per unit volume at a standardized temperature of 23°C of a base resin prior to compounding with additives and modifiers.

Density, Pipe Compound - The mass per unit volume of a compound at standardized temperature of 23°C. Note this is pipe compound density, not base resin density.

Design Coefficient (DC) - A number greater than 1.00 that when divided into the Minimum Require Strength (MRS) establishes the maximum design stress of the product for the application. The DC takes into consideration the variables in resin and processing involved in the production of plastic pipe. The user needs to also consider other variables such as: shipping, handling, installation and service of properly installed thermoplastic pressure piping systems.

Design Factor (DF) - A number less than 1.00 that takes into consideration the variables in resin and processing as well as the variables involved in the shipping, handling, installation and service of properly installed thermoplastic pressure piping systems.

Design Stress – (ISO12162) Allowable stress (MPa) for a given application. It is derived by dividing the MRS by the design coefficient C then rounding to the next lower value in the R-20 series (ISO 3). (For HDB rated materials see Hydrostatic Design Stress).

Dimension Ratio (DR) - The ratio of pipe diameter to wall thickness. It is calculated by dividing the specified outside diameter of the pipe, in inches, by the minimum specified wall thickness, in inches. Specifying PE pipes with the same DR regardless of O.D. assures all pipes will have the same design pressure assuming the PEs have the same HDB rating. The <u>standard</u> dimension ratio (SDR) is a common numbering system that is derived from the ANSI preferred number series R-10.

Dimple - A term used in tight fitting pipeline reconstruction, where the new plastic pipe forms an external departure or a point of expansion slightly beyond the underlying pipe wall where unsupported at side connections. The dimples are used for location and reinstatement of lateral sewer service.

Ductile Failure - A failure mode that exhibits material deformation (stretching, elongation, or necking down) in the area of the break.

Easement - A liberty, privilege, or advantage without profit which the owner of one parcel of land may have in the hand of another. In this agreement, all land, other than public streets, in which the owner has sewer system lines or installations and right of access to such lines or installations.

Easement Access - Areas within an easement to which access is required for performance of work.

Effluent - Outflow or discharge from a sewer us sewage treatment equipment.

EHMWHD - Extra High Molecular Weight High Density as originally noted in ASTM D1248, Grade P34 materials were specifically EHMW high-density polyethylene materials.

Elastic Modulus - A measure of the stress buildup associated with a given strain.

Electrofusion - A heat fusion joining process where the heat source is an integral part of the fitting.

Elevated Temperature Testing - Tests on plastic pipe above 23°C (73°F) for HDB rated materials and 20°C (68°F) for MRS rated materials.

Elongation – (strain) The increase in length of a material stressed in tension.

Embankment (or fill) - A bank of earth, rock or other material constructed above the natural ground surface.

Embrittlement - Loss of ductility of a material resulting from a chemical or physical change.

Emergency Repair - A repair that must be made while the main is pressurized, or flowing.

End Section - Flared attachment on inlet and outlet of a culvert to prevent erosion of the roadbed improve hydraulic efficiency, and improve appearance.

Endurance Limit - The maximum stress that a material can withstand for an infinitely large number of fatigue cycles (See Fatigue Strength).

Energy Gradient - Slope of a line joining the elevations of the energy head of a stream.

Energy Head - The elevation of the hydraulic gradient at any section, plus the velocity head.

Engineer - The person, firm or corporation named as such in the contract documents; the "Engineer of Record".

Environment - The surroundings or conditions (physical, chemical, mechanical) in which a material exists.

Environmental Stress Crack Resistance (ESCR) - The resistance to crack or craze under the influence of specific chemicals and stress and/or mechanical stress.

Environmental Stress Cracking – Under certain conditions of temperature and stress in the presence of certain chemicals, polyethylene may begin to crack sooner than it would at the same temperature and stress in the absence of those chemicals. The susceptibility to crack or craze under the influence of specific chemicals, stress, and/or mechanical stress.

Epoxy - Resin formed by the reaction of bisphenol and epichlorohydrin.

Equalizer - A culvert placed where there is no channel but where it is desirable to have standing water at equal elevations on both sides of a fill.

Erosion - Deterioration of a surface by the abrasive action of moving fluids. This is accelerated by the presence of solid particles or gas bubbles in suspension. When deterioration is further increased by corrosion, the term "Corrosion-Erosion" is often used.

Erosion Corrosion - A corrosion reaction accelerated by the relative movement of the corrosive fluid and the metal surface.

ESCR - Environmental Stress Crack Resistance. The ability to resist environmental stress cracking when tested under standards such as ASTM F 1248 and F 1473.

Ethylene Plastics - Plastics based on polymers of ethylene or copolymers of ethylene with other monomers, the ethylene being in greatest amount by mass.

Exfiltration - The leakage or discharge of flows being carried by sewers out into the ground through leaks in pipes, joints, manholes, or other sewer system structures; the reverse of "infiltration."

Existing Linear Feet - The total length of existing sewer pipe in place within designated sewer systems as measured from center of manhole to center of manhole from maps or in the field.

Experimental Grade (E) - A PPI HSB recommended rating that is valid for a limited duration, given to those materials covered by data that do not yet comply with the full requirements of the Standard Grade, but satisfy the applicable minimum preliminary data requirements which are detailed in TR-3.

Extrusion - A process whereby heated or unheated plastic forced through a shaping orifice becomes one continuously formed piece.

Fabricated Fittings – Large diameter polyethylene fittings fabricated by fusing together special shapes to create reducer fittings, tees, ells and bends.

Fatigue - The phenomenon leading to fracture under repeated or fluctuating stresses having a maximum value less than the tensile strength of the material.

Fatigue Strength - The stress to which a material can be subjected for a specified number of fatigue cycles.

Field Orders - A written order issued by the engineer clarifies or interprets the contract documents in accordance with the terms of the contract or orders minor changes in the work in accordance with the terms of the contract.

Filter - Granular material or geotextile placed around a submarine pipe to facilitate drainage and at the same time strain or prevent the admission of silt or sediment.

Flash Point - Temperature at which a material begins to vaporize.

Flexible - Readily bent or deformed without permanent damage.

Flexural Modulus - The ratio, within the elastic limit, of the applied stress in the outermost fibers of a test specimen in three point static flexure, to the calculated strain in those outermost fibers (ASTM D 790).

Flexural Strength – (Flexural Modulus of Rupture) – The maximum calculated stress in the outermost fibers of a test bar subjected to three point loading at the moment of cracking or breaking (ASTM D 790). The maximum stress in the outer fiber of a test specimen at rupture.

Flow Attenuation - The process of reducing the peak flow rate, in a sewer system, by redistributing the same volume of flow over a longer period of time.

Flow Control - A method whereby normal sewer flows or a portion of normal sewer flows are blocked, retarded, or diverted (bypassed) within certain areas of the sewer collection system.

Flow Reduction - The process of decreasing flows into a sewer system or of removing a proportion of the flow already in a sewer system.

Flow Simulation - The modeling of flow in surface water or combined sewer systems using a dynamic digital model.

Fold and Form Pipe - A pipe rehabilitation method where a plastic pipe manufactured in a folded shape of reduced cross-sectional area is pulled into an existing conduit and subsequently expanded with pressure and heat. The reformed plastic pipe fits snugly to and takes the shape of the ID of the host pipe.

Fouling - An accumulation of deposits. This term includes accumulation and growth of marine organisms on a submerged metal surface and also includes the accumulation of deposits (usually inorganic) on heat exchanger tubing.

Foundation Drain - A pipe or series of pipes which collect groundwater from the foundation or footing of structures and discharge it into sanitary, storm, or combined sewers, or to other points of disposal for the purpose of draining unwanted waters away from such structures.

Fracture Mechanics - A quantitative analysis for evaluating structural reliability in terms of applied stress, crack length, and specimen geometry.

Fractures - Cracks visibly open along the length and/or circumference of the conduit with the pieces still in place.

Galvanic Cell - A cell consisting of two dissimilar metals in contact with each other and with a common electrolyte (sometimes refers to two similar metals in contact with each other but with dissimilar electrolytes; differences can be small and more specifically defined as a concentration cell).

General Corrosion - Corrosion in a uniform manner.

Glass Transition Temperature - The temperature below which a plastic is more brittle and glassy.

Gradation - Sieve analysis of aggregates.

Grade - Profile of the center of a roadway, or the invert of a culvert or sewer. Also refers to slope, or ratio of rise or fall of the grade line to its length. (Various other meanings.)

Gradient - See Grade.

Grain - A portion of a solid metal (usually a fraction of an inch in size) in which the atoms are arranged in an orderly pattern. The irregular junction of two adjacent grains is known as-a grain boundary; also a unit of weight, 1/7000th of a pound; also used in connection with soil particles i.e. = grain of sand.

Granular - Technical term referring to the uniform size of grains or crystals in rock.

Graphitization (graphitic corrosion) - Corrosion of gray cast iron in which the metallic constituents are converted to corrosion products, leaving the graphite flakes intact, Graphitization is also used in a metallurgical sense to mean the decomposition of iron carbide to form iron and graphite.

Groin - A-jetty built at an angle to the shoreline, to control the waterflow and currents or to protect a harbor or beach.

Ground Water Table (or level) - Upper surface of the zone of saturation in permeable rock or soil. (When the upper surface is confined by impermeable rock, the water table is absent.)

Grout - A fluid mixture of cement, and water (and sometimes sand), that can be poured or pumped easily; also encompasses chemical mixtures recognized as stopping water infiltration through small holes and cracks.

Grouting - (1) The joining together of loose particles of soil in such a manner that the soil so grouted becomes a solid mass which is impervious to water, (see also PIPE JOINT SEALING) (2) The process of flowing a cement/water grout (without sand) into the annular space between a host pipe and a slipline pipe.

Haunch - That portion of the pipe barrel extending below the pipe springline.

Haunching - The act of placing embedment material below the springline.

Head (Static) - The height of water above any plane or point of references (the energy possessed by each unit of weight of a liquid, expressed as the vertical height through which a unit of weight would have to fall to release the average energy posed). The standard inch-pound unit of measure is feet of water. The relation between pressure in psi and feet of head at 68°F is 1 psi = 2.310 ft of head.

Headwall - A wall (of any material) at the end of a culvert or, drain to serve one or more of the following purposes: protect fill from scour or undermining; increase hydraulic efficiency, divert direction of flow, and serve as a retaining wall.

Height Of Cover (HC) - Distance from crown of a culvert or conduit to the finished road surface, or ground surface, or the base of the rail.

High-Density Polyethylene (HDPE) - A plastic resin made by the copolymerization of ethylene and a small amount of another hydrocarbon. The resulting base resin density, before additives or pigments, is greater than 0.941 g/cm.

Holiday - Any discontinuity or bare spot in a coated surface.

Hoop Stress - The circumferential force per unit areas, psi, in the pipe wall due to internal pressure.

Hydraulic Cleaning - Techniques and methods used to clean sewer lines with water e.g. water pumped in the form of a high velocity spray and water flowing by gravity or head pressure. Devices include high velocity jet cleaners, cleaning balls, and hinged disc cleaners.

Hydraulic Gradient or Hydraulic Grade Line - An imaginary line through the points to which water would rise in a series of vertical tubes connected to the pipe. In an open channel, the water surface itself is the hydraulic grade line.

Hydraulic Radius - The area of the water prism in the pipe or channel divided by the wetted perimeter. Thus, for a round conduit flowing full or half full, the hydraulic radius is d/4. Another term sometimes used for this quantity is hydraulic mean depth.

Hydraulics - That branch of science or engineering which treats water or other fluid in motions.

Hydrocarbon, Gaseous - An organic compound made up of the elements of carbon and hydrogen that exists as a gas at ambient conditions (14.7 psi, 73.4T).

Hydrocarbon, Liquid - An organic compound made up of the elements of carbon and hydrogen that exists as a liquid at ambient conditions (14.7 psi, 73.4"F).

Hydrogen Blistering - Subsurface voids produced in a metal by hydrogen absorption in (usually) low strength alloys with resulting surface bulges.

Hydrogen Induced Cracking (HIC) - A form of hydrogen blistering in which stepwise internal cracks are created that can affect the integrity of the metal.

Hydrogen Ion (pH) - Refers to acidity or alkalinity of water or soil. An ion is a charged atom or group of atoms in solution or in a gas. Solutions contain equivalent numbers of positive and negative ions.

Hydrogen Stress Cracking - A cracking process that results from the presence of hydrogen in a metal in combination with tensile stress. It occurs most frequently with high strength alloys.

Hydrostatic Design Basis (HDB) – One of a series of established stress values specified in Test Method D 2837 "Standard Test Method for Obtaining Hydrostatic Design Basis for Thermoplastic Pipe Materials" for a plastic compound obtained by categorizing the LTHS determined in accordance with Test Method D 2837. HDB refers to the categorized LTHS in the circumferential, or hoop direction, for a given set of end use conditions. Established HDBs are listed in PPI TR-4.

Hydrostatic Design Stress HDB (HDS$_{HDB}$) – The estimated maximum tensile stress (psi) in the wall of the pipe in the circumferential orientation due to internal hydrostatic pressure that can be continuously applied with a high degree of certainty that failure of the pipe will not occur.
HDS_{HDB} = HDB X DF

Hydrostatic Design Stress MRS (HDS$_{MRS}$) – The estimated maximum tensile stress (psi) in the wall of the pipe in the circumferential orientation due to internal hydrostatic pressure that can be continuously applied with a high degree of certainty that failure of the pipe will not occur. HDS_{MRS} = MRS/C

I. D. –Inside diameter of pipe or tubing.

Ignition Temperature - Temperature at which the vapors emitted from a material will ignite either without exposure to a flame (self-ignition) or when a flame is introduced (flash ignition).

Impact - Stress in a structure caused by the force of a vibratory, dropping, or moving load. This is generally a percentage of the live load.

Impact Strength - The ability of a material to withstand shock loading.

Impervious - Impenetrable. Completely resisting entrance of liquids.

Inert Material - A material which is not very reactive, such as a noble metal or plastic.

Infiltration - The water entering a sewer system, including building sewers, from the ground, through such means as defective pipes, pipe joints, connections, or manhole walls. Infiltration does not include, and is distinguished from inflow.

Infiltration/Inflow - A combination of infiltration and inflow wastewater volumes in sewer lines, with no way to distinguish the basic sources, and with the effect of usurping the capacities of sewer systems and facilities.

Inflow - The water discharged into a sanitary sewer system, including service connections from such sources as roof leaders, cellar, yard, area drains, foundation drains, cooling water discharges, drains from springs and swampy areas, manhole catch basins, storm waters, surface runoff, street washwaters and/or drainage. Inflow does not include and is distinguished from infiltration.

Ingredient – Any chemical, mineral, polymer or other ingredient that has been added to a resin composition for the purpose of imparting certain desired processing or product performance properties.

Inhibitor - (1) A chemical substance or combination of substances which, when present in the environment, prevents or reduces corrosion without significant reaction with the components of the environment. (2) A substance which sharply reduces corrosion, when added to water, acid, or other liquid in small amounts. (3) A chemical additive that delays the chemical reaction in epoxy resin systems.

Injection Molding - The process of forming a material by melting it and forcing it, under pressure, into the cavity of a closed mold.

Insert Stiffener - A length of tubular material, usually metal, installed in the ID of the pipe or tubing to reinforce against OD compressive forces from a mechanical compression type fitting.

Inspector - The owner's on-site representative responsible for inspection and acceptance, approval, or rejection of work performed as set forth in these specifications.

Inspector (Construction Observer, Resident Inspector, Construction Inspector, Project Representative) - An authorized representative of the engineer assigned to observe the work and report his findings to the engineer.

Interaction - The division of load carrying between pipe and backfill and the relationship of one to the other.

Intercepting Drain - A ditch or trench filled with a pervious filter material around a subdrainage pipe.

Interceptor Sewer - A sewer which receives the flow of collector sewers and conveys the wastewaters to treatment facilities.

Intergranular Stress Corrosion Cracking (IGSCC) - Stress corrosion cracking in which the cracking occurs along grain boundaries.

Internal Corrosion - Corrosion that occurs inside a pipe because of the physical, chemical, or biological interactions between the pipe and the water as opposed to forces acting outside the pipe, such as soil, weather, or stress conditions.

Internal Erosion - Abrasion and corrosion on the inside diameter of the pipe or tubing due to the fluid or slurry that is being transported.

Internal Pipe Inspection - The television inspection of a sewer line section. A CCTV camera is moved through the line at a slow rate and a continuous picture is transmitted to an above ground monitor. (See also PHYSICAL PIPE INSPECTION.)

Inversion - The process of turning a fabric tube inside out with water or air pressure as is done at installation of a cured in place pipe.

Invert - That part or a pipe or sewer below the spring line - generally the lowest point of the internal cross section.

Invert Level (elevation) - The level (elevation) of the lowest portion of a liquid-carrying conduit, such as a sewer, which determines the hydraulic gradient available for moving the contained liquid.

Ion - An electrically charged atom (e.g., Na^+, Al^{+3}, Cl^-, S^{-2}) or group of atoms known as "radicals" (e.g. NH_4, SO_4, PO_4).

Ionization - Dissociation of ions in an aqueous solution (e.g., $H2CO3 \longrightarrow H+ + HCO3$ or $H20 \longrightarrow H+ + OH-$).

IPS – Iron pipe sizing convention for PE pipe.

Jacking (for conduits) - A method of providing an opening for drainage or other purposes underground, by cutting an opening ahead of the pipe and forcing the pipe into the opening by means of horizontal jacks.

Joint, Butt-Fused - A thermoplastic pipe connection between two pipe ends using heat and force to form the bond.

Joint, Electrofused – A joint made with an Electrofusion fitting in which the heating source is an integral part of the fitting.

Joint, Flanged - A mechanical joint using pipe flanges, a gasket, and bolts.

Joint, Heat-Fused - A thermoplastic pipe connection made using heat and usually force to form the fusion bond.

Joint, Mechanical - A connection between piping components employing physical force to develop a seal or produce alignment.

Joint, Saddle-Fused - A joint in which the curved base of the saddle fitting and a corresponding area of the pipe surface are heated and then placed together to form the joint.

Joint, Socket-Fused - A joint in which the joining surfaces of the components are heated, and the joint is made by inserting one component into the other.

Joints - The means of connecting sectional lengths of sewer pipe into a continuous sewer line using various types of jointing materials. The number of joints depends on the lengths of the pipe sections used in the specific sewer construction work.

Kip - A force unit equal to 1000 pounds.

Lateral - Any pipe connected to a sewer.

Linear Foot - Being one foot to the length of a sewer line.

Long-Term Strength - The hoop stress in the wall of the pipe is sufficiently low that creep (relaxation) of the materials is nil and assures service life in excess of 50 years.

Long-Term Hydrostatic Strength (LTHS) - The hoop stress that when applied continuously, will cause failure of the pipe at 100,000 hours (11.43 years). This is the intercept of the stress regression line with the 100,000-h coordinate as defined in ASTM D 2837. Note –The typical condition uses water as the pressurizing fluid at 23°C (73°F).

Low-Density Polyethylene Plastics (LDPE) - Polyethylene plastics, having a standard density of 0.910 to 0.925 g/cm^3.

Lower Confidence Limit (LCL) - A calculated statistical value used in ASTM D 2837 to determine the suitability of a data set for use in determining LTHS and HDB.

Lower Confidence Limit of the Predicted Hydrostatic Strength (σLPL)(ISO 9080) - A quantity in MPA, with the dimension of stress, which represents the 97.5% lower confidence limit of the predicted hydrostatic strength at temperatures T and time t.

LP-Gas – Liquid petroleum gas, permitted to be piped in PE piping, in vapor phase, Maximum Allowable Operating Pressure only at pressures ≤ 30 psig.

MAG PIPE – Magnetically detectable polyethylene pipe.

Major Blockage - A blockage (structural defect, collapse, protruding service connection, debris) which prohibits manhole-to-manhole cleaning, TV inspections, pipe flow, or rehabilitation procedures.

Manhole Section - The length of sewer pipe connecting two manholes.

Manning's Formula - An equation for the value of coefficient c in the Chezy Formula, the factors of which are the hydraulic radius and a coefficient of roughness: an equation itself used to calculate flows in gravity channels and conduits.

Maximum Allowable Operating Pressure - The highest working pressure expected and designed for during the service-life of the main.

Maximum Allowable Operating Pressure (MAOP) – In USA Regulation for gas piping, the highest allowed pressure, in psig, as determined in accordance with US CFR, Title 49, Part 192.121 and as represented in the following: MAOP= 2 x HDB$_T$ x 0.32 / (DR-1)

Mechanical Cleaning - Methods used to clean sewer lines of debris mechanically with devices such as rodding machines, bucket machines, winch-pulled brushes, etc.

Mechanical Fitting - Fitting for making a mechanical joint to provide for pressure integrity, leak tightness, and depending on category, as defined in ASTM F 1924, resistance to end loads and pull-out.

Median Barrier - A double-faced guardrail in the median or island dividing two adjacent roadways.

Medium Density Polyethylene Plastics (MDPE) - Those branched polyethylene plastics, having a standard density of 0.926to 0.940 g/cm^3.

Melt Flow - A measure of the molten material's fluidity.

Melt Flow Rate - The quantity of thermoplastic material in grams that flows through an orifice during a 10-minute time span under conditions as specified by ASTM D 1238.

Melt Index - A measurement of a polymer's molten flow properties (ASTM D 1238), is related to molecular weight, or the length of the individual polymer chains. Generally, lower melt indices represent higher molecular weights while higher values indicate lower molecular weights. For any given PE resin, a lower melt index (higher molecular weight) will normally have superior physical properties.

Melt Viscosity - The resistance of the molten material to flow.

Melting Point - That temperature at which the plastic transitions to a completely amorphous state.

Minimum Required Pressure (MRP) – One of a series of established pressure values for a plastic piping component (multilayer pipe, fitting, valve, etc.) obtained by categorizing the long-term hydrostatic pressure strength in accordance with ISO 9080.

Minimum Required Strength (MRS) – (ISO 12162) The lower confidence limit in accordance with ISO 9080 at 20°C for 50 years with internal water pressure, rounded down to the next smaller value of the R-10 series or of the R-20 series conforming to ISO 3 and ISO 497, and categorized in accordance with ISO 12162, "Thermoplastic materials for pipes and fittings for pressure applications – Classification and designation – Overall service (design) coefficient."

Modification - (1) A written amendment of the contract documents signed by both parties. (2) A change order. (3) A written clarification or interpretation issued by the engineer in accordance with the terms of the contract. (4) A written order for a minor change or alteration in the work issued by the engineer pursuant to the terms of the contract. A modification may only be issued after execution of the agreement.

Modulus of Elasticity (E) - ASTM D 638 The ratio of stress (nominal) to corresponding strain below the proportional limit of a material.

Molecular Weight Distribution - The ratio of the weight average molecular weight (M_w) to the number average molecular weight (M_n). This gives an indication of the distribution.

Molecular Weight, Number Average (abbreviation M_n) - The total weight of all molecules divided by the number of molecules.

Molecular Weight, Weight Average (abbreviation M_w) - The sum of the total weight of molecules of each size multiplied by their respective weights divided by the total weight of all molecules.

Moment of Inertia - Function of some property of a body or figure - such as weight, mass, volume, area, length, or position, equal to the summation of the products of the elementary portions by the squares of their distances from a given axis.

Moment, Bending - The moment which produces bending in a beam or other structure. It is measured by the algebraic sum of the products of all the forces multiplied by their respective lever arms.

Multilayer Pipe – (Composite Pipe). **TYPE 1:** A pressure rated pipe having more than one layer (bonded together) in which at least 60% of the wall thickness is polymeric material that has an HDB (Hydrostatic Design Basis) or MRS (Minimum Required Strength), from which the pressure rating of the pipe is determined by pipe size and pipe wall construction.

Multilayer Pipe – (Composite Pipe). **TYPE 2:** A pressure rated pipe having more than one layer (bonded together) where at least 60% of the wall thickness is polymeric material, where the pipe pressure rating is determined by pipe size and pipe wall construction, and this pipe rating is listed by a PDB (Pressure Design Basis) or MRP (Minimum Required Pressure).

Multilayer Pipe – (Composite Pipe). **TYPE 3:** Non-pressure rated pipe comprising more than one layer in which at least 60% of the wall thickness is polymeric material.

Neutral Axis - An axis of no stress.

Nominalize - To classify a value into an established range or category.

Non-Pressure Pipe - Pipe designed for gravity-conveyed medium which must resist only intermittent static pressures and does not have a pressure rating.

Non-Uniform Corrosion - Corrosion that attacks small, localized areas of the pipe. Usually results in less metal loss than uniform corrosion but causes more rapid failure of the pipe' due to pits and holes.

Notch Sensitivity - The extent to which an inclination to fracture is increased by a notch, crack, scratch, or sudden change in cross-section. **NOTE:** The SDB is used only for a material intended for molding applications. The SDB shall not be used for pipe applications.

Notice of Award - The written notice by owner to the apparent successful bidder stating that upon compliance with the conditions precedent to be fulfilled by him within the time specified, the owner will execute and deliver the agreement to him.

Notice to Proceed - A written notice given by the owner to the contractor (with a copy to the engineer) fixing the date on which the contract time will commence to run and on which contractor shall start to perform his obligations under the contract documents.

Nylon (Polyamides) - Plastics based on resins composed principally of a long-chain synthetic polymer amide, which has recurring amide groups as an integral part of the main polymer chain.

O.D. – Outside diameter of pipe or tubing.

Odorants - To enhance safety, the fuel gas industries add chemical compounds to their gases, with a unique odor, to alert the user if a leak occurs. This odor is designed to be readily detectable when the fuel gas mixes with the atmosphere at low concentrations. The compounds used as odorants usually consist of aliphatic mercaptans, such as propyl and tertiary butyl mercaptan, and sulfides, such as thiopane or dimethyl sulfide at ordinary temperatures. Most gas odorants are liquids at full concentrations, and, in this state, might be harmful to some plastic pipe materials. However, in the small amounts sufficient to odorize a gas they are in the vapor state and cause no harm to plastic piping.

Outfall (or outlet) - In hydraulics, the discharge end of drains and sewers.

Out-of-Roundness - The allowed difference between the maximum measured diameter and the minimum measured diameter (stated as an absolute deviation).

Ovality - (%) - ((max_measured_O.D.) - (min_measured_O.D.) ÷ (average_O.D.)) x 100

Overflow - (1) The excess water that flows over the ordinary limits of a sewer, manhole, or containment structure. (2) An outlet, pipe, or receptacle for the excess water.

Owner - A public body of authority, corporation as partnership, or individual for whom the work is to be performed.

Oxidation - Loss of electrons, as when a metal goes from the metallic state to the corroded state.

Parapet - Wall or rampart, breast high. Also, the wall on top of an abutment extending from the bridge seat to the underside of the bridge floor and designed to hold the backfill.

Pascal's Law - Pressure exerted at any point upon a confined liquid is transmitted undiminished in all directions.

Pavement, Invert - Lower segment of a corrugated metal pipe provided with a smooth bituminous material that completely fills the corrugations, intended to give resistance to scour, erosion, and to improve flow.

PE - Polyethylene

PE 2406 – Medium-density polyethylene with ESCR in accordance with ASTM D1693 equal to or greater than 600 hours or a PENT value per ASTM D1473 equal to or greater than 10 hours and a hydrostatic design basis of 1250 psi.

PE 3408 – High-density polyethylene with ESCR in accordance with ASTM D1693 equal to or greater than 600 hours or a PENT value per ASTM D1473 equal to or greater than 10 hours and a hydrostatic design basis of 1600 psi.

PE 80 – A polyethylene classified by the ISO MRS system as having a minimum required strength of 8.0 MPa (1160 psi) in accordance with ISO 12162.

PE 100 - A polyethylene classified by the ISO MRS system as having a minimum required strength of 10.0 MPa (1450 psi) in accordance with ISO 12162.

PENT - The common name given for a test to determine the slow crack resistance of PE materials by placing a razor-notched tensile bar under a constant tensile load of 2.4 MPa at 80°C in accordance with ASTM F 1473.

Perched Water Table - In hydrology, the upper surface of a body of free ground water in a zone of saturation, separated by unsaturated material from an underlying body of ground water in a differing zone of saturation.

Periphery - Circumference or perimeter of a circle, ellipse, pipe-arch, or other closed curvilinear figure.

Permeability - Penetrability

PEX – Crosslinked polyethylene

pH - A measure of the acidity or alkalinity of a solution. A value of seven is neutral. Numbers lower than seven are acid, with the lower numbers more acid. Numbers greater than seven (up to 14) indicate alkalinity, with the higher numbers more alkaline.

Physical Pipe Inspection - The crawling or walking through manually accessible pipelines. The logs for physical pipe inspection record information of the kind detailed under TELEVISION INSPECTION. Manual inspection is only undertaken when field conditions permit this to be done safely. Precautions are necessary.

Pile, Bearing - A member driven or jetted into the ground and deriving its support from the underlying strata and/or by the friction of the ground on its surface.

Pipe - Nominal Weight - The pipe or tubing weight, expressed in pounds per 100 feet, calculated in accordance with PPI TN-7 by using the nominal diameter, and the nominal wall thickness of the pipe.

Pipe Joint Sealing - A method of sealing leaking or defective pipe joints which permit infiltration of groundwater into sewers by means of injecting chemical grout into and/or through the joints from within the pipe.

Pipeline Reconstruction - The insitu repair of an existing pipeline that has suffered loss of pressure integrity or has been structurally damaged. The liner becomes the principal pressure containment or structural element of the insitu composite pipe structure.

Pipeline Rehabilitation - The insitu repair of an existing pipeline, which has become corroded or abraded, by insert renewal of a liner which rehabilitates the bore of the pipeline but does not contribute significantly to increased pressure capability or increased structural strength, yet does improve flow efficiency/hydraulics.

Pitting - Highly localized corrosion resulting in deep penetration at only a few spots.

Pitting Factor - The depth of the deepest pit divided by the "average penetration" as calculated from weight loss.

Planting Piping - Installation procedure that digs a trench and lays the pipe in one step.

Plastic - A polymeric material that contains as an essential ingredient one or more organic polymeric substances of large molecular weight, is solid in its finished state, and, at some stage in its manufacture or processing into finished articles (See Thermoplastic and Thermoset).

Plastic Pipe - A hollow cylinder of a plastic material in which the wall thicknesses are usually small when compared to the diameter and in which the inside and outside walls are essentially concentric and which follows the O.D. sizing convention of steel pipe (IPS) or the sizing convention of ductile iron pipe (DIPS).

Plastic Tubing - A particular size of smooth wall plastic pipe in which the outside diameter is essentially the same as the corresponding size of copper tubing (CTS) or other tubing sizing conventions.

Plate - A flat-rolled iron or steel product.

Plough-in Piping - Installation procedure that splits the earth and pulls the pipe into position.

Poly (Vinyl Chloride) (PVC) - A polymer prepared by the polymerization of vinyl chloride as the sole monomer.

Polyester - Resin formed by condensation of polybasic and monobasic acids with polyhydric alcohols.

Polyethylene - A ductile, durable, virtually inert thermoplastic composed of polymers of ethylene. It is normally a translucent, tough solid. In pipe grade resins, ethylene-hexene copolymers are usually specified with carbon black pigment for weatherability.

Polymer - A substance consisting of molecules characterized by the repetition (neglecting ends, branch junctions, and other minor irregularities) of one or more types of monomeric units.

Polymerization - A chemical reaction in which the molecules of a monomer are linked together to form polymers. When two or more different monomers are involved, the process is called copolymerization.

Ponding - (1) Jetting or the use of water to hasten the settlement of an embankment - requires the judgment of a geotechnical engineer. (2) In hydraulics, ponding refers to water backed up in a channel or ditch as the result of a culvert of inadequate capacity or design to permit the water to flow unrestricted.

PPI (Plastic Pipe Institute) - A trade organization whose Membership is composed of manufacturers of plastic pipe, fittings, and valves; plastic materials for piping; metallic fittings for plastic piping: and equipment that is used for fabricating, joining or installing plastic piping.

Precipitation - Process by which water in liquid or solid state (rain, sleet, snow) is discharged out of the atmosphere upon a land or water surface.

Pressure Class (PC) – (AWWA C906) The design capacity to resist working pressure up to 80°F (27°C) maximum service temperature, with specified maximum allowances for reoccurring positive surges above working pressure.

Pressure Design Basis (PDB) – One of a series of established pressure values for a plastic piping component (multilayer pipe, fitting, valve, etc.) obtained by categorizing the long-term hydrostatic pressure strength determined in accordance with an industry test method that uses linear regression analysis. Although ASTM D 2837 does not use "pressure values", the PPI Hydrostatic Stress Board uses the principles of ASTM D2837 in plotting log pressure vs. log time to determine a "long-term hydrostatic pressure strength" and the resulting "Pressure Design Basis" for multilayer pipe that is listed in PPI TR-4.

Pressure Pipe - Pipe designed to resist continuous pressure exerted by the conveyed medium.

Pressure Rating - Estimated maximum internal pressure that allows a high degree of certainty that failure of the pipe will not occur.

Pressure Rating, HDB (PR_{HDB}) - The estimated maximum pressure (psig) that the medium in the pipe can exert continuously with a high degree of certainty that failure of the pipe will not occur. PR_{HDB} = 2 (HDB) (DF)/(DR-1)

Pressure Rating, MRS (PR_{MRS}) - The estimated maximum pressure (bar) that the medium in the pipe can exert continuously with a high degree of certainty that failure of the pipe will not occur. PR_{MRS} = 20 (MRS)/(DR-1) C

Pressure, Surge - The maximum positive transient pressure increase (commonly called water hammer) that is anticipated in the system as the result of a change in velocity of the water column.

Pressure, Working - The maximum anticipated sustained operating pressure, in pounds per square inch gauge, applied to the pipe or tubing, exclusive of surge pressures.

Primary Properties - The properties used to classify polyethylene materials.

Profile - Anchor pattern on a surface produce by abrasive, blasting or acid treatment.

Project - The entire construction to be performed as provided in the contract documents.

PSF - Pounds per square foot. PSF= lb/in^2 x 144

PSI - Pounds per square inch.

PSIG – Pounds per square inch gauge.

Pull-in Piping - Also referred to as insert renewal; installation procedure whereby pipe is pulled inside old mains and service lines to provide the new main or service line.

Quality Assurance Test - A test in a program that is conducted to determine the quality level. DISCUSSION— Quality assurance includes quality control, quality evaluation, and design assurance. A good quality assurance program is a coordinated system, not a sequence of separate and distinct steps.

Quality Control Test - A production, in-plant test that is conducted at a given test frequency to determine whether product is in accordance with the appropriate specification(s).

Quick Burst Test - (ASTM D 1599) An internal pressure test designed to produce rupture (bursting) of a piping component in 60-70 seconds determined in accordance with ASTM D 1599.

Radian - An arc of a circle equal in length to the radius; or the angle at the center measured by the arc.

Radius of Gyration - The distance from the reference at which all of the area can be considered concentrated that still produces the same moment of inertia. Numerically it is equal to the square root of the moment of inertia, divided by the area.

Rainfall - Precipitation in the form of water (usage includes snow).

Rapid Crack Propagation (RCP) – A running-crack failure associated with lower temperatures and compressed gas media, initiated by a significant impact. Cracks, once initiated, run at high speed (300 to 1400 ft/sec) and result in cracks many feet in length.

Rate Process Method (RPM) – A three coefficient mathematical model for calculating plastic piping performance projections at use conditions – see TN-16.

Reduction - Gain of electrons, as when copper is electro-plated on steel from a copper sulfate solution (opposite of "Oxidation").

Regression Analysis - An evaluation of the long-term hoop stress data. A linear curve is calculated using the least Squares method to fit the logarithm of hoop stress versus the logarithm of the resulting hours-to-failure.

Regulator - A device for controlling the quantity of sewage and storm water admitted from a combined sewer collector line into an interceptor, pump station or treatment facility, thereby determining the amount and quality of the flow discharged through an overflow device to receiving waters or other points of disposal.

Rehabilitation - All aspects of upgrading the performance of existing sewer systems. Structural rehabilitation includes repair, renovation and renewal. Hydraulic rehabilitation covers replacement, reinforcement, flow reduction or attenuation and occasionally renovation.

Reinforcement - The provision of an additional sewer which in conjunction with an existing sewer increases overall flow capacity.

Renewal - Construction of a new sewer, on or off the line of an existing sewer. The basic function and capacity of the new sewer being similar to those of the old.

Renovation - Methods by which the performance of a length of sewer is improved by incorporating the original sewer fabric, but excluding maintenance operations such as isolated local repairs and root or silt removal.

Repair - Rectification of damage to the structural fabric of the sewer and the reconstruction of short lengths, but not the reconstruction of the whole of the pipeline.

Replacement - Construction of a new sewer, on or off the line of an existing sewer. The function of the new sewer will incorporate that of the old, but may also include improvement or development work.

Reprocessed Plastic - A thermoplastic prepared from usually melt processed scrap or reject parts by a plastics processor, or from non-standard virgin material or non-uniform virgin material.

Resin Impregnation (wet-out) - A process used in cured-in-place pipe installation where a plastic coated fabric tube is uniformly saturated with a liquid thermosetting resin while air is removed from the coated tube by means of vacuum suction.

Resins - An organic polymer, solid or liquid: usually thermoplastic or thermosetting.

Retaining Wall - A wall for sustaining the pressure of earth or filling deposited behind it.

Revetment - A wall or a facing of wood, willow mattresses, steel units, stone or concrete placed on stream banks to prevent erosion.

Reworked Plastic - A plastic from a manufacturer's own production that has been reground or pelletized for reuse by that same manufacturer.

Reynolds Number - A dimensionless quantity named after Osbourne Reynolds who first made know the difference between laminar and turbulent flow. The practical value of the Reynolds Number is that it indicated the degree of turbulence in a flowing liquid. It depends on the hydraulic radius of the conduit, the viscosity of the water and the velocity of flow. For a conduit of a given size, the velocity is generally the major variable and the Reynolds Number will increase as the velocity of flow increases.

Right Bank - That bank of a stream which is on the right when one looks downstream.

Ring Compression - The principal stress in a confined thin circular ring subjected to external pressure.

Rip Rap - Rough stone of various large sizes placed compactly or irregularly to prevent scour by water or debris.

Roadway (highway) - Portion of the highway included between the outside lines of gutters or side ditches, including all slopes, ditches, channels and appurtenance necessary to proper drainage, protection and use.

Roof Leader - A drain or pipe that conducts storm water from the roof of a structure downward and thence into a sewer for removal from the property, or onto the ground for runoff or seepage disposal.

Roughness Coefficient - A factor in the Kutter, Manning, and other flow formulas representing the effect of channel (or conduit) roughness upon energy tosses in the flowing water.

Runoff - That part of precipitation carried off from the area upon which it falls. Also, the rate of surface discharge of the above. That part of precipitation reaching a stream, drain or sewer. Ratio of runoff to precipitation is a "coefficient" expressed decimally.

Saddle Fitting - A fitting used to make lateral connection to a pipe in which a portion of the fitting is contoured to match the OD of the pipe to which it is attached.

Samples - Physical examples which illustrate materials, equipment or workmanship and establish standards by which the work will be judged.

Sanitary Sewer - A sewer intended to carry only sanitary and industrial wastewaters from residences, commercial buildings, industrial parks, and institutions.

Scaling - (1) High temperature corrosion resulting in formation of thick corrosion product layers. (2) Deposition of insoluble materials on metal surfaces, usually inside water boilers or heat exchanger tubes.

SDR (Standard Dimension Ratio) - The ratio of the average outside diameter to the minimum wall thickness. A common numbering system that is derived from the ANSI preferred number series R-10.

Secondary Stress - Forces acting on the pipe in addition to the internal pressure such as those forces imposed due to soil loading and dynamic soil conditions.

Section Modulus - The moment of inertia of the area of a section of a member divided by the distance from the center of gravity to the outermost fiber.

Sectional Properties - End area per unit of widths, moment of inertial, section modulus, and radius of gyration.

Seepage - Water escaping through or emerging from the ground along a rather extensive line or surface, as contrasted with a spring, the water of which emerges from a single spot.

Serviceability of The Piping System - Continued service life with a high degree of confidence that a failure will not occur during its long-term service.

Sewer Cleaning - The utilization of mechanical or hydraulic equipment to dislodge, transport, and remove debris from sewer lines.

Sewer Interceptor - A sewer which receives the flow from collector sewers and conveys the wastewaters to treatment facilities.

Sewer Pipe - A length of conduit, manufactured from various materials and in various lengths, that when joined together can be used to transport wastewaters from the points of origin to a treatment facility. Types of pipe are: Acrylonitrile-butadiene-styrene (ABS): Asbestos-Cement (AC); Brick Pipe (BP); Concrete Pipe (CP); Cast Iron Pipe (CIP): Polyethylene (PE); Polyvinylchloride (PVC); Vitrified Clay (VC).

Sewer, Building - The conduit which connects building wastewater sources to the public or street sewer, including lines serving homes, public buildings, commercial establishments and industry structures. The building sewer is commonly referred to in two sections: (1) the section between the building line and the property line, frequently specified and supervised by plumbing or housing officials; and (2) the section between the property line and the street sewer, including the connection thereto, frequently specified and supervised by sewer, public works, or engineering officials. (Referred to also as "house sewer," "building connection," or "service connection.")

Shaft - A pit or wall sunk from the ground surface into a tunnel for the purpose of furnishing ventilation or access to the tunnel.

Sheeting - A wall of metal plates or wood planking to keep out water, soft or runny materials.

Shop Drawings - All drawings, diagrams, illustrations, brochures, schedules, and other data which are prepared by the contractor, a subcontractor, manufacturer supplier or distributor and which illustrate the equipment, materials or some portion of the work as required by the contract documents.

Siphon (Inverted) - A conduit or culvert with a U or V shaped grade line to permit it to pass under an intersecting roadway, stream or other obstruction.

Site - Any location where work has been or will be done.

Site Access - An adequately clear area of a size sufficient to accommodate personnel and equipment required at the location where work is to be performed, including roadway or surface sufficiently unobstructed to permit conveyance of vehicles from the nearest paved roadway to the work location.

Skew (or Skew Angle) - The acute angle formed by the intersection of the line normal to the centerline of the road improvement, with the centerline of a culvert or other structure.

Slide - Movement of a part of the earth under force of gravity.

Slit-Type Failure - A form of brittle failure that exhibits only a very small crack through the wall of the pipe with no visible material deformation in the area of the break.

Slow Crack Growth (SCG) – the slow extension of the crack with time.

Smooth Radius Bend - A contoured sweep or bend with no sharp or angular sections.

Social Costs - Costs incurred by society as a result of sewerage works and for which authorities have no direct responsibility. These include unclaimed business losses due to road ensures and the cost of extended journey times due to traffic diversions.

Socket Fusion Joint - A joint in which the joining surfaces of the components are heated, and the joint is made by inserting one component into the other.

Softening Temperature - There are many ways to measure the softening temperature of a plastic. The commonly reported Vicat Softening Temperature method is to measure the temperature at which penetration of a blunt needle through a given sample occurs under conditions specified in ASTM D 1525.

Solar Radiation - The emission of light from the sun, including very short ultraviolet wavelengths, visible light, and very long infrared wavelengths.

Solubility - The amount of one substance that will dissolve in another to produce a saturated solution.

Spalling - The spontaneous chipping, fragmentation, or separation of a surface or surface coating.

Span - Horizontal distance between supports, or maximum inside distance between the sidewall of culverts.

Special Conditions - When included as a part of the contract documents. Special conditions refer only to the work under this contract.

Specific Gravity - The density of a material divided by the density of water usually at 4°C. Since the density of water is nearly 1 g/cm, density in g/cm and specific gravity are numerically nearly equal.

Specifications - Those portions of the contract documents consisting of written technical descriptions of materials, equipment, construction systems, standards and workmanship as applied to the work.

Spillway - (1) A low-level passage serving a dam or reservoir through which surplus water may be discharged; usually an open ditch around the end of a dam, a gateway, or a pipe in a dam. (2) An outlet pipe, flume or channel serving to discharge water from a ditch, ditch check, gutter or embankment protector.

Spring Line - A line along the length of the pipe at its maximum width along a horizontal plane. The horizontal midpoint of a sewer pipe.

Springing Line - Line of intersection between the intrados and the supports of an arch. Also the maximum horizontal dimension of a culvert or conduit.

Spun Lining - A bituminous lining in a pipe, made smooth or uniform by spinning the pipe around its axis.

Stabilizer - An ingredient used in the formulation of some plastics to assist in maintaining the physical and chemical properties of the materials at their initial values throughout the processing service life of the material.

Standard Dimension Ratio (SDR) - A specific ratio of the average specified outside diameter to the minimum specified wall thickness for outside diameter-controlled plastic pipe, the value of which is derived by adding one to the pertinent number selected from the ANSI Preferred Number Series 10. Specifying PE pipes with a given SDR regardless of O.D. assures all pipes will have the same design pressure assuming the PEs have the same HDB rating.

Standard Grade (S) - A PPI HSB recommended rating that is valid for a five-year period, given to those materials that comply with the full data requirements of TR-3.

Standard Thermoplastic Material Designated Code - In this designation system, which is widely used by major national product standards, the plastic is identified by its standard abbreviated terminology in accordance with ASTM D 1600, "Standard Terminology Relating to Abbreviations, Acronyms, and Codes for Terms Relating to Plastics", followed by a four or five digit number. The first two or three digits, as the case may be, code the material's ASTM classification (short-term properties) in accordance with the appropriate ASTM standard specification for that material. The last two digits of this number represent the PPI recommended HDS (0.5 design factor) at 73°F (23°C) divided by one hundred. For example, PE 2406 is a grade P24 polyethylene with a 630-psi design stress for water at 73.4°F (23°C). The hydrostatic design stresses for gas are not used in this designation code.

Strength Design Basis (SDB) - Refers to one of a series of established stress values (specified in Test Method D 2837) for a plastic molding compound obtained by categorizing the long-term strength determined in accordance with ASTM Test Method F 2018, "Standard Test Method for Time-to-Failure of Plastics Using Plane Strain Tensile specimens".

Stress Crack - An internal or external crack in a plastic caused by tensile or shear stresses less than the short-term tensile strength of the material. The development of such cracks is frequently related to and accelerated by the environment to which the material is exposed. More often than not, the environment does not visibly attack, soften or dissolve the surface. The stresses may be internal, external, or a combination of both.

Stress Relaxation - The decay of stress with time at constant strain.

Sustained Pressure Test - A constant internal pressure test for an extended period of time.

Tensile Strength at Break - The maximum tensile stress (nominal) sustained by the specimen during a tensile test where the specimen breaks.

Tensile Strength at Yield - The maximum tensile stress (nominal) sustained by the specimen during a tensile test at the yield point.

Thermal Stabilizers - Compounds added to the plastic resins when compounded that prevent degradation of properties due to elevated temperatures.

Thermoplastic - A plastic, such as PE, that can be repeatedly softened by heating and hardened by cooling through a temperature range characteristic of the plastic, and that in the softened state can be shaped by flow into articles by molding or extrusion.

Thermoset - A material, such as epoxy, that will undergo or has undergone a chemical reaction by the action of heat, chemical catalyst, ultraviolet light, etc., leading to an infusible state.

Thermosetting - Resins that are composed of chemically cross-linked molecular chains, which set at the time the plastic is first formed; these resins will not melt, but rather disintegrate at a temperature lower than its melting point, when sufficient heat is added.

Threading - The process of installing a slightly smaller pipe or arch within a failing drainage structure.

Toe Drain - A subdrain installed near the downstream toe of a dam or levee to intercept seepage.

Toe-in - A small reduction of the outside diameter at the cut end of a length of thermoplastic pipe.

Tuberculation - Localized corrosion at scattered locations resulting in knob-like mounds.

Ultraviolet Absorbers (Stabilizers) - Compounds that when mixed with thermoplastic resins selectively absorb ultraviolet rays protecting the resins from ultraviolet attack.

Underdrain - See subdrain.

Uniform Corrosion - Corrosion that results in an equal amount of material loss over an entire pipe surface.

UV Degradation - Sunlight contains a significant amount of ultraviolet radiation. The ultraviolet radiation that is absorbed by a thermoplastic material may result in actinic degradation (i.e., a radiation promoted chemical reaction) and the formation of heat. The energy may be sufficient to cause the breakdown of the unstabilized polymer and, after a period of time, changes in compounding ingredients. Thermoplastic materials that are to be exposed to ultraviolet radiation for long periods of time should be made from plastic compounds that are properly stabilized for such conditions.

Velocity Head - For water moving at a given velocity, the equivalent head through which it would have to fall by gravity to acquire the same velocity.

Vinyl Plastics – Compositions of polymers and ingredients that are based on polymers of vinyl chloride, or copolymers of vinyl chloride with other monomers, the vinyl chloride being in the greatest amount by mass.

Virgin Plastic - A plastic material in the form of pellets, granules, powder, floc, or liquid that has not been subjected to use or processing other than that required for its initial manufacture.

Voids - A term generally applied to paints to describe holidays, holes, and skips in the film. Also used to describe shrinkage in castings or welds.

Wale - Guide or brace of steel or timber, used in trenches and other construction.

Water Table - The upper limit of the portion of ground wholly saturated with water.

Watershed - Region or area contributing to the supply of a stream or lake; drainage area, drainage basin, catchment area.

Weatherability - The properties of a plastic material that allows it to withstand natural weathering; hot and cold temperatures, wind, rain and ultraviolet rays.

Wetted Perimeter - The length of the perimeter in contact with the water. For a circular pipe of inside diameter "d", flowing full, the wetted perimeter is the circumference, d. The same pipe flowing half full would have a wetted perimeter of d/2.

Work - Any and all obligations, duties and responsibilities necessary to the successful completion of the project assigned to or undertaken by contractor under the contract documents, including all labor, materials, equipment and other incidentals, and the furnishing thereof.

Working Pressure (WP) - The maximum anticipated, sustained operating pressure applied to the pipe exclusive of transient pressures.

Working Pressure Rating (WPR) - The capacity to resist Working Pressure (WP) and anticipated positive transient pressure surges above working pressure.

Written Notice - The term "notice" as used herein shall mean and include all written notices, demands, instructions, claims, approvals, and disapproval required to obtain compliance with contract requirements. Written notice shall be deemed to have been duly served if delivered in person to the individual or to a member of the firm or to an officer of the corporation for whom it is intended, or to an authorized representative of such individual, firm or corporation, or if delivered at or sent by registered mail to the last business address known to him who gives the notice. Unless otherwise stated in writing, any notice to or demand upon the owner under this contract shall be delivered to the owner through the engineer.

Yield Point (ASTM D 638) - The stress at which a material exceeds its elastic limit. Below this stress, the material will recover its original size and shape on removal of the stress. The first point on the stress-strain curve at which an increase in strain occurs without an increase in stress.

Yield Strength (ASTM D 638) – The stress at which a material exhibits a specified limiting deviation form the proportionality of stress to strain. Unless otherwise specified, this stress will be the stress at the yield point, and when expressed in relation to the tensile strength, shall be designated as Tensile Strength at Yield or Tensile Stress at Yield.

Organizations and Associations

AASHTO   **American Association of State Highway & Transportation Officials**
444 N. Capitol St., N.W., Suite 225
Washington, DC 20001
(202) 624-5800
www.aashto.org

ACS   **American Chemical Society**
1155 Sixteenth Street NW
Washington, DC 20036
(800) 333-9511
www.acs.org

AGA/PMC   **American Gas Association**
Plastic Materials Committee
400 N. Capitol Street NW
Washington, DC 20001
(202) 824-7336
www.aga.com

ANSI   **American National Standards Institute**
11 W. 42nd St., 13th Floor
New York NY 10036
(212) 642-4900
www.ansi.org

APC   **American Plastics Council**
1300 Wilson Blvd.
Arlington, VA 22209
1-800-2-HELP-90
www.americanplasticscouncil.org

API   **American Petroleum Institute**
1220 L St., N.W.
Washington, DC 20005
(202) 682-8000
www.api.org

APGA   **American Public Gas Association**
Suite 102
11094-D Lee Highway
Fairfax, VA 22030
(703) 281-2910
www.apga.org

APWA   **American Public Works Association**
Mark Twain Building
06 W. 11th Street Suite 1080
Kansas City, MO 64105
www.apwa.net

ASAE   **American Society of Agricultural Engineers**
2950 Niles Road
St. Joseph, MI 49085
(616) 429-0300
www.asae.org

ASCE   **American Society of Civil Engineers**
345 East 47th St.
New York NY 10017
(212) 705-7496
www.asce.org

ASDWA	**Association of State Drinking Water Administrators**	

ASDWA   **Association of State Drinking Water Administrators**
1120 Connecticut Avenue, NW Suite 1060
Washington, DC 20036
(202) 293-765
(202) 293-7656
www.asdwa.org

ASHRAE   **American Society of Heating, Refrigerating and Air-Conditioning Engineers**
1791 Tullie Circle, N.E.
Atlanta, GA 30329
(404) 321-5478
(404) 636-8400
(800) 527-4723
www.ashrae.org

ASME   **American Society of Mechanical Engineers**
345 East 47th St.
New York, NY 10017
(212) 705-7722
www.asme.org

ASTM   **ASTM International**
100 Barr Harbor Drive
West Conshohocken, PA 19428-2959
(610) 832-9500
www.astm.org

ASTPHLD   **Association of State and Territorial Public Health Laboratory Directors**
1211 Connecticut Avenue, NW, Suite 608
Washington, DC 20036
(202) 822-5227
(202) 887-5098
www.astphld.org

AWWA   **American Water Works Association**
6666 West Quincy Ave.
Denver. CO 80235
(303) 794-7711
www.awwa.org

AWWRF   **American Water Works Research Foundation**
6666 West Quincy Avenue
Denver, CO 80235
(303) 347-6118
www.awwarf.org

BOCA   **Building Officials and Code Administrators**
4051 West Flossmoor Road
Country Club Hills, IL 60478
(708) 799-2300
(708) 799-4981
www.bocai.org

CABO   **The Council of American Building Officials**
5203 Leesburg Pike, Suite 708
Falls Church, VA 22041
www.cabo.org

CMA   **Chemical Manufacturers Association**
2501 M Street NW
Washington, DC 20037
(202) 887-1378

CERF	**Civil Engineering Research Foundation**	

CERF   **Civil Engineering Research Foundation**
1015 15th St., NW
Washington, DC 20005
(202) 789-2200
(202) 289-6797
www.cerf.org

CSA   **Canadian Standards Association**
178 Rexdale Blvd.
Rexdale, Ont. M9W 1R3, Canada
(416) 747-4000
www.csa.ca

DOT/OPS   **U.S. Department of Transportation Office of Pipeline Safety**
400 7th Street SW
Washington, DC 20590
www.opts.dot.gov

DOT/TSI   **U.S. Department of Transportation Transportation Safety Institute**
P.O. Box 25082
Oklahoma City, OK 73125
(405) 686-2466
tsi.dot.gov

FHA   **Federal Housing Authority**
820 First Street, NE
Washington, DC 20002-4205
(202) 275-9200
(202) 275-9212
www.hud.gov/fha/fhahome.ht

FM   **Factory Mutual**
1151 Boston Providence Turnpike
P.O. Box 688
Norwood, MA 02062
(617) 762-4300

GPTC   **Gas Piping Technology Committee**
400 N. Capitol Street NW
Washington, DC 20001
(202) 824-7335

GTI   **Gas Technology Institute**
1700 South Mount Prospect Road
Des Plaines, IL 60018
(847) 768-0500
www.gastechnology.org

GRI   **Geosynthetic Research Institute at Drexel University**
33rd and Lancaster Walk
Rush Bldg. - West Wing
Philadelphia, PA 19104
(215) 895-2343
www.drexel.edu

HSB   **Hydrostatic Stress Board**
Plastics Pipe Institute
1825 Connecticut Ave. NW, Suite 680
Washington, DC 20009
(202) 462-9607
www.plasticpipe.org

HUD   **Department of Housing and Urban Development**
451 7th Street, SW
Washington, DC 20410
(202) 708-4200
(202) 708-4829
(800) 347-3735
www.hud.gov

IAPMO   **International Association of Plumbing and Mechanical Officials**
20001 S. Walnut Drive
Walnut, CA 91789
(714) 595-8449
www.iapmo.org

ICBO   **International Conference of Building Officials**
5360 S. Workman Mill Road
Whittier, CA 90601
(213) 699-0541
www.icbo.org

IGSHPA   **International Ground Sourced Heat Pump Association**
374 Cordell South
Stillwater, OK 74078
(405) 744-5175
www.igshpa.okstate.edu

ISO   **International Standard Organization**
11 West 42nd Street
New York, NY 10036
(212) 642-4900
(212) 398-0023
www.ansi.org

ISO/SC4   **International Standards Organization**
Secretariat for Subcommittee SC4 "Gas"
GASTEC
Postbus 137, 7300 Ac Apeldoorn
Wilmersdorf 50
Apeldoorn Netherlands
055-494 949

NACE   **National Association of Corrosion Engineers**
P.O. Box 218340
Houston. TX 77218
(713) 492-0535
www.nace.org

NACO   **National Associations of Counties**
440 First Street, N.W.
Washington, DC 20001
(202) 393-6226
www.naco.org

NASSCO   **National Association of Sewer Service Companies**
101 Wymore Rd., Suite 501
Altamonte, FL 32714
(407) 774-0304
www.nassco.org

NASTT   **North American Society for Trenchless Technology**
435 N. Michigan Ave., Suite 1717
Chicago. IL 60611
(312) 644-0828
www.bc.irap.nrc.ca/nodig

NCSL	**National Conference of State Legislatures** 1560 Broadway, Suite 700 Denver, CO 80202 (303) 830-2200 www.ncsl.org
NEMA	**National Electrical Manufacturers Association** 2101 L Street NW Washington, DC 20037 (703) 841-3200 (703) 841-3300 www.nema.org
NFPA	**National Fire Protection Association** 1 Batterymarch Park Quincy, MA 02269 (617) 770-3000 www.nfpa.org
NGA	**National Governors' Association** 444 North Capitol Street Washington, DC 20001 (202) 624-5300 www.nga.org
NRWA	**National Rural Water Association** 2915 S. 13th Street Duncan, OK 73533 (405) 525-0629 (405) 255-4476 www.nrwa.org
NSF	**NSF International** NSF Bldg. P.O. Box 130140 Ann Arbor, MI 48113 (313) 769-8010 (313) 769-0109 (800) NSF-MARK www.nsf.org
NTSB	**National Transportation Safety Board** 800 Independence Ave., S.W., Room 820A Washington, DC 20594 (202) 382-6600 www.ntsb.gov
NUCA	**National Utility Contractors Association** 4301 N. Fairfax Drive Suite 360 Arlington, VA 22203 (703) 358-9300 www.nuca.com
NWRA	**National Water Resources Association** 3800 N. Fairfax Drive, Suite 4 Arlington, VA 22203 (703) 524-1544 www.nwra.org
NWWA	**National Well Water Association** 6375 Riverside Drive Dublin, OH 43017

PCGA   **Pacific Coast Gas Association**
1350 Bayshore Highway, Suite 340
Burlingame, CA 94010
(415) 579 7000

PHCC-NA   **Plumbing, Heating, Cooling Contractors Association**
180 S. Washington Street
P.O. Box 6808
Falls Church, VA 22040
(703) 237-8100
(703) 237-7442
(800) 533-7694
www.naphcc.org

PPFA   **Plastic Pipe and Fittings Association**
800 Roosevelt Road
Building C, Suite 200
Glen Ellyn, IL 60137
(708) 858-6540
www.ppfahome.org

PVRC   **Pressure Vessel Research Council of the Welding Research Council**
345 East Fifty-Seventh Street
New York, NY 10017
www.forengineers.org/pvrc

RCAP   **Rural Community Assistance Program**
602 South King St., Suite 402
Leesburg, VA 20175
(703) 771-8636
www.rcap.org

SBCCI   **Southern Building Codes Council International**
900 Montclair Road
Birmingham, AL 35213
(205) 591-1853
www.sbcci.org

SCA   **Standards Council of Canada**
45 O'Connor Street, Suite 1200
Ottawa, ON K1P6N7
(613) 238-3222
www.scc.ca

SwRI   **Southwest Research Institute**
6220 Culebra Rd.
P.O. Drawer 28510
San Antonio, TX 78284
(512) 522-3248
www.swri.org

UL   **Underwriters Laboratories, Inc.**
333 Pfingsten Road
Northbrook, IL 60062
(847) 272-8800
www.ul.com

Uni-Bell   **Uni-Bell PVC Pipe Association**
2655 Villa Creek Drive, Suite 155
Dallas, Texas 75234
(214) 243-3902
www.uni-bell.org

VI	**The Vinyl Institute**	
	1300 Wilson Blvd., Suite 800	
	Arlington, VA 22209	
	(703) 741-5670	
	(703) 741-5672	
	www.vinylinfo.org	
WEF	**Water Environment Federation**	
	601 Wythe St.	
	Alexandria, VA 22314	
	(703) 684 2492	
	(703) 684-2452	
	www.wef.org	

Abbreviations

ABS	Acrylonitrile Butadiene Styrene
AGA	American Gas Association
ANSI	American National Standards Institute, (formerly USASI, formerly ASA)
ASA	American Standards Association (see ANSI)
ASME	American Society of Mechanical Engineers
ASTM	American Society for Testing and Materials; now just ASTM International.
AWWA	American Water Works Association
CPPA	Corrugated Plastic Pipe Association
CTS	Copper Tubing Size
DIPS	Ductile Iron Pipe Size
DOT	Department of Transportation, a bureau of the Federal Government
DR	Dimension Ratio
ESC	Environmental Stress Cracking
ESCR	Environmental Stress Cracking Resistance
HDB	Hydrostatic Design Basis
HDBC	Hydrostatic Design Basis Category
HDPE	High Density Polyethylene
HDS	Hydrostatic Design Stress
IPS	Iron Pipe Size
ISO	International Standards Association
LPG	Liquefied Petroleum Gas
LTHS	Long-Term Hydrostatic Strength
MDPE	Medium Density Polyethylene
MRS	Minimum Required Strength
MSS	Manufacturers Standardization Society of the Valve and Fitting Industry
NFPA	National Fire Protection Association
NPGA	National Propane Gas Association
NSF	National Sanitation Foundation
NTSB	National Transportation Safety Board
OPS	Office of Pipeline Safety, a branch of the U.S. Department of Transportation
PA	Polyamide (nylon)
PB	Polybutylene
PE	Polyethylene
PEX	Crosslinked Polyethylene
PPI	Plastics Pipe Institute
RCP	Rapid Crack Propagation
RTRP	Reinforced Thermosetting Resin Pipe
SCG	Slow Crack Growth
SDR	Standard Dimension Ratio
SPE	Society of Plastic Engineers
SPI	Society of the Plastics Industry, Inc.
TSI	Transportation Safety Institute

Index

A

AASHTO 155, 195, 196, 197, 211, 212, 221

Above Ground 243, 483

Above Ground Piping Systems 243

Abrasion Resistance 57, 80, 106, 151

Acceptable Diameter 299

Acceptance Deflection 265

Access 401, 405, 406, 476

Advantages 41, 466

Aging 101, 189

Air Entrapment 359

Allowable Operating Pressures 455

American Association of State Highway and Transportation Officials 195

American Society for Testing and Materials (ASTM) 129, 144, 327, 412

American Water Works Association (AWWA) 34, 45, 129, 146, 245, 327, 349

Anchor 243, 319, 325, 347, 385

Anchoring 317, 326, 327, 357

ANSI 21, 26, 139, 140, 144, 145, 245, 340, 343, 344, 349

Apparent Modulus 60, 209, 241, 308, 427, 462

Apparent Tensile Strength 126

Appearance 124

Applications 1, 138, 151, 152, 153, 155, 221, 267, 305, 346, 398, 429, 442, 451, 458, 466, 494

Arching 193, 225, 226, 249, 266, 423

Armored 497

Army Corps of Engineers 271

ASME 22, 245, 349

Asphalt Dipped 165

Assembly 5, 341, 354, 364

ASTM 18, 22, 26, 28, 29, 31, 34, 35, 39, 50, 54, 65, 66, 67, 68, 73, 74, 75, 76, 78, 80, 81, 82, 84, 86, 88, 99, 100, 101, 105, 107, 108, 109, 110, 117, 118, 122, 124, 125, 126, 129, 131, 132, 133, 134, 135, 136, 137, 138, 142, 143, 144, 145, 146, 148, 149, 150, 152, 153, 155, 158, 159, 161, 162, 192, 197, 201, 208, 209, 211, 213, 214, 215, 216, 221, 241, 245, 246, 260, 267, 268, 269, 273, 290, 293, 295, 296, 297, 302, 306, 326, 327, 333, 337, 338, 347, 349, 364, 376, 385, 393, 401, 412, 413, 418, 419, 429, 433, 437, 454, 455, 456, 463, 465, 468, 469, 470, 472, 475, 476, 478, 486, 487, 494, 498

ASTM D1586 213

ASTM D1598 125

ASTM D2321 221, 267, 327, 412

ASTM D2513 107, 158, 162, 338, 349

ASTM D2683 108

ASTM D2774 221, 364, 412

ASTM D2837 65, 66, 67, 68, 136, 161, 306, 327

ASTM D3035 107, 158, 327

ASTM D3212 214

ASTM D 1248 131, 133, 293

ASTM D 1693 133, 454

ASTM D 2683 454, 487

ASTM F1055 108

ASTM F894 107, 159, 208, 209, 216, 349, 401, 412

Aude 178, 245

Average Inside Diameter 135

AWWA 26, 34, 39, 45, 107, 129, 146, 147, 148, 149, 154, 158, 159, 190, 191, 221, 244, 245, 327, 337, 340, 343, 349, 360, 412, 419

Axial Bending Stress 448

Axial Tensile Stress 450

B

Backfill 267, 272, 284, 285, 290, 295

Ballast 355, 361, 365, 367, 374, 385, 386, 387

Barbed Mechanical Fittings 486

Barrel 127, 287

Bead Removal 331

Beam Analysis 323, 324

Beam Deflection 322

Bedding 208, 248, 263, 267, 282, 285, 289, 290, 294, 295, 296, 297, 364

Bell and Spigot 406

Below Ground Installations 474

Bend-Back Test 126

Bending Radii 371, 479

Bending Strain Development 320

Bending Stress 440, 448

Bend Radius 479

Bent Strap Test 18

Berm 318

Biological Resistance 104

Blowing 488, 489

Bolting 341, 362

Bolt Type Mechanical Flange Adapter 343

Bore 387, 472, 474, 477

Borehole 420, 422, 423, 424, 442, 445

Bowles 229, 246

Box 287, 336

Bracing 280

Branch Saddles 345

Breaker Plate/Screen Pack 113

Bridge Structures 495

Broken or Damaged Fittings 32

Buckling 218, 220, 231, 232, 236, 246, 426, 428, 430, 432, 461

Building Code 497

Bulk Pack 6, 7

Bundling 118

Buoyant Force 376, 449

Burial of HDPE Fabricated Fittings 286
Buried PE Pipe Design 190
Buried Piping Systems 243
Burst Pressure 126
Butt Fusion 17, 37, 142, 151, 330, 331, 349, 402, 487

C

Cable 466, 467, 471, 477, 484, 488, 489, 490, 498, 499
California Bearing Ratio 302
Canadian Standards Association 129, 148, 154
CANDE 207, 246, 264
Capstan 435
Casing 150, 297
Cast Iron 165, 397
Category 206, 222, 224, 232, 293, 338, 456
Catenary 319
Cell Classification 132, 468
Cement Mortar Relined 165
Cement Stabilized Sand 272
Center-to-Center Span 322
Chemical Resistance 39, 57, 88, 89, 90, 91, 92, 93, 94, 95, 96, 126, 130, 310, 469
CIC 467
Class 150 340
Class II Material 297
Class III 272, 276
Clay 229, 234, 300, 397
Cleaning 14, 27
Cleanup 300
Coarse Grain Soils 301
Coefficient of Linear Expansion 81
COF 468, 490, 491, 492
Cohesion 300
Cohesionless Soils 273
Cold Weather Handling 13
Collapse 236, 407, 412, 423, 424, 426, 432, 440
Color 13, 111, 133
Colorants 470

Combustion Toxicity 88, 106
Compaction 211, 212, 269, 273, 274, 275, 276, 284, 285, 286, 301
Compressed Air 327
Compressible Gas Flow 182
Compression Fittings 154, 486
Compression Nut 338
Compressive Ring Thrust 215, 225
Compressive Strength 75, 213
Compressive Stress 217, 432
Concrete 165, 187, 287, 317, 319, 363, 383, 384, 385, 386, 387, 397
Conduit Entry Issues 497
Conduit Wall Determination 472
Connections 14, 296, 297, 337, 339, 390, 400, 409, 410
Connection to Rigid Structure 318
Constrained Pipe Repair 33
Construction of the Land-to-Water Transition 372
Continuous Beam Analysis 323
Continuous Support 317, 326
Continuous Trench 474, 479
Contraction 81, 151, 309, 313
Contract Documents 401
Controlling Surge Effects 176
Cooling 117, 126, 334
Copper Tubing Sizes 138, 337
Corrugated Duct 494
Cracking 99, 469, 498
Cradle 319
Creep Recovery 62, 63
Creep Rupture 60, 64, 105
Critical Velocity 178
Crown 214, 263, 408
Crush 191, 477, 478
Crystallinity 51
CSA 129, 148, 154, 161
CTS 122, 138, 147, 158, 333, 337, 338, 344
Culvert Analysis and Design 207
Cumulative Deflection Effects 323
Currents 378
Curvature 415, 448, 493
Cyclic Fatigue 70

D

D3350 101, 109, 133, 134, 401, 412
Damage 9, 18, 32, 152, 155, 466, 474
Deep Fill Installation 191, 194, 216, 224
Definitions of Symbols 499
Deflection 19, 20, 84, 85, 165, 207, 208, 211, 213, 214, 215, 227, 246, 248, 263, 265, 288, 303, 319, 322, 323, 324, 392, 427, 428, 429, 442, 443, 475
Deformation Factor 228, 229, 230
Density 50, 57, 61, 63, 80, 81, 105, 130, 132, 133, 153, 155, 184, 211, 221, 245, 273, 276, 283, 284, 290, 291, 292, 301, 322, 465, 466, 468, 472, 498
Design 1, 45, 57, 65, 66, 105, 113, 114, 127, 133, 134, 135, 136, 137, 138, 139, 140, 143, 150, 151, 153, 154, 155, 156, 157, 158, 160, 161, 162, 165, 169, 175, 190, 191, 192, 207, 209, 215, 221, 222, 232, 240, 242, 244, 245, 246, 247, 248, 265, 306, 307, 308, 311, 313, 325, 327, 328, 353, 355, 360, 361, 370, 376, 387, 390, 398, 412, 413, 419, 420, 429, 433, 442, 447, 455, 463, 469, 471, 484, 487
DF 137, 138, 140, 160, 161, 228, 229, 230, 242, 247, 307, 308, 455
Diameters 116, 138, 284, 405, 471
Dielectric Strength 86
Digging 480
Dimensions 124, 150, 151, 258, 384, 385, 454
Dimension Considerations 471
Dimension Ratio 139, 140, 150, 160, 171, 173, 192, 208, 216, 219, 228, 237, 247, 248, 403, 429, 442, 447, 479
DIPS 138, 147, 158, 337, 439
Directional Bores 482
Directional Drilling 40, 150, 155, 192, 413, 414, 416, 417, 442, 482
Direct Plow 474, 475
Disinfecting Water Mains 26, 34
Disinfection 151
Dissipation Factor 86, 87
Distributed Loads 203, 204

Index | 537

DR 12, 19, 28, 30, 35, 138, 139, 158, 159, 160, 169, 170, 171, 173, 175, 188, 190, 192, 193, 194, 208, 209, 214, 215, 216, 217, 218, 219, 221, 222, 223, 226, 227, 228, 237, 238, 245, 247, 258, 259, 260, 293, 307, 308, 327, 354, 355, 356, 357, 370, 371, 376, 392, 393, 394, 403, 419, 426, 429, 430, 433, 434, 435, 437, 438, 439, 442, 444, 445, 446, 447, 449, 454, 455, 462, 481, 482

Drag Resistance 434

Drawn Tubing 165

Dredging 374

Drilling Mud 415

Drilling Process 414

Ductile 66, 138, 337, 339, 340, 397, 439

Ductility 42, 58, 352, 466

Duncan 210, 212, 246

E

E' 39, 105, 126, 133, 192, 193, 195, 201, 202, 206, 207, 208, 209, 210, 211, 212, 213, 214, 219, 220, 221, 222, 225, 226, 227, 228, 237, 241, 245, 246, 247, 269, 270, 275, 285, 293, 295, 302, 314, 315, 327, 328, 391, 412, 428, 429, 430, 442, 444, 445, 454, 462

E'N 212, 213, 214

Earth and Groundwater Pressure 422, 425, 428

Earth Load 193, 428

Earth Pressure 227

Effluent 367

Electrical 76, 78, 80, 85, 86, 466, 488, 497, 498

Electrofusion 39, 108, 122, 142, 150, 153, 334, 335, 336, 346, 347, 488, 498

Elevation Change 167

Elongation 130

Embankment 191, 206, 412

Embedment 211, 212, 263, 265, 266, 267, 268, 269, 270, 273, 276, 290, 295, 300

End Restrained Thermal Effects 240

Engineering Properties 43, 47, 135, 161

Enveloping Soil 264

Environmental Stress Crack Resistance 55, 99, 100, 150

Equipment 9, 10, 17, 118, 151, 333, 405, 417

Equivalent Length of Pipe 167

ESC 88, 89, 99

ESCR 55, 99, 100, 101, 108, 110, 124, 126, 130, 132, 133, 150, 469

Existing Pipe 401, 408

Expansion Joints 486

Explosive Failure 25

Exposure to UV and Weather 13

External Hydraulic Load

External Load 245, 420

External Pressure 439

Extruded Profile Pipe 13

Extruder 109, 111

Extrusion 50, 54, 108, 109, 111, 112, 126, 127, 498

F

Fabricated Fittings 108, 121, 123, 153, 286

Factory Mutual Research 148

Fanning Formula 162

Fat Clays (CH) 300

FDA 103

Features 38, 458, 467

Fibrillation 75

Field Effects of Friction 492

Field Fusion Joining 15

Field Handling 15

Field Joining 14

Fine Grain Soil 300

Finish 124

Fire Protection 148

Fittings 65, 107, 108, 118, 119, 120, 121, 122, 123, 124, 125, 131, 134, 141, 142, 146, 148, 150, 151, 152, 153, 154, 166, 167, 245, 248, 286, 336, 338, 344, 349, 412, 454, 485, 486, 498

Flame Retardant 495

Flammability 88

Flanges 349

Flange Adapter 120, 339, 340, 342, 343

Flange Assembly 341

Flange Gasket 341

Flat-Bottomed Ballasts 361

Flattening 29, 126

Flexibility 40, 351

Flexible Pavement 197, 198

Flexible Pipe 245, 246, 262, 288, 300, 302, 412

Flexural Modulus 132, 133, 468

Flexural Strength and Modulus 76

Flex Pipe Restraint 347

Float 353, 357, 369

Flotation 191, 207, 232, 233, 235, 238

Flows for Slipliners 187, 188

Flow Capacity 158, 395

Flushing 457

FM 122, 148

Fold and Form 411

Food and Drug Administration 103

Forklift 10

Foundation 103, 105, 246, 263, 267, 288, 412

Fracture Mechanics 69, 105

Freeze Tolerant 459

Friction 87, 168, 169, 247, 396, 490, 491, 492

Frictional Forces 498

Fuel Oil 99

Fusion 15, 16, 17, 37, 108, 120, 142, 151, 153, 329, 330, 331, 332, 333, 336, 349, 402, 410, 456, 461, 487, 498

G

Galvanized Iron 165

Gas Flow 182, 183

Gas Permeation 81, 183

Gaube 191, 225, 227, 228, 229, 230, 246

General Guidelines 151, 349, 416

Geotechnical 246, 288, 417, 418

Geotextile Separation Fabrics 272

Geothermal Heat Pump System 463

Glass Transition Temperature 83

Grade Beam 317

Gravel 234

Gravity 22, 26, 34, 151, 155, 177, 180, 184, 185, 188, 245, 267, 291, 396, 398, 400, 410

Grip Ring 338

Groundwater Pressure 422, 425, 428

Ground Heat Exchangers 451, 452, 455

Ground Source Heat Pump Systems 451

Ground Water 218, 232, 233, 238

H

H2O 183, 199, 248, 396, 432

Handling 6, 9, 13, 15, 110, 351

Hardness 57, 79, 80, 130

Haunching 267, 282, 285, 290

HDB 65, 127, 134, 135, 136, 137, 138, 140, 143, 151, 160, 161, 242, 293, 307, 308, 455

HDD 192, 413, 414, 416, 417, 418, 419, 421, 422, 423, 424, 425, 426, 428, 433, 434, 440, 441, 442

HDPE 1, 2, 32, 37, 38, 39, 40, 41, 42, 43, 52, 64, 81, 86, 89, 90, 91, 92, 93, 94, 95, 96, 97, 99, 105, 139, 144, 148, 153, 155, 157, 176, 208, 209, 214, 217, 220, 238, 241, 243, 246, 266, 267, 272, 286, 287, 291, 292, 293, 295, 296, 297, 308, 367, 374, 375, 379, 380, 382, 385, 391, 418, 427, 432, 433, 434, 437, 438, 439, 440, 441, 442, 444, 447, 452, 453, 454, 455, 456, 465, 466, 467, 468, 469, 472, 473, 474, 476, 492, 498

Head Loss 162, 166, 167

Health Effects 103, 105, 146, 149

Heat Pump Systems 451, 463

High Density Polyethylene Pipe 245

High Pressure 25, 182, 489

High Temperature 459

Hooke's Law 58

Hoop 68, 135, 215, 226, 249

Horizontal Directional Drilling 150, 155, 191, 192, 413, 414, 416, 442, 482 See also HDD

Howard 210, 211, 213, 245, 246, 264, 288, 412

HSB 143

HVAC 2, 451

Hydrokinetic 436, 450

Hydrostatic Design Basis 65, 66, 105, 127, 134, 135, 136, 138, 150, 160, 161, 242, 248, 327, 471 See also HDD

Hydrostatic Design Stress 65, 105, 134, 137, 138, 140, 327

Hydrostatic Load 412

Hydrostatic Pressure 150, 155

Hydrostatic Strength 65, 105, 132, 134, 136, 469

Hydrostatic Stress Board 65, 110, 143, 306

Hydrostatic Testing 23

I

IAPMO 147, 349

ICC 147

ID Controlled Pipe 159

IGSHPA 454, 455, 456, 458

IGT Distribution Equation 182

Immersion Testing 89

Impact 57, 77, 78, 79, 126, 150, 195, 196, 197, 275, 311, 469, 474, 498

Improperly Made Fusion 32

Infiltration 299, 485

Influence Value 202, 203, 204, 205, 248

Infrastructure 414

Initial Service Testing 26

Injection Molded Fittings 118

Innerducts 471

Inspection 6, 8, 16, 19, 23, 265, 288, 298

Installation 11, 14, 15, 20, 22, 26, 34, 40, 41, 45, 84, 105, 151, 154, 155, 156, 191, 194, 195, 206, 207, 216, 221, 222, 224, 232, 245, 246, 261, 262, 264, 267, 268, 272, 278, 282, 288, 289, 290, 294, 296, 297, 300, 302, 306, 316, 327, 353, 359, 360, 366, 369, 373, 401, 406, 412, 413, 433, 437, 439, 440, 442, 444, 455, 456, 458, 460, 463, 467, 473, 474, 475, 476, 477, 478, 479, 482, 483, 484, 486, 488, 498

Intermittent Support 317

Internal Pressure 65, 105, 125, 134, 135, 150, 248

International Ground Source Heat Pump Association 454, 458, 463

International Plumbing Code (IPC) 147

Iowa Formula 191, 207, 208, 209, 213, 428

IPC 147

IPS 17, 28, 31, 122, 138, 147, 158, 169, 251, 329, 337, 338, 382, 384, 385, 438, 442, 444

Iron Pipe Size 138, 337, 438, 439

ISO 65, 96, 132, 134, 135, 140, 143, 144, 145, 147, 148, 155

Izod 78, 469

J

Janson 246, 358, 362, 376, 379, 442

Jetting 489

Joining 14, 15, 34, 37, 134, 141, 142, 150, 151, 153, 329, 334, 336, 338, 339, 349, 363, 364, 390, 456, 461, 484, 487, 488, 498

Joint 17, 287, 330, 332, 333, 335, 336, 343, 344, 390, 409

K

Katona 207, 246

KBULK 171, 248

Kinetic COF 491

Kinked Pipe 33

Knight 428, 442

L

L/D Ratio 113

Lag Factor 209, 211, 213

Land-to-Water Transition 364, 372

Landfill Gas 184

Lateral Connections 409

Lateral Deflection 319

Lateral Service Connection 400

Launching 354, 363, 366

Loading 63, 105, 130, 191, 195, 197, 200, 201, 207, 222, 311, 395, 412, 426, 442, 474

Load Per Unit Length 322

Long-Term Deflection 213

Long-Term Hydrostatic Strength See LTHS

Long-Term Loading 130, 474

Longitudinal Force vs. Temperature Change 314
Longitudinal Stress vs. Temperature Change 314
Loose Storage Stacking Heights 12, 13
Love's Equation 391, 392
Low Pressure 26, 34, 183, 489
Low Temperature 57, 130, 309
LPG 161, 162
LTHS 65, 66, 67, 134, 135, 136, 151, 161, 306, 307
Lubricants 105

M

Make-up Water 23
Manhole Placements 399
Manning 185, 186, 187, 395, 396, 397
Manual 55 147, 154
Marking 145
Marston 225, 246
Material Selection 131, 468, 483
Maxwell Model 59
MDPE 39, 81, 86, 105, 209, 427
Measuring Density 301
Mechanical Connections 337, 390
Mechanical Couplings 155, 486, 498
Mechanical Fittings 152, 338, 485, 486
Mechanical Impact 311
Mechanical Properties 57, 71
Mechanical Protection 497
Mechanical Pulling Heads 407
Mechanical Stress 478
Medium Density PE (MDPE) *See* MDPE
Melt Index 54, 55, 57, 130, 132, 133, 468
Mesh 179
Migration 271
Mining 306
Model Specifications 134, 143, 149, 463

Modulus 60, 61, 62, 63, 73, 75, 76, 132, 133, 208, 209, 210, 213, 226, 241, 245, 246, 247, 269, 288, 308, 427, 430, 447, 462, 468
Modulus of Soil Reaction Values 246, 269
Molded Fittings 118
Moody Diagram 163, 166
Moore-Selig Equation 191, 218, 231
Mounting 354, 365
MRS 127, 135, 143, 151, 160
Ms 226, 227, 248
Mueller Equation 182, 183
Multi-Cell Conduit 497

N

Native Soil 213, 284

O

Open Trench 474, 479
Organic Silts 234
Original Line 405
Ovality 237, 238, 247, 391, 392, 429, 431, 443, 444, 446, , 479
Ovality Compensation Factor 238, 431
Ovality Correction Factor 237, 247, 392
Ovalization 427, 428, 429, 479

P

Packing List 8
Particle Size 177, 179, 181, 271
Passive Resistance 245
Path Profile 447
Pavement (Rigid) 197
PC 170, 172, 173, 174, 175, 207, 248
PDB 127, 135, 143, 151, 160

PE 34, 35, 39, 40, 41, 42, 43, 45, 47, 81, 101, 105, 106, 107, 108, 125, 131, 133, 134, 136, 137, 141, 142, 143, 144, 145, 146, 147, 148, 150, 151, 152, 153, 154, 155, 161, 165, 168, 169, 171, 190, 191, 192, 193, 194, 207, 208, 209, 215, 216, 217, 221, 225, 227, 228, 230, 240, 241, 243, 245, 246, 247, 248, 262, 263, 265, 267, 276, 278, 281, 282, 289, 291, 293, 298, 302, 303, 312, 327, 339, 343, 346, 347, 349, 351, 352, 353, 354, 355, 356, 357, 358, 360, 361, 362, 363, 367, 368, 369, 372, 373, 374, 375, 376, 378, 390, 396, 401, 410, 412, 414, 416, 420, 421, 423, 424, 425, 426, 427, 428, 429, 431, 432, 435, 437, 442, 443, 451, 452, 454, 455, 456, 458, 459, 460, 461, 462, 465, 466, 467, 468, 469, 470, 474, 475, 480, 482, 483, 484, 485, 486, 487, 492, 497
PE2406 28, 30, 307, 308
PE3408 28, 30, 162, 171, 172, 173, 175, 223, 251, 307, 308, 313, 403, 433, 440
Performance Limits 192, 426
Permanent Repair 33
Permeability 57, 81, 130, 184
Permeation 55, 81, 106, 183
Personal Protection 21
PEX 143, 151, 154
Physical Properties 77, 184, 468, 469
Physical Property Tests 125
Pilot Hole 414, 415
Pipeline Being Floated Out 367
Pipeline Research Council 433
Pipe Bursting 411
Pipe, Damaged 30, 33
Pipe Deflection 165, 246, 288
Pipe Deformation 264
Pipe Directly Beneath a Surcharge Load 202
Pipe Embedment 211, 212, 265, 266, 290
Pipe Extrusion 108, 126
Pipe Hanger 326
Pipe ID for Flow Calculations 158
Pipe Installation 84, 105, 262, 300, 456
Pipe Joining Methods 456
Pipe Laying 293, 297

Index

Pipe Length vs. Temperature Change 313
Pipe Liner Diameter 391
Pipe Restraint 21, 347
Pipe Sizing Operation 115
Pipe Specifications and Requirements 453
Pipe Splitting 411
Pipe Support in Transition 284
Pipe Surface Condition, Aging 189
Pipe Surface Damage 18
Pipe Trench 263
Pipe Tunnels and Casing 297
Pipe Wall Buckling 218, 236
PL 199, 200, 201, 202, 205, 206, 208, 216, 223, 248, 377
Placement Planning 493
Plastic Pipe to Steel 344
Plastometer 54, 498
Plenum 496
Plow 474, 475, 481
Plowing 475, 481
Plumbing Codes 147
Pneumatic Testing 25
Poisson's Ratio 75, 229, 231, 250, 430, 443, 462
Poisson's Ratio 75, 229, 231, 250, 430, 443, 462
Polyethylene Grade 133
Polyethylene Plastics 47, 50, 131, 150, 498, 515
Polyethylene Resin Properties 130
Polymeric water-soluble 491
Polymer Characteristics 49
Portable Trench Shield 281
Post-Installation Survey 373
Pre-Installation Storage 11
Preferred Number Series 139, 140, 521
Preparing an Installation 290
Preparing the Land-to-Water Transition Zone 354
Pressure Capability 306, 311
Pressure Flow 168, 176
Pressure Pipe 31, 34, 84, 105, 107, 146, 152, 154, 155, 175, 245, 288, 349, 391, 395, 400, 516, 518
Pressure Rating 134, 151, 159, 170, 173, 174, 249, 307, 518, 522

Pressure Surge 172, 175, 176
Pressure Test 21, 22, 125, 290, 521
Primary Initial Backfill 267
Prism 194, 424, 445
Proctor 211, 214, 221, 226, 227, 269, 273, 275, 283, 284, 285, 286, 287, 290, 295, 296, 297, 301, 302
Product Design 419
Profile Pipe 13, 246
Profile Wall 107, 117, 153, 209, 224, 245, 296, 297, 349, 404, 412
Proofing 482
Properties of Gases 184
Properties of Water 168
Proper Transition 284
Pull-line Construction 492
Pulling 371, 402, 403, 406, 407, 442, 449, 471, 476, 477, 478, 483, 488, 498, 499
Purchase Order 8
Purging 457
Push 407, 471, 480, 493
Pushing Force 404
Pushing Length 404
Pushing Technique 408
Pylon 319

Q

Quality Assurance 124, 126, 291, 470, 518
Quality Control 68, 105, 124, 127, 518
Quick Burst Test 518

R

Radius of Curvature 448
Railroad Loading 201
Rapid Crack Propagation 105, 519, 531
Rate Process Method 67, 105, 106, 151, 519
Ravine Crossing 317
Raw Materials 108, 110
RCP 309, 510, 519, 531
Reaming 415
Receiving Report 9
Recurring Surge Pressure 170, 174, 176, 248

Reduction Factor 444
Reel Method 484
Rehabilitation Methods 410
Repairs 32, 33, 345
Repair Clamps 345
Repair Operations 488
Repair Saddle 346
Reporting Damage 9
Required Strength 127, 132, 135, 143, 151, 160, 515, 516, 531
Required Weighting 355, 359
Resin Parameters 130
Resistance to External Collapse 440
Restrained Joints 39
Restrained Pipelines 318
Restraining 346
Reynolds Numbers 178
Rigidity Factor 228, 229, 230
Rigid Pavement 426
Ring Compression 217, 519
Ring Stiffness 208, 209, 249, 262, 292, 394
Riser Raceways 496
River Crossings 155, 192, 413, 418
Riveted steel 165
Roadway 519
Rock Trench Bottom 284
Rodent and Mechanical Protection 497
Rolldown 411

S

Saddle Fusion 120, 153, 332, 349
Safety 5, 15, 88, 329, 349, 393, 394, 417, 430, 443, 445, 447, 462, 526, 528, 531
Safety Factor 393, 394, 447,
Safe Deflection Limits for Pressurized Pipe 215
Safe Pull Load 438, 439
Safe Pull Stress 427, 447
Safe Pull Tensile Stress 427
Sand Stabilized with Cement 287
Saturated and Dry Soil Unit 234
Saturated and Dry Soil Unit Weight 234
Scale 179

Scrapes 32
Screen Mesh 179
Screen Pack 113
Screw 111, 112, 119, 121, 127
SDB 127, 135, 143, 151, 160, 516, 521
SDR 28, 101, 107, 118, 139, 140, 153, 154, 159, 215, 220, 230, 232, 245, 251, 252, 253, 254, 255, 256, 257, 260, 262, 265, 266, 273, 282, 288, 292, 312, 313, 320, 353, 354, 356, 357, 412, 429, 440, 441, 454, 462, 463, 472, 474, 475, 477, 479, 494, 510, 520, 521, 531
Secondary Initial Backfill 267
Selection of an Appropriate Pipe Diameter 353, 354
Selection of an Appropriate Site for Staging, Joining and Launching the Pipe 353
Select a Pipe Liner Diameter 391
Service and Lateral Connections 409
Service Connections 400
Service Loads 426, 442
Service Temperatures 84, 161
Service Temperature Design Factors 162
Service Temperature vs. Modulus of Elasticity 308
Service Testing 26
Shallow Cover 191, 195, 207, 222, 223, 232
Shear Properties 76
Sheathing 280
Sheeting 520
Shield 281
Shipping Claims 9
Shop Drawings 292, 520
Short-Term Mechanical Properties 71
SIDR 35, 107, 139, 153, 159, 245, 337, 472, 494
Silo 6, 9
Silts 234
Sizing 115, 116
Slipliners 187, 188
Sliplining 389, 400, 401, 405, 406
Sloping of Trench Walls 280

Slow Crack Growth 101, 105, 132, 133, 134, 150, 469, 498, 520, 531
Slurry 180, 245, 449
Smooth Pipes 165
Softening 57, 130, 521
Soft pigs 27
Soil Arching 266
Soil Classification 300
Soil Group Symbol 269
Soil Pressure 193, 197, 198, 201
Soil Reaction 210, 213, 246, 269
Soil Support Factor 208, 212, 213, 247
Soil Tests 18
Soil Type 211, 229, 234
Soil Unit Weight 234
Solar 151, 451, 458, 459, 521
Solid Wall 151, 155, 392, 402, 408, 468, 472, 498
Spacing of Ballasts 360
Span 323, 324, 521
Spangler 207, 213, 215, 245, 246, 264, 288, 412, 426, 442, 475
Special Applications 494
Specifications 20, 129, 134, 141, 142, 143, 149, 155, 197, 327, 349, 412, 453, 454, 463, 465, 509, 521
Specific Gravity 177, 180, 184, 521
Specific Heat 82, 83
Spitzglass Equation 182, 183
Springline 408
Stability 131, 246, 278, 279, 280, 385
Stabilization 102, 279, 319, 470
Stabilize 409
Stabilizer 133, 521
Stable 277, 289, 422
Stab Type Fitting 338
Stacking Heights 12, 13, 292
Staging 363
Standard Dimension Ratio See SDR
Standard Mesh 179
Standard Practice 142, 144, 150, 153, 155, 267, 327, 349, 412, 498
Standard Proctor 214, 221, 227, 269, 273, 275, 283, 284, 285, 286, 287, 290, 295, 296, 297, 302

Standard Specification 34, 35, 50, 107, 108, 131, 141, 142, 144, 151, 152, 153, 154, 155, 209, 245, 327, 349, 412, 465, 468, 486, 498
Standard Test Method 22, 26, 34, 54, 73, 75, 76, 78, 80, 81, 84, 88, 99, 100, 124, 125, 144, 150, 154, 155, 498, 513, 521
Static Charge 87
Static COF 490
Stationary Reel 484
Steel 154, 165, 327, 339, 344, 397, 410, 412
Stiffener 338, 514
Stiffness 57, 130, 151, 208, 209, 226, 249, 262, 292, 394, 405, 477
Storage 6, 11, 12, 13, 15, 292, 363
Strain 43, 58, 60, 71, 74, 77, 88, 320, 328, 521
Strength Design Bases See SDR
Strength of Embedment Soil 267
Stress Distribution 266
Stress Rating 134
Strip Load Truckload 8
Stub Ends 340
Submarine Outfalls 376
Submerged 357, 360, 376, 378
Submersion 354, 369
Sudden Velocity Change 172, 173
Sulphuric Acid 99
Sunlight 522
Support 208, 212, 213, 247, 284, 287, 317, 322, 325, 326
Surcharge Load 202, 205
Surface Damage 18
Surface Resistivity 86
Surface Roughness 165
Suspended from Rigid Structure 317
Suspended Pipelines 321
Suspension Spacing 322
Sustained Pressure 126, 521
Swagelining 410
Swelling 89
Swivels 441
Symbols 499

T

Take-off Equipment 118
Tamping 274, 290
Temperature 42, 57, 59, 60, 75, 83, 84, 85, 105, 130, 151, 152, 160, 161, 162, 168, 173, 217, 241, 247, 306, 307, 308, 309, 313, 314, 320, 432, 452, 459, 467, 508, 511, 512, 513, 521
Temporary Floating Lines 375
Temporary Repair 34
Tensile Creep 59, 60, 61, 63
Tensile Properties 58, 71, 73
Tensile Strength 57, 73, 75, 126, 130, 133, 469, 507, 509, 521, 523
Tensile Stress 58, 427, 437, 450, 523
Terminal Connections 409
Testing 16, 20, 21, 22, 23, 25, 26, 30, 34, 35, 50, 70, 72, 88, 89, 105, 106, 124, 126, 129, 144, 149, 150, 151, 155, 245, 290, 298, 299, 325, 327, 412, 457, 463, 467, 507, 511, 531
Test Duration 22
Test Pressure 22
Test Procedure 26
Thermal Conductivity 82
Thermal Expansion 81, 151, 509
Thermal Heat Fusion Methods 329
Thermal Properties 81
Thermal Stability 131
Thermal Stresses 441
Thermoformed Fittings 122
The International Organization for Standardization (ISO) 145
  See also International Organization for Standardization (ISO);
  See also ISO
Threaded Mechanical Fittings 486
Thrust Anchor 243
Thrust Block 319
Time-Dependent Behavior 427
Timoshenko 198, 199, 201
Titeliner 411
Torsion Stress 441
Toxicological 103

Traffic 405
Transition 83, 152, 284, 344, 364, 372, 410, 512
Transport 15, 34, 89, 106, 151, 327, 469, 498
Trench 191, 206, 263, 272, 276, 277, 278, 279, 280, 281, 282, 284, 289, 290, 294, 296, 297, 474, 479, 480
Trenching 289, 474, 479
Truckload 6, 7, 8
Tunnels 297
Turbulent 177
Typical Electrofusion Join 335
Typical Impact Factors for Paved Roads 196

U

Ultraviolet 101, 306, 310, 522
Unconfined Compressive Strength 213
Unconstrained Buckling 430
Unconstrained Pipe Wall Buckling 236
Undercut 480
Underwater Bedding 364
Uniform Particle Size 181
Uniform Plumbing Code 147
Unloading 9, 10, 427
Unpaved 197, 198
Unrestrained Thermal Effects 240
UPC 147
USBR 211, 246, 288
Use of Embedment Materials 270
UV 13, 101, 102, 103, 109, 133, 311, 454, 459, 468, 470, 483, 495, 522

V

Vacuum 115, 116, 176, 432, 451, 461, 463
Valve 167, 287, 342, 531
Vehicles 198
Vehicle Loading 197
Vehicle Loads 196
Vehicular Load 191, 195, 207, 222
Velocity 172, 173, 176, 178, 181, 188, 522
Venting 23, 31
Vertical Arching Factor 225, 226, 249
Vertical Deflection 214
Vertical Soil Pressure 193
Vertical Surcharge 204, 206
Viscoelastic 105, 327
Viscoelasticity 225
Voids 119, 522
Volume Resistivity 86

W

Wall Buckling 218, 236
Wall Compressive Stress 432
Wall Determination 472
Wall Thickness 135, 160, 251, 252, 253, 254, 255, 256, 257, 258, 259, 260, 391
Water-Base Slurry Specific Gravities 180
Water-Soluble Lubricants 491
Water Jetting 276
Water Pressure Pipe Design Example 175
Water Service 146, 154, 245, 349
Water Test 23, 24, 299
Watkins 190, 191, 207, 222, 225, 227, 228, 229, 230, 245, 246, 264, 288, 412, 428, 442
Watkins and Spangler 264
Weatherability 101, 102, 105, 151, 522
Weld Lengths of Polyethylene Pipe 402
Working Pressure Rating 170, 173, 174, 249, 522
Workmanship 124
WPR 170, 173, 174, 175, 176, 249, 522

Y

Yield Strength 469, 523
Young's Modulus 73